D0936812

Practical Thermodynamic Tools for Heat Exchanger Design Engineers

Practical Thermodynamic Tools for Heat Exchanger Design Engineers

Henri Soumerai

A WILEY-INTERSCIENCE PUBLICATION

JOHN WILEY & SONS

New York • Chichester • Brisbane • Toronto • Singapore

Library of Congress Cataloging in Publication Data:

Soumerai, Henri.
Practical thermodynamic tools for heat exchanger
design engineers.

"A Wiley-Interscience publication."
Bibliography: p.
Includes index.
1. Heat exchangers—Design and construction.
2. Thermodynamics. I. Title.

TJ263.S62 1987 621.402′5 86-34028
ISBN 0-471-81854-2

Printed in the United States of America

10 9 8 7 6 5 4 3 2 1

To Brigid Eithne Bourke-Soumerai (BEBS), who sparked this project and sustained it through from beginning to end

FOREWORD

This will be interesting reading for professional engineers as well as for students with a rudimentary background in thermal engineering sciences. The author's extensive experience as a practicing thermal engineer provides this work with practical recommendations useful to every engineer. The many references to the historical basis for the fundamental laws and the empirical relations make it valuable study material for practicing engineers and teachers of thermal science.

The author introduces a number of guidelines to be satisfied by physical laws in thermal engineering, treating thermodynamics, fluid mechanics, and heat transfer as a unified body of knowledge.

A detailed discussion of the Bernoulli theorem and friction factor with a comparison of the many equations proposed is followed by an excellent presentation on the complexities of two-phase flow, illustrated by equipment for the refrigeration industry, in which the author has extensive experience.

A unified treatment of the interrelationship between heat transfer and friction is refreshing. Reduced properties are introduced to unify results.

The book contains detailed examples of a variety of heat exchangers including introduction of reduced properties for boiling and condensation.

The text concludes with a discussion of departures from ideal analytical conditions in heat exchangers. It discusses the entropy generation concept and relates thermodynamics and heat transfer.

This book presents a refreshing look in great detail at our knowledge of heat transfer and pressure drop and relates it to thermodynamics. It introduces reduced properties to unify many presentations. The emphasis is on practical engineering design, which stems from the author's extensive experience in the processing, refrigeration, and power generation industries.

Experienced heat transfer engineers and teachers will find this book well worth studying in depth.

WARREN M. ROHSENOW
Professor of Mechanical Engineering
Massachusetts Institute of Technology

PREFACE

The key words in the title of the book highlight its *utilitarian objectives*, central novel feature, and scope. The proposed analytical techniques or tools were derived and effectively applied in the course of my research, development, and engineering activities in the electrical heating, chemical and food processing, air-conditioning and refrigerating, and power generation industries. *The central novel feature is the application of thermodynamic techniques* in the practical solution of complex heat transfer engineering problems. Deliberately narrowing the scope to focus on thermodynamically efficient heat exchangers (except in the last section of Chapter 9) made it possible to reduce the coverage of well-known topics, such as thermal conduction and radiation, to a bare minimum and thus allocate adequate space to novel methods not covered in standard thermal engineering textbooks. Although the main emphasis is on the specific problems faced by heat exchanger engineers, it will become clear to the attentive reader that the proposed analytical techniques can also be put to practical use by system and components engineers in many other engineering applications.

The detailed *table of contents* provides a fairly comprehensive description of the *scope and organization* of the book. The main topics addressed have been grouped in nine chapters, starting with the most elementary concepts and gradually moving up to more advanced and unconventional methods. However, I have made a special effort (at the price of some minor redundancies) to make each chapter a self-contained unit so that experienced engineers, teachers, or advanced engineering students should be able to select their own chapter sequence more or less "à la carte" according to their individual needs. To further lighten the text and yet provide a fully self-contained treatment in a single volume, less essential and/or well known material has been relegated to the general appendices (see Appendices A, B, F) along with less widely known relevant back up material (see Appendices C, D, E). For similar reasons, the many footnotes appearing in the text (for additional clarifications, historical perspective, etc.) have been consolidated in chapter Notes at the end of each chapter.

The most *significant and/or novel features* addressed in each chapter are highlighted below:

Chapter 1: The main objectives and philosophy adopted for the whole book are defined in this introductory chapter. The useful concept of *methodological guidelines*, which is particularly important in a text dedicated to new methodologies, is introduced. Relevant fundamental concepts, definitions, and frequently forgotten theoretical assumptions are formulated as numbered guidelines, starting with **Guideline 1**, the definition of a physical law in Chapter 1, and ending with **Guideline 29** in Chapter 9, which identifies the significant thermodynamic characteristics of low thermal lift heat exchangers. For quick reference the guidelines are adequately cross indexed in the general index.

Chapter 2: The practical validity limits of idealized (inviscid) Bernoulli flows are assessed and the concept of friction dissipation, expressed as a dimensionless "number of velocity heads," is introduced. The chapter concludes with a broad assessment of idealized (inviscid) and real (viscid) fluid flows in the context of classical thermodynamics. The basic philosophy adopted throughout the book is already apparent in this succinct overview of inviscid fluid mechanics.:

(a) The extensive use of *numerical analyses* as an integral part of the text, primarily to provide the rationale for recommended analytical procedures and additionally to illustrate more concretely some of the novel and/or abstract concepts presented here. Although the numerical calculations were all performed on a simple engineering hand calculator (Texas Instruments), it is evident that many of the proposed solutions could be further refined through the use of modern computerized techniques.

(b) The systematic application of *dimensionless (or reduced properties) analytical methods to generalize important results*.

Chapter 3 focuses on *single-phase Newtonian fluid mechanics*, except for a brief overview of entrance effects in two-phase flows. Recognized correlations for laminar and turbulent flow regimes are reviewed and quantitatively assessed in the light of the end engineering objective of fluid flow analysis formulated as a guideline at the beginning of the chapter. This leads to an adequately accurate universal Blasius type friction factor correlation suitable for both laminar and turbulent flows in channels of any natural wall roughness. This correlation is later put to practical use in a universal Chilton–Colburn type of heat transfer correlation and two-phase flow thermodynamic generalizations. The *emphasis on "real world" engineering applications* is evidenced by detailed discussions of such topics as: the statistical nature of the natural wall roughness, the impact of inlet effects and nonisothermal conditions, and marginally stable flow regimes. The chapter concludes with a qualitative thermodynamic assessment of flow regime bifurcations anticipating the entropy maximizing principle methodology highlighted in Chapter 9 and Appendix E.

Chapter 4: *Multiphase fluid flows* are addressed very broadly in order to obtain a preliminary "organic" view of the whole field and provide the necessary background for fundamental concepts and techniques especially useful in two-phase flow heat exchangers. Departing from tradition, reverse power cycles rather than steam power generation cycles are emphasized. A number of pragmatic and didactic reasons justifying this departure from the standard "steam water menu" are indicated in this chapter. The most significant single argument is highlighted here because it explains the major emphasis on thermodynamically efficient heat exchangers throughout this book (except in the last section of Chapter 9):

> Out of necessity the refrigeration industry has a long tradition in "fighting second law irreversibilities" and has developed the know-how to cope with the challenge of designing compact and economical heat exchangers for operation at low temperature differentials. These traditional refrigeration design techniques have become quite relevant in other industrial applications since the first oil shocks of the 1970s and the general worldwide emphasis on energy conservation. By focusing on *normally efficient fluid-to-fluid surface heat exchangers rather than atypical nuclear reactor* applications, it is possible to greatly simplify the analysis of complex two-phase flow heat exchangers on fundamental grounds without any compromise in prediction accuracy.

The first purely thermodynamic guideline is introduced in this chapter; it emphasizes the *necessity* in evaporator and/or condenser design optimization studies of a *unified treatment* involving all three branches of thermal sciences. A qualitative assessment of the current state of the art in the refrigeration and nuclear power generation industries brings to light the extreme complexity of two-phase flow phenomena and provides a rationale for the renewed interest in the corresponding states principle thermodynamic generalization methods addressed in later chapters.

Chapter 5 is essentially a critical review and a *comparative quantitative assessment of recognized correlations for the prediction of the pressure drop in two-phase flows*, focusing first on a "naive all vapor" flow model, then on the zero slip flow model, and finally on the more sophisticated Martinelli type slip flow model. Although two-component two-phase flows are also considered in view of their importance in the chemical process industries, the main emphasis is on the single-component two-phase flow situations encountered in forced convection evaporators and condensers.

Chapter 6: The *thermodynamic generalization techniques* based on reduced pressures initially proposed in the late 1950s (and to my knowledge never addressed in standard heat transfer textbooks) are reviewed and updated to reflect the current state of the art. The practical application of this methodology is illustrated in several numerical analyses with special emphasis on single-component two-phase flow applications. It is shown that this thermodynamic generalization method provides the *fluid flow system engineer* and the *heat exchanger designer* with an additional tool that is simple, effective, and more reliable than conventional semiempirical two-phase flow correlations. The significant fundamental and practical advantages of this "*universal*" *thermodynamic tool* are summarized at the end of this chapter.

Chapter 7: The *fundamental aspects* of the older discipline, the science of heat transfer, are addressed first with special emphasis on the convection mode of heat transfer, then classical equilibrium thermodynamics (ET), which is treated here as the (idealized) limiting case of the new twentieth century thermodynamics of (real) irreversible processes, or nonequilibrium thermodynamics (NET), a topic that has not yet been given the attention it deserves in modern thermal engineering textbooks. The fundamental basis of conventional or first law heat exchanger analytical methods is examined to yield a very general methodological guideline which is particularly useful in heat exchangers with phase changes, as shown in the next chapter.

My succinct and admittedly rudimentary overview (in Section 7.5) of extremely abstract concepts of NET theory is adequate for its intended purpose, namely, a pragmatic assessment (expressed as a fundamental guideline) of the *practical value of modern NET theory* in the "art" of heat transfer. Professor P. Glansdorff, one of the earlier pioneers of NET theory, was kind enough to review critically the final manuscript of Section 7.5 to make sure that my personal interpretation of this complex new branch of mathematical physics is correct. This bird's eye overview of NET theory, which has so far remained a restricted domain in mathematical physics, also provides the theoretical foundation for the central novel feature proposed in this book, a "unified thermodynamic treatment of momentum, heat, and mass transfer," a *fully interdisciplinary* approach, which is further justified on practical engineering grounds outlined in this chapter. Friction dissipation (the "hidden variable" systematically ignored in conventional heat exchanger analytical methods) is the key parameter of novel variation analyses. Based exclusively on the conservation laws and recognized empirical correlations, these analyses demonstrate the existence, under specified design conditions, of a *common relationship* between the total rate of friction dissipation and total rate of heat exchange, which *surprisingly*,[1] *is valid for both turbulent and laminar* single-phase tube flow regimes. From this primitive fundamental empirical relationship a universal Chilton–Colburn type of Reynolds analogy is formulated, which is applicable to laminar and turbulent flows in circular as well as noncircular channels of *any natural wall roughness*. The same variation meth-

[1] Professor P. Grassmann, who critically reviewed several sections of Chapter 7, strongly felt it was a "pity to bury" these variations analyses in a textbook without prior publication in a recognized international scientific/engineering journal!

odology is put to practical use to quantify the significant economical advantages of turbulent flow regimes in forced convection heat exchangers.

The emphasis on convective heat transfer engineering, *the art*, rather than the theory, is evident by the questions addressed in several sections of this key chapter, such as: the nonlinear relationships between heat flux and heat transfer driving forces; the validity/accuracy limits of the conventional linear relationships in so-called "linear heat flux heat exchangers," and the resulting symmetry between the fluid heating and cooling modes (the theoretical basis for the mirror image concept); the introduction of the useful chemical engineering concept of mean residence time; the experimental methods of determining the heat transfer coefficients, including an assessment (expressed as a practical guideline) of the impact of temperature measurement deviations on the empirical coefficient as well as the end effect on the overall heat exchanger performance of heat transfer coefficient deviations; the major impact of fouling resistance in water-cooled condensers and unavoidable manufacturing tolerances; the formulation of a pragmatic general guideline defining realistic quantitative accuracy targets for the elusive heat transfer coefficient, the engineering end objective of convective heat transfer research.

In the remaining two chapters the proposed *unified thermodynamic approach is systematically applied*. In contrast to the concluding chapter which focuses on second law considerations, an understanding of the rather abstract concept of entropy is not necessary in Chapter 8.

Chapter 8 is broken down into three main sections focusing on single-phase turbulent channel flows, single-component two-phase channel flows with heat exchange, and thermodynamic generalizations of heat transfer data without and with phase changes (evaporation and condensation) in heat exchangers of any arbitrary geometry including external as well as internal flows. The most significant features are highlighted below section by section.

Section 8.1 focuses on normally effective low thermal lift applications (after a practical assessment of nonisothermal and inlet effects) to determine the Prandtl exponent β in a *universal form of the Chilton–Colburn Reynolds analogy*. This leads to quantitative recommendations for β in terms of the generalized Blasius friction factor exponent which are applicable (in contrast to conventional correlations) to channels of *any* natural wall roughness. An integrated form of the modified Reynolds analogy is proposed for multicircuit tubular heat exchangers. This novel analytical method emphasizes the significant *impact of individual friction losses* (typically ignored in conventional local optimization studies!) and the constructive role of the wall friction component of the total friction dissipation allocated to the heat exchanger by the thermal system engineer. The significant advantages of small hydraulic diameters in "compact heat exchangers" due to the resulting higher *wall friction effectiveness* is quantified in a numerical analysis for single-phase flows, then briefly highlighted and illustrated for single-component two-phase flows with heat addition (evaporation) or removal (condensation).

In Section 8.2 the emphasis is on convective *heat transfer with phase changes* in *low thermal lift* power cycles and heat pumping systems. The vague notion of low thermal lift heat exchangers is quantified on the basis of extremely simple Carnot cycle analyses (in lieu of more abstract second law methods) to demonstrate the negligible role of nucleate boiling regimes in thermodynamically efficient forced convection evaporators and the resulting symmetry between the evaporating and condensing modes. The absence of nucleate boiling is further confirmed on the basis of published experimental data and an excellent survey paper on heat transfer with boiling by a world renowned authority Prof. Warren M. Rohsenow. Some of the results derived in these *nucleate boiling suppression analyses* are generalized on the basis of reduced pressures. The most widely accepted empirical heat transfer correlation proposed by the Swedish researcher Bo Pierre is analyzed, with special emphasis on the more effective fully wet wall flow regimes, to derive the so-called "Bo Pierre paradox." This paradox is verified empirically and a plausible simple thermodynamic explanation offered, based on alternative enthalpy heat transfer driving forces. As a spin-off from this thermodynamic interpretation a useful extension of the conventional single-phase *NTU methodology is proposed for the forced convection evaporating* mode in the fully wet wall regimes *and* for the mirror image *condensing mode*.

In Section 8.3 the *corresponding states principle* generalization methods introduced in Chapter 6 are applied to *generalize theoretical and/or empirical heat transfer data* in both thc fluid heating and cooling modes, with and without phase changes in any geometric configuration. The extremely broad validity, simplicity, and accuracy of this methodology is illustrated in a numerical analysis and a brief overview of its application in high flux nuclear reactor applications. This section concludes with a summary of the main advantage of this thermodynamic generalization method, namely, its quasi-universal validity and, most importantly, in the context of the alternative entropy-based heat transfer driving forces addressed in Chapter 9, the empirical fact that the *molar entropy* is practically a *universal function* of reduced pressure for most pure substances and azeotropic mixtures.

Chapter 9: This concluding chapter focuses on the key variable of thermodynamics: entropy. Linear heat flux heat exchangers operating under conditions "near equilibrium" in the linear domain of NET are first considered. The useful *concept of a mirror image heat exchanger system* (MIHES) is defined and its key characteristics first analyzed according to conventional (first law) methods with the help of several numerical analyses to provide a more concrete basis for otherwise rather abstract nonequilibrium thermodynamic considerations. In addition to its value as a didactic tool, the physical model of the MIHES provides the basis for an experimental test set-up to determine empirically heat transfer coefficients in single-phase and (by extension) two-phase flows under truly isothermal wall boundary conditions as opposed to constant heat fluxes. The thermodynamic analyses of the simple MIHES set a practical bridge with the highly theoretical concepts of twentieth century NET and extend, in effect, the validity of linear NET to include nonlinear turbulent flow regimes, thus demonstrating the *relevancy of the new nonequilibrium branch of thermodynamics* in practical momentum, heat, and mass transfer applications.

The concept of linear second law heat transfer driving forces is extended on the basis of a "reversible work" driving force to deal with more complex situations in the nonlinear domain of NET theory. The *advantages of second law driving forces* are summarized and a number of potentially fruitful applications *beyond the current* state of the art are suggested. This is supplemented by excerpts from new papers shown in Appendix D, an extension of the mirror image concept to heat exchangers with phase changes, and Appendix E addressing flow regime bifurcations on the basis of an entropy maximizing principle.

To make this text fully self-contained, and realizing that most practicing heat transfer engineers have forgotten abstract second law thermodynamic concepts (usually with no regrets!), I have incorporated in this chapter several sections on relevant second law nonlinear relationships applicable in high thermal lift systems as well as the simpler linear solutions applicable in low thermal lift systems, topics *not addressed in current thermodynamic textbooks* (see Sections 9.2 and 9.3).

HENRI SOUMERAI

ACKNOWLEDGMENTS

I would like to express my sincere gratitude to Professor Warren M. Rohsenow for his encouragement and critical reviews in the early phases of this work and in particular for writing the Foreword to this book.

I am also greatly indebted to all those who through their support and critical review of the manuscripts for the book and related papers helped to finalize this work, and more specifically to the following individuals:

Mogens Andersen, Brown Boveri & Cie., Baden, Switzerland
Carlisle Ashley, Carrier Corporation, Syracuse, New York
Professor A. Bejan, Duke University, Durham, North Carolina
Cecil Boling, Naples, Florida (former President, Dunham-Bush Inc.)
B. Eithne Bourke-Soumerai, HED Soumerai & Associates, Multinational Consultants, Fislisbach, Switzerland
Frank Cerra, Editor, Wiley-Interscience Division, and supporting production staff at John Wiley & Sons, New York
Professor P. Glansdorff, Université Libre de Bruxelles, Belgium
Professor P. Grassmann, Eidg. Technische Hochschule, Zurich, Switzerland
Professor G. G. Haselden, University of Leeds, Leeds, England
William Holladay, Private Consultant, Altadena, California
Bertrand Jacobsen, Private Consultant, Zurich, Switzerland
Dr. Tamami Kusuda, National Bureau of Standards, Washington, D.C.
Professor G. Lorentzen, Norges Tekniske Hogskole, Trondheim, Norway
Dr. Michel Nicolet, Eidg. Institut fuer Reaktorforschung, Wuerenlingen, Switzerland
Hiroshi Ohba, Daiwa (Switzerland) SA, Zurich
David N. Soumerai, HED Soumerai & Associates, Multinational Consultants, West Hartford, Connecticut

CONTENTS

LIST OF SYMBOLS

Symbol	Name / Definition	Units
A	area	m^2
a	area per unit length, or wetted perimeter	m
AF	approach factor	
$(AF)_{tp}$	two-phase approach factor	
BF	bypass factor	
C	heat capacity ratio $= (\dot{m}c_p)_{min}/(\dot{m}c_p)_{max}$	
C	empirical constants	
C_{st}	constant in generalized Reynolds analogy	
c	acoustic velocity	m/s
c_p	constant pressure specific heat	J/kg · K
c_v	constant volume specific heat	J/kg · K
CF	contact factor	
$(CF)_{tp}$	two-phase contact factor	
COP	coefficient of performance, heat pump	
$D = D_h$	circular tube internal diameter	m
$D_h = 4r_h$	hydraulic or volumetric diameter	m
$\bar{e}$	effective wall roughness, statistical average	m
e	effective wall roughness, actual	m
$\bar{e}/D_h$	effective relative wall roughness, statistical average	
e/D_h	effective relative wall roughness, actual	
$(ENVH)_{tp}$	effective number of velocity heads, two-phase flow	

f	friction factor based on $D = D_h$	
f_{rh}	friction factor based on $r_h = 1/4D_h$	
G	mass velocity	$kg/m^2 \cdot s$
g	gravitational acceleration	m/s^2
$\Delta H = \Delta P/\rho$	friction head loss (dissipation)	J/kg
H, h	enthalpy, specific enthalpy	J, J/kg
h_{lg}	latent heat of vaporization	J/kg
K_{in}, K_{ex}	head loss coefficient, inlet, exit	
K_{indiv}	individual head loss coefficient	
K_n	Blasius friction factor constant, generalized	
K_p	sum of individual head loss coefficients per pass	
k	thermal conductivity	$W/m \cdot K$
L	length of travel	m
L/D_h	dimensionless length of travel or aspect ratio	
L_c	characteristic length in dimensionless groups	m
L_p	length per pass	m
l	axial position	m
M	molecular "weight"	
$M_{\Delta T}$	heat transfer coefficient, non-isothermal correction factor	
$\overline{M}_g$	integrated two-phase pressure drop multiplier based on all gas flow	
$M_h = \alpha_{tp}/\alpha_L$	two-phase heat transfer coefficient multiplier	
$M_{L/D}$	heat transfer coefficient, inlet effects correction factor	
Ms_{lg}	molar entropy change from liquid to vapor state	$J/mol \cdot K$ $= kJ/kmol \cdot$
$\dot{m}$	mass flow rate	kg/s
n	generalized friction factor Blasius exponent	
N_c	number of circuits in parallel per pass	
N_p	number of passes	
NTU	number of thermal units	
NVH	number of velocity heads	
P	pressure (absolute)	$Pa = N/m^2 = J/m^3$
ΔP	pressure drop	$Pa = N/m^2 = J/m^3$
P_{cr}	critical pressure (absolute)	$Pa = N/m^2 = J/m^3$

q''	heat flux	W/m^2
Q_e	external heat transfer	J
$\dot{Q}_e$	rate of external heat transfer	W
$\dot{Q}_f$	rate of internal heat generation due to wall friction	W
R	overall thermal resistance	K/W
R	specific gas constant	$J/kg \cdot K$
R, r	radius	m
r_h	hydraulic or volumetric radius	m
S	entropy	J/K
$\dot{S}$	total rate of entropy production (irreversible)	$J/K \cdot s = W/K$
s	specific entropy	$J/kg \cdot K$
s_{lg}	specific entropy change from liquid to vapor state	$J/kg \cdot K$
ΔS_e	entropy change due to external heat exchange	J/K
Δs_e	specific entropy change due to external heat exchange	$J/kg \cdot K$
$(\Delta s_e)_n$	normalized specific entropy change due to external heat exchange	
Δs_f	specific entropy change due to wall friction	$J/kg \cdot K$
$(\Delta s_f)_n$	normalized specific entropy change due to wall friction	
ΔS_i	entropy change due to internal dissipation	J/K
Δs_i	specific entropy change due to internal dissipation	$J/kg \cdot K$
T, t	absolute, common temperature	K, °C,
ΔT	temperature change or differential	K or °C
ΔT_w	wall to fluid temperature differential or conventional heat transfer driving force	K or °C
U	overall heat transfer coefficient, conventional	$W/m^2 \cdot K$
U, u	internal energy, specific internal energy	J, J/kg
V, v	volume, specific volume	m^3, m^3/kg
$\mathbf{V}, V_x, V_y, V_z$	velocity vector, components in x, y, z directions	m/s
W, w	work, specific work	J, J/kg
$\dot{W}$	rate of energy transfer as work	W
$\dot{w}$	rate of energy transfer as work per unit mass flowing	W/kg

x	mass flow fraction of gas phase, or vapor quality	
X	entrance distance to diameter ratio	
$X_{tt}, X_{tv}, X_{vt}, X_{vv}$	Martinelli two-phase flow parameters	
y	mass flow fraction of liquid phase	
Z	compressibility factor	

Greek

$\alpha, \bar{\alpha}$	heat transfer coefficient, local, average	$W/m^2 \cdot K$
β	Prandtl exponent in generalized Chilton-Colburn correlation	
δ	thickness	m
δ_{bl}	effective boundary layer thickness	m
δ_l	liquid film thickness	m
ε	heat exchanger effectiveness	
ε_η	normal total emmisivity, thermal radiation	
η	efficiency or effectiveness	
η_C	Carnot cycle efficiency	
η_{wf}	wall friction effectiveness	
κ	c_p/c_v = isentropic exponent	
μ	dynamic viscosity	$N \cdot s/m^2 = kg/m \cdot s$
ν	kinematic viscosity	m^2/s
θ	time	s
$\overline{\Delta\theta}$	mean residence time	s
Ω	cross-section free flow area	m^2
π_i	arbitrary property term	
$\pi_i R$	arbitrary liquid to vapor phase property ratio	
φ	flow rate ratio $= \dot{m}/\dot{m}_{virt}$	
$\varphi(\ )$	function of terms in ( )	
$\varphi(P^+)$	function of reduced pressure	
Φ_g^2, Φ_l^2	two-phase flow multiplier based on, gas flow, liquid flow	
$\Phi_{tt}, \Phi_{tv}, \Phi_{vt}, \Phi_{vv}$	Martinelli two-phase flow parameters	
$\Psi(\ )$	universal function of terms in ( )	
$\Psi(P^+)$	universal function of reduced pressure P^+	
$\rho, g\rho$	mass density, specific weight	kg/m^3, N/m^3
τ	shear stress	N/m^2

Script

$\mathscr{C}$	molar specific heat	J/mol · K = kJ/kmol · K
$\mathscr{P}$	thermodynamic probability	
$\mathscr{P}_i$	internal entropy production rate (irreversible)	J/K · s = W/K
$\mathscr{R}$	universal gas constant	J/mol · K = kJ/kmol · K
$\mathscr{V}$	volume of thermodynamic system	m^3
$\Delta\mathscr{V}$	elementary volume within system	m^3
$\dot{\mathscr{V}}$	volumetric flow rate	m^3/s

Subscripts

1, 2,	fluid inlet, outlet
a	acceleration
act	actual
am	arithmetic mean
avg	average
b	bulk
bl	boundary layer
BP	boiling point, normal
C	colder fluid
cond	condensing
cr	critical thermodynamic state, also flow regime transition
des	design
eff	effective
evap	evaporating
est	estimated
f	wall friction
fd, h	fully developed, hydrodynamically
fr	fully rough
g, go	gas phase, gas phase only
H	hotter fluid, high temperature sink or source
h	enthalpy based
i	internal dissipation, inside
indiv	individual
in/out	inlet/outlet

irr	irreversible
L	low temperature sink or source
l, lo	liquid phase, liquid phase only
lam	laminar
lg	change from liquid to vapor state
max	maximum
meas	measured
min	minimum
n	normalized,
nom	nominal
o, O	outside, origin of entropy scale
om	overall mean
P, p	constant pressure
q	constant heat flux
r	reversible mixing effluent state
rad	radiation
red	reduced
ref	reference
rel	relative
rev	reversible
s	entropy based
sat	saturation
sk	sink
so	source
sup	superheat
T	constant temperature
t	temperature based, also triple point
tot	total
tp	two-phase
tp, a	two-phase acceleration component
tp, f	two-phase wall friction component
tt, tv, vt, vv	two-phase flow regimes, Martinelli
turb	turbulent
v	constant volume
virt	virtual
VH	one velocity head
w	wall, wall interface
zero	zero individual head losses

Overbars, Dots, and Superscripts

$-$	average
$\cdot$	per unit time (rate expressions)
$+$	reduced properties as in $T^+ = T_{sat}/T_{cr}$,

Dimensionless Groups

Symbol	*Name/Definition*
Ma	Mach: $\mathbf{V}/c$
Nu	Nusselt: $\alpha L_c/k = \text{St Re Pr} = \text{St Pe}$
Pe	Peclet: $L_c \rho c_p \mathbf{V}/k = \text{Re Pr}$
Pr	Prandtl: $c_p \mu/k$
Re	Reynolds: $L_c \rho \mathbf{V}/\mu$
St	Stanton: $\alpha/(c_p \rho \mathbf{V}) = \text{Nu}/\text{Re Pr} = \text{Nu}/\text{Pe}$

Acronyms

ADP	apparatus dew point
AF	approach factor
AIChE	American Institute of Chemical Engineers
AMTD	arithmetic mean temperature difference
ARI	Air-conditioning Refrigeration Institute
ASHRAE	American Society of Heating, Refrigerating and Air-Conditioning Engineers
ASME	American Society of Mechanical Engineers
AWL	available work loss
BF	bypass factor
CF	contact factor
COP	coefficient of performance
CPE	condition preceding entrance
CSP, ECSP	corresponding states principle, extended CSP
DF	driving force
ECC	external Carnot cycle
ES	equilibrium state
ET	equilibrium thermodynamics
ETL	external thermal lift
FCM	fluid cooling mode
FFSHES	fluid-to-fluid surface heat exchanger system

FHM	fluid heating mode
GL	guideline
HTL, HTLS	high thermal lift, HTL system
ICC	internal Carnot cycle
IIR	International Institute of Refrigeration
In/Out SSSF	inlet/outlet steady state steady flow
ITL	internal thermal lift
LD	larger difference
LDF	larger driving force
LHFHE	linear heat flux heat exchanger
LMTD	logarithmic mean temperature difference
LTL, LTLS	low thermal lift, LTL system
MHMT	momentum, heat, and mass transfer
MIHES	mirror image heat exchanger system
NE	non-equilibrium
NET	nonequilibrium thermodynamics
NIC	nonisothermal conditions
NSTR	number of stagnation temperature rise
NTU	number of thermal units
NVH, ENVH	number of velocity heads, effective NVH
RAWL	relative available work loss
REW	relative excess work
RW	reversible work
SD	smaller difference
SI	système international
SRRW	square root reversible work
SSSF	steady state steady flow
STR	stagnation temperature rise
TDF	terminal driving force
TPC-MIHES	two-phase counterflow mirror image heat exchanger system
TTD	terminal temperature difference
UT-MHMT	unified treatment of momentum, heat, and mass transfer
UTT-MHMT	unified thermodynamic treatment of momentum, heat, and mass transfer
VDI	Verein Deutscher Ingenieure

Practical Thermodynamic Tools for Heat Exchanger Design Engineers

CHAPTER 1

PHYSICAL LAWS IN THERMAL ENGINEERING

According to the scientific methods adopted by experimental physicists and researchers since the publication of Galileo's *Essayer* [1.1] in the seventeenth century, the mathematical expressions of observed physical phenomena, or "physical laws," must satisfy a few generally accepted criteria. These are summarized in the first section of this chapter to emphasize clearly their central significance in heat transfer engineering.

1.1 PHYSICAL LAWS AS METHODOLOGICAL GUIDELINES

Relevant fundamental concepts generally recognized by experimental physicists and engineers will be expressed as methodological guidelines (GL) throughout this text, starting with the definition of a physical law.

> ***GUIDELINE 1*** *A physical law is expressed quantitatively, either analytically or graphically, in the simplest possible form.*

Because the mathematical expressions of physical laws are never totally accurate, or valid outside certain empirical limits, Guideline 1 is dangerously incomplete for design purposes. Therefore the following additional guideline is of vital importance, particularly in the case of fluid flow and heat transfer correlations, which are notoriously inaccurate:

> ***GUIDELINE 2*** *The validity limits of a physical law or any correlation of empirical data must be documented clearly and include a quantitative estimate of the expected accuracy of the proposed equation or correlation within the full range of the stated validity limits.*

Unfortunately, the practical significance of this second guideline is overlooked by too many "pure" scientists engaged in fundamental heat and fluid flow research. This may explain why practicing heat transfer design engineers continue to favor the fully documented simpler correlations recommended in old reliable "bibles" of the 1950s such as McAdams [1.2] and Kern [1.3] in preference to the more sophisticated theoretical correlations proposed in recent years with inadequately documented proof of their superior accuracy in "real world" heat exchangers. The constructive role of Guideline 2 will be illustrated in a number of practical engineering design cases beginning in Chapter 2 where the oldest of the three formerly separate branches of thermal sciences—fluid mechanics—is addressed.

As the name implies, the laws of fluid mechanics are based on a generalization of the classical Newtonian laws of motion to continuous media. We know that even such "exact" fundamental laws as the classical (nonrelativistic) conservation laws of mass and energy have their own finite validity limits beyond which they are no longer applicable, as in the case of micro- and astrophysics. Fortunately, relativistic effects in normal earthbound thermal engineering applications (excluding nuclear reactor-core physics) are so insignificant [1.4] that they could not be measured even with the most refined instruments available to date.[1] Consequently, the mass and energy conservation laws of classical thermodynamics may be applied confidently within the full application range of normal macroscopic (human scale) thermal engineering systems, and it is not necessary to specify any validity/accuracy limits in conjunction with these "exact" laws. However, this constitutes a notable exception to the general applicability of the fundamental Guideline 2 in thermal engineering sciences when we recall that the most exact heat transfer or fluid flow correlations are accurate to within 5–10% at best in the simplest heat exchanger design applications. Even such generally reliable thermodynamic relations as the ideal gas laws can become significantly inaccurate outside relatively narrow empirical validity limits, falling well within the normal operating range of many process heat exchangers.

1.2 THERMAL ENGINEERING SCIENCES

Concise definitions of the three branches of thermal sciences [1.5, 1.6] are listed below in the chronological order according to which each discipline was officially recognized as a full-fledged branch of physics, to highlight the fact that thermodynamics is the "youngest" and fluid mechanics the "oldest" of these formerly separate branches of physics. These simple definitions will be refined as needed in subsequent chapters.

1. **Fluid Mechanics: Eighteenth Century.** This branch of thermal sciences, alternatively called fluid dynamics or fluid flow, deals with the following:

GUIDELINE 3 *The transportation of energy and the resistance to motion associated with flowing fluids.*

2. **Heat Transfer: Early Nineteenth Century.** Also commonly known under the name of heat flow or heat transmission [1.2], this discipline describes the following:

GUIDELINE 4 *The transfer of a specific form of energy as a result of the existence of a temperature difference or gradient.*

3. **Classical Thermodynamics: Mid-nineteenth Century.** This most fundamental and broadest discipline in thermal sciences encompasses the following:

GUIDELINE 5a *The study of heat and work and those properties of substances that bear a relation to heat and work.*

It is interesting to note that the first part of Guideline 5a, the study of heat and work,[2] closely reflects the title and actual content of the epoch-making memoire by Sadi Carnot: *Reflexions sur la Puissance Motrice du Feu et sur les Machines Propres a Developper Cette Puissance* [Reflections on the Motive Power of Fire and on the Machines Best Suited to Develop this Power] [1.7] published in 1824. A French military engineer and graduate of the Paris Ecole Polytechnique, Sadi Carnot was then only 28 years old (1796–1832). He has rightfully been recognized as the legitimate father of classical equilibrium thermodynamics by such eminent scientists as William Thompson (Lord Kelvin) and Rudolph Clausius who formalized this new branch of physics around 1850 in the light of the "first law" of thermodynamics (the energy conservation law) on the basis of Carnot's work. In actual fact, Carnot formulated the substance of the "second law" almost two decades before the first law of classical thermodynamics [1.8, 1.9].

A most compact alternative definition of engineering thermodynamics [1.6] simply states the following:

GUIDELINE 5b *The science of energy and entropy.*

This succinct definition recognizes the central importance of the thermodynamic state property called *entropy* (from the Greek for transformation, change, or evolution) by Rudolph Clausius. The rather abstract but extremely useful and powerful concept of entropy[3] was first presented at a scientific conference in Zurich, Switzerland, on April 24, 1865 by Clausius [1.10] who was a professor at the University and at the newly founded Institute of Technology, now known as the Swiss Federal Institute of Technology [Eidgenössische Technische Hochschule, Zurich (ETHZ)], during the period 1855–1867.

1.3 HEAT TRANSFER ENGINEERING AND NONEQUILIBRIUM THERMODYNAMICS

Equilibrium thermodynamics, the science of energy and transformation, has undergone significant changes since its official recognition as a separate branch of physics, outside the scope of classical Newtonian mechanics, in the middle of the nineteenth century. According to twentieth century concepts of nonequilibrium thermodynamics, the arbitrary compartmentalization of thermal sciences in three tightly separated disciplines has become archaic, at least on fundamental grounds. However, we are more concerned here with the application of thermal sciences as useful engineering tools in heat transfer engineering, the art, rather than purely academic aspects of the evolution of thermal sciences. Consequently, we systematically assess the scientific tools available to heat transfer engineers from a strictly utilitarian point of view, that is, their practical value in real world heat exchanger analyses. By the time we address the central heat exchanger design problem—the methods of predicting the illusive heat transfer coefficients—we shall have come to two major conclusions that provide a rationale for the title and the organization of this book.

Conclusion 1: The Art Usually Precedes the Science. Motivated by the need to invent new technologies or improve existing analytical methods, pragmatic thermal engineers (and application-oriented physicists) simply ignored the official arbitrary boundaries between the three branches of thermal sciences by combining thermodynamics and fluid flow to design better steam or gas turbines and compressors, while heat transfer engineers, following the lead of Osborne Reynolds and his disciple T. E. Stanton [1.11], improved the methods of predicting convective heat transfer coefficients by coupling the laws of fluid and heat flow. In other words, thermal and heat transfer engineers anticipated, more than a century ago, fundamental concepts of nonequilibrium thermodynamics and the unity of thermal sciences, thus confirming the well-known adage that the art usually precedes the science.

Conclusion 2: Practical Advantages of a Unified Treatment in Heat Transfer Engineering. Fully recognizing the arbitrariness of the barriers between the three traditional branches of thermal sciences and the fruitful results of the symbiosis between (1) fluid flow and thermodynamics on the machinery side (turbo machinery) and (2) fluid flow and forced convection on the appliance side (steady-state heat exchangers) of thermal systems, we conclude the following:

- The early *nineteenth century barriers* between the three traditional branches of thermal sciences are both *unnecessary* and totally *counterproductive* in the case of heat exchangers designed to operate under steady-state conditions.

- The unified thermodynamic treatment of momentum, heat, and mass transfer (UTT-MHMT), the central feature of this book, solves complex convective heat exchanger design problems more effectively.

The text is structured with this prime utilitarian objective in mind.

1.4 SYSTEMS OF UNITS AND SYMBOLS

The international system of units (SI) is used almost exclusively. However, since a large amount of the available thermophysical data is still published in the (U.S.) technical system of units, it is occasionally necessary to make use of the technical system of units. For the benefit of practicing heat transfer engineers who are more comfortable with the old technical system or who are not familiar with the international system of units, the following steps have been taken: The SI units for fundamental and derived quantities are summarized in Tables 1.1 and 1.2, respectively, and the SI unit prefixes are summarized in Table 1.3. In addition, all the conversion factors relevant in thermal engineering applications are listed for convenience with the main physical constants in Appendix F.

It should be emphasized, however, that most of the novel design correlations recommended here are in *dimensionless form* and therefore *applicable in any coherent system of units*.

The symbols used throughout the book generally conform with U.S. and, where feasible, International Engineering Standards.

TABLE 1.1. SI Units for Fundamental Quantities

Fundamental Quantity	Unit Name	Unit Symbol
Length	Meter	m
Mass	Kilogram	kg
Time	Second	s
Temperature	Kelvin	K
Electric current	Ampere	A

Source: F. W. Schmidt et al., Introduction to Thermal Sciences, p. 15, Wiley, New York, 1984. Reproduced with permission.

TABLE 1.2. SI Units for Derived Quantities

Derived Quantity	Unit Name	Unit Symbol	Relationship to Other Units
Force	Newton	N	$m \cdot kg / s^2$
Pressure or stress	Pascal[a]	Pa	N / m^2
Energy	Joule	J	$N \cdot m$
Power	Watt	W	J / s
Electric charge	Coulomb	C	$A \cdot s$
Electric potential	Volt	V	W / A
Electric resistance	Ohm	Ω	V / A

Source: Adapted from F. W. Schmidt et al., Introduction to Thermal Sciences, p. 15, Wiley, New York, 1984. Reproduced with permission.

[a]By definition, the pascal $Pa \equiv N / m^2 \equiv Nm / m^3 \equiv J / m^3$ and also represents a specific energy per unit volume. Therefore, the ratio P / ρ which appears frequently in fluid mechanics, where ρ is the specific mass in kg / m^3, has the dimensions $N \cdot m / kg \equiv J / kg$, representing a specific energy per unit mass.

TABLE 1.3. SI Unit Prefixes

Multiplication Factor	Prefix Name	Prefix Symbol
10^{-12}	Pico	p
10^{-9}	Nano	n
10^{-6}	Micro	μ
10^{-3}	Milli	m
10^{3}	Kilo	k
10^{6}	Mega	M
10^{9}	Giga	G
10^{12}	Tera	T

Source: F. W. Schmidt et al., Introduction to Thermal Sciences, p. 16, Wiley, New York, 1984. Reproduced with permission.

NOTES

1. Grassmann [1.4] estimates that the internal energy increase of 1 kg of water heated from 0 to 100°C would correspond to a mass increase of only 4.7×10^{-12} kg or a relative change of 4.7×10^{-10}%!
2. His use of a somewhat archaic language (science of fire) and obsolescent concepts (the "caloric fluid") may explain in part why Sadi Carnot was ignored by his compatriots until 1878 when his pioneering work was republished and widely disseminated in France [1.8, 1.9].
3. However, the fundamental concept of the existence of a new thermodynamic state property (later called entropy by Clausius) was derived earlier from an extension of the maximum Carnot cycle efficiency principle stating that for any fully reversible closed thermodynamic cycle, the algebraic sum $\Sigma Q/T = 0$ generalized to the integral form $\oint Q/T = 0$, which is designated the *Clausius integral* in the German language scientific literature. The equation $\oint Q/T = 0$ was in fact independently derived by R. Clausius and W. Thompson (Lord Kelvin), who communicated their findings, respectively, in December and May of the same year, 1854. Therefore, as indicated by Rudolf Plank in a footnote on p. 45 of his *Thermodynamische Grundlagen* [1.12], Thompson communicated his demonstration of this most fundamental law of classical thermodynamics a few months ahead of Clausius. This comment is corroborated in Thompson's biography [1.13].

CHAPTER 2

INVISCID FLUID MECHANICS

A number of eighteenth century mathematicians and physicists laid down the foundations of hydrodynamics, the precursor of modern fluid mechanics, by extending with rigorous mathematical logic Newton's laws of motion to continuous media for hypothetical ideal frictionless (inviscid) incompressible liquids. The names of three mathematical physicists from Basel, Switzerland, figure prominently among the early pioneers of hydrodynamics or hydraulics: two members of the illustrious Bernoulli family [2.1], Johann Bernoulli (1667–1748) and his son Daniel Bernoulli (1700–1782), and their close friend Leonhard Euler (1707–1783).

2.1 THE BERNOULLI THEOREM

Like his friend Euler, Daniel Bernoulli had the distinction of having won the prize of the French Academy of Sciences several times and has been called the founder of mathematical physics [2.2, 2.3]. His pioneering works in fluid flow were consolidated in his famous book *Hydrodynamica*, published in 1733 during his stay in St. Petersburg, Russia. More practice oriented than his father Johann or Euler, Daniel Bernoulli actually carried out experiments to validate several of his many novel analytical solutions in the field of hydraulics. He was the first to derive practical engineering solutions for fluid flow inside long conduits of arbitrary cross sections and to propose the concept of ship jet propulsion. Besides his research work in hydrodynamics he was also very productive and innovative [2.2–2.5] in other fields of physics, including the *kinetic* theory of *gases*, as evidenced by the fact that all his publications are to be consolidated in a series of eight volumes [2.6] with a 1987 completion target date for the two volumes on hydraulics.

According to Bernoulli's theorem, the sum of the pressure head, the velocity head, and the potential head, defined as the total head, remains constant between two points. The total head is therefore conserved, or

$$H_{tot} = \frac{P}{\rho g} + \frac{\mathbf{V}^2}{2g} + z = \text{constant} \tag{2.1}$$

where $\mathbf{V}$ = a vector in the general three-dimensional formulation
H_{tot} = total head
$P/\rho g$ = static head
$\mathbf{V}^2/2g$ = velocity head
z = potential head with reference to an arbitrary horizontal reference plane

Equation (2.1) is incomplete without a clarification of its validity limits per Guideline 2. In the case of purely analytical solutions this can best be approached by a full statement of the simplifying axiomatic assumptions made to derive Eq. (2.1) as shown below.

Theoretical Basis of Bernoulli's Theorem

1. Incompressible fluid (ρ = constant).
2. Inviscid fluid (zero friction dissipation).
3. No work is done by or on the fluid.
4. Steady-state steady flow SSSF (i.e., time independent).
5. Validity for any points along a streamline.

Note that by definition the velocity vectors $\mathbf{V}$ at a given instant of time must be tangent to the *streamline*. In the special SSSF case under consideration the streamline also represents the actual path of the fluid particles.

By multiplying all the terms of Eq. (2.1) by a constant (ρg), Bernoulli's equation can also be written in terms of pressures:

$$P_{tot} = P + \rho\frac{V^2}{2} + \rho g z = \text{constant} \tag{2.2}$$

where P_{tot} = total pressure
P = static pressure
$\frac{1}{2}\rho V^2$ = dynamic pressure
$\rho g z$ = potential pressure (due to gravity field)

Equation (2.2) simply states that when the values of P, $\frac{1}{2}\rho V^2$, and $\rho g z$ are known at one point in the flow, the sum P_{tot} of these three individual pressure terms will remain constant at all other points *along the same streamline*.

Johann Bernoulli's contributions to the field of hydraulics have long been blemished by the stigma of plagiarism against his son as a result of the publication in 1742 of Johann's treatise on hydraulics entitled *Hydraulica ... anno 1732*! By predating *Hydraulica* 10 years to 1732, or one year earlier than the 1733 publication of Daniel's *Hydrodynamica*, he caused a bitter conflict of priority with his son, who as a result gave up further research work in the field of hydraulics [2.5–2.7]. More recent investigations seem to indicate that the stigma of plagiarism may not be fully justified since, as Leonhard Euler pointed out, Johann's *Hydraulica* not only contains some novel concepts but actually extends the generality of Daniel Bernoulli's theorem to nonstationary flow conditions!

Leonhard Euler was strongly influenced initially by his teachers at the Basel University, the senior Bernoullis, but the student soon surpassed his masters. Euler's original research work, spanning the field of pure mathematics, theoretical physics, engineering, astronomy, and philosophy, is so extensive that 74 volumes have already been published since 1911 under the general title *Leonhard Euler—Opera Omnia* [2.8] and the end of this series is not yet in sight!

Euler's most significant contribution to the field of hydrodynamics is his publication in 1755 of the equations known in the scientific literature as *Euler's equations of motion* or Euler's differential equations. These and Euler's generalized form of the continuity equations provide a complete analytical tool for the solution of incompressible inviscid fluid flow problems. From these equations it is possible to derive more rigorously Bernoulli's theorem for a streamline as demonstrated by Euler [2.9]. This may explain why Euler is also given credit for this theorem instead of Daniel or Johann Bernoulli in the scientific literature [2.4]. Although reputed to be more a pure mathematician than a physicist, let alone an engineer [2.1], Euler nevertheless published a memoire (1754) on a conceptually new and workable reaction turbine design.[1]

On the basis of Euler's more exact equations of motion, one can demonstrate that the static pressure P remains constant in a flow whose streamlines are straight (zero curvature) and horizontal and one can finally derive rigorously the validity of the special and quite useful case of Bernoulli's theorem discussed below.

2.2 SPECIAL BERNOULLI EQUATION: HORIZONTAL STRAIGHT STREAMLINE

In this special case the elevation term in Eq. (2.2) disappears and we obtain, using the alternative modern definition $P_0 \equiv P_{\text{tot}} \equiv$ stagnation pressure,

$$P_0 = P + \tfrac{1}{2}\rho V^2 = \text{constant} \tag{2.3}$$

Equation (2.3) shows that while the stagnation pressure remains constant on a horizontal streamline, the static pressure P increases when the velocity decreases and conversely. The static pressure P therefore reaches a maximum

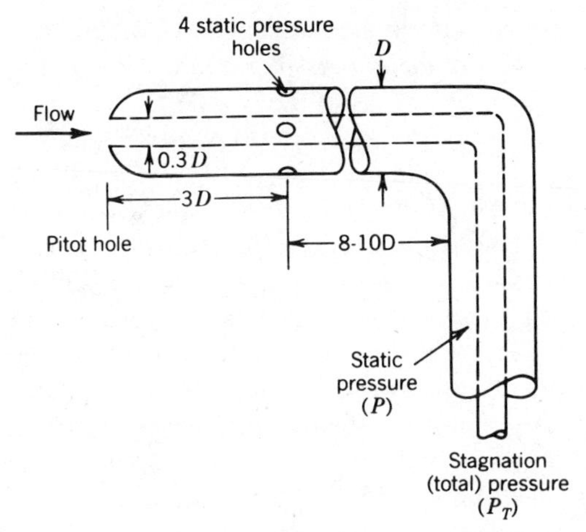

FIGURE 2.1. Velocity measurement by Pitot tube technique. [From F. W. Schmidt et al., "Introduction to Thermal Sciences," p. 177, Wiley, New York, 1984. Reproduced with permission.]

value $P = P_0$ at that point in the flow field where the fluid is brought to rest (reversibly) and $\frac{1}{2}\rho V^2 = 0$. This is the basis for the measurement of fluid velocity with the Pitot tube. This instrument (Fig. 2.1) measures the difference between the stagnation and static pressures $(P_0 - P)$ from which the fluid velocity at a point in the flow stream is computed according to Eq. (2.3):

$$V = \left(\frac{2(P_0 - P)}{\rho} \right)^{1/2} = \sqrt{\frac{2(P_0 - P)}{\rho}} \tag{2.4}$$

The Pitot tube must be properly aligned parallel to a streamline and carefully calibrated in a known flow to compensate for the effect of its intrusion into the flow field.

The Bernoulli equation (2.4) also provides the theoretical basis for the nozzle, orifice, and venturi tube flow measurement methods described in standard engineering handbooks [2.10, 2.11].

2.3 ZERO VELOCITY LIMITING CASE: HYDROSTATICS

Incompressible fluids at rest satisfy all the theoretical validity requirements of Bernoulli's theorem (points 1–5, p. 9) for the limiting case of a vanishing velocity field, including real (viscous) fluids, since there are no tangential stresses without fluid motion. For this special limiting case, the dynamic velocity terms in Eqs. (2.1) and (2.2) vanish and Bernoulli's theorem converges to

$$\frac{P_1}{\rho g} + z_1 = \frac{P_2}{\rho g} + z_2 \tag{2.5a}$$

Regrouping to express the hydrostatic pressure at different levels below the horizontal surface of any incompressible liquid, we obtain

$$P_1 = P_2 + \rho g(z_2 - z_1) \tag{2.5b}$$

where P_2 is the constant pressure at the surface and z_2 and z_1 are the elevations at and below the surface, respectively. This relationship clearly shows that the pressure P_1 at any point below the surface increases linearly with the depth $\Delta z = z_2 - z_1$ for any fluid of constant specific weight (ρg). For any liquid in a container opened to atmospheric pressure, P_2 is the constant atmospheric pressure P_{atm} and the hydrostatic pressure increase $(P_1 - P_{atm})$ at any point z_1 below the surface is equal to the product $\rho g(z_2 - z_1)$, that is, the weight of a column of fluid above a unit area, in accordance with well-known laws of hydrostatics. Thus, Bernoulli's theorem converges, as it should, to the older laws of hydrostatics in the limiting case of a vanishing velocity ($\mathbf{V} = 0$).

2.4 NUMERICAL ANALYSES

In contrast to Daniel Bernoulli's more pragmatic attitude, the mathematical physicists Leonhard Euler and Johann Bernoulli adopted the dogmatic position that their analytical solutions required no further empirical verification. This standpoint, based entirely on pure mathematical logic, fails to recognize the impact of frictional effects in many engineering applications where the viscosities of real fluids may play a significant role.

We assess the practical value and limitations of the Bernoulli equations in a few typical engineering fluid flow problems which also provide a rational basis for real (i.e., viscid) fluid flow analysis with energy dissipation.

NUMERICAL ANALYSIS 2.1

An ideal liquid flows from a large open reservoir through a well-rounded smooth circular nozzle as shown in Fig. 2.2.

Problem Statement

Determine analytically from Bernoulli's theorem (a) the fluid velocity at nozzle discharge, (b) the volumetric flow rate, (c) the mass flow rate, (d) the effect of fluid density, and (e) the virtual velocity and mass flow rate for water with a specific mass $\rho = 1000\ \text{kg/m}^3$ for $\Delta z = 10$ m and an orifice diameter of 2 cm $= 2 \times 10^{-2}$ m. (f) Compare the above theoretical results with those obtained from empirically validated methods described in standard engineering handbooks.

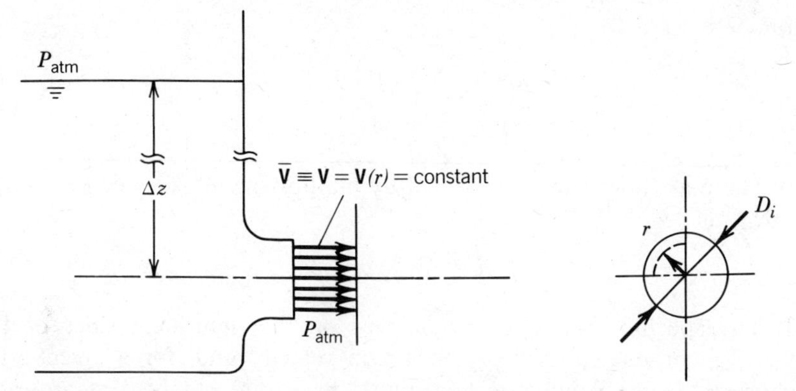

FIGURE 2.2. Ideal inviscid fluid flow from a large reservoir through a circular nozzle.

Solution

(a) Because the nozzle exits to the atmosphere, the static pressure term in Eq. (2.1) at the water surface and nozzle outlet cancel out, and since the fluid velocity in the large reservoir is negligibly small compared to the nozzle velocity, Eq. (2.1) yields

$$\frac{V_1^2}{2g} + z_1 = 0 + z_2 \tag{2.6}$$

Solving for $V_1 \equiv V_{\text{virt}}$, we obtain

$$V_{\text{virt}} = (2g\,\Delta z)^{1/2} \propto (\Delta z)^{1/2} \tag{2.7}$$

where V_{virt} is the virtual velocity and $\Delta z = z_2 - z_1$.

The term *virtual velocity* has been selected to emphasize the fact that V_{virt} represents an *idealization valid only for quasifrictionless fluid flow applications*. Since the effect of friction is to resist motion, it can logically be expected that the value of V_{virt} obtained from Eq. (2.7) will overestimate the real V—hence the use of "virtual" in the sense of a maximum potential velocity in the absence of frictional effects.

(b) For ideal frictionless liquids, the velocity is uniform throughout the nozzle cross section (i.e., there is no "sticking" at the wall) and the volumetric flow rate $\dot{m}/\rho$ is simply

$$\frac{\dot{m}}{\rho} = V_{\text{virt}}\Omega \propto \Omega \tag{2.8}$$

where Ω is the nozzle cross section, or in the special case of a circular cross section ($\Omega = \pi D^2/4$),

$$\frac{\dot{m}}{\rho} = V_{\text{virt}}\left(\frac{\pi D^2}{4}\right) \propto D^2 \tag{2.9}$$

(c) The mass flow rate $\dot{m}$, obtained by multiplying all sides of Eq. (2.8) or (2.9) by ρ, is

$$\dot{m} = \Omega V_{\text{virt}}\rho \propto \rho \tag{2.10}$$

(d) The specific mass does not appear on the right-hand sides of Eqs. (2.7)–(2.9); consequently, the same nozzle velocity and, for a given orifice diameter, the same volumetric flow rates ($\dot{m}/\rho$) are obtained regardless of fluid density for a given Δz. On the other hand, Eq. (2.10) shows that the mass flow rate is directly proportional to the specific mass.

(e) From Eqs. (2.7) and (2.10) we obtain with $g = 9.81\ \text{m/s}^2$ and $\Omega = (\pi/4)(2 \times 10^{-2})^2$

$$V_{\text{virt}} = (2 \times 9.81 \times 10)^{1/2} = 14.01\ \text{m/s}$$

$$\dot{m}_{\text{virt}} = 1000 \times 14.01 \times 3.14 \times 10^{-4} = 4.40\ \text{kg/s}$$

(f) The actual flow rate according to Refs. 2.10 and 2.11 is

$$\dot{m} = C\dot{m}_{\text{virt}} \tag{2.11}$$

where C is an empirical factor allowing for friction and flow contraction effects. For smooth well-rounded nozzles, the empirical constant C has a typical value of 0.97 for the stated conditions. Therefore, the water flow rate computed from inviscid fluid flow theory in the case under consideration overstates, as expected, the actual flow rate by some 3%. Thus, $\dot{m} = 0.97\,\dot{m}_{\text{virt}}$ or, stated along the lines suggested in Guideline 2,

$$\dot{m} = \dot{m}_{\text{virt}} \tag{2.12}$$

with an estimated deviation of $+0$ to -3%.

Comments

1. Although the above conclusions may seem somewhat surprising at first, they make sense when one recalls that we are dealing with idealized "frictionless" fluids. Bernoulli's theorem is essentially an extension to continuous media of the classical theorem of dynamics stating that a solid body initially at rest at an elevation $\Delta z = z_2 - z_1$ above ground level at z_1 will accelerate in a free fall to the same end velocity $V = [2g(z_2 - z_1)]^{1/2}$ regardless of its density when it reaches ground level in the *absence* of any air *frictional resistance* or dissipation.[2]

2. The inviscid flow theory provides quite a good estimate of fluid flow rates in the particular instance of a well-rounded short nozzle where viscous and contraction effects are negligibly small compared to the available potential pressure differential $\rho g \Delta z$.

NUMERICAL ANALYSIS 2.2

The smooth orifice described in Numerical Analysis 2.1 now forms the inlet of a straight horizontal pipe of identical internal diameter D as shown in Fig. 2.3. In fluid flow and heat transfer analyses, the total length of the pipe is conveniently expressed as a dimensionless ratio L/D. We shall assume here values ranging from $L/D = 100$ to 1000 typically encountered in thermodynamically efficient fluid to fluid tubular heat exchangers.

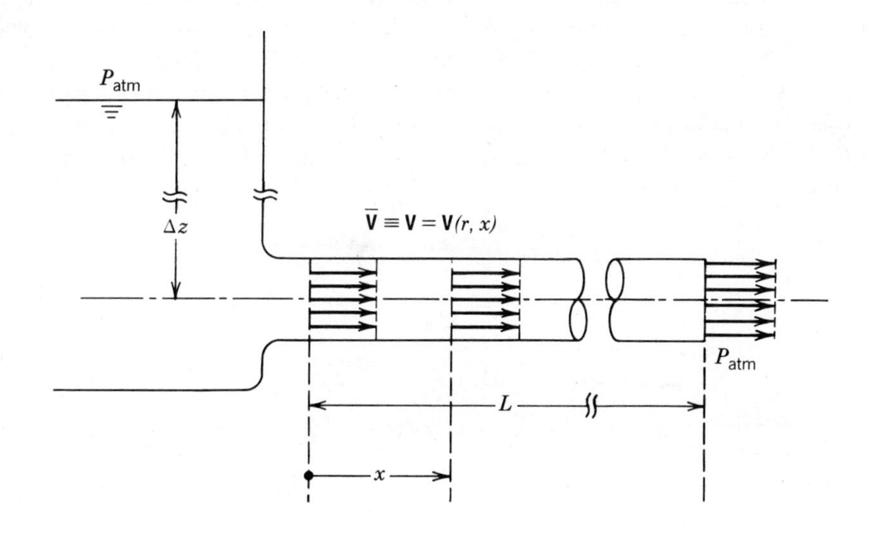

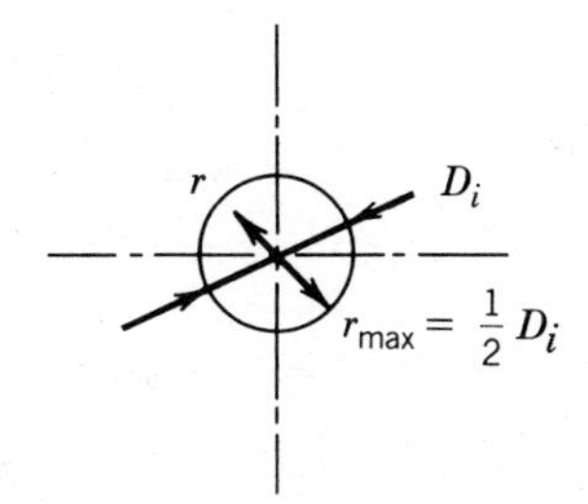

FIGURE 2.3. Ideal inviscid fluid flow from a large reservoir through a circular nozzle into a constant diameter horizontal pipe.

Problem Statement

Under otherwise identical conditions as in the previous examples, (a) determine the velocity distribution within the tube, (b) find the mass flow rate, (c) assess the plausibility of the results obtained on the basis of inviscid fluid flow theory under (b), and (d) anticipating nineteenth century thermodynamics concepts (one-dimensional energy equation), derive a suitable analytical method to account for the resistance to flow of real fluids. The analytical method is to be independent of the types of flow regime (streamline or turbulent in the case of single-phase fluids) and therefore generally applicable, regardless of flow regimes, as long as steady-state steady-flow (SSSF) conditions are possible.

Solution

(a) In the absence of viscous effects, the ideal fluid "slips" by the wall boundary at bulkstream velocity, the velocity $\mathbf{V}(x, r)$ is uniform throughout the full tube cross section Ω (no boundary layer), and the constant local velocity $\mathbf{V}(x, r)$ coincides with the average velocity $\overline{\mathbf{V}}$ as sketched in Fig. 2.3. Since we are dealing with incompressible fluids (ρ = constant) flowing in tubes of constant cross section Ω, we obtain from the mass conservation law under SSSF conditions, $\dot{m}_{\text{inlet}} = \dot{m}_{\text{outlet}}$ = constant, the following continuity equation:

$$\dot{m} = G\Omega = \overline{V}\rho\Omega = \text{constant} \tag{2.13a}$$

where by definition

$$G \equiv \frac{\dot{m}}{\Omega} = \text{mass velocity} \tag{2.13b}$$

Referring to Fig. 2.3, we can summarize the above conclusions as follows:

$$\overline{V} \equiv \frac{G}{\rho} \equiv \frac{\dot{m}}{\rho\Omega} = \text{constant} \equiv |\mathbf{V}(x, r)| \tag{2.14a}$$

and the local velocity $\mathbf{V}(x, r)$ is independent of the radial and axial position in the tube

Validity: Ideal inviscid incompressible fluid under SSSF conditions (2.14b)

(b) Since frictional effects are not considered (axiomatic assumptions of eighteenth century hydrodynamics theory) the values of V_{virt} and $\dot{m}_{\text{virt}}$ computed in the previous example with $L/D = 0$ remain unchanged at any value of L/D from 0 to ∞!

(c) Empirical evidence clearly shows that the actual mass flow rate $\dot{m}$ decreases as L/D increases in the case of real (viscous) incompressible

(ρ = constant) fluids or, expressed more generally in dimensionless terms,

$$\varphi \equiv \frac{\dot{m}}{\dot{m}_{\text{virt}}} = \frac{G}{G_{\text{virt}}} = \frac{\overline{V}}{V_{\text{virt}}} \leq 1 \tag{2.15}$$

where V_{virt} is computed according to Eq. (2.7). In the limiting case $L/D \to 0$, φ approaches unity and $\dot{m} = \dot{m}_{\text{virt}}$ per Eq. (2.11) when $C \cong 1$ and the frictional effects are negligible compared to the available head Δz. However, for extremely long tubes when $L/D \gg 1$, or at the limit $L/D \to \infty$, $\varphi \to 0$ and the results obtained from *inviscid hydrodynamics* theory become *totally inaccurate*.

(d) The effect of irreversible energy dissipation can be estimated by recognizing the empirical fact that the net head available to accelerate the fluid from rest in the large reservoir to $\overline{V}$ at tube outlet must be less than Δz assumed in Eq. (2.7) by an amount equivalent to Δz_f because of the dissipation of energy from tube inlet to outlet: that is,

$$(\Delta z)_{\text{net}} = \Delta z - \Delta z_f \tag{2.16}$$

Introducing Eq. (2.16) into Eq. (2.7), we obtain

$$\frac{\overline{V}^2}{2} = g\,\Delta z - g\,\Delta z_f \tag{2.17a}$$

or

$$\frac{\overline{V}^2}{2} + g\,\Delta z_f = g\,\Delta z \tag{2.17b}$$

Without making any assumption (until Chapter 3) for the analytical prediction of the term $g\,\Delta z_f \equiv \Delta H_f \equiv \Delta P_f/\rho$, it is *permissible* and *convenient* to introduce the concept of *number of velocity heads* $(\text{NVH})_f$ defined as follows:

$$g\,\Delta z_f \equiv \Delta H_f \equiv \Delta P_f/\rho \equiv (\text{NVH})_f(\overline{V}^2/2) \tag{2.18a}$$

or

$$(\text{NVH})_f \equiv \frac{\Delta H_f}{\overline{V}^2/2} \equiv \frac{\Delta P_f}{\rho\overline{V}^2/2} \equiv \frac{g\,\Delta z_f}{\overline{V}^2/2} \tag{2.18b}$$

Substituting the expression of $g\,\Delta z_f = \Delta H_f = \Delta P_f/\rho$ per Eq. (2.18a) in Eq. (2.17b), solving for $\overline{V}$, and recalling that $V_{\text{virt}} = (2g\,\Delta z)^{1/2}$ per Eq. (2.7), we obtain

$$\overline{V} = \frac{(2g\,\Delta z)^{1/2}}{[1 + (\text{NVH})_f]^{1/2}} = \frac{V_{\text{virt}}}{[1 + (\text{NVH})_f]^{1/2}} \tag{2.19}$$

and with Eq. (2.15) the following general dimensionless expression in terms of $(\mathrm{NVH})_f$

$$\varphi = \frac{\dot{m}}{\dot{m}_{\mathrm{virt}}} = \frac{G}{G_{\mathrm{virt}}} = \frac{\overline{V}}{V_{\mathrm{virt}}} = \frac{1}{\left[1 + (\mathrm{NVH})_f\right]^{1/2}} \tag{2.20a}$$

or alternatively

$$\varphi^2 = \left(\frac{\dot{m}}{\dot{m}_{\mathrm{virt}}}\right)^2 = \left(\frac{G}{G_{\mathrm{virt}}}\right)^2 = \left(\frac{\overline{V}}{V_{\mathrm{virt}}}\right)^2 = \frac{1}{1 + (\mathrm{NVH})_f} \tag{2.20b}$$

Equations (2.20a) and (2.20b) are useful to determine either the actual $\dot{m} = \varphi(\dot{m}_{\mathrm{virt}})$ when the term $(\mathrm{NVH})_f$ is known or the required $(\mathrm{NVH})_f$ for a prescribed velocity ratio φ by solving Eq. (2.20a) or (2.20b) for $(\mathrm{NVH})_f$: that is,

$$(\mathrm{NVH})_f = \left(\frac{1}{\varphi}\right)^2 - 1 = \frac{1}{\varphi^2} - 1 \tag{2.21}$$

which converges to

$$(\mathrm{NVH})_f = \frac{1}{\varphi^2} \tag{2.22}$$

when $\varphi^2 \ll 1$ at $L/D \gg 1$.

Comments

It is clear from Eqs. (2.19)–(2.21) that φ converges, as it should, toward unity and $\overline{V}/V_{\mathrm{virt}} = \dot{m}/\dot{m}_{\mathrm{virt}} = 1$ when $(\mathrm{NVH})_f \to 0$. Conversely, as $\varphi \to 0$, $\overline{V}/V_{\mathrm{virt}} = \dot{m}/\dot{m}_{\mathrm{virt}} \to 0$ when $(\mathrm{NVH})_f \to \infty$. Thus, the term $(\mathrm{NVH})_f$ provides a convenient means of estimating the real mass flow rate $\dot{m}$ for a given total head when $(\mathrm{NVH})_f$ is known or prescribed. We shall take full advantage of the dimensionless friction dissipation term $(\mathrm{NVH})_f$ in subsequent fluid flow and forced convection heat exchanger analyses.

2.5 FROM REVERSIBLE TO IRREVERSIBLE FLUID FLOW

Anticipating nineteenth century thermodynamic concepts not known to the Bernoullis and Euler, we see that the theory of inviscid fluid flow represents a special case within a broader class of idealized physical processes defined as *reversible processes* in classical thermodynamics. This conclusion is obvious when one recalls that the isothermal (constant temperature) theory of hydrodynamics is based solely on one universal gravitational constant g and Newton's classical laws of dynamics which are inherently reversible[3] in the absence of frictional effects [2.12 to 2.14].

To illustrate this point in a more concrete fashion, consider the ideal hydrodynamic system sketched in Fig. 2.4. According to inviscid hydrodynamic theory, the power produced under SSSF conditions by a perfect (100% efficient) hydraulic turbine (motor) is exactly equal to $\dot{m}g\,\Delta z$ and therefore exclusively a function of Δz for a given $\dot{m}$. In other words, the higher the value of Δz the higher the specific work output, and in the limiting case when $\Delta z \to 0$ the potential work output vanishes.

Conversely, if we conceptually reverse the flow direction by substituting a perfect (100% efficient) pump for the turbine as indicated by the dashed lines in Fig. 2.4, the power input required to lift the same mass flow rate $\dot{m}$ against the same Δz is exactly equal, in absolute value, to the turbine power output $\dot{m}g\,\Delta z$. We can therefore visualize two identical (twin) systems operating concurrently in parallel: one in the turbine-operating mode supplying exactly the amount of power required in the second system to pump "uphill" the same amount of water $\dot{m}$ flowing "downhill" through the hydraulic turbine. Thermodynamically, this combined system is completely *reversible* [2.15].

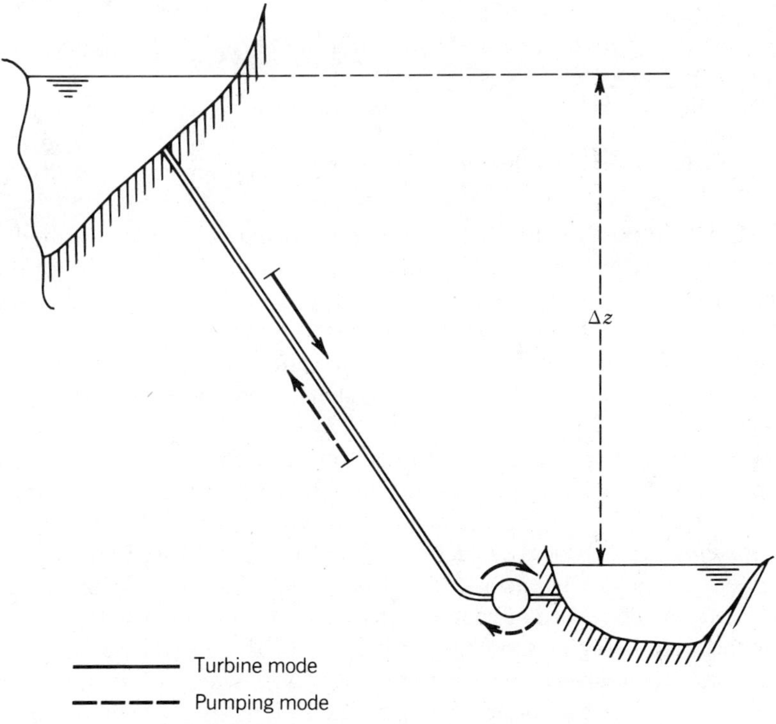

FIGURE 2.4. Twin turbine/pump scheme.

The situation changes radically when real (viscid) fluid dissipation effects are considered by accounting for the unavoidable friction head loss in the system $g\,\Delta z_f = \Delta H_f$ defined in Numerical Analysis 2.2. Since the end effect of frictional resistance is to resist flow, as the name implies, the available net head in the turbine mode is reduced and the ideal turbine output is only $\dot{m}g(\Delta z - \Delta z_f)$. Conversely, the hydraulic pump has to overcome a higher net head requiring a correspondingly higher power input $\dot{m}g(\Delta z + \Delta z_f) > \dot{m}g(\Delta z - \Delta z_f)$. Consequently, in the case of real fluids, the pump power consumption exceeds the turbine output by an amount

$$\dot{m}g(\Delta z + \Delta z_f) - \dot{m}g(\Delta z - \Delta z_f) = 2(\dot{m}g\,\Delta z_f)$$

This power dissipation can be expressed by the following equivalent equations:

$$\dot{m}(2g\,\Delta z_f) = \dot{m}(2\,\Delta H_f) = \Omega\left(\frac{G}{\rho}\right)(2\,\Delta P_f) = \Omega\bar{V}(2\,\Delta P_f) \qquad (2.23)$$

where

$2g\,\Delta z_f = 2\,\Delta H_f =$ total system friction dissipation in $\mathrm{N \cdot m/kg = J/kg}$

$2\,\Delta P_f =$ total system friction pressure drop, in $\mathrm{N/m^2 = N \cdot m/m^3 = J/m^3}$

$\dot{m} = \dot{m}_{\text{turbine}} = \dot{m}_{\text{pump}}$ in Kg/s

$\Omega G/\rho = \Omega G v = \Omega\bar{V} =$ volumetric flow rate in m^3/s

Thus, it is impossible to lift the same rate of water, $\dot{m}$, back to the higher reservoir without an additional external source of power supply to offset the irreversible friction losses $2\dot{m}\,\Delta H_f$ expressed in J/s = W. Thermodynamically the real (viscid) combined fluid flow system[4] is typically *irreversible* even in the idealized case of 100% efficient turbines and pumps because of internal friction dissipation within the complete system schematically described in Fig. 2.4.

By the seemingly innocuous introduction of a dissipative term ΔH_f, we have in effect made a momentous conceptual step from the simpler idealized (utopian) world of *reversible* phenomena to the vastly more complex physics of *irreversible* real processes [2.13, 2.14]. This is discussed more fully in subsequent chapters where some useful new analytical heat exchanger design tools based on such thermodynamic considerations are proposed.

For the time being it is sufficient to note that the dissipative term $\Delta H_f = (\mathrm{NVH})_f V^2/2$ can always be expressed as $T\Delta s_f$, where T is the constant absolute temperature (isothermal) and $\Delta s_f > 0$ is the exclusive result of internal dissipation. This internal nonnegative entropy production term, Δs_f, makes it possible to differentiate between idealized hydrodynamics and real fluid flow, since Δs_f is always positive in real (viscous) flow phenomena, whereas $\Delta s_f = 0$ applies only in the limiting idealized case of inviscid fluid flow. Because T and Δs_f are two of the most important thermodynamic state properties, a direct connection or "coupling" is thereby established between real fluid flow mechanics and thermodynamics, at least in principle. The practical advantages of this "coupling" will become more evident later, particularly when we address two-phase flow applications.

If one accepts literally the definition that *entropy is thermodynamics*, then it follows logically that real (viscid) fluid flow may (or must) be considered an integral part of thermodynamics [2.16–2.18]. This is indeed the central theme of Ackeret's 1961 Guggenheim Memorial lecture entitled "The Role of Entropy in the Aerospace Sciences" [2.16]. Equally at home in the fields of aerodynamics and thermodynamics, Ackeret was one of the early disciples of L. Prandtl [2.19], the leading pioneer of *boundary layer* fluid flow theory. He was greatly influenced, like all mechanical engineering students and educators at the Zurich Institute of Technology, by the traditional emphasis on the second law of thermodynamics under such world reknowned teachers as Clausius, the "inventor" of the state property "entropy," and Stodola,[5] who is considered by German speaking[6] thermal engineers as the "father" of the steam and gas turbine theory. In the introduction of his 1961 Guggenheim lecture on the role of entropy in aerospace sciences, Ackeret makes a special point to stress how appropriate it is to lecture on entropy in Zurich, the "cradle of entropy."

NOTES

1. A working model shown at the Swiss Federal Institute of Technology in Zurich on the occasion of the 200th anniversary celebration of Leonhard Euler's death indicated that the hydraulic turbine performed right on target with an acceptable efficiency, notwithstanding the inherent limitations of inviscid fluid flow theory! A similar comparison between the early steam engines and modern steam turbines would not be nearly as favorable and this may explain the rather condescending attitude of the early hydraulics pioneers regarding the first technological applications of the "science of fire" in eighteenth century England with Newcomen's and Watt's steam engines [2.20] before Sadi Carnot had laid down the scientific foundation of thermodynamics in 1824. It is noteworthy that in the case of the steam engine the art again preceded the science by more than a century, as shown in the interesting article on Watt [2.20]. [The first concept of the steam engine came from the Marquis of Worcester around 1663, while large-scale production was carried out by such pragmatic pioneers as Captain Savary, Newcomen (an ironmonger), Crawley (a glazier), and Watt in the eighteenth century.]
2. This is usually shown in introductory physics courses by free-fall experiments in vacuum to demonstrate that a very low density ball of paper, or even a feather, can acquire the same velocity **V** as a steel ball for a given vertical drop in a drag-free medium.
3. Classical (reversible) dynamics can be formulated in a compact and elegant way in terms of a single function, called the *Hamiltonian* in honor of Ireland's greatest mathematician, William Rowan Hamilton. This is one of the main topics of the interesting account "An Irish Tradegy—Hamilton" in Ref. 2.1, see also "Mathematics and Theoretical Physics" by T. D. Spearman in Ref. 2.21, published on the occasion of the bicentennial (1785–1985) of the Royal Irish Academy.
4. This concept is of more than academic interest. It is in principle the scheme used in hydropower pumped storage stations as shown for instance in paragraph 1.9.6 of Ref. 2.22 entitled Pumped Storage. However, in actual pumped storage systems, as opposed to Fig. 2.4, the surplus power available from a lower-elevation stream is

used to divert some water from this stream to *store* it in a higher-elevation reservoir for *future* use.

5. In one of his papers Stodola addresses the following problem: measurement of temperature variation in the nozzle:

   > If the temperature at a certain point, say, at the mouth of the nozzle, is known, then in the case of saturated steam, the pressure is thereby determined, and the velocity can be obtained... The procedure, however, is purposeless, because no method is known that will determine the true temperature in flowing fluids. The friction of the fluid against the thermometer or thermo-couple generates heat affecting the readings to an extent as yet unknown.

   The question raised by Stodola is still very relevant, as shown in a recent (1984) paper by Eckert [2.23]. This problem provides another illustration of the intimate relationship and symbiose between classical engineering thermodynamics, fluid and heat flow, the central theme of this book.
6. Although "physical laws" know no national borders, the questions raised by scientists can certainly be influenced by their cultural heritage and environment. Di Francia [1.1] makes the interesting observation that nineteenth century English and German engineers/scientists were culturally better equipped to accept the new abstract thermodynamic theories pioneered by the French engineer Sadi Carnot in 1824 and his disciple E. Clapeyron in 1834 than the leaders of the French scientific community with their Cartesian mathematical logic! In this context it is important to make a sharp distinction between common language and nationality. A common language certainly created a scientific bond between German speaking thermal engineers of different nationalities regardless of the changing national boundaries. Thus, whenever we refer to the German literature in this book, we really mean publications in the German language.

CHAPTER 3

SINGLE-PHASE NEWTONIAN FLUID MECHANICS

A fluid is defined in contemporary textbooks as a substance that deforms continuously when it experiences a shearing or tangential stress and thereby flows. This broad definition includes the flow of single-phase fluids such as liquids (e.g., water and oil) or real gases (e.g., air and saturated or superheated vapors) as well as multiphase flow situations such as the technologically important two-phase flow of a liquid and its vapor phase in evaporators and condensers or two-phase flow of two different chemical species such as oil and air.

The prime engineering end objective of fluid flow analysis reflecting the definition of fluid mechanics can be translated into the following two distinct problem statements:

> ***GUIDELINE 6*** *Determine under* SSSF *conditions for single-phase or multiphase fluids and any configuration either* (a) *the flow rate* $\dot{m}$ *for a prescribed total* $(\Delta H)_{\mathrm{tot}} = (\Delta P/\rho)_{\mathrm{tot}}$ *or* (b) *the total* $(\Delta H)_{\mathrm{tot}} = (\Delta P/\rho)_{\mathrm{tot}}$ *for a prescribed flow rate* $\dot{m}$.

The problem statements expressed by Guideline 6 are simpler than their solutions, which become more and more complex as we move away from single-phase Newtonian fluids in laminar isothermal flow regimes to the more chaotic turbulent flow regimes typical of many heat exchanger applications, and entirely different semianalytical approaches are required for multiphase flow situations with and without external heat flow.

In this chapter we focus mainly on single-phase Newtonian fluids (except for a brief overview of inlet effects in two-phase flow) flowing in straight channels of constant cross section. The highly complex multiphase flow situations are addressed in Chapters 4 and 5. Novel thermodynamic tools, which are particularly useful in two-phase flow applications with evaporation and condensation in any type of geometric configuration, are described in Chapter 6.

3.1 NEWTONIAN AND NON-NEWTONIAN FLUIDS

By definition *Newtonian fluids* satisfy the law known today as Newton's equation of viscosity outlined in the second book of *Principia* (1687), in which Isaac Newton laid down the principles of the motion of bodies in resisting media and fluid motion [3.1]. This oldest of all phenomenological rate equations used in the study of transport phenomena simply postulates [3.2, 3.3] that a Newtonian fluid exhibits a linear relationship between the applied shear stress τ_{xy} and the fluid strain rate, expressed as $\partial V_x/\partial y$ for one-dimensional flow:

$$\tau_{xy} \propto \frac{\partial V_x}{\partial y} \tag{3.1}$$

The dynamic or absolute viscosity is by definition the proportionality factor in the linear expression

$$|\tau_{xy}| = \mu \frac{\partial V_x}{\partial y} \tag{3.2}$$

Newtonian fluids include all gases and most common liquids. The absolute viscosities of a number of Newtonian fluids are presented in Appendix A together with other key thermophysical properties.

A cursory review of the absolute viscosity data shown in Appendix A clearly shows that the impact of temperature variation is more significant for liquids than for gases. In the case of pure *liquids*, such as water, the viscosity *decreases* significantly with an *increase* in temperature from the triple (or freezing) point to the boiling point and ultimately the critical thermodynamic temperature, the upper existence limit of the liquid phase. By contrast, the viscosity of *gases increases* with *increasing* temperature. We will come back to this important difference when we make a preliminary assessment of the impact of nonisothermal conditions in laminar and turbulent flow regimes with gases and liquids in the last two sections of this chapter.

In the case of non-Newtonian fluids, such as molten plastics, the linear relationship according to Eq. (3.1) is not valid. Therefore, it is important to keep in mind that all theoretical *fluid flow equations derived from the axiomatic assumption of a constant viscosity are a priori not applicable to strongly non-Newtonian fluids* in accordance with the fundamental Guideline 2. We defer considerations of the more complex flow of two-phase mixtures exhibiting typical non-Newtonian characteristics [3.4] and focus our attention mainly on the flow of single-phase Newtonian fluids for the balance of the present chapter with special emphasis on isothermal (T = constant) applications.

3.2 LAMINAR AND TURBULENT FLOW REGIMES

On February 22, 1880 Osborne Reynolds [3.5] ushered in a radically new period in fluid flow mechanics and physics [3.6] when he presented the results of his research together with a physical demonstration of the "... experimental investigation on the circumstances which determine whether the motion of water shall be direct or sinuous... and the law of resistance in parallel channels... ." To visualize the "sinuous" or unsteady motion, Reynolds introduced a small amount of colored dye into the water at the entrance of a transparent glass tube to observe what he called a color band as it moved through the tube as highlighted in Fig. 3.1.

Reynolds clearly demonstrated [3.5, 3.7] that below a critical flow rate the streak of colored water is completely straight and steady, indicating an orderly streamline or laminar flow. He observed that this streamline flow persisted until a critical velocity $\overline{V}_{cr}$ was reached when the colored streak suddenly changed, indicating that the color band started fluctuating and dispersing into the main flow. Today the scientific literature refers to this type of flow regime as turbulent flow. In the years between 1880 and 1883, Reynolds carried out a large number of experiments with tubes of different sizes and different liquids and established that the transition to turbulence occurred in a relatively narrow range of a dimensionless group, now called the Reynolds number, defined for tube flow as follows:

$$\mathrm{Re} \equiv \frac{DG}{\mu} \equiv \frac{D\overline{V}\rho}{\mu} \equiv \frac{D\overline{V}}{\nu} \tag{3.3}$$

Experience has shown that at Re below a certain value, the critical Reynolds number $(\mathrm{Re})_{cr}$, the flow regime remains laminar regardless of the conditions prevailing at the tube inlet. It is generally accepted today that this critical Reynolds number occurs at $(\mathrm{Re})_{cr} = 2200$ in the case of a circular tube. However, the transition from laminar to turbulent flow can occur in a relatively broad band of Reynolds numbers. Under extremely careful test conditions (absence of vibrations, excellent surface conditions, etc.) stable

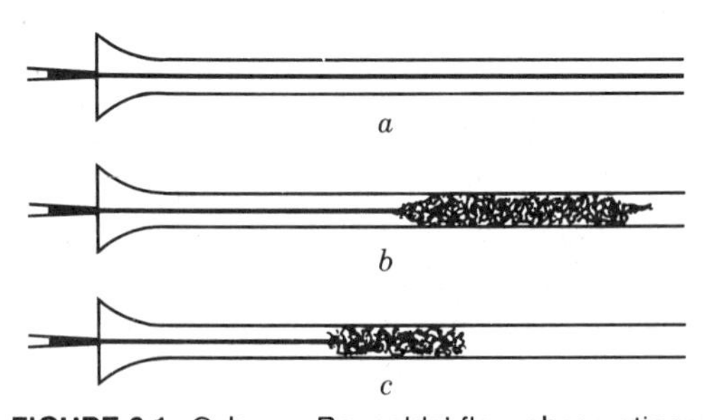

FIGURE 3.1. Osborne Reynolds' flow observations.

laminar flow regimes have been observed at a Reynolds number of the order of 70,000! However, under conditions typically encountered in normal heat exchanger design practice it is safe to assume that stable turbulent flow regimes occur at a Reynolds number in excess of 10,000. Most of the fluid flow and heat transfer coefficient correlations recommended in heat exchanger design practice for turbulent flow regimes are based on a conservative limit Re > 10,000 in the case of tube flow. The transition zone extending roughly from Reynolds number 2000 to 10,000 represents an uncertain operation band discussed later in this chapter.

The key characteristics of these two radically different stable flow regimes can be summarized as follows:

Laminar Regimes

- The streamlines are parallel and the momentum transfer (friction) results from molecular diffusion only.
- The velocity, pressure, temperature, and specific mass ρ are completely independent of time.
- The central fluid flow problems expressed by Guideline 6 can be solved fully analytically under isothermal conditions, at least in principle.

Turbulent Regimes

- It is no longer possible to talk of streamlines. The motion of macroscopic "packets" of fluid is three dimensional and exhibits nondeterministic (randomness) characteristics.
- The flow patterns within the tube are unsteady on a macroscopic scale even under SSSF conditions at the channel inlet and outlet.
- It is recognized that the general momentum conservation equations are valid at any given instant. However, the *nonlinearity* of the Navier–Stokes equations in the case of turbulent flow [3.8–3.11] excludes a fully analytical solution and therefore it is necessary to resort to experimentation to solve the central fluid flow problems expressed by Guideline 6.

Because of the radically different characteristics of the two basic single-phase Newtonian fluid flow regimes, the first step in any fluid flow analysis is to identify the types of flow regime as illustrated in the following numerical analysis.

NUMERICAL ANALYSIS 3.1

Necessary $(NVH)_f$ for turbulent and laminar regimes in long circular tubes $(L/D \gg 1)$.

Problem Statement

A smooth orifice forms the inlet of a straight horizontal pipe of identical internal diameter $D = 1$ cm, as shown in Fig. 2.3 of Chapter 2, and $\Delta z = 10$ m. For water at a constant temperature of 285 K (12°C) and corresponding constant fluid properties (from Table A.1 of Appendix A), $\rho = 1000\ \mathrm{kg/m^3}$; $\mu = 1225 \times 10^{-6}\ \mathrm{N \cdot s/m}$; $\nu = \mu/\rho = 1.225 \times 10^{-6}\ \mathrm{m^2/s}$:

(a) Find the maximum number of velocity head $(\mathrm{NVH})_{f,\max}$ that may be dissipated in the pipe to ensure stable turbulent flow regimes at $\mathrm{Re} \geq 10{,}000$.

(b) Find the minimum number of velocity heads $(\mathrm{NVH})_{f,\min}$ that must be dissipated to ensure stable laminar flow regimes at $\mathrm{Re} \leq 2000$.

(c) For ratios $100 < L/D < 1000$ typical of heat exchanger design practice, verify that the flow regimes are turbulent on the basis of the following approximate empirical correlations already known to nineteenth century hydraulic engineers prior to the more scientific methods pioneered by Osborne Reynolds:

$$\Delta P_f = \bar{f} \frac{L}{D} \rho \frac{\bar{V}^2}{2} \tag{3.4}$$

or in dimensionless form and recalling the definition of $(\mathrm{NVH})_f$ per Eq. (2.18),

$$\bar{f} \frac{L}{D} \equiv (\mathrm{NVH})_f \equiv \frac{\Delta H_f}{\bar{V}^2/2} = \frac{\Delta P_f}{\rho \bar{V}^2/2} = \frac{\Delta z_f}{\bar{V}^2/2g} \tag{3.5}$$

Note that Eq. (3.5) is independent of the system of units selected and therefore applicable in the U.S. or European technical as well as the International System (SI) of units with experimental values of the friction factor falling within the following limits:

$$\bar{f} = 0.025 \pm 30\% \quad \text{for } \textit{turbulent} \text{ flow in } \textit{smooth} \text{ tubes} \tag{3.6}$$

(d) Determine the minimum ratio $(L/D)_{\min}$ required for stable laminar flow regimes at $\mathrm{Re} < 2000$ on the basis of the well-known expression for the fully developed friction factor f:

$$f = 64/\mathrm{Re} = 64\ \mathrm{Re}^{-1} \quad \text{for } \textit{laminar} \text{ flow} \tag{3.7}$$

Solution

(a) Solving Eq. (3.3) for $\bar{V}$ at $\mathrm{Re} \geq 10{,}000$ yields $\bar{V} \geq 1.23$ m/s. With $\Delta z = 10$ m we obtain from Eq. (2.7) a virtual velocity $V_{\mathrm{virt}} = (2 \times 10 \times 9.81)^{1/2} = 14.01$ m/s and the corresponding maximum allowable number of velocity heads can be calculated with Eq. (2.20) from Eq. (2.21) as

$$(\mathrm{NVH})_{f,\max} = \frac{1}{(1.23/14.01)^2} - 1 = 128.7 \tag{3.8a}$$

(b) Following exactly the same procedure as above with $\text{Re} \leq 2000$ yields $\bar{V} \leq 0.25$ m/s and

$$(\text{NVH})_{f,\min} = \frac{1}{(0.25/14.01)^2} - 1 = 3139 \tag{3.8b}$$

(c) From Eq. (3.5) with $f = 0.025$ and $L/D = 100\text{–}1000$, we obtain

$$(\text{NVH})_f = 0.025 \times (100\text{–}1000) = 2.5\text{–}25$$

Since $(\text{NVH})_f \ll (\text{NVH})_{f,\max}$ the flow regimes are certainly turbulent even allowing for the $\pm 30\%$ uncertainty in $\bar{f}$.

(d) Solving Eq. (3.5) for L/D yields, at $\text{Re} = 2000$ with $(\text{NVH})_{f,\min} = 3139$ per item (b) and with $f = 64/2000 = 0.032$ from Eq. (3.7),

$$(L/D)_{\min} > (\text{NVH})_{f,\min}\left(\frac{1}{f}\right) = (3139)\left(\frac{1}{0.032}\right) = 98{,}406! \tag{3.9}$$

Comments

1. The required L/D for laminar SSSF regimes under the stated conditions are several orders of magnitude larger than $100 < L/D < 1000$ typical of normal heat exchanger design practice. It is clear that turbulent flow regimes are more typical in forced-convection heat exchangers than streamline flows.

2. Although the above analysis is rather rudimentary in the case of turbulent flow owing to the $\pm 30\%$ uncertainty in $\bar{f}$, the results thus obtained are sufficiently accurate in many practical fluid flow applications.

3. It is necessary, however, to predict the friction factor $\bar{f}$ more accurately in heat transfer analyses because of the direct "coupling" between the convective heat transfer coefficient and friction factor in single- and two-phase flow heat exchanger design applications. This explains the rather detailed review of the best friction factor correlations in the following sections.

3.3 ISOTHERMAL LAMINAR FLOW REGIMES IN STRAIGHT CHANNELS

We first consider flow in long straight *circular* tubes and pipes, then streamline flow in *noncircular* channels of constant cross section.

3.3.1 Isothermal Laminar Flow in Long Circular Pipes ($L/D \gg 1$)

The more rigorous and general mathematical analyses based on the well-known three-dimensional Navier–Stokes equations can be solved fully analytically only in simple situations as shown in standard texts [3.2, 3.3]. Because the more complex Navier–Stokes equations yield the same results in the case of streamline circular tube flow as the simpler analytical solution independently derived by Hagen in 1839 [3.12] and Poiseuille in 1840 [3.13], we confine ourselves here to the Hagen–Poiseuille expression of the friction pressure drop gradient:

$$\frac{-dP_f}{dx} = \left|\frac{dP_f}{dx}\right| = 32\frac{\mu}{D^2}\overline{V} \tag{3.10}$$

Validity: Constant fluid properties (ρ and μ) fully developed laminar flow regimes with a parabolic *axial velocity distribution* V_r within the circular tube cross section: (3.11)

$$\frac{V_r}{\overline{V}} = 2\left[1 - \left(\frac{r}{r_0}\right)^2\right] \tag{3.12}$$

where $r_0 = D/2$
r = radial distance from the tube center line
$\overline{V}$ = average velocity $\equiv \dot{m}/\Omega\rho \equiv G/\rho$
Ω = flow cross section
$\dot{m}$ = mass flow rate
G = mass velocity $\equiv \dot{m}/\Omega$

Equation (3.12) shows that at $r = 0$ the velocity ratio $V_r/\overline{V} = 2$. In other words, V_r reaches a maximum value $(V_r)_{\max} = 2\overline{V}$, that is, twice the average velocity at the center line of the circular tube. At the wall–fluid interface, $r = r_0 = D/2$, Eq. (3.12) shows that the axial velocity $V_r = 0$, as it should in the case of classical Newtonian fluid flows.

Since the pressure gradient $|dP_f/dx|$ is constant in *fully developed* incompressible fluid flow

$$\frac{|dP_f|}{dx} = \frac{|\Delta P_f|}{L} = \text{constant} \tag{3.13}$$

Introducing Eq. (3.13) into Eq. (3.10) and solving for ΔP_f, we obtain for the total pressure drop in long tubes ($L/D \gg 1$)

$$\Delta P_f = 32\frac{L}{D^2}\overline{V} \propto \overline{V} \tag{3.14}$$

The above analytical expression has been verified for the isothermal flow of hydrocarbon oils in cylindrical tubes ranging from capillary tubes up to 12 in. ($\simeq$ 0.30 m) standard U.S. steel pipes [3.14].

The most significant and unique characteristic of laminar flow regimes according to Eq. (3.14) is the linear relationship between the *driving force* ΔP_f

and the average velocity $\overline{V}$. For a given fluid and channel geometry, that is, constant μ and D, the mass velocity G and the mass flow rate $\dot{m}$ are proportional to $\overline{V}$ and we obtain the more general linear relationships,

$$\Delta P_f \propto \overline{V} \propto G \propto \dot{m} \tag{3.15}$$

that uniquely characterize fully developed laminar flow regimes.

Instead of applying the analytical Hagen–Poiseuille expression (3.14), it is frequently more convenient in fluid and heat flow design practice to reduce Eq. (3.14) to the standard format used in turbulent regimes according to Eq. (3.4) or (3.5) by setting ΔP_f, per Eq. (3.14) $\equiv \Delta P_f$ per Eq. (3.4), and solving for the friction factor f, thus obtaining Eq. (3.7) introduced in Numerical Analysis 3.1.

It should be noted that Eq. (3.7) and/or (3.14) can also be applied with fully compressible (elastic) fluids such as gases and superheated single-phase vapor under isothermal conditions when $\Delta P_f/P \ll 1$. According to the Boyle–Mariotte ideal gas law $\rho \propto P$ when T = constant, or

$$\Delta\rho/\rho = \Delta P/P \quad \text{at } T = \text{constant} \tag{3.16}$$

therefore

$$\Delta\rho/\rho \rightarrow 0 \quad \text{and} \quad \rho = \text{constant} \quad \text{when } \Delta P/P \rightarrow 0 \tag{3.17}$$

3.3.2 Isothermal Laminar Flow in Noncircular Channels

Generally, the equations for laminar flow in straight constant cross-section channels of various shapes *do not* coincide with Eq. (3.14), which is valid specifically for circular pipes, even when expressed in terms of an equivalent hydraulic diameter $D_h = 4\ \Omega/a$, where a is the *wetted perimeter*, which is also equivalent to the internal tube heat transfer area per unit length in the case of constant cross-section channels.

Several theoretical equations have been derived by many researchers for various shapes of channel cross sections [3.14–3.20]. Experimental data have been presented to verify some of the proposed theoretical analytical solutions, particularly in the case of straight long channels of rectangular and annular constant cross section.

The important point to stress is that the characteristic linear relationship, Eq. (3.15), remains valid for all the proposed analytical and empirical correlations in fully developed laminar flow regimes in straight long channels $(L/D_h \gg 1)$ of constant cross section regardless of geometry.

3.4 THE CENTRAL ROLE OF DIMENSIONAL ANALYSIS

The powerful concepts of similarity, model theory, and dimensional analysis provide extremely effective tools in the complex thermal sciences and these topics form an integral part of classical fluid flow and heat transfer textbooks [3.21–3.26].

One of the most rigorous algebraic methods of establishing the correct form of an equation by dimensional analysis was proposed in 1914 by Buckingham [3.27]. The Buckingham π-theorem has proved particularly useful in predicting the most general expression for the friction pressure drop ΔP_f or head loss $\Delta P_f/\rho$ in cylindrical tubes as shown in Ref. 3.28. On the assumption that ΔP_f is exclusively a function of six significant independent variables—ρ, μ, $\bar{e}$, $\bar{V}$, L, and D, where $\bar{e}$ is a statistical measure of the pipe average roughness and has the dimension of length—the application of the Buckingham π-theorem leads to an equation of the following form, *applicable* a priori to fully developed *turbulent and laminar* SSSF *regimes* when $L/D \gg 1$:

$$\frac{\Delta P_f}{(L/D)\rho\bar{V}^2} = \Psi_1(\mathrm{Re}, \bar{e}/D) \tag{3.18}$$

where Ψ_1 is a yet unknown function of the Reynolds number and the relative wall roughness $\bar{e}/D$. Multiplying both sides of Eq. (3.18) by a constant factor (2) to make the dynamic pressure terms $\rho(\bar{V}^2/2)$ appear, and since $2\Psi_1(\mathrm{Re}, \bar{e}/D)$ is another universal function $\Psi_2(\mathrm{Re}, \bar{e}/D)$, we obtain

$$\frac{\Delta P_f}{\rho(L/D)(\bar{V}^2/2)} = \Psi_2(\mathrm{Re}, \bar{e}/D) \tag{3.19a}$$

Comparing Eq. (3.19a) with the nineteenth century *Darcy–Weissbach* empirical correlation, introduced earlier as Eq. (3.4), we see that the unknown function $\Psi_2(\mathrm{Re}, \bar{e}/D)$ represents the dimensionless friction factor in Eq. (3.4) for fully developed SSSF regimes which we shall denote by f_∞, or

$$f_\infty = \Psi_2(\mathrm{Re}, \bar{e}/D) \tag{3.19b}$$

It is important to recognize the fundamental limitations and strengths of the dimensional analysis techniques from which Eqs. (3.19a) and (3.19b) were derived.

- The weakness of dimensional analysis is that it sheds no light on the physical nature of a phenomenon, nor does it provide any explicit formulation for the unknown dimensionless function $f_\infty = \Psi_2(\mathrm{Re}, \bar{e}/D)$. In the absence of any analytical theory to predict the unknown function f_∞ in turbulent regimes, the only recourse is experimental information.
- Conversely, the strength of this technique is its generality, since it does not depend on any particular physical model of the phenomenon, a factor that is most important in complex situations such as the chaotic analytically intractable turbulent flow regimes. Once the conclusions derived

from dimensional analysis have been empirically validated, as is actually the case for the friction factor per Eqs. (3.19a) and (3.19b), a universal function of only two independent variables Re and $\bar{e}/D$ can be obtained from relatively few experiments. The semiempirical correlation is perfectly general and therefore not restricted to either the kind of fluid used in the test, or the range of velocities measured, or even the particular kind of pipe employed in the experiments within the stated test range of the two key parameters Re and $\bar{e}/D$.

The importance of dimensionless analysis in the highly complex thermal sciences, and a fortiori the "art" of momentum, heat, and mass transfer (MHMT), cannot be overstated.

3.5 SENSITIVITY TO CHANNEL ENTRANCE EFFECTS

In engineering design practice, as opposed to pure fluid and heat flow research, it is not practical to ignore the so-called entrance effects in the tube section preceding the hydrodynamically fully developed region. By definition (see Fig. 3.2), the fully developed friction factor f_∞ = constant is valid only for $x > x_{fd,h}$, or expressed in a more general dimensionless form $X \equiv x/D > x_{fd,h}/D \equiv X_{fd,h}$.

In the entrance region at $X < X_{fd,h}$, the local friction factor $f_X > f_\infty$ is a function of the relative axial position X and, most generally, of the conditions preceding the channel entrance (CPE). It is evident that the fluid's "prior history" must have an impact on the velocity distribution at $X = 0$ and, consequently, on the local friction factor and on the value $X_{fd,h}$ beyond which the inlet effects become so insignificant that $f_X = f_\infty$.

By considering the two additional independent variables $X = x/D$ and, at least in a formalistic fashion, the effects of CPE, it is easy to show on the basis

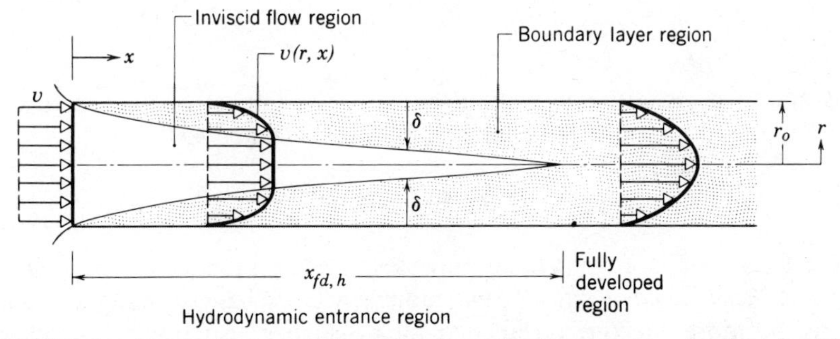

FIGURE 3.2. Laminar hydrodynamic boundary layer development in a circular tube. [From F. P. Incropera and D. P. de Witt, *Fundamentals of Heat and Mass Transfer*, 2nd ed., Wiley, New York, 1985. Reproduced with permission.]

of dimensional analysis (for instance Buckingham's π-theorem) that the most general expression of f_X must be

$$f_X = \Psi_3(\mathrm{Re}, \bar{e}/D, X, \mathrm{CPE}) \tag{3.20}$$

For fixed values of Re, $\bar{e}/D$, and similar CPE, f_X in Eq. (3.20) reduces to

$$f_X = \Psi_4(X) \quad \text{for fixed Re}, \bar{e}/D, \text{and CPE} \tag{3.21}$$

indicating that the local friction factor f_X is a unique function of $X \equiv x/D$ *only*!

Experience indicates that f_X converges more rapidly toward f_∞—that is, at lower values of $X_{fd,h}$—in turbulent than in pure laminar flow regimes. This is expected because any uncertain inlet effects are likely to be more quickly "dampened" or "erased" in the chaotic motion associated with turbulent flow regimes. This intuitive qualitative conclusion is quantified by the following approximate correlations for the relative entrance length $X_{fd,h} = x_{fd,h}/D$ recommended in some recent textbooks [3.29–3.31] for single-phase turbulent and laminar flow regimes.

3.5.1 Single-Phase Turbulent Regimes

$$10 < (X_{fd,h})_{\mathrm{turb}} \leq 60 \quad \text{and independent of Re} \tag{3.22a}$$

and many authors assume fully developed turbulent flow at

$$X_{fd,h} \geq 10 \tag{3.22b}$$

The average friction factor $\bar{f}$ used with Eqs. (3.4) and (3.5) in normal fluid and heat flow engineering analyses can be evaluated on the basis of Eq. (3.21) as follows for any total channel length L and corresponding total L/D ratio:

$$\bar{f} = \frac{1}{L/D}\int_{X=0}^{X=L/D} \Psi_4(X)\, dX = \Psi_5(f_\infty, L/D) \tag{3.23a}$$

This allows us to estimate, with Eq. (3.20), the impact of inlet effects on $\bar{f}$ according to the ratio $\bar{f}/f_\infty$ at prescribed Re, $\bar{e}/D$, and CPE, or

$$\bar{f}/f_\infty = \Psi_6(L/D) \tag{3.23b}$$

where $\Psi_6(L/D)$ is, for practical purposes, a universal function of L/D *only*.

Since *thermodynamically efficient* tubular heat exchangers designed to operate in the more effective turbulent regimes (higher heat transfer coefficients) require large L/D ratios in the range of $100 < L/D < 1000$ and $f_X \to f_\infty$ at $X_{fd,h} = 10$ per Eq. (3.22b), the inlet effects are negligibly small compared to normal accuracy bands of ± 5–10% for f_∞ in the case of ideal smooth tubes. Therefore, it is permissible to set

$$\bar{f} = f_\infty$$

$$\text{for turbulent flows at } 100 < L/D < 1000 \tag{3.24}$$

$$\text{within the accuracy of } f_\infty \text{ correlations}$$

Because the heat transfer coefficients α_∞ and $\bar{\alpha}$ are directly "coupled" to the fluid flow friction factors f_∞ and $\bar{f}$, it is also possible to simplify the correlations for α, without loss of rigor, by setting $\bar{\alpha} = \alpha_\infty$ at $L/D > 100$ as shown in subsequent chapters. In fact, the more effective the heat exchanger (i.e., the larger the aspect ratio L/D), the simpler the correlations for $\bar{f}$ and $\bar{\alpha}$ because one can dispense with the complexities associated with highly uncertain entrance effects in single-phase stable turbulent flow tubular heat exchangers.

3.5.2 Single-Phase Laminar Regimes

True to the general pattern, the situation is more complex in the case of the theoretically simpler laminar flow regimes since the recommended empirical correlation for $(X_{fd,h})_{\text{lam}}$ indicates a significant effect of the Reynolds number according to

$$(X_{fd,h})_{\text{lam}} \cong 0.05\,\text{Re} \propto \text{Re} \tag{3.25}$$

Thus, $(X_{fd,h})_{\text{lam}}$ can vary within the following limits at typical values of Reynolds numbers in stable laminar flow regimes:

$$5 \leq (X_{fd,h})_{\text{lam}} \leq 115 \tag{3.26a}$$

at corresponding Re of

$$100 \leq \text{Re} \leq 2300 \cong (\text{Re})_{cr} \tag{3.26b}$$

Since heat exchangers operating in pure laminar flow regimes may frequently require L/D ratios substantially below 115, the inlet effects and associated impact of the uncertain CPE can become quite significant. This confirms what every experienced heat transfer engineer knows: The design of heat exchangers is inherently more difficult in actual practice for laminar than for turbulent flow regimes, because of the former's greater sensitivity to entrance and nonisothermal effects.

3.5.3 Two-Phase Flow With and Without Heat Addition or Removal

The very significant impact of entrance effects in adiabatic and diabatic two-phase channel flow is emphasized in Refs. 3.4 and 3.32 (and references cited therein). The exact extent of the relative entrance length $X_{fd,h}$ is not known even in the simpler case of two-component or adiabatic single-component (without vapor quality change, which implies constant average mixture velocity) two-phase flow. It is known, however, that entrance effects may persist much further downstream than in single-phase flow at $X_{fd,h}$ in excess of 300. Based on this observation, two-phase flow studies for the petroleum industry have been carried out with extremely long tubes to minimize entrance effects and thus produce correlations applicable to actual pipeline design practice [3.33].

The situation is further complicated in single-component two-phase flow with heat addition (vapor generation in evaporators) or removal (vapor condensation in condensers) where fully developed flow may actually never occur [3.32] and entirely new techniques are required to cope effectively with such complex two-phase flow phenomena. One relatively new technique introduced in the 1960s is based on the extended corresponding states principle (ECSP) of thermodynamics. The ECSP thermodynamic generalization method, based on universal reduced fluid properties in terms of reduced pressures, is a logical extension of the dimensionless generalization techniques highlighted previously. Like dimensional analysis techniques, the ECSP thermodynamic generalization method does not explain the detailed mechanisms of two-phase flow but it does provide the heat transfer engineer with simple additional tools to cope effectively with otherwise hopelessly complex heat exchanger design problems. The ECSP generalization methods are described in Chapters 6 and 8, which address the hydrodynamic and heat transfer aspects, respectively, with special emphasis on two-phase heat exchangers.

3.6 PROLIFERATION OF FRICTION FACTOR DEFINITIONS

The total dissipation $\Delta H_f = \Delta P_f/\rho$, which results from wall friction according to Eq. (3.4) or the equivalent dimensionless expression $(\mathrm{NVH})_f$ per Eq. (3.5), obviously does *not* depend on the *arbitrary* choice of the circular tube diameter D as the key *reference length* l. Equation (3.5) can be expressed most generally as

$$(\mathrm{NVH})_f = \bar{f}\left(\frac{L}{l}\right) \tag{3.27a}$$

or

$$\frac{(\mathrm{NVH})_f}{L} = \frac{\bar{f}}{l} \tag{3.27b}$$

using any convenient characteristic dimension l as a reference.

At least three different reference lengths have been proposed for tube flow by European and American researchers. Those discussed below are still found in contemporary scientific publications:

$$l \equiv r \equiv \tfrac{1}{2}D \tag{3.28}$$

$$l \equiv r_h = \frac{\text{channel internal volume}}{\text{channel internal surface}} = \text{volumetric or hydraulic radius} \tag{3.29a}$$

with

$$l \equiv r_h \equiv \frac{\Omega L}{aL} \equiv \frac{\Omega}{a} \tag{3.29b}$$

where a = wetted perimeter = internal tube surface per unit length of tube in $m^2/m = m$
Ω = constant channel cross section in m^2
L = total channel (circuit) length in m

In the case of *circular* channels with $\Omega = (\pi/4)D^2$ and $a = \pi D$ Eq. (3.29b) reduces to

$$l = r_h = \tfrac{1}{4}D \tag{3.29c}$$

The third choice of reference length l, conforming to currently accepted (U.S.) engineering practice, for flow in channels is simply

$$l \equiv D \tag{3.30}$$

The last characteristic length has the advantage of being consistent with the generally accepted definition of the Reynolds number Re, which is also based on $l = D$ for internal channel flow.

Since the left-hand side of Eq. (3.27b) is independent of the arbitrary choice of the characteristic length l, it is evident that the right-hand side term $\bar{f}/l$ must also be independent of l and the *friction factor* must be *inversely proportional* to l. Using as a standard reference the friction factor based on $l = D$, we obtain the following relationships:

$$\text{For } \bar{f}_r \text{ based on } l = r = \tfrac{1}{2}D \quad \bar{f}_r = \tfrac{1}{2}\bar{f} \tag{3.31a}$$

$$\text{For } \bar{f}_{r_h} \text{ based on } l = r_h = \tfrac{1}{4}D \quad \bar{f}_{r_h} = \tfrac{1}{4}\bar{f} \tag{3.31b}$$

where $\bar{f}$ is based on $l = D$.

A side benefit of the dimensionless dissipation term $(\text{NVH})_f$ per Eq. (3.5) (besides its practical advantages in actual design analyses that involve several individual resistances as shown in Chapter 8) is that it practically eliminates any possibility of confusion owing to the proliferation of friction factor definitions!

Because circular tube correlations for the friction factor f_∞ (or $\bar{f}$) and the "coupled" heat transfer coefficient α_∞ (or $\bar{\alpha}$) are applicable to noncircular channels of constant cross section Ω in turbulent regimes, when $l \equiv D_h \equiv 4r_h$ is used as the characteristic length for the Reynolds number, we have adopted for the balance of this text the following widely accepted terminology:

$$\mathrm{Re} = \frac{\rho \bar{V} D_h}{\mu} = \frac{G D_h}{\mu} = \frac{\bar{V} D_h}{\nu} = \frac{4\dot{m}}{\mu(a)} \tag{3.32a}$$

Validity: Internal flow in channels of *arbitrary constant* cross section Ω and constant wetted perimeter a = wall surface per unit length (3.32b)

It is noteworthy that Eq. (3.32) is fully consistent with the previous definition of Re per Eq. (3.3) since $D_h = 4(\pi D^2/4)/\pi D \equiv D$ in the case of circular channels where $\mathrm{Re} = (4/\pi)\dot{m}/\mu D = 1.273\dot{m}/\mu D$.

3.7 CHANNEL FLOW FRICTION FACTOR: GRAPHICAL SOLUTIONS

The conclusions derived from dimensional analysis in Section 3.4 have been validated by prolific nineteenth and twentieth century fluid flow researchers. For an "organic" overview of the whole field we focus first on the thoroughly documented recommendations by William H. McAdams in the third edition (1954) of the great American classic: *Heat Transmission* [3.34] and then on the well-known Moody (1944) graph.

It should be noted that McAdams' original graph (Fig. 1.4 of Ref. 3.34) and all his recommended equations have been modified here to conform with the now widely accepted definition of f according to Eq. (3.30), that is, using $f = 4f_{rh}$ instead of McAdams' values of f_{rh}. Furthermore, to simplify the writing we now use the term f to designate the friction factor since $\bar{f} = f_\infty$ per Eq. (3.24) at $L/D > 100$ as is typically encountered in turbulent forced-convection heat exchangers.

3.7.1 McAdams' (1954) Graph

Figure 3.3 shows the analytical and semiempirical friction factors for isothermal flow of Newtonian fluids in long pipes ($L/D \gg 1$) having two degrees of relative wall roughness. It reflects the most significant findings by Osborne Reynolds: the existence of two radically different stable flow regimes below and above a critical band of Re extending roughly from 1000–2000 up to 5000–10,000. We briefly review laminar regimes at the lower end of the Reynolds number scale, then focus at length on the turbulent flow regimes in "smooth" tubes and commercial pipes in view of their central importance in forced-convection heat exchanger design applications.

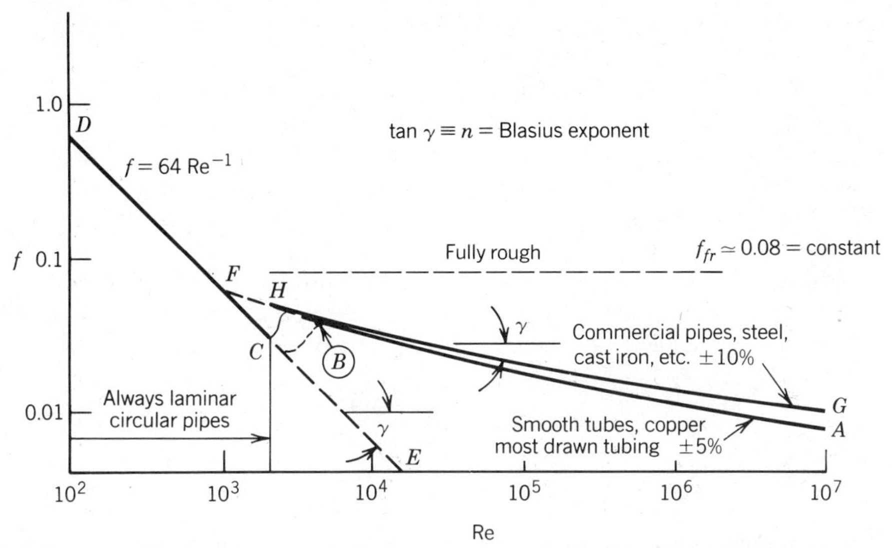

FIGURE 3.3. Friction factors for isothermal flow of Newtonian fluids in pipes having two typical degrees of roughness and for very large ratios of wall roughness to pipe diameter.

For streamline flow regimes the single straight line *DE* represents, in the log–log scale used in Figure 3.3, Eq. (3.7) derived from the Hagen–Poiseuille analytical solution. The fact that a single correlation is applicable for smooth ($\bar{e}/D = 0$) as well as nonsmooth ($\bar{e}/D \gg 0$) steel pipes indicates that the wall roughness effects suggested by dimensional analysis according to Eq. (3.20) are suppressed in highly viscous laminar flow regimes, at least within the normal range of natural wall roughness encountered in commercial pipes of substantially constant cross section.

For turbulent flow regimes Fig. 3.3 confirms what was predicted solely on the basis of dimensional analysis, namely, the impact of the relative wall roughness $\bar{e}/D$ on the friction factor, in as much as two different empirical curves are shown for f in terms of Re. The lower, slightly steeper curve marked *AB* is valid for ideal smooth drawn tubes ($\bar{e}/D \cong 0$), whereas the higher, slightly flatter curve marked *GH* is applicable to typical commercial steel and cast iron pipes ($\bar{e}/D \gg 0$).

According to McAdams, curve *AB* in Fig. 3.3 is "... based on the data of a number of investigators for the flow of air, water, and oils in clean smooth tubes, mostly drawn tubing made of brass, copper, lead and glass with diameters ranging from 0.5 to 5.0 in. (1.25 to 12.5×10^{-2} m). It is found that the empirical values of f, when plotted versus the Re number lie in a relatively narrow band..." of ±5% around the curve *AB*. This conclusion is based on the correlation of 1380 experimental points by Koo and other MIT researchers [3.35a] in the 1930s. Curve *AB* is therefore backed up by a statistically significant number of reliable individual tests by leading researchers to draw some meaningful conclusions for practical fluid flow and heat transfer applications which are still relevant today.

Curve *HG* in Fig. 3.3 correlates (within a stated deviation of $\pm 10\%$ in f) experimental data of a number of observers for clean pipes of steel and cast iron [3.35b]. Empirical results with compressible gases at higher (acoustic) velocities are also in reasonable agreement with this correlation [3.36, 3.37]. For fully rough walls, McAdams recommends an approximate constant value $f = 0.08$ independent of Re.

When the same scales are used for the abscissa $X = \log \text{Re}$, and the ordinate $Y = \log f$, as is the case here, the local slope of any curve, that is, the tangent of the angle γ sketched in Fig. 3.3, quantifies the local exponent n in the expression

$$f \propto (\text{Re})^{-n} \tag{3.33}$$

or

$$Y \equiv \log(f) \propto -n \log(\text{Re}) \equiv -nX \tag{3.34}$$

We can therefore draw the following general conclusions from Fig. 3.3 for laminar and turbulent regimes, including the fully rough limit $f_{\text{fr}} = \text{constant}$.

1. For laminar regimes at Re < 2300 the exponent is constant, or

$$n = 1 \text{ and is independent of Re and } \bar{e}/D \tag{3.35}$$

2. For turbulent regimes at Re > 3000–5000, the exponent n varies with Re and $\bar{e}/D$ roughly as follows:

(a) For ideal smooth tubes ($\bar{e}/D \cong 0$) the empirical curve marked *AB* in Fig. 3.3 has a slightly decreasing slope (concave side up) varying from roughly

$$n_{\text{max}} \cong 0.28 \text{ at } (\text{Re})_{\text{min}} = 3000 \quad \text{to} \quad n_{\text{min}} \cong 0.15 \text{ at } (\text{Re})_{\text{max}} = 3{,}000{,}000 \tag{3.36}$$

(b) For commercial steel and cast iron pipes ($\bar{e}/D > 0$) the line *HG* in Fig. 3.3 is less steep with values

$$n_{\text{max}} \text{ and } n_{\text{min}} \text{ below those for ideal smooth tubes shown above} \tag{3.37}$$

(c) For the fully rough limit $f_{\text{fr}} = \text{constant}$

$$n = 0 \tag{3.38}$$

3.7.2 Moody's (1944) Graph

Since the universal empirical curve for ideal smooth tubes in turbulent regimes and the analytically derived straight line for laminar flow are identical in the Moody (Fig. 3.4) [3.38] and the McAdams graphs, we confine this discussion to the friction factor curves in turbulent regimes at Re > 3000 and $\bar{e}/D \gg 0$.

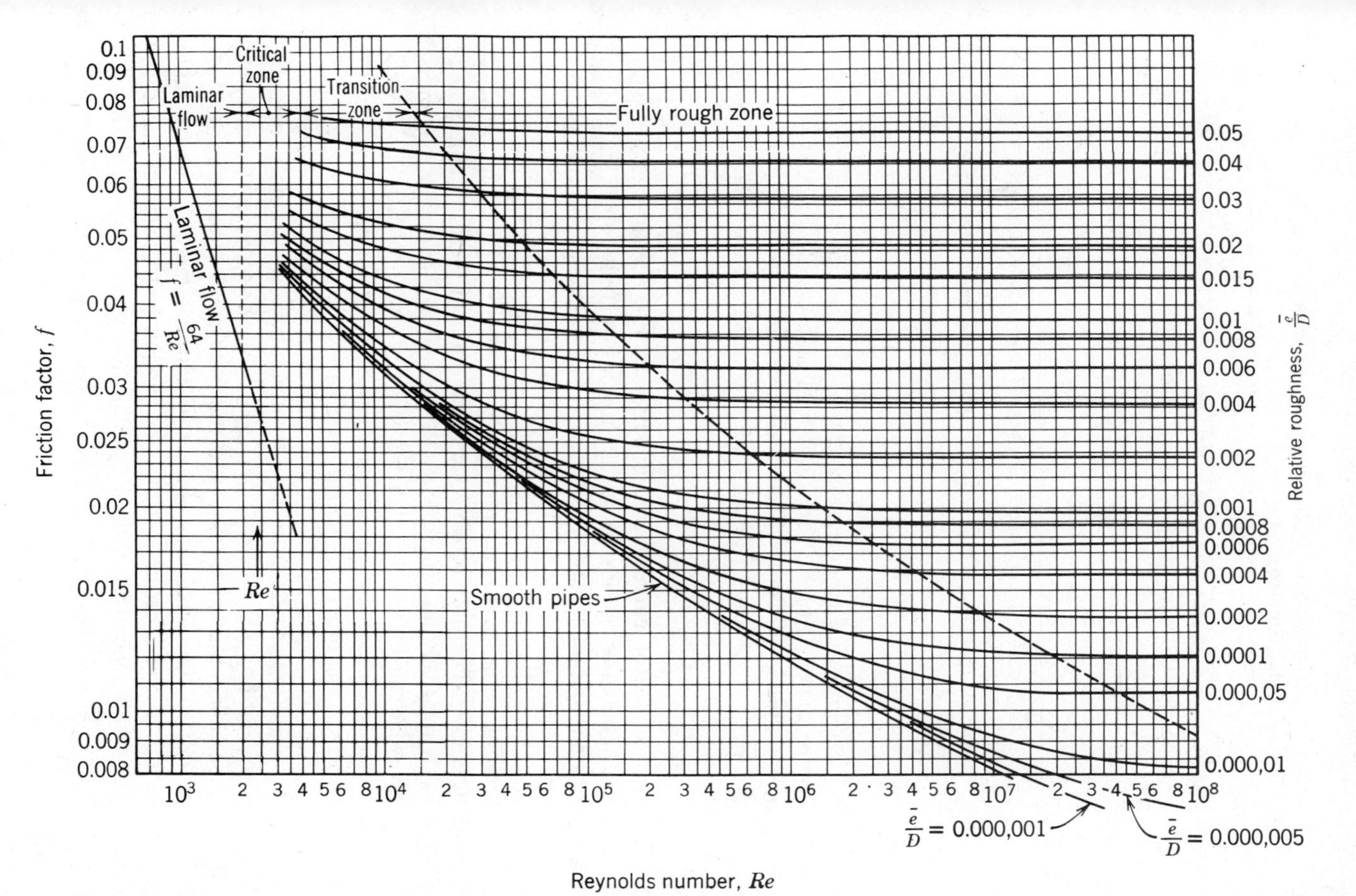

FIGURE3.4. Moody's chart for the friction factor for fully developed flow in circular tubes. [From F. P. Incropera and D. P. de Witt, *Fundamentals of Heat and Mass Transfer*, 2nd ed., Wiley, New York, 1985. Reproduced with permission.]

TABLE 3.1. Effective Average Roughness $\bar{e}$ and Ratio $\bar{e}/\bar{e}_{smooth}$ of Some Conduit Surfaces

Material	$\bar{e}$ (mm)	$\bar{e}/\bar{e}_{smooth}$
Commercially ***smooth*** brass, lead, copper, transite, or plastic pipe	0.0015	1
Steel and wrought iron	0.045	30
Galvanized iron or steel	0.15	100
Cast iron	0.25	167
Concrete	0.3 – 3.0	200 – 2000

Source: Adapted with permission from ASHRAE Handbook, Fundamentals, 1985, Table 3, p. 2.10, and from F. P. Incropera and D. P. de Witt, Fundamentals of Heat and Mass Transfer, Wiley, New York, 1985, Fig. 8.3, p. 373.

Figure 3.4 is convenient to determine graphically the friction factor f for any circular pipe diameter (D) of any statistical relative wall roughness $\bar{e}/D$ in conjunction with the estimated effective average wall roughness $\bar{e}$ shown in Table 3.1. A cursory overview of Fig. 3.4 and Table 3.1 leads to the following general conclusions:

1. The friction factor curves become flatter in the transition zone with increasing $\bar{e}/D$ and converge asymptotically toward the fully rough limit when $f \to f_{fr}$ = constant and independent of Re, therefore independent of $\bar{V}$, G, and $\dot{m}$ for a fixed D and μ.

2. The limiting value f_{fr} = constant is only a function of the pipe relative wall roughness $\bar{e}/D$. The fully rough zone to the right of the dropping dotted line begins at a relatively low Re $\leq 15 \times 10^4$ at high relative wall roughness $\bar{e}/D \geq 0.050$, whereas extremely high Re on the order of 10^7–10^8 (well beyond the usual design range of normal heat exchangers) are necessary to reach f_{fr} at $\bar{e}/D$ or the order of 0.00020–0.00001.

3. In contrast to the critical transition or bifurcation from orderly laminar to chaotic turbulent flow regimes at Re > 2300, there is no physical transition between the so-called transition zone and the fully rough zone indicated in Fig. 3.4; the dotted line separating these two zones is merely a convenient demarcation line to highlight the fact that the friction factor remains substantially constant and independent of the Reynolds number at values of Re above those shown by the dotted line at corresponding values of the relative wall roughness $\bar{e}/D$.

4. The practical advantage of Moody's graph is that it is applicable in all engineering applications, including large-diameter pipes used in civil engineering canalizations and made of various materials as shown in Table 3.1. By

contrast, McAdams' Fig. 3.3 is specifically tailored to normal heat exchanger materials and pipe sizes.

5. Figure 3.4 quantifies the basic conclusions derived theoretically from dimensional analysis, namely, the significance of the relative wall roughness $\bar{e}/D$. The absolute roughness being substantially independent of the pipe diameter for a given wall material, the relative wall roughness $\bar{e}/D$ *increases* with *decreasing* channel diameter and this results in higher friction factors for the smaller pipe sizes at a given Re than for the larger pipes made of the same material.

6. Moody's graph is in agreement with earlier tabular and/or graphical representations of reliable empirical friction factor correlations with typical wall materials such as those shown in Ref. 3.28. The results obtained from the Russian version of Moody's graph (shown as Fig. 47 of Ref. 3.39) also concur within reading accuracy with those of Fig. 3.4.

7. It should be noted, however, that the results obtained from curve *GH* in Fig. 3.3 are in line with those obtained from Moody's chart within McAdams' stated validity and accuracy limits of $\pm 10\%$. This may explain why McAdams did not feel it necessary to modify his recommendations in the 1954 revision of *Heat Transmission* even though he was evidently aware of Moody's 1944 paper to which he makes a specific reference (see section entitled "Friction Factors for Rough Pipes," p. 157 of Ref. 3.34).

In closing this overview of incompressible Newtonian fluid flow in circular tubes, it is useful to reemphasize that *all* the recommended explicit correlations for $f = \Psi(\text{Re}, \bar{e}/D)$ in turbulent flow regimes are based on experimentation and not on any fundamental theory as in the case of laminar flow. This statement remains true for the most sophisticated correlations based on various theoretical models, although their empirical base is shifted to more abstract concepts such as the various "laws of the wall" [3.38–3.44].

The recommended semiempirical correlations for smooth as well as rough wall circular tubes and pipes are systematically assessed in the next sections because of their central importance in forced-convection heat exchanger analyses due to the direct link between f (or $\bar{f}$) and the forced-convection heat transfer coefficient α (or $\bar{\alpha}$).

3.8 TURBULENT FRICTION FACTOR EQUATIONS FOR SMOOTH CIRCULAR PIPES

McAdams recommends three different equations for long straight smooth circular tubes and pipes. The first correlation is the mathematical expression for the curve AB in Fig. 3.3 proposed by Koo, expressed below in terms of $f = 4f_{rh}$

$$f = 0.0056 + 0.5(\text{Re})^{-0.32} \tag{3.39a}$$

Validity: 3000 < Re < 3,000,000; $(L/D \gg 1)$; ideal smooth $(\bar{e}/D \simeq 0)$ circular tubes made of glass and copper and most drawn tubing with diameters in the range of 12.7 mm (0.5 in.) < D < 127 mm (5 in.) (3.39b)

Deviation: ±5% within the above validity limits, based on 1380 reliable experimental points. (3.39c)

A simpler mathematical approximation of line AB is recommended for the narrower range of Re typically encountered in "normal" heat exchanger design applications:

$$f = 0.184(\text{Re})^{-0.2} \tag{3.40a}$$

Validity: 5000 < Re < 200,000; otherwise the same as Eq. (3.39b) (3.40b)

Deviation: Not explicitly stated but totally adequate in normal heat exchanger design applications as shown in Numerical Analysis 3.2. (3.40c)

The third semianalytical equation recommended by McAdams [3.34], which is based on boundary layer theory [3.40, 3.41], fits the data for smooth pipes and is recommended at extremely high Re:

$$\left(\frac{1}{f}\right)^{1/2} = 2\log_{10}\left[\text{Re}(f)^{1/2}\right] - 0.802 \tag{3.41a}$$

Validity: Recommended for extrapolation to Re > 3,000,000; otherwise same as Eq. (3.39b) (3.41b)

Deviation: *Not stated*[1] but presumably better than Koo's correlation per Eq. (3.39a) when extrapolated beyond the stated validity limits, Eq. (3.39b), at Re > 3,000,000 (3.41c)

For good measure we shall assess two more widely accepted correlations.
The first one is particularly relevant because it is used in the most advanced correlations for forced-convection heat transfer correlations in tube flow

proposed by Petukhov [3.42, 3.43]:

$$f = (1.82 \log_{10}\text{Re} - 1.64)^{-2} \tag{3.42a}$$

Validity/Deviation: Not stated by Petukhov [3.42]; presumably comparable to Eq (3.41) since it is based on the same basic boundary layer theory (see Numerical Analysis 3.2) (3.42b)

The second correlation is the old Blasius equation still recommended in the 1985 edition of the respected "bible" of the refrigeration and air conditioning industry [3.45]:

$$f = 0.3164(\text{Re})^{-0.25} \tag{3.43a}$$

Validity: Re < 100,000 (3.43b)

Deviation: Not stated in Ref. 3.45 (see Numerical Analysis 3.2). (3.43c)

3.8.1 Generalized Blasius Type Equations

Noting the similar form of the Blasius type equations for laminar and turbulent flow regimes—Eqs. (3.7), (3.40), and (3.43)—it is evident that all these correlations can be expressed in a general form for laminar as well as for turbulent flow in circular tubes as

$$f = K_n(\text{Re})^{-n} \tag{3.44a}$$

with suitable pairs of constants K_n and n. Therefore, under otherwise fixed conditions, the total friction head loss $\Delta H_f = \Delta P_f/\rho$ varies with $\dot{m}$ according to

$$\Delta H_f = \Delta P_f/\rho \propto f\bar{V}^2 \propto f\dot{m}^2 \propto \dot{m}^{2-n} \tag{3.44b}$$

in view of Eq. (3.5).

Equations (3.44a) and (3.44b) yield the following useful general relationships.

For laminar flow,

$$K_n = 64 \quad \text{and} \quad n = 1, \quad \text{then } 2 - n = 1 \tag{3.45a}$$

for Re < 2300 per Eq. (3.7).

For turbulent flow in ideal smooth tubes,

$$K_n = 0.3164 \quad \text{and} \quad n = 0.25, \quad \text{then } 2 - n = 1.75 \tag{3.46}$$

for Re < 100,000 per Eq. (3.43). Alternatively,

$$K_n = 0.184 \quad \text{and} \quad n = 0.20, \quad \text{then } 2 - n = 1.8 \tag{3.47}$$

for 5000 < Re < 200,000 per Eq. (3.40).

For turbulent flow in fully rough tubes,

$$K_n = \text{constant} \quad \text{and} \quad n = 0, \quad \text{then } 2 - n = 2 \tag{3.48}$$

when K_n is mainly a function of $\bar{e}/D$ (see Numerical Analysis 3.3).

The main advantage of the generalized Blasius type of correlation per Eq. (3.44) is its simplicity compared to the more complex semianalytical expressions according to Eqs. (3.41) and (3.42). Furthermore, since the log of a product is the sum of the logs, Eq. (3.44) can be expressed logarithmically as

$$\log(f) = \log(K_n) + \log(\text{Re})^{-n} = \log K_n - n \log(\text{Re}) \tag{3.49}$$

Introducing two new variables defined as

$$Y \equiv \log(f) \quad \text{and} \quad X \equiv \log(\text{Re}) \tag{3.50}$$

we can reduce Eq. (3.49) to

$$Y = \log(K_n) - n(X) \tag{3.51}$$

which represents a straight line in an $X = \log(\text{Re})$ versus $Y = \log(f)$ graph. It can be fully defined by a single point, for example, $f = K_n$ at Re = 1 (since log 1 = 0), and the negative slope of the straight line in the log–log graph, which is exactly equal to n if the same logarithmic scales are used for the ordinate and the abscissa. It is therefore evident that any of the curves shown in Figs. 3.3 and 3.4 can be represented with a totally adequate accuracy of better than $\pm 5\%$, within a selected range of Re, by Eq. (3.44) with a suitable pair of Blasius constants K_n and n, keeping in mind that all the correlations for turbulent flow (even in ideal smooth circular tubes) have at best an accuracy of $\pm 5\%$.

The fluid flow and heat transfer design engineer is now faced with a sophisticated problem: an overabundance of semianalytical or purely empirical solutions for turbulent regimes! Depending on temperament, sense of aesthetics, cultural environment, and language, the design engineer may choose either the simpler solution (the usual choice of pragmatic engineers) or the more complex solution (frequent choice of pure mathematicians) or simply the favorite national correlation. Ironically, it does not really matter what choice is made in normal heat exchanger design practice since any of these equations will yield the same results within the stated validity limits and best accuracy band of $\pm 5\%$ per Eq. (3.39) for the smooth circular tubes in stable turbulent flow regimes, even if we disregard additional statistical deviations due to unavoidable manufacturing tolerances in wall quality, $\bar{e}/D$, and wall thickness.

The prime purpose of the following numerical analysis is to make a *rational and pragmatic* choice between all the recommended correlations for ideal smooth tubes for stable turbulent flow regimes on the basis of the end engineering objective defined in Guideline 6.

NUMERICAL ANALYSIS 3.2

Selection of adequately accurate friction factor correlations for turbulent flow in ideal smooth tubes.

Problem Statement

(a) Compare the results obtained with Eqs. (3.39), (3.40), (3.42), and (3.43).

(b) Based on the end engineering objectives defined by Guideline 6, select an adequately accurate correlation for f, keeping in mind the stated accuracy of $\pm 5\%$ for the best documented Eq. (3.39).

(c) The heat transfer coefficient $\bar{\alpha}$ is directly related to the friction factor according to the modified Reynolds analogy: $\bar{\alpha} = \frac{1}{8} c_p G(\text{Pr})^{-\beta}(\bar{f}) \propto \bar{f}$. Therefore, $\bar{\alpha}$ is directly proportional to the friction factor $\bar{f}$ under otherwise fixed conditions. On the basis of this fundamental relationship, estimate the best accuracy one can ever hope to achieve for $\bar{\alpha}$ in the case of turbulent flow regimes in long ideal smooth tubes.

(d) Determine the pair of constants K_n and n per Eq. (3.44) to duplicate Eq. (3.39) with a maximum deviation of *substantially less than* $\pm 5\%$.

(e) Predict the variation of the heat transfer coefficient $\bar{\alpha}$ with the flow rate $\dot{m}$ under otherwise fixed conditions on the basis of the results obtained under items (c) and (d) above.

Solution

(a) The results are summarized in Table 3.2 using Eq. (3.39) as a reference (since it is backed up by 1380 reliable empirical tests and not based on any particular theory, it is therefore, totally unbiased) to compare the values of f computed according to different recommended equations. All calculated values of f falling within the $\pm 5\%$ accuracy band of Eq. (3.39) are shown in Table 3.2 without parentheses, whereas all numerical results deviating by more than $\pm 5\%$ are shown in single parentheses for deviations between ± 5 and $\pm 10\%$ and double parentheses for deviations between ± 10 and $\pm 15\%$.

(b) Except for the approximate Blasius type Eq. (3.43) recommended by McAdams, which underestimates the friction factor in the lower Re range by less than 13 and 10% at Re of 5000 and 10,000, respectively, *all* the

TABLE 3.2. Friction Factor $f \times 100$ in Terms of Re Computed According to Four Different Recommended Correlations[a] for Ideal Smooth Tubes

Reynolds Number	Koo Eq. (3.39)	Petukhov Eq. (3.42)	Blasius Eq. (3.43)	McAdams Eq. (3.40)
3,000	4.417	NA[b]	4.275	NA
5,000	3.836	NA	3.763	((3.349))
10,000	3.184	3.143	3.164	(2.916)
20,000	2.662	2.611	2.661	2.539
40,000	2.244	2.204	2.237	2.210
80,000	1.909	1.885	1.881	1.924
160,000	1.641	1.630	NA	1.675
200,000	1.566	1.55	NA	1.601
320,000	1.426	1.424	NA	NA
3,000,000	0.983	0.971	NA	NA

[a]All values shown without parentheses fall within a $\pm 5\%$ uncertainty band of the most thoroughly documented correlation [Eq. (3.39)] recommended by McAdams [3.34].
[b]NA = Not applicable; outside stated validity limits.

recommended equations yield results at Re > 15,000, falling well within the $\pm 5\%$ tolerance band of the most accurately documented correlations recommended by McAdams [3.34].

It should also be noted that a deviation of $\pm 5\%$ represents an uncertainty of $\pm 5\%$ in $\Delta H_f = \Delta P_f/\rho$ for a prescribed flow rate $\dot{m}$ but only a $\pm 2.5\%$ uncertainty in $\dot{m}$ for a prescribed $\Delta H_f = \Delta P_f/\rho$ since $\dot{m} \propto 1/f^{1/2}$.

This analysis provides a rationale for the continued use of the simpler Blasius type correlations such as Eqs. (3.40) and (3.43) in actual heat exchanger design practice according to the recommendations of heat transfer pioneers such as McAdams and his colleagues who, in effect, followed the pragmatic philosophy stated below as a corollary to Guideline 2 of Chapter 1.

> ***GUIDELINE 7*** *For practical design purposes it is permissible and sensible to select the simplest equation yielding results falling within the accuracy band of the most exact empirically validated correlation.*

(c) For a Pr $\simeq$ 1 fluid such as saturated water vapor in the neighborhood of the normal boiling point (see Table A.1 in Appendix A) the expression for $\bar{\alpha}$ reduces to the simple Reynolds analogy $\bar{\alpha} = \frac{1}{8}\, c_p G \bar{f}$ and under otherwise prescribed conditions $\bar{\alpha} \propto \bar{f}$. Therefore, the heat transfer coefficient can at best be as accurate as the friction factor $\bar{f}$. Thus, Eq. (3.39c) also defines the best accuracy of the heat transfer coefficient $\bar{\alpha}$ that can ever be expected for turbulent flow in ideal smooth circular tubes in the simplest case of Pr = 1 fluids and a fortiori for Pr $\neq$ 1 fluids!

(d) Suitable pairs of constants K_n and n can be determined for any Re interval on the basis of Eq. (3.39) as follows: For $(\text{Re})_1 = 3000$ and $(\text{Re})_2 = 5000$ with corresponding values of the friction factor f_1 and f_2 computed from

Eq. (3.39), we set

$$f_1 = K_n(\text{Re})_1^{-n} \tag{3.52}$$

$$f_2 = K_n(\text{Re})_2^{-n} \tag{3.53}$$

Dividing Eq. (3.52) by (3.53) we obtain

$$\frac{f_1}{f_2} = \left(\frac{(\text{Re})_1}{(\text{Re})_2}\right)^{-n} = \left(\frac{(\text{Re})_2}{(\text{Re})_1}\right)^{n} \tag{3.54}$$

which yields a specific value for the exponent n according to

$$n = \frac{\log(f_1/f_2)}{\log[(\text{Re})_2/(\text{Re})_1]} \tag{3.55}$$

for any selected interval

$$(\text{Re})_1 < \text{Re} < (\text{Re})_2 \tag{3.56}$$

and a corresponding constant K_n, computed with the above value of n per Eqs. (3.52) and (3.53)

$$K_n = f_1(\text{Re})_1^n = f_2(\text{Re})_2^n \tag{3.57}$$

The pairs of constants K_n and n obtained in this manner over the full validity range of Koo's Eq. (3.39) are shown in Table 3.3, together with the estimated maximum percentage deviation in any interval, $(\text{Re})_i < \text{Re} < (\text{Re})_{i+1}$. It is evident that any desired level of precision can be achieved by selecting suitably small Re intervals. However, the percent deviations shown in Table 3.3 are already negligibly small compared to a $\pm 5\%$ accuracy band.

As expected from the shape of the friction curves in Moody's chart (Fig. 3.4) and what was said in Section 3.7, the exponent n decreases monotonously from a value $n = 0.28$ at $\text{Re} = 3000$ to $n = 0.15$ at $\text{Re} = 3{,}000{,}000$, although it does not converge toward $n = 0$ as quickly, in terms of Re, as in the case of rough pipes. This has interesting implications in connection with the coupled heat transfer coefficient as highlighted below.

(e) For a prescribed tube geometry and under otherwise fixed conditions since $G \propto \dot{m}$ and $\text{Re} \propto \dot{m}$, the modified Reynolds analogy yields the following relationship between $\bar{\alpha}$ and $\dot{m}$, with $\bar{f} \propto (\text{Re})^{-n} \propto \dot{m}^{-n}$:

$$\bar{\alpha} \propto \bar{f}\dot{m} \propto \dot{m}^{1-n} \tag{3.58}$$

with values of the exponent $1 - n$ increasing from about 0.77 to 0.85 with increasing Re from 3000 to 3,000,000 in the case of ideal smooth tubes as shown in Table 3.3. Equations (3.58) and (3.44b), indicating that $\Delta H_f =$

TABLE 3.3. Blasius Constants K_n and n for Ideal Smooth Circular Tubes

Reynolds Number		100 × f Eq. (3.39)	Constants in Eq. (3.44) for Interval $(\text{Re})_i - (\text{Re})_{i+1}$: K_n Eq. (3.57)	n Eq. (3.55)	%Deviation in f Eq. (3.44a) Versus Eq. (3.39)	Exponent $2-n$ for $\Delta H_f = \Delta P_f/\rho$ Eq. (3.44b)	Exponent $1-n$ for $\bar{\alpha}$ Eq. (3.58)
$(\text{Re})_1$	3,000	4.417	—	—	0		
	4,000	4.078	0.4029	0.2761	+0.05	1.7239	0.7239
$(\text{Re})_2$	5,000	3.836	—	—	0		
	7,500	3.437	0.3786	0.2688	+0.1	1.7312	0.7312
$(\text{Re})_3$	10,000	3.184	—	—	0		
	20,000	2.662	0.3255	0.2524	+0.4	1.7476	0.7476
$(\text{Re})_4$	40,000	2.244	—	—	0		
	80,000	1.909	0.2454	0.2257	+0.6	1.7743	0.7743
$(\text{Re})_5$	160,000	1.641	—	—	0		
	320,000	1.426	0.1689	0.1946	+0.5	1.8054	0.8054
$(\text{Re})_6$	640,000	1.253	—	—	0		
	1,820,000	1.1155	0.0905	0.1484	+0.8	1.8516	0.8516
$(\text{Re})_7$	3,000,000	0.983	—	—	0		

$\Delta P_f/\rho \propto \dot{m}^2\bar{f} \propto \dot{m}^{2-n}$ with an exponent $2-n$ increasing with Re as quantified in Table 3.3, strongly infer the following trend: As $n \to 0$, the exponent $2-n \to 2$; therefore $\Delta H_f = \Delta P_f/\rho \to \dot{m}^2$ and the exponent $1-n$ in Eq. (3.58) tends toward unity, or $\bar{\alpha} \propto \dot{m}^1$. Thus, as the fully rough turbulent flow limit is approached when the friction dissipation ΔH_f characteristically varies with the second power of the flow rate, one expects that the coupled heat transfer coefficient $\bar{\alpha}$ will vary in almost direct proportion to $\dot{m}^{1-n\to 1}$. This is indeed the case in typical two-phase flow forced-convection applications discussed in subsequent chapters.

3.9 FRICTION FACTOR FOR ROUGH CIRCULAR PIPES

McAdams [3.14] recommends a correlation for line *HG* of Fig. 3.3 that reduces in terms of f (instead of f_{rh}) to

$$(1/f)^{1/2} = 1.6\log_{10}(\text{Re}\,f^{1/2}) - 0.1184 \tag{3.59a}$$

Validity: Commercial pipes made of steel and cast iron ($\bar{e}/D > 0$); other wise same as Eq. (3.39b) (3.59b)

Estimated Deviation: $\pm 10\%$ (3.59c)

McAdams concludes his recommendation for rough pipes with the following statement:

> For badly corroded or tuberculated pipes, due especially to a decrease in diameter [f_{r_h}] based on the observed friction and the original diameter may rise to a value as high as 0.02 [therefore $f = 4f_{rh} = 0.08$ as shown in Fig. 3.3] depending primarily on the average actual diameter of the opening in the tuberculated pipe and the ratio of wall roughness to pipe diameter [3.38, 3.46, 3.47].

The various friction factor curves shown on Moody's chart (Fig. 3.4), with $\bar{e}/D$ as an independent parameter, can be represented by Colebrook's *natural roughness* functions [3.45] as follows:

For the *fully rough* zone where f_{fr} = constant

$$\left(\frac{1}{f_{\text{fr}}}\right)^{1/2} = 1.14 + 2\log_{10}\left(\frac{D}{\bar{e}}\right) \tag{3.60a}$$

For the *transition zone* to the left of the fully rough region in Fig. 3.4.

$$\left(\frac{1}{f}\right)^{1/2} = \left(\frac{1}{f_{\text{fr}}}\right)^{1/2} - 2\log_{10}\left[1 + \frac{9.3}{\text{Re}(\bar{e}/D)f^{1/2}}\right] \tag{3.60b}$$

Validity: Presumably for Re and $\bar{e}/D$ values shown in Fig. 3.4. (3.60c)

Deviation: Not stated in Ref. 3.45 but certainly in *excess of* $\pm 5\%$ *for smooth tube correlations* (3.60d)

Thus, f can be determined from Eq. (3.60b), in principle, at any arbitrary value of the natural relative wall roughness $\bar{e}/D$ in terms of Re with the constant value $(1/f_{\text{fr}})^{1/2}$ computed from Eq. (3.60a), provided $\bar{e}/D$ is known exactly.

3.9.1 Statistical Nature of the Effective Wall Roughness

All engineering oriented fluid and heat flow textbooks [3.44, 3.45, 3.48] rightfully emphasize the statistical nature of the mean effective roughness $\bar{e}$ of typical conduit surfaces shown in Table 3.1 and the fact that $\bar{e}$ is evaluated indirectly from empirical and theoretical boundary flow considerations. Thus, $\bar{e}$ "...is not some easily calculated height, as from mechanical surface-roughness measurement, and it is liable to increase with conduit use or aging" [3.45]. The essential information conveyed by Table 3.1 is that *on average* the clean new steel or wrought iron pipes used in conjunction with Eqs. (3.60a) and

(3.60b) are effectively 30 times rougher than clean new smooth pipes and that this ratio increases to values ranging from 200 to 2000 for materials used in civil engineering. If we focus exclusively on normal heat exchanger design applications and accept a statistical variation of $\pm 5\%$ for the friction factor f in ideal smooth tubes with Eq. (3.39) (which is quite reasonable since this empirical correlation is validated by 1380 reliable individual tests obtained within the usual range of D and Re encountered in actual heat exchanger design practice), then it is easy to show[2] that e can indeed vary from the nominal value $\bar{e} = 0.0015$ mm for smooth tubes by at least ± 33–50%. A similar statistical variation can certainly be expected for steel and wrought iron pipes with a nominal $\bar{e} = 0.0450$ mm.

Although wall roughness variations of the order of ± 33–50% around the mean value $\bar{e}$ have little effect in extremely large-diameter pipes used in civil engineering applications, the same percentage variation can have a significant impact with small hydraulic diameter channels typically encountered in heat exchanger design practice as highlighted in Numerical Analysis 3.3.

NUMERICAL ANALYSIS 3.3

Impact of statistical wall roughness variations for turbulent flow in clean commercial steel pipes.

Problem Statement

Quantify the impact of a statistical wall roughness variation for commercial steel pipes of nominal internal diameters extending from 1 to 1000 mm with special emphasis on heat transfer applications, that is, D_{h} ranging from 1 mm in compact heat exchangers up to 50 mm in shell-and-tube heat exchangers:

(a) On f_{fr} in the fully rough zone under the highly optimistic assumption that Eq. (3.60a) is totally exact.

(b) On the flow rate $\dot{m}$, $\Delta H_f = \Delta P_f/\rho$, and the heat transfer coefficient $\bar{\alpha}$ as in Numerical Analysis 3.2 and compare the results with those obtained with ideal smooth tubes.

Solution

(a) Let e and $\bar{e}$ designate the actual and average effective wall roughness, respectively, with $\Delta e = e - \bar{e}$ and the ratio $(\Delta e/\bar{e})100$ representing the percentage variation from the "typical" average value of $\bar{e}$ shown in Table 3.1. The ratio $(f_{\mathrm{fr}})_e/(f_{\mathrm{fr}})_{\bar{e}}$ can be calculated directly from Eq. (3.60a) for any D_h and percentage wall roughness variation. The results thus obtained are summarized in Table 3.4 for $(\Delta e/\bar{e})100 = +50\%$ and -50% in terms of D_h. The

TABLE 3.4. Impact of Statistical Wall Roughness Variations for Turbulent Flow in Clean Commercial Steel Pipes

Hydraulic Diameter (mm)	(m)	$100 \times f_{fr}$ at $\bar{e} = 0.045$ mm	Percent Change in f_{fr} at $100\,(\Delta e/\bar{e})$ +50%	−50%	Total % Variation
1[b]	0.001	6.804	+21.3	−25.3	46.6
5[b]	0.005	3.654	+15.0	−19.6	34.6
10[a]	0.010	2.939	+13.3	−17.8	31.1
50[a]	0.050	1.912	+10.5	−14.8	25.4
100	0.100	1.630	+9.6	−13.8	23.4
1000	1.000	1.034	+7.6	−11.2	18.8

[a]10–50 mm: typical of heat exchanger design applications; 100–1000 mm: typical of civil engineering and air conditioning duct work.
[b]In compact heat exchangers such as the internally finned tubes discussed in Chapter 4 (Fig. 4.6).

theoretical "average" value $(f_{fr})_{\bar{e}}$ at $e = \bar{e} = 0.045$ mm for various hydraulic diameters D_h calculated from Eq. (3.60a) are also listed in Table 3.4.

(b) Table 3.4 clearly shows that the "cosmetic" accuracy of Moody's charts and Colebrook's Eq. (3.60a) can be dangerously misleading if the statistical nature of $\bar{e}$ is not fully recognized. The impact of unavoidable wall roughness variations from the "typical effective" $\bar{e}$ used in conjunction with Moody's chart is indeed significant in the range of diameters used in heat transfer engineering as opposed to civil engineering applications.

In retrospect, a $\pm 10\%$ deviation band per Eq. (3.59) is still reasonable for many bare tube heat exchangers ($D_h = 25$ mm $\cong 1$ in.) but rather optimistic for smaller hydraulic diameters of the order of a few millimeters typically used in compact tubular heat exchangers [3.4, 3.32, 3.49].

In any case, one can safely conclude (for the reasons stated in Numerical Analysis 3.2) that in the case of *turbulent flow in clean commercial steel pipes the best accuracy* one can ever hope to achieve *for the heat transfer coefficient is* $\pm 10\%$ *and more likely* $\pm 20\%$ as correctly stated in engineering oriented heat transfer textbooks.

3.10 FRICTION FACTOR FOR NONCIRCULAR CHANNELS

For turbulent flow regimes the friction factor can be evaluated with the same correlations recommended for circular tubes by simply using the *equivalent hydraulic diameter* $D_h = 4r_h = 4\Omega/a$ within a $\pm 5\%$ accuracy, except for extreme cross sections [3.45]. This conclusion has been verified by numerous laboratory, production, and field tests performed with straight constant cross section internally finned channels in single- and two-phase flow heat exchangers as highlighted in Ref. 3.49 (and references cited therein).

By contrast, the prediction of the friction factor for laminar flow regimes on the basis of D_h for noncircular channels can be off by a factor of 2 to 1 [3.50]. This illustrates again the important general conclusion emphasized earlier, namely, that the seemingly more complex chaotic flow structures we call turbulent regimes are in fact more predictable in actual engineering fluid and heat flow applications than the theoretically simpler laminar flow regimes.

3.11 SENSITIVITY TO NONISOTHERMAL CONDITIONS

All fluid mechanics equations reviewed so far are purely "mechanistic" interpretations of fluid flow phenomena in the sense that the absolute temperature, a key thermodynamic state property, never appears explicitly in any of the fluid flow equations.

Thermal effects are ignored "axiomatically" in reversible fluid hydrodynamics which is based on ideal inviscid incompressible fluids, whereas they are at least "inferred" in irreversible real fluid mechanics since the effect of temperature on the viscosity is recognized. It is important to recall, however, that conventional analytical solutions (the Navier–Stokes equations) are usually based on the fundamental assumption of constant viscosity (μ) and constant density (ρ) fluids. Since according to classical thermodynamics all fluid state properties, including the key fluid flow transport property μ, are uniquely defined by two independent intensive thermodynamic state properties, such as the absolute temperature T and pressure P, the assumption of a constant μ *and* ρ infers most generally that T and P are substantially constant. For common liquids and diluted gases or vapors, the empirical thermophysical properties shown in Appendix A clearly indicate that the effect of P is normally negligible compared to the significant impact of T on the absolute viscosity. Thus, in the case of constant density liquids (not too near the thermodynamic critical point) and quasi-incompressible gases (when $\Delta P/P \ll 1$ and $\rho \cong$ constant) *the key independent parameter* is the *temperature*. Consequently, all the *conventional fluid flow analytical solutions* reviewed here are strictly *valid only under isothermal conditions*, that is, when the fluid temperature T is substantially constant throughout the whole conduit cross section and length, or more generally the full system under consideration.

The sensitivity to nonisothermal conditions (NIC) can roughly be assessed for any fixed geometry on the basis of the generalized Blasius friction factor correlation, Eq. (3.44), by estimating the impact of a small relative temperature change $\Delta T/T$ on the flow rate $\dot{m}$ for a given $\Delta H_f = \Delta P_f/\rho$ or conversely [per Guideline 6(a) and (b)] as follows.

From Eq. (3.44) for a fixed geometry Ω and L/D_h at $\rho =$ constant,

$$\Delta H_f = \frac{\Delta P_f}{\rho} = f\left(\frac{L}{D_h}\right)\left(\frac{\dot{m}}{\Omega\rho}\right)^2 \propto f(\dot{m})^2 \tag{3.61}$$

with Re $\propto \dot{m}/\mu$ and the general Blasius type correlation for f,

$$f = K_n(\text{Re})^{-n} \propto \left(\frac{\mu}{\dot{m}}\right)^n \tag{3.62}$$

Introducing the above expression for f into Eq. (3.61) yields

$$\Delta H_f = \frac{\Delta P_f}{\rho} \propto \dot{m}^{2-n}\mu^n \tag{3.63}$$

Using the subscript ΔT to identify all parameters at $T + \Delta T$, Eq. (3.63) yields for a *prescribed mass flow rate* $\dot{m}$ the following variations:

$$\frac{(\Delta H_f)_{\Delta T}}{\Delta H_f} = \frac{(\Delta P_f)_{\Delta T}}{\Delta P_f} = \left(\frac{\mu_{\Delta T}}{\mu}\right)^n \quad \text{for } \dot{m} \equiv \dot{m}_{\Delta T} \tag{3.64}$$

and, conversely, for a *prescribed dissipation* $(\Delta H_f) = (\Delta H_f)_{\Delta T}$

$$\dot{m}^{2-n}\mu^n = \left(\dot{m}^{2-n}\mu^n\right)_{\Delta T} \tag{3.65}$$

Solving Eq. (3.65) for the ratio $\dot{m}_{\Delta T}/\dot{m}$ yields

$$\left(\frac{\dot{m}_{\Delta T}}{\dot{m}}\right)^{2-n} = \left(\frac{\mu}{\mu_{\Delta T}}\right)^n, \tag{3.66}$$

therefore

$$\frac{\dot{m}_{\Delta T}}{\dot{m}} = \left(\frac{\mu}{\mu_{\Delta T}}\right)^{n/(2-n)} \tag{3.67}$$

at $\Delta H_f = \Delta P_f/\rho \equiv (\Delta H_f)_{\Delta T} = (\Delta P_f/\rho)_{\Delta T}$.

The above results are summarized in Table 3.5 for Blasius exponents n ranging from $n = 1$ for laminar flow regimes to $n = 0.28$ down to 0.15 for turbulent flow regimes in smooth tubes according to Numerical Analysis 3.2, as well as the limiting case of fully rough wall turbulent flows.

3.11.1 Laminar Versus Turbulent Regimes

It is clear from Table 3.5 and Eq. (3.67) that laminar flows are inherently more sensitive to nonisothermal conditions (NIC) than turbulent flow regimes. For the limiting case of fully rough turbulent regimes when $n = 0$, the system becomes *totally insensitive to NIC* since the friction factor remains constant and independent of Re, and therefore of $\dot{m}$ (as could be expected from Moody's chart, Fig. 3.4).

TABLE 3.5. Exponents in Eqs. (3.44), (3.67), and (3.69) for Stable Laminar and Turbulent Flow Regimes in Ideal Smooth to Fully Rough Tubes

	n Eq. (3.44)	$\frac{n}{2-n}$ Eq. (3.67)	$0.68\left(\frac{n}{2-n}\right)$ Eq. (3.69)	Wall Characteristic
Laminar				
Re < 2,000	1	1	0.68	$\bar{e}/D \cong 0$ and $\bar{e}/D \gg 0$
Turbulent				
Re > 10,000	0.25[a]	0.143	0.097	Ideal smooth $\bar{e}/D \cong 0$
Re ~ 3,000,000	0.15[a]	0.081	0.055	
Re → ∞	0	0	0	
Turbulent				
Re ≥ 15,000	0	0	0	Fully rough $\bar{e}/D \geq 0.05$[b]

[a]See Table 3.3.
[b]From Moody's chart and / or Colebrook's equations.

3.11.2 Single-Phase Liquids

The absolute *viscosity* of liquids may *decrease* several orders of magnitude with *increasing* T within the "domain of existence" of a pure liquid, that is, between the triple point T_t (about equal to T_f = freezing point at atmospheric pressure) and the thermodynamic critical point T_{cr} beyond which the liquid phase vanishes.

As an illustration, Table A.1 of Appendix A shows for water a viscosity $\mu = 1750 \times 10^{-6}$ N · s/m² at $T_f = 273.15$ K decreasing to a value $\mu_{cr} = 45 \times 10^{-6}$ at $T_{cr} = 647.3$ K. This corresponds to a maximum ratio $\mu_f/\mu_{cr} = 38.9$ for a temperature change $T_{cr} - T_f = 374$ K.

For a typical (clean) engine oil, Table A.2 shows that an increase from 273 to 430 K, or $\Delta T = 157$ K, causes a decrease in μ from 385×10^{-2} to 0.470×10^{-2} N · s/m² or a ratio $\mu_{273}/\mu_{430} = 819$. This indicates that highly viscous liquids are more sensitive (by about two orders of magnitude) to thermal effects than water with a lower viscosity ratio over the same temperature range: $\mu_{273}/\mu_{430} = 1750/173 \cong 10$.

3.11.3 Single-Phase Gases and Vapors

In contrast to liquids, the *absolute viscosity* μ of most common gases and vapors *increases slightly* with *increasing* T, as shown in Table A.3 of Appendix A, and is about one or two orders of magnitude *smaller* than a relatively inviscid reference liquid such as water, the most popular classical reference liquid in engineering. Since tubular gas heat exchangers are deliberately designed to operate in stable turbulent regimes, gases are generally much less sensitive to NIC than liquids on two counts: first, because the exponent $n \ll 1$ in turbulent regimes (as shown in Table 3.5) and second, the absolute viscosity of gases at normal absolute P and absolute T (thus excluding ultralow cryogenic temperatures at $T \to 0$) vary very weakly with T according to gas kinetic theories [3.51] and empirical data show [3.52] that μ is approximately proportional to T^a with $a = 0.68$. Therefore, Eqs. (3.64) and (3.67) reduce to

$$\frac{(\Delta H_f)_{\Delta T}}{\Delta H_f} = \frac{(\Delta P_f)_{\Delta T}}{\Delta P_f} = \left(1 \pm \frac{|\Delta T|}{T}\right)^{0.68n} \quad \text{for } \dot{m} \equiv (\dot{m})_{\Delta T} \tag{3.68}$$

and

$$\frac{(\dot{m})_{\Delta T}}{\dot{m}} = \left(\frac{T}{T \pm |\Delta T|}\right)^{0.68n/(2-n)} = \frac{1}{(1 \pm |\Delta T|/T)^{0.68n/(2-n)}} \quad \text{for } \Delta H_f \equiv (\Delta H_f)_{\Delta T} \tag{3.69}$$

Note that the *positive* sign in the above equations corresponds to temperature *increase* and conversely. Equations (3.68) and (3.69) clearly indicate a *very low sensitivity to NIC for gases in turbulent regimes* at absolute temperature $T \gg 0$ even in the case of ideal smooth tubes. This is further quantified in Table 3.5.

3.11.4 Quantitative Comparison

Table 3.6 provides a quantitative comparison of the sensitivity to NIC for two liquids—water and clean engine oil—as well as quasi-ideal gases such as air or hydrogen with a viscosity proportional to T^a and $a = 0.68$ (note that the numerical values shown in Table 3.6 could easily be adjusted for exponents $a \neq 0.68$). The sensitivity to NIC can be estimated in a similar manner for different single-phase liquids and gases with the help of Tables A.1-A.3 in Appendix A.

The key data summarized in Tables 3.5 and 3.6 highlight once more why streamline flows are inherently more complex in actual heat exchanger design practice (notwithstanding their theoretical simplicity) than the chaotic turbulent flow regimes, particularly when one recalls that laminar flow patterns are also more sensitive to entrance effects and uncertain conditions preceding the tube entrance. As a matter of fact, the more chaotic (i.e., the more turbulent) the flow regimes, the easier the solution of the basic fluid and heat flow engineering problems in forced-convection heat exchangers. In the limiting case of fully rough wall turbulent flows, the most chaotic and analytically most intractable regimes, it is only necessary to know or find empirically one single constant, f_{fr}, to solve with the greatest degree of confidence the basic engineering fluid and heat flow problems in both single- and two-phase heat exchanger design applications.

TABLE 3.6. Impact of Small Relative Temperature Changes $(\Delta T/T) \times 100$ on Mass Flow Rate Expressed as a Percentage of the Flow Rate at 310 K and Computed[a] from Eqs. (3.67) and (3.69) for Liquids[b] and Gases[c] in Stable Laminar and Turbulent Flow Regimes

ΔT (°C / K) $(\Delta T/T) \times 100\%$	−30 −9.7	−10 −3.2	0 0	+10 +3.2	+30 +9.7	Flow Regime
Engine oil	16.7	52.1	100	179.4	476.5	Laminar: $n = 1$
Water	48.9	81.3	100	120.5	165.5	Re < 2000
Gases	107.2	102.3	100	97.9	93.9	
Engine oil	73.5	91.1	100	108.7	125.0	Turbulent: $n = 0.25$
Water	90.3	97.1	100	102.7	107.5	$\bar{e}/D = 0$
Gases	101.0	100.3	100	99.7	99.10	10,000 < Re < 100,000
Engine oil	84.0	94.5	100	104.9	113.5	Turbulent: $n = 0.15$
Water	94.4	98.3	100	101.5	104.2	$\bar{e}/D = 0$
Gases	100.6	100.2	100	99.8	99.5	Re ≈ 3,000,000
Engine oil Water Gases	100	100	100	100	100	Turbulent: $n = 0$ Re → ∞ for $\bar{e}/D = 0$ Re > 15,000 for $\bar{e}/D > 0.05$

[a]Based on incompressible fluid (ρ = constant).
[b]Viscosity μ obtained from Tables A.1 and A.2 in Appendix A.
[c] Based on $\mu \propto T^a$ with $a = 0.68$, per Ref. 3.52. Note that essentially identical results are obtained with μ from Table A.3 of Appendix A for common gases.

3.12 UNSTEADY OR MARGINALLY STABLE FLOW REGIMES

By demonstrating the existence of a critical transition band in terms of Re, which is inherently *unstable* or only marginally stable and rather unpredictable, Osborne Reynolds showed that the tacit assumption of steady-state steady-flow (SSSF) conditions cannot be taken for granted.

In the simplest and best known case of single-phase Newtonian isothermal incompressible fluids flowing in ideal smooth circular conduits, this uncertain grey zone falls roughly within a Reynolds number band of 2000 < Re < 5000–10,000.

At Reynolds numbers safely below the lower transition limit, Re < 2000, laminar SSSF regimes can be expected. These regimes are characterized by the linear relationship $\Delta H_f = \Delta P_f/\rho \propto \bar{V} \propto G \propto \dot{m}$ for a given geometry at constant μ and ρ (i.e., isothermal conditions and $\Delta P/P \ll 1$). Within this stable streamline region the fluid behaves in an orderly manner and fully analytical solutions are possible on the basis of the general Navier–Stokes equations when the trajectories of fluid particles can be *predicted mathematically* under *isothermal* conditions, at least in principle.

Above the uncertain grey zone at Re > 5000–10,000, the fluid reorganizes itself from an orderly streamline flow to a new type of stable flow structure called turbulent, which is characterized by the *nonlinear relationships* $\Delta H_f = \Delta P_f/\rho \propto \dot{m}^{2-n} \rightarrow \dot{m}^2$ for $n \rightarrow 0$ in the limiting case of fully rough flow regimes.

The critical Reynolds number $(\mathrm{Re})_{cr}$ that controls the transition from turbulent to laminar regimes (or vice versa) is inversely proportional to μ. It is clear from Tables 3.5 and 3.6 that the flow regimes can be affected more seriously by small relative temperature changes in the case of liquids, particularly highly viscous liquids, than gases. For example, an absolute change of only $-7\,^{\circ}\mathrm{C}$ from $T = 280$ down to 273 K or a relative change $\Delta T/T = -7/280 = -0.025 = -2.5\%$ corresponds roughly to a 2:1 increase in viscosity for a clean engine oil or glycerine (see Table A.2 in Appendix A). Such a viscosity increase could be sufficient to trigger a switch from one regime to another under otherwise fixed conditions. By comparison, a 2:1 viscosity increase would require, in the case of gases, a temperature ratio $[(T + \Delta T)/T]^{0.68} = 2$ or $(T + \Delta T)/T = 2.772$. This would require an increase from $T = 280$ K to $(T + \Delta T) = 776.2$ K or a temperature increase of close to $500\,^{\circ}\mathrm{C}$ instead of a $7\,^{\circ}\mathrm{C}$ drop! This explains why gases are generally more predictable than viscous fluids and quite insensitive to the temperature variations typically encountered in thermodynamically efficient heat exchangers designed to operate at low relative wall to fluid temperature differentials.

Because the heat transfer coefficients α_{turb} are significantly higher than α_{lam}, pragmatic heat transfer engineers deliberately circuit their tubular heat exchangers (if at all possible) so that Re is significantly above $(\mathrm{Re})_{cr}$ and typically in the range $10{,}000 < \mathrm{Re} < 120{,}000$, where the friction factor and the coupled heat transfer coefficient are more accurately predictable. Indeed, the most popular design correlations for $\bar{\alpha}$ [3.42, 3.53] are usually based on $\mathrm{Re} > 10{,}000$.

Thus, heat transfer engineers have learned to cope with the uncertainties associated with regime transitions by avoiding this potentially unstable zone under specified design conditions. However, heat exchangers also are expected to operate stably well below nominal full load conditions, and the possibility of operating within an unstable critical zone at part loads cannot always be ruled out. Consequently, a better understanding of the grey zone $2000 < \mathrm{Re} < 10{,}000$, in the case of tubular heat exchangers, is of considerable scientific and practical interest.

It is quite a sobering experience, when searching for rational analytical solutions for the prediction of flow regime transitions, to realize how little more we know today, over a century after Reynolds' discovery of "sinuous" flows, regarding the exact nature of turbulence, let alone the key factors that may trigger a switch from an unstable flow pattern to another more probable and therefore stabler flow structure.

The thermodynamic implications of flow instabilities and regime transitions become intuitively more evident when one recalls that entropy is a measure of probability, according to Boltzmann,[3] and assesses the flow crises associated with regime transitions from a thermodynamic point of view. Whereas the sudden switch from laminar to turbulent flow is depicted in classical fluid mechanics as a transition from an orderly analytically predictable streamline flow to an unpredictable disorganized turbulent pattern, this self-reorganization can be viewed in exactly the opposite fashion from a strictly thermodynamic standpoint: the critical bifurcation from laminar to turbulent flow can be visualized instead as a rearrangement from a highly heterogeneous fragile (except at Re $\ll$ 2000) flow pattern to a thermodynamically more homogeneous and *more probable structure associated with a maximum entropy dissipation*. This qualitative conclusion is not inconsistent with the recent recognition that turbulent flows possess an orderly structure [3.6]. The central role of entropy is also evident in the context of the new (twentieth century) irreversible thermodynamics theory overviewed in Chapter 7 and further supported by the older classical thermodynamic methods of predicting critical supersonic to subsonic flow transitions on the basis of the well-known *Fanno curve* in a temperature–entropy diagram [3.54–3.57]. One analytical method of flow regime predictions based on nonequilibrium thermodynamic considerations is highlighted in Chapter 9 and Appendix E.

NOTES

1. One can detect a subtle trend in thermal sciences with the appearance of theoretically more refined computer-aided correlations: the *longer* and *more complex* the mathematical expressions, the *shorter* and the *more cryptic* the accompanying statements on accuracy/validity limits, compared to those found in the pragmatic American heat transfer textbooks of the 1950s by McAdams [1.2], Kern [1.3], and later disciples.
2. f is computed at various values of Re according to Koo's Eq. (3.39), allowing for a $\pm 5\%$ variation in f and solving Colebrook's Eqs. (3.60a) and (3.60b) for e/D with these extreme values of f at the corresponding Re.
3. The abstract concept of entropy can readily be explained in terms of *thermodynamic probability* [3.58–3.62] on the basis of the statistical mechanics concepts pioneered by Ludwig Boltzmann (University of Vienna, Austria) and Josiah Willard Gibbs (Yale University, United States) according to

$$s_2 - s_1 = k(\ln \mathscr{P}_2 - \ln \mathscr{P}_1) \tag{3.70a}$$

where $\mathscr{P}$ = thermodynamic probability defined as the number of possible microstates for a given macroscopic state
k = Boltzmann's constant $\equiv \mathscr{R}/\mathscr{N} \equiv 1.380 \times 10^{-23}$ J/K · molecule
$\mathscr{R}$ = universal gas constant = 8.315 J/mol · K
$\mathscr{N}$ = Avogadro's number = 6.024×10^{23} molecules/mol

This well-known equation (first proposed by Boltzmann in the nineteenth century) states that when an isolated thermodynamic system moves from a less probable state (1) to a more probable state (2), the increase in entropy, $s_2 - s_1 > 0$, is directly proportional to the increase in the natural logarithm of the probability of the macrostate of the system. Thus, the stable end equilibrium state of an isolated system is always associated with a maximum entropy s_2, that is, a maximum thermodynamic probability.

CHAPTER 4

OVERVIEW OF MULTIPHASE FLUID FLOW

So far we have dealt mainly with the simplest fluid flow phenomena encountered in nature and in engineering systems: single-phase internal flows of Newtonian fluids. We now address multiphase flow situations very broadly to obtain an "organic" view of the whole field and to identify the salient features of these complex phenomena. The prime purpose of this broad overview is to highlight qualitatively the unique characteristics of multiphase flows and provide the necessary background for two-phase pressure drop correlations and the effective fluid and heat flow thermodynamic generalization methods presented in Chapters 5, 6 and 8, respectively.

4.1 NATURAL MULTIPHASE FLOW PHENOMENA

Multiphase flows are a common occurrence in nature. Clouds are good examples of multiphase systems consisting of two, sometimes three, different phases: a gas phase (air plus water vapor), a liquid phase (droplets or mist), and/or a solid phase (ice, snow). The sirocco, a violent North African desert wind, can readily whip up dust clouds to create solid–gas two-phase systems easily transported across the Mediterranean Sea to Europe where the fine desert sand finally separates from the carrier, to settle down as red dust on the European mainland. A murky turbulent stream of water flowing down a steep sandy hill after a heavy rain is essentially a solid–liquid two-phase system and it takes a long time (to the despair of fishermen) for the water to become crystal clear again.

All these natural multiphase flow phenomena have one important feature in common: It is relatively easy, if enough potential energy is available, to whip, mix up, or emulsify a mixture of finely divided solid particles in a fluid (gas or liquid) or a heavier liquid phase in a lighter gas phase ($\rho_g \ll \rho_l$). By contrast, it is not as easy, and much more time is required, for the small entrained particles to *separate again*, particularly at high carrier velocities in the turbulent flow regimes more frequently encountered in nature and technological applications. The reason for the extremely slow separation of heavier particles, such as very fine liquid droplets or solid particles of diameter d, is that the drag (frictional) forces are proportional to the surface exposed (i.e., $\propto d^2$), whereas the gravity forces are proportional to the volume (i.e., $\propto d^3$); thus, the ratio drag/gravity $\propto 1/d \to \infty$ when $d \to 0$ and this explains why aerosols never settle out of an air stream [4.1, 4.2].

4.2 MULTIPHASE FLOW IN ENGINEERING SYSTEMS

From the above rudimentary overview of natural multiphase flow phenomena it would appear that nature favors the more intimately *mixed* or *homogeneous* flow structures associated with turbulent regimes over the *heterogeneous separated* flow patterns, and many technological multiphase flow situations display the same characteristic.

Instead of illustrating the unique features of complex multiphase flow phenomena on the basis of steam and water, as is customary in most engineering heat transfer textbooks (traditional mechanical engineers' emphasis on steam power plants), we highlight here *reverse power cycles*, particularly the simple mechanical refrigeration cycles. Several practical and didactic reasons justify this departure from the standard two-phase water–steam flow "menu":

1. Mechanical refrigeration systems are a much more common occurrence in household, commercial and industrial technological applications than large fossil and/or nuclear steam power generating plants!

2. Notwithstanding the vital importance of nuclear two-phase flow phenomena and research results, they are in many respects atypical of conditions encountered in normally efficient fluid to fluid exchangers and more akin to electric heating. Indeed, the prime motivation for nuclear two-phase flow research, the avoidance of a nuclear channel physical burnout at some critical heat flux, is based on the *boiling crises* demonstrated in 1934 by Nukiyama [4.3] with water boiling on an electrically heated horizontal platinum wire. Such wide temperature excursions to the point of a physical metal burnout are *not* conceivable in normally efficient fluid to fluid convective heat exchangers, particularly in refrigeration systems. Contrary to electric and/or nuclear heating, the maximum tube wall temperature has a finite upper bound: the absolute temperature of the heat source. Furthermore, it has been shown [4.4–4.6] that two-phase fluid flow and heat transfer results obtained by

evaporating a fluid inside electrically heated tubes or channels are not necessarily applicable to forced-convection evaporators where the evaporating refrigerant (the coolant) is heated by an external fluid, or source, at temperature T_{so}, which is cooled as the refrigerant evaporates. Thus, T_{so} must decrease monotonously or, at the limit, remain constant but never increase. This

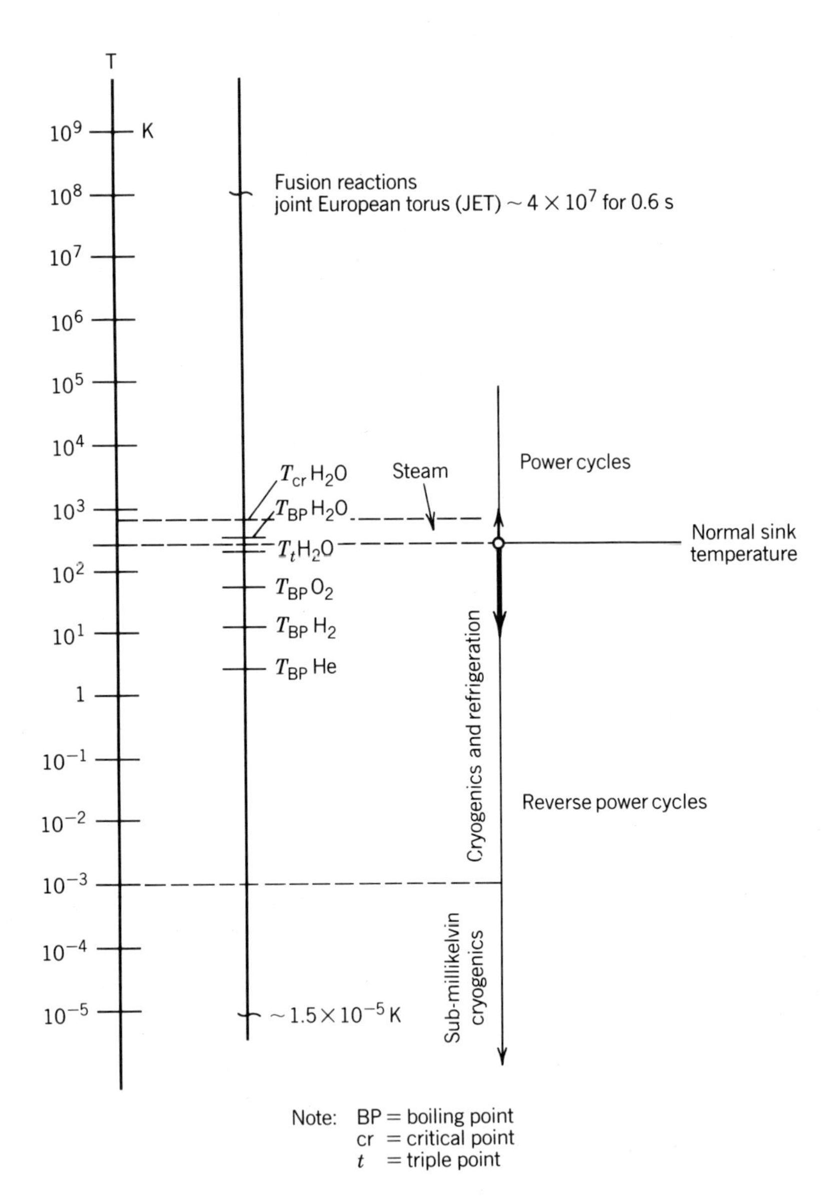

FIGURE 4.1. Absolute temperature range.

characteristic has a profound effect on the possible types of stable flow regime and consequently on the thermal performance of forced-convection evaporators. These topics are discussed more fully by Soumerai [4.4–4.6].

3. In contrast to steam power generation cycle analysis, which focuses on the single chemical species H_2O, many different refrigerants are used in refrigeration systems [4.7–4.9] as illustrated by Table A.5 in Appendix A, which lists nearly 100 different refrigerants. In this context typical refrigeration two-phase flow applications have more in common with general process heat exchangers.

4. Thermodynamically, the refrigeration/heat pump industry covers *a much larger logarithmic absolute temperature range* as highlighted in Fig. 4.1.

5. From the beginning of this relatively new industry, refrigeration engineers were compelled to select significantly lower temperature differentials than was customary in general process heat exchanger design practice to keep the power consumption within acceptable limits. Thus, the refrigeration industry has a long tradition in fighting "second law irreversibilities" and has developed the know-how to cope with the challenge of designing heat exchangers for *low fluid-to-fluid temperature differentials*. These traditional refrigeration and cryogenic heat exchanger design techniques have become quite *relevant in general process heat transfer applications since the oil crises of the 1970s* and the renewed emphasis on energy conservation or entropy production minimization [4.10–4.13].

When the International Institute of Refrigeration was founded in 1908, refrigeration engineers and scientists had pioneered a new trend: the internationalization of the scientific community. ASHRAE, the American Society of Heating, Refrigerating and Air-Conditioning Engineers, also has a long tradition in multinational cooperation with local chapters on all continents. Readers interested in the history and the significant scientific/engineering contributions from this vast multidisciplinary branch of engineering called refrigeration are referred to the brief overview given by Thevenot [4.14] or the more detailed historical review by the same author [4.15] published under the auspices of the International Institute of Refrigeration (IIR).

4.3 SIMPLE VAPOR COMPRESSION REFRIGERATION CYCLE

The purpose of a refrigeration cycle or heat pump is to *transfer heat from* a low-temperature level, the *heat source*, *to* a high-temperature level, the *heat sink*. As demonstrated in 1824 by Sadi Carnot [1.7], who conceptually invented[1] the mechanical refrigeration cycle, power must be expended in any continuous refrigeration cycle since heat of itself cannot flow up a temperature gradient (the second law of classical thermodynamics).

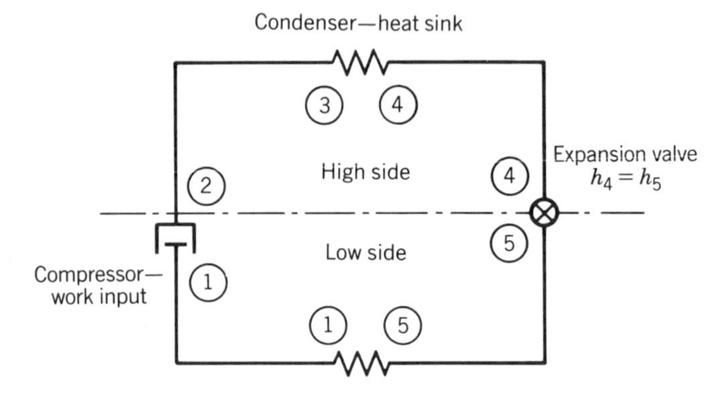

FIGURE 4.2. Simple saturation compression cycle. [Adapted with permission from H. Soumerai, Compression Refrigeration, in *Air Conditioning Refrigerating Data Book*, Design Volume, 10th ed., pp. 4-01 to 4-25, ASRE (now ASHRAE), 1957.]

In a *vapor* compression cycle [4.7] a liquifiable refrigerant is employed which changes phase from liquid to vapor as it absorbs heat in the evaporator. A compressor withdraws the vapor generated in the evaporator at a low pressure, raises its pressure, and discharges the vapor to the condenser. The refrigerant changes phase from vapor to liquid in the condenser as heat is discarded to the sink. High-pressure liquid refrigerant is expanded and fed to the evaporator to complete the refrigeration cycle.

All vapor refrigeration systems (regardless of the type of compressor used) operate because of a difference in pressure which permits the *collection* by the evaporating refrigerant (two-phase flow with heat addition) at a *low* saturation temperature T_L and *disposal* at a *higher* saturation temperature T_H by the condensing refrigerant as shown schematically in Fig. 4.2.

4.3.1 Ideal Reference Cycle Analyses

Idealized cycles can be set up to evaluate the effects of various system modifications on overall performance, efficiency, and first cost. The analysis of ideal cycles is extended to *real* compression cycles, usually on the basis of pressure–enthalpy diagrams which are more suitable than temperature–entropy graphs in practical design analyses, as shown in standard refrigeration texts and handbooks [4.7–4.9]. For didactic/heuristic purposes, however, the temperature–entropy diagram as shown in Fig. 4.3 is more useful because it provides a simple means of quantifying the effects of deviations from the ideal reference Carnot cycle requiring the least work input per unit of refrigeration due to the fact that irreversible losses appear as areas in a *T-s* chart.

Referring to points 1–5 in Figs. 4.2 and 4.3, we can identify the following single- and two-phase fluid flow situations for a full refrigerant cycle.

Points

1	Saturated vapor at low-pressure side: *single-phase* (gas).
2	Superheated vapor at high-pressure side: *single-phase* (gas).
3	Saturated vapor at high-pressure side: This is the *theoretical* transition point between single-phase dry vapor to two-phase flow in the condenser
3–4	*Condensation* with *heat rejection* from refrigerant to sink at high-pressure level: *two-phase flow* regimes of refrigerant (liquid, vapor) at $T_H = T_{sink}$.
4	Saturated liquid at high-pressure side: *single-phase* (liquid).
4–5	*Adiabatic irreversible throttling* (all potential energy is dissipated as internal friction heat under assumed adiabatic change) from the high- to the low-pressure side: *two-phase flow* (liquid, vapor). At point 5 the mixture has a vapor quality $x_5 \equiv$ kg vapor per kg total mixture and a corresponding liquid content $(1 - x_5)$ kg liquid per kg total mixture. Point 5 is uniquely defined in the *T-s* diagram at the intersection of the constant enthalpy line $h_5 = h_4$ and the constant prescribed low-side saturation temperature T_L corresponding to a known saturation pressure P_1 for a given refrigerant. In ideal cycle analyses the refrigerant is assumed to evaporate exactly at the source temperature $T_{source} = T_L$ without any irreversible internal losses, that is, zero two-phase pressure drop or $\Delta P_{tp} = P_5 - P_1 = 0$ and $T_5 - T_1 = 0$.
5–1	*Evaporation* of refrigerant from inlet quality x_5 to saturated outlet $x_1 = 1$, or *two-phase flow with heat addition* at constant low-pressure side $P_5 = P_1$ and corresponding constant saturation temperature $T_5 = T_1 = T_{source} = T_L$. This implies a vanishingly small temperature differential $(T_{source} - T_L = 0)$, that is, zero entropy production, and an infinitely large heat exchanger surface in the absence of any heat transfer driving force.

4.3.2 Real Mechanical Refrigeration Cycle Analyses

The *idealized* refrigeration cycle portrayed by Figs. 4.2 and 4.3 departs from *real* cycles in several respects, which are highlighted below as far as is necessary for a better understanding of the unique features typically encountered in two-phase refrigeration heat exchangers. For more detailed real cycle analyses the interested reader is referred to the specialized technical literature [4.7–4.9].

Item 1: Finite Heat Transfer Driving Forces and Internal Friction Dissipation. It is tacitly inferred, as a result of the assumption of zero irreversible

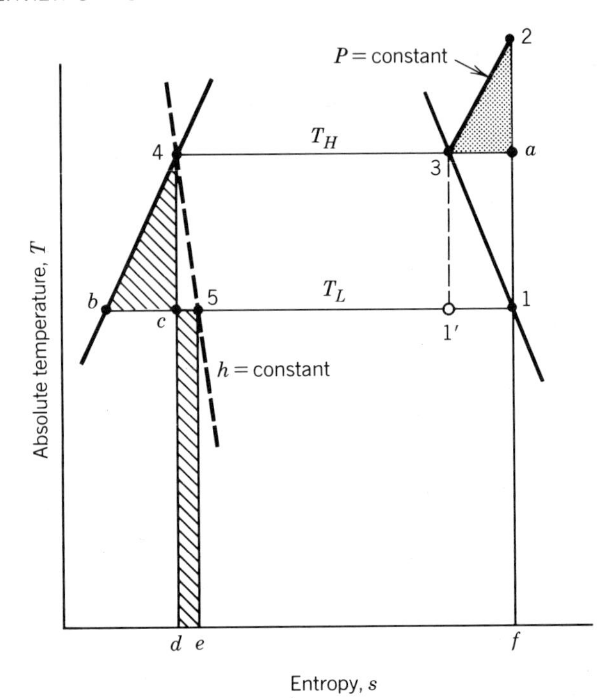

FIGURE 4.3. Simple saturation compression cycle in a *T-s* diagram. [Adapted with permission from H. Soumerai, Compression Refrigeration, *Air Conditioning Refrigerating Data Book*, Design Volume, 10th ed., pp. 4–01 to 4–25, ASRE (now ASHRAE), 1957.]

losses in the heat exchangers, that the condenser and evaporator heat exchange surfaces are infinitely large since the temperature gradient necessary to drive the heat transfer is postulated to be infinitely small. By contrast, real world condensers/evaporators must operate at *finite temperature differentials* and finite two-phase flow pressure drop $\Delta P_{tp} > 0$ associated with *finite saturated temperature drops* ΔT_{sat}.

Item 2: Superheat Horn Loss. Area $a23a$ in Fig. 4.3, the *superheat horn*, represents one cause of increase in specific work input compared to the ideal Carnot cycle work represented by area $a4c1$. The only theoretical way to eliminate this loss fully in a single-stage refrigeration cycle would be to compress the refrigerant vapor from a wet compressor intake at point $1'$ to 3 isentropically, as indicated in Fig. 4.3. However, wet suction will actually increase irreversible losses due to enhanced (irreversible) heat transfer within the compressor and this may lead to practical lubrication problems in positive displacement or erosion in turbo compressors as highlighted in Ref. 4.17.

Item 3: Direct Coupling of Thermofluid Phenomena. The ideal cycle assumes that no irreversible friction losses occur in any part of the system except for the irreversible throttling ($s_5 - s_4 > 0$) between points 4 and 5 in the expansion valve. In principle, the expansion valve could be replaced by an ideal 100% efficient engine to let the high-pressure saturated liquid expand isentropically from point 4 to point c at $s_c = s_4$ as shown in Fig. 4.3. Such an expander would improve the refrigeration cycle performance on two counts:

1. By yielding an additional positive amount of external work represented on the *T-s* chart by the nearly triangular area $4bc4$, thereby reducing the net specific amount of work required per unit mass of refrigerant flowing through the system.
2. By increasing the net refrigerating effect per unit mass of refrigerant by an amount equivalent to area $cde5c$ = area $4bc4$ in the *T-s* diagram since in an adiabatic (zero external heat exchange) throttling process the heat of friction is fully absorbed by the refrigerant; therefore, $h_4 = h_5 > h_c$.

The reason for this deliberate waste of energy in mechanical refrigeration systems is a matter of economics and reliability. The expansion valve is much simpler, more reliable, and less costly than a sufficiently efficient two-phase flow expander. Although the elimination of this throttling irreversible loss is being considered more seriously since the energy crisis of the 1970s and the resulting escalating energy costs [4.17], we nevertheless focus here on the conventional systems equipped with expansion valves to reflect the current state of the art of mechanical refrigeration.

The fact that the available energy in the high-pressure liquid leaving the condenser is in any case fully dissipated during the throttling process occurring in the expansion valve means that *no additional pumping power* needs to be spent to *overcome* the flow resistance $|\Delta P_{tp}|$ between evaporator inlet and outlet in contrast to normal heat exchanger applications. The refrigerant will simply be expanded irreversibly from condenser pressure to a slightly higher expansion valve exit pressure by an amount $\Delta P_{tp} > 0$ when $\Delta P_{tp} \neq 0$. Consequently, the heat transfer engineer could, in principle, circuit the evaporator for any ΔP_{tp} because the *pumping power* required to force the refrigerant through the evaporator channels is available practically "free of charge." There is nevertheless a penalty for an excessively high refrigerant pressure drop because ΔP_{tp} *is always associated with a corresponding saturation temperature drop* ΔT_{sat} and the saturation temperature at the evaporator inlet will *increase proportionally* to ΔP_{tp} at a *prescribed suction pressure* (the usual case in refrigeration applications) thus reducing the net driving force available for heat transfer. As a result, the total pressure drop ΔP_{tp} has a major impact on the optimum design of the evaporator.

Item 4: Presence of Lubrication Oil in the System. Like all machinery, normal mechanical refrigeration compressors require good lubrication (disregarding highly specialized oil-free compressor applications) and a certain amount of liquid oil "mist" is unavoidably entrained with the high-velocity dense gas beyond the compressor into the system. Consequently, in real mechanical refrigeration systems, one has to contend *not* with a *surgically clean* oil-free single-component refrigerant but rather with a *multicomponent refrigerant and oil system*. The important positive and/or negative impacts of the presence of oil can best be highlighted by referring to the typical modern refrigeration system described below.

4.3.3 Large Hermetic Package Chillers

Figure 4.4 illustrates the physical layout of a complete R-22 hermetic screw package chiller with a nominal capacity of 226 tons of refrigeration (795 kW) under the following standard (ARI) conditions for airconditioning applications: water inlet/outlet temperatures of 54/44°F (12.22/6.66°C) for the cooler, and 85/95°F (35/29.44°C) for the condenser with a specified standard fouling factor allowance in both the cooler and the condenser. Other units with nominal capacities from 120 to 350 tons (422–1231 kW) designed along the same lines differ only in overall dimensions, as indicated in Fig. 4.4. The

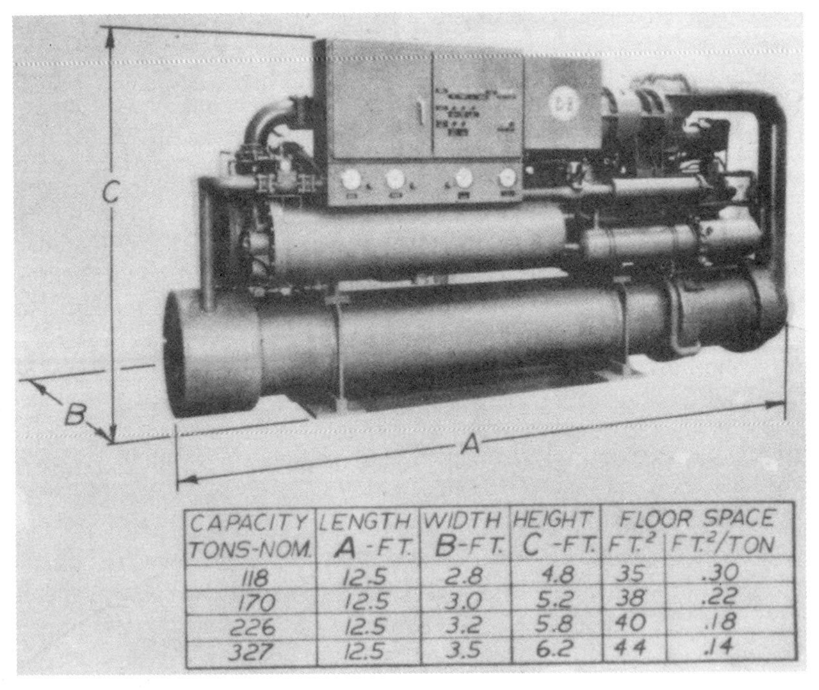

CAPACITY TONS-NOM.	LENGTH A-FT.	WIDTH B-FT.	HEIGHT C-FT.	FLOOR SPACE FT.^{2}	FT.^{2}/TON
118	12.5	2.8	4.8	35	.30
170	12.5	3.0	5.2	38	.22
226	12.5	3.2	5.8	40	.18
327	12.5	3.5	6.2	44	.14

FIGURE 4.4. Second-generation hermetic package chiller design. [From H. Soumerai "Design and Operation of Modern Two-Pole Hermetic Screw Package Chillers," X11th IIR Congress, Madrid, Spain, 1967, reproduced with permission.]

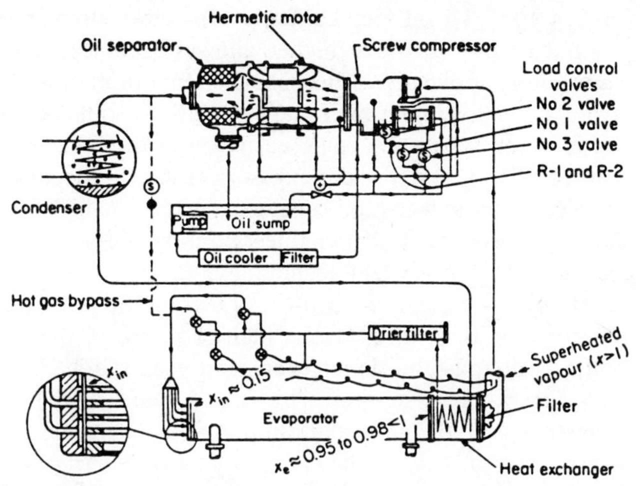

FIGURE 4.5. Schematic refrigeration cycle and flow diagram of compact screw package chiller with incomplete evaporation and post evaporator heat exchanger. [Adapted from H. Soumerai, "Design and Operation of Modern Two-Pole Hermetic Screw Package Chillers," XIIth International Institute of Refrigeration Congress, Madrid, Spain, 1967. [4.18] reproduced with permission.]

basic components—refrigerant, oil, and water circuits—are shown schematically in Fig. 4.5. This figure highlights the design and operating characteristics that are most relevant for a broad overview of multiphase flow phenomena in reverse power cycles.

Motor-Compressor Unit. In these oil-flooded rotary screw compressors, oil is injected in large quantities of the order of 20–40% of the refrigerant mass flow rate (but only about 1% of the volumetric flow rate compressor inlet conditions) inside the compression chamber. It is this profuse oil injection that made this type of positive-rotary compressor highly competitive [4.19] in medium to large refrigeration systems ranging from a nominal capacity in the air conditioning range of 100 to over 600 kW when this new line was introduced on the American market in the 1960s. The popularity of the oil-flooded refrigeration screw compressor systems has increased steadily since then, and they are now being used over a much wider capacity range extending from 15 to 1125 kW and above in multiple sets according to recent manufacturers' literature and the 1983 edition of ASHRAE equipment volume [4.20].

The reverse side of the coin is that oil entrainment rates of the order of $x_{oil} \sim$ 20–40% are totally unacceptable in the refrigeration system for practical and theoretical reasons. Thus, adequately effective oil separators are mandatory with oil-flooded rotary compressors (as in most lubricated high-speed compressors for critical applications) to reduce the oil entrained into the system to an acceptably low level, varying several orders of magnitude from traces—$x_{oil} = 5 \times 10^{-6}$ kg oil/per kg refrigerant circulating in critical low temperature applications—to much higher allowable values, $x_{oil} = 2$–3% in normal airconditioning applications. Because the separation of finely dispersed oil droplets from a gas flowing at high mass velocities is difficult, the size and cost of conventional oil separators could easily offset the inherent advantages of the compact oil-flooded screw compressors. Indeed, the design of adequately effective compact low-cost oil separators was probably the single most important challenge faced by design engineers during the research and development phases of this relatively new technology.

An important point to note in the context of this multiphase flow discussion is that the compressor is located at the highest point of the unit and the segmented baffle shell-and-tube direct expansion evaporator is at the lowest point of the packaged system as shown in Figs. 4.4 and 4.5. The residual oil entrained downstream of the oil separator into the refrigerant circuit must return upward, against gravity, to the compressor (since refrigeration systems must operate several years without any oil addition!) even at the minimum gas flow rates (5–10% of nominal). The fact that the oil entrained from the high-pressure to the low-pressure side of the refrigerant, which is at the lowest point of the physical package (see Fig. 4.4), automatically returns to the compressor illustrates concretely the comments made at the beginning of this chapter: Small liquid droplets are easily entrained in a fast dense (high mass velocity) gas stream. Conversely, it is generally more difficult to separate fine droplets entrained in gases flowing at substantial mass velocities as highlighted below.

Oil-Separation Systems. The high side gas–oil-mist cooled[2] hermetic motor sketched in Fig. 4.5 serves as an effective first-stage oil separator, removing over 95% of the oil entrained with the discharge gas at full gas flow and up to practically 100% at part load. The residual fine oil mist is separated in a special demister section downstream of the motor. One of the unique design features of the annular demister is the unrestricted inner section as shown in Fig. 4.5. Without this free center core, a much larger demister face area, and therefore larger separator shell diameter, would be required to prevent oil carryover at maximum flow rates. (Above a certain critical average gas velocity, the oil separated in a full flow demister is reentrained and the liquid separation efficiency drops suddenly to practically zero!) The combination of changes of directions (arrows in Fig. 4.5) and lower gas velocity through the annular mesh (the gas chooses the path of least resistance in the center free area) together with the natural tendency of the heavier more viscous phase to "hug" the surfaces (in this case the internal diameter of the annular demister core) results in a maximum carryover of $x_{oil} < 1\%$ that is well below the design target of 2–3% at maximum load. The percentage of oil entrainment drops off very quickly at reduced load, and below 50% of full load the oil entrainment (traces of $x_{oil} < 0.1\%$) is no longer measurable with normal laboratory and standard performance test code instrumentation techniques.

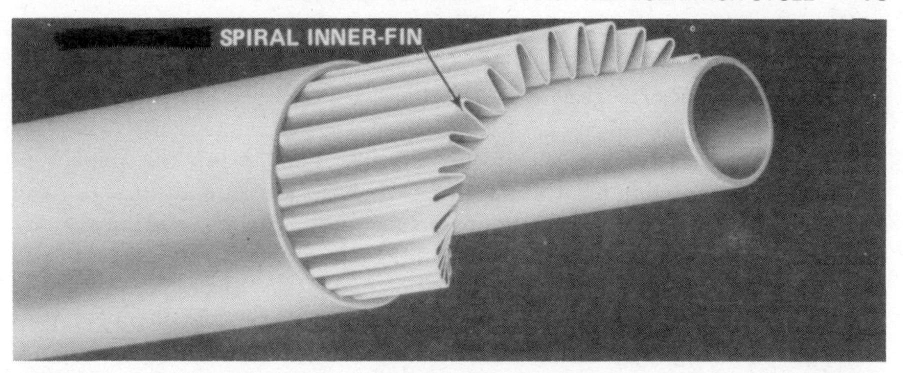

FIGURE 4.6. Inner-fin configuration. [From Dunham Bush, Inc., West Hartford, CT. Reproduced with permission.]

This operating characteristic is highly desirable since proper oil return from an evaporator to a compressor located at a higher elevation could be a serious problem during prolonged periods of operation at refrigeration loads below 5–10% of nominal capacity at extremely low suction pressures and correspondingly low mass velocities in low-temperature refrigeration systems.

Water Cooler System. The direct-expansion cooler, postevaporator heat exchanger, refrigerant feed control, and distribution system are shown schematically in Fig. 4.5. Locating the "bulbs" of the expansion valves (which automatically control the refrigerant flow rate to maintain a specified superheat $\equiv T_g - T_{sat} > 0$ at the bulb location) downstream of the postevaporator heat exchanger ensures dry superheated gas at compressor inlet for maximum compressor[3] and system efficiency with *incomplete evaporation* of the refrigerant at evaporator outlet. It has been shown that operation with refrigerant vapor qualities at evaporator outlet ranging from 95% to as high as 99% with the internally finned tubes used in these compact chillers in the so-called wet wall boiling regimes significantly improves the heat transfer, thus reducing the total surface required per unit of refrigeration [4.19–4.22].

Figure 4.6 shows qualitatively the design of the type of proprietary inner fin used in compact direct expansion evaporators (and other applications in single-phase and two-phase flow with condensation) described in greater detail in several of the above-mentioned references.

Effects of Oil in Forced-Convection Evaporators. The presence of the *second chemical species*, the entrained lubricating *oil*, has two significant effects on the thermohydrodynamic performance of forced-convection evaporators [4.5, 4.6, 4.21–4.27] as summarized in the following excerpt from the 1985 edition of the *ASHRAE Handbook* [4.24]:

> The effect of oil on forced convection evaporation has not been clearly determined. *Increases* occur in the *average heat transfer coefficient* for R-12 up to

10% oil by weight, with a maximum of about 4% [4.25, 4.26]. Oil quantities greater than 10% cause reduction in heat transfer. Oil can *increase the pressure drop*, offsetting possible gains in the heat transfer coefficient.

Although not clearly determined theoretically, [4.23–4.34], the impact of various oil flow fractions x_{oil} on the two-phase pressure drop ΔP_{tp} and corresponding saturation temperature penalty ΔT_{sat} can be, and have been, quantified with sufficient accuracy for design purposes in the case of bare circular tubes [4.27] and internally finned channels [4.5] of the type shown in Fig. 4.6. By a judicious choice of refrigerant mass velocity and channel aspect ratios L/D_h (i.e., proper circuiting of multicircuit forced-convection evaporators and/or condensers), it is possible to capitalize[4] on the positive aspects of the presence of traces of oil ($0 < x_{oil} <$ 1–2% in airconditioning applications) without incurring any measurable penalty due to excessive pressure drop since a minimum amount of friction dissipation is necessary to avoid the *endemic* problem of all multicircuit heat exchangers, particularly evaporators and condensers, namely, *maldistribution*.

4.4 MALDISTRIBUTION IN MULTICIRCUIT HEAT EXCHANGERS

The rather sophisticated design features incorporated in the horizontal water cooler to ensure equal refrigerant flow in each circuit of the multicircuit direct expansion evaporator sketched in Fig. 4.5 emphasize what is probably the most important single problem in heat exchanger design engineering: *maldistribution*. The net chiller capacity obtained in laboratory tests with the first full size prototype direct-expansion chiller operating in horizontal position, as intended, was more than 33% below design targets in the absence of the special distribution system in Fig. 4.5.

A 33% deviation in net overall chiller system capacity actually corresponds to an *effective* evaporating *heat transfer coefficient reduction of over 50%* because a fixed maximum displacement compressor will tend to "rebalance" the system at a lower saturation suction pressure and corresponding lower evaporating temperature. A massive 50% drop below reliable heat transfer coefficients obtained in numerous elementary single-circuit chiller tests under practically identical conditions as those of the final multicircuit design (same channel, L/D_h, external fluid, and external heat transfer coefficient) could only be attributed to uneven refrigerant distribution. This assumption was very quickly verified by simply repeating the tests with the direct expansion cooler in a *vertical position* and *bottom fed* to ensure *even distribution*, under otherwise identical conditions. In the vertical position the total net capacity immediately came up to par (well within the standard test code allowable tolerance band) without any other physical change. This clearly confirmed that the loss of capacity was *entirely* due to maldistribution. Without adequate means of distributing the refrigerant evenly among the many parallel circuits with the cooler in the horizontal position, the system apparently settled down under SSSF conditions in such a manner that the lower tubes were "liquid logged" and the upper tubes "liquid starved". Whatever the true reasons, the significant fact is that the *effective* average boiling *heat transfer coefficient* could be *increased* by a factor of over *two to one* solely through improved *distribution*. This illustrates vividly the central importance of even distribution, which is essentially an *overall system fluid flow problem*, in heat transfer engineering.

Although illustrated above for a particular type of refrigeration design application, this maldistribution problem is typically encountered, to various degrees, in *all types* of *multicircuit* surface heat exchangers both in two-phase (evaporation and/or condensation) and single-phase flow applications for the refrigeration [4.35], power generation [4.36 to 4.41], and general process industries [4.10, 4.11, 4.42–4.46].

Figure 4.7 highlights the major research efforts invested by a leading European[5] manufacturer in the field of power generation to improve the

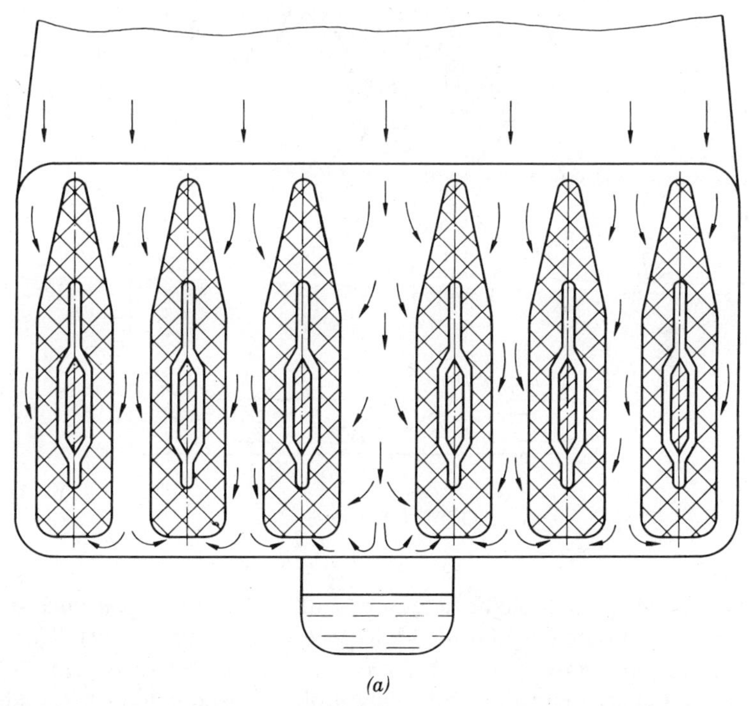

(*a*)

FIGURE 4.7. "Church-window" condensers for large steam turbine: (*a*) The "church-window" condenser concept. (*b*) Steam flow in a model. (*c*) "Church-window" condenser being assembled in the workshop prior to tube fitting. [From G. Oplatka and H. Lang, Theory and Design of 'Church-Window' Condensers for Large Steam Turbines, pp. 316–336, Figs. 2, 5, and 11, *Brown Boveri Review*, 7/8 Vol. 60, Brown Boveri & Co., Baden, Switzerland, 1973. Reproduced with permission.]

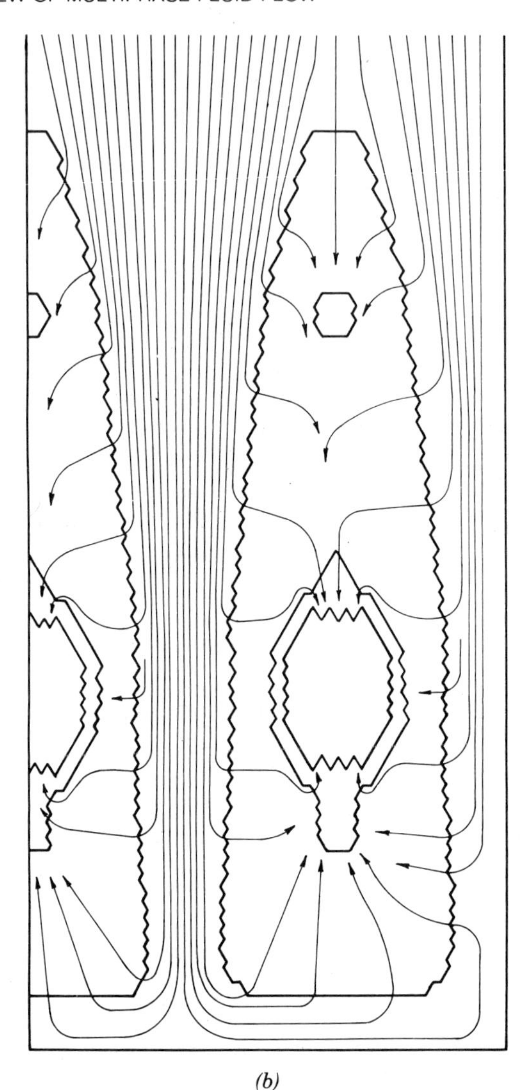

(b)

FIGURE 4.7. Continued.

performance of water cooled condensers suitable for the largest nuclear steam turbines built to date (> 1,000,000 kW generator power output). To increase the effective two-phase heat transfer coefficient on the steam side of these condensers, Oplatka and Lang [4.38] list in their introduction (italics added) a series of design objectives starting as follows:

> The demands made of the new condenser can be summarized as follows: the primary aim is good heat transfer. This is to be achieved mainly by smooth steam flow, promoting *even distribution* of steam to all tubes.

FIGURE 4.7. Continued.

The prime emphasis on even distribution shows clearly that maldistribution is typically encountered in all *large* heat exchangers regardless of the types of industrial application. The following additional excerpts [4.38] are intended to provide some perspective on the physical size of these large nuclear power plant condensers:

> These three parameters, i.e., type of "church-windows," number of "church-windows" and length of tube, enable a virtually unlimited assortment of condensers to be built.
>
> A condenser with 16 "church-windows" each containing 2925 tubes with a total heat transfer surface of about 37,000 m^2, is shown in Fig. 11 [our Fig. 4.7*c*] which also gives an impression of the size.
>
> The condenser height varies between 3.5 and 4.5 m depending on the type of "church-window." The width depends on the number of "church-windows" arranged side by side. Condensers built to date vary between 5 and 23 m.

As maldistribution is generally one of the most significant problems in real world heat transfer engineering and since it is essentially a fluid flow problem that is particularly critical in two-phase flow applications, the above conclusions are summarized below as a guideline.

GUIDELINE 8 *The achievement of even distribution represents the most important common problem/opportunity in heat exchanger design practice. The effective heat transfer coefficient can be improved massively through sound fluid flow engineering alone.*

Of course, Guideline 8 is equally applicable to old *flooded evaporators* which have been aptly described by Lorentzen [4.47] as "disorganized forced convection evaporators." Indeed, the significant cost and size reductions achieved in modern compact external flow evaporators are due in large part to improved distribution. As air conditioning system engineers know, a simple air heating or cooling coil can perform significantly below par inside a package unit compared to ideal wind tunnel test conditions because of poor air distribution over the entire coil area. Finally, uneven penetration and numerous flow by-passes represent the main design challenges in shell-and-tube heat exchangers [4.44, 4.48, 4.49].

4.5 PROLIFERATION OF FLOW REGIMES

It is clear that stable overall flow conditions and even distribution are highly improbable even in "simple" single-phase heat exchangers if the Reynolds number in each circuit falls within, or near, the inherently unstable transition region between laminar and turbulent regimes, as illustrated in Rohsenow [4.45] and Putnam and Rohsenow [4.46]. The same basic problem can occur in single-component multicircuit *two-phase* flow heat exchangers with heat addition (boiling) or removal (condensation) except that the problem is *compounded several orders of magnitude* because of the occurrence of not just one but about *ten different flow regimes* and uncertain potentially unstable transitions.

This is illustrated in Fig. 4.8. Several different flow patterns, or regimes, are described and identified by such labels as bubble, plug, churn (or semiannular), annular, spray-annular, mist, and dry flows. Additional modes of gas–liquid flows are identified in Chapter 4 of the *ASHRAE Handbook*, *Fundamentals* [4.23], grouped in *eight major classes* of flow patterns. This clearly shows that flow pattern visualization and labeling techniques are rather subjective and two-phase flow phenomena are certainly far more complex than single-phase Newtonian fluid flows, where one has to contend only with a single uncertain transition region.

In this context it is quite understandable why so *many different types* of empirical *correlation* have been proposed to predict ΔP_{tp} with a *significantly lower level of accuracy than in single-phase* laminar or turbulent SSSF regimes.

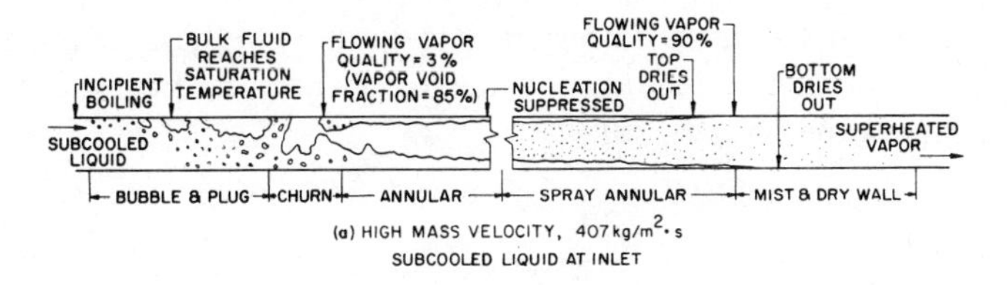

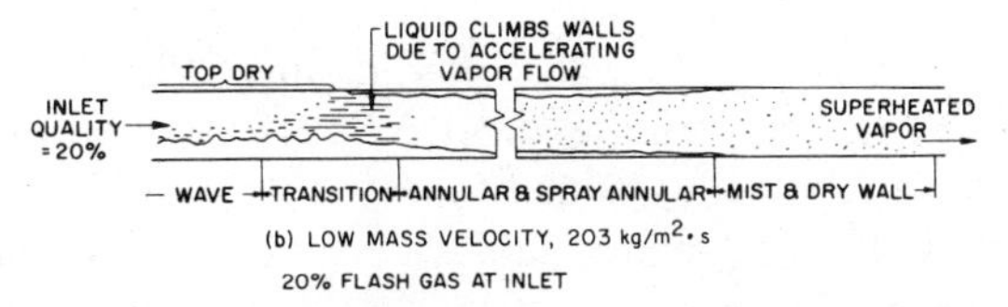

FIGURE 4.8. Flow regimes in a horizontal tube evaporator. [From *ASHRAE Handbook, Fundamentals*, Chap. 4, *Two-Phase Flow Behavior,* p. 4.5, ASHRAE, Atlanta, 1985. Reproduced with permission.]

4.6 THERMODYNAMIC COUPLING OF HEAT AND FLUID FLOW

If we assume SSSF conditions are possible (regardless of the particular type of flow regime), it follows that the two-phase pressure drop ΔP_{tp} must be most generally a function of *all* relevant *independent variables*. These are the total mass flow rate $\dot{m}$, the effective consistency of the flowing mixture defined by the average (or local) vapor quality $\bar{x}$ (or x) of the flowing mixture, which in turn uniquely defines the liquid mean $\bar{y}$ (or local y) flowing fraction as well as the two relevant gas- and liquid-phase properties μ_g, ρ_g and μ_l, ρ_l, respectively, the channel geometry (D_h, $\bar{e}/D_h$, L/D_h), wall material, and the conditions preceding the channel entrance (CPE).

For single-component two-phase flow situations (in contrast to multicomponent such as oil and gas), *all* the liquid- and vapor-phase *equilibrium state and transport properties* are *uniquely defined* at any pressure $P = P_{\text{sat}}$ since there exists a unique relationship between P_{sat} and the corresponding equilibrium[6] saturation temperature T_{sat}. Therefore, the thermodynamic state of the mixture is fully defined by either x and T_{sat} or x and P_{sat}.

In two-phase heat exchanger design practice it is convenient to select the more easily measured variable, the *pressure* $P = P_{\text{sat}}$ as the *independent variable* [4.4–4.6, 4.42]. Thus, the most general expression for the length integrated total two-phase pressure drop ΔP_{tp} can be stated as follows:

$$\Delta P_{\text{tp}} = \varphi_1(\dot{m}, \bar{x}, P, D_h, \bar{e}/D_h, L/D_h, \text{wall material, and CPE}) \quad (4.1)$$

where φ_1 represents a yet unknown function of all the above independent variables. For a fixed total mixture flow rate $\dot{m}$, prescribed channel geometry, material, and CPE, Eq. (4.1) reduces to a specific function, $\varphi_2(\bar{x}, P)$ or

$$\Delta P_{\text{tp}} = \varphi_2(P, \bar{x}) \quad (4.2)$$

for *prescribed* $\dot{m}$, D_h, $\bar{e}/D_h$, L/D_h, wall material, and CPE. For an infinitesimal channel length dl and corresponding local two-phase friction pressure drop dP_{tp} at the prevailing local value of P and x, Eq. (4.2) yields the following expression for dP_{tp}/dl:

$$\frac{dP_{tp}}{dl} = \varphi_3(P, x) \tag{4.3}$$

for *prescribed* $\dot{m}$, D_h, $\bar{e}/D_h$, wall material, and CPE.

The saturation temperature drop ΔT_{sat} associated with a known (either empirically or analytically) value of $\Delta P_{tp} = \Delta P_{sat}$ could readily be determined from standard thermodynamic tables or charts for the particular fluid under consideration.

Fortunately, ΔT_{sat} can also be calculated most generally and with a totally adequate level of accuracy for two-phase flow and heat transfer engineering applications from the well-known Clapeyron–Clausius[7] equation of classical thermodynamics derived quite simply from a *differential Carnot cycle*, as shown in Fig. 4.9.

The slope of the saturation temperature curve in a T_{sat} versus P_{sat} graph, that is, the first derivative dT_{sat}/dP_{sat}, can be expressed most generally according to the Clapeyron–Clausius equations:

$$\frac{dT_{sat}}{dP_{sat}} = \frac{v_{lg}}{s_{lg}} \tag{4.4}$$

where

$$v_{lg} = v_g - v_l = \text{specific volume change from } x = 0 \text{ to } x = 1 \tag{4.5}$$

$$s_{lg} = s_g - s_l = \text{specific entropy change from } x = 0 \text{ to } x = 1 \tag{4.6}$$

and with

$$h_{lg} = h_g - h_l = \text{specific enthalpy change from } x = 0 \text{ to } x = 1 \tag{4.7}$$

the specific entropy change at a fixed T_{sat} corresponding to P_{sat} is expressed by

$$s_{lg} = \frac{h_{lg}}{T_{sat}} \tag{4.8}$$

The Clapeyron–Clausius equation is valid for a constant saturation pressure sufficiently below ($\sim$ 10%) the critical pressure P_{cr} at the thermodynamic critical point of the particular pure substance under consideration; or with

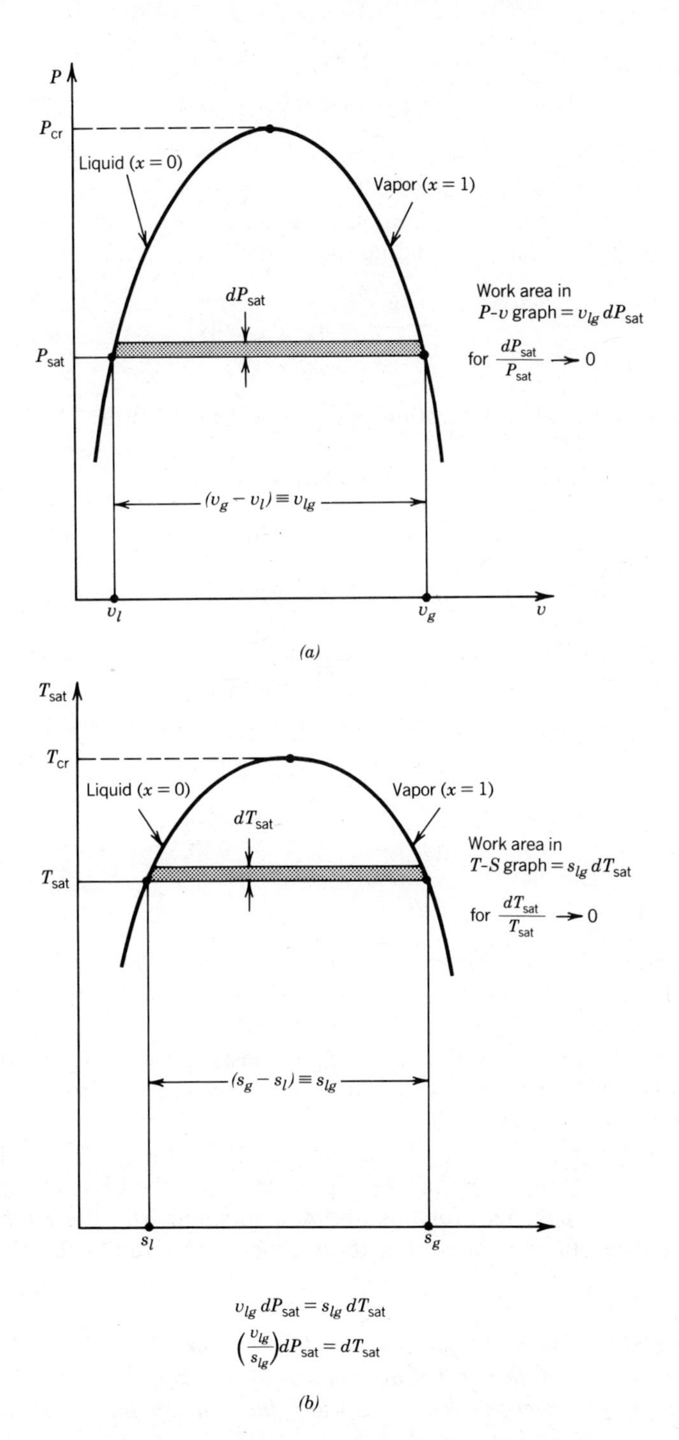

FIGURE 4.9. Derivation of Clapeyron – Clausius equation from differential Carnot cycle: (*a*) pressure versus volume diagram (*b*) temperature versus entropy diagram.

$P^+ \equiv P_{sat}/P_{cr} \equiv$ reduced pressure,

$$\textit{Validity}: \quad P^+ = \frac{P_{sat}}{P_{cr}} \lesssim 0.9 \tag{4.9}$$

It is important to note that the term on the right-hand side of Eq. (4.4), the ratio of two state properties uniquely defined at P_{sat}, is itself uniquely defined in terms of P_{sat}, that is, a unique function φ_4 of P_{sat} only or

$$\frac{v_{lg}}{s_{lg}} = \varphi_4(P_{sat} \text{ only}) \tag{4.10}$$

For small relative saturation pressure and temperature changes, $\Delta P_{sat}/P_{sat} \ll 1$ and $\Delta T_{sat}/T_{sat} \ll 1$, respectively, it is permissible to substitute $\Delta T_{sat}/\Delta P_{sat}$ for dT_{sat}/dP_{sat} in Eq. (4.4):

$$\frac{\Delta T_{sat}}{\Delta P_{sat}} = \frac{v_{lg}}{s_{lg}} \tag{4.11}$$

Thus, the saturation temperature drop associated with a known pressure drop ΔP_{tp} can be computed from Eq. (4.11) according to

$$\Delta T_{sat} = \left(\frac{v_{lg}}{s_{lg}}\right) \Delta P_{tp} \propto \Delta P_{tp} \tag{4.12a}$$

$$\textit{Validity}: \quad \frac{\Delta P_{tp}}{P_{sat}} \ll 1 \quad \text{and} \quad P^+ = \frac{P_{sat}}{P_{cr}} \lesssim 0.9 \tag{4.12b}$$

Equation (4.12a) shows that the saturation *temperature drop penalty* in two-phase forced-convection evaporators or condensers varies in *direct proportion* with the two-phase fluid flow *pressure drop* ΔP_{tp}, since the term v_{lg}/s_{lg} is fully defined at a prescribed practically constant absolute saturation pressure P_{sat} when $\Delta P_{tp}/P_{sat} \ll 1$, a condition typically satisfied in *thermodynamically efficient two-phase flow heat exchangers*.

Although Eq. (4.12a) can be simplified further at $P^+ \ll 1$ by setting $v_{lg} = v_g\,(1 - v_l/v_g) \cong v_g$ and by expressing v_g on the basis of the well-known gas laws, it is sufficient for this overview to summarize the essential point of this analysis in the form of a most general and fundamental guideline as follows:

GUIDELINE 9 *In forced-convection evaporators and condensers, the heat and fluid flow problems cannot be uncoupled since the saturation temperature drop penalty* ΔT_{sat}, *which has a major impact on the net heat* *transfer temperature driving force in thermodynamically efficient heat exchangers, is directly proportional to the two-phase pressure drop ΔP_{tp} according to the universal relationship, Eq. (4.12a), derived from a differential Carnot cycle, that is, the second law of classical thermodynamics. Thus, the practical solution of forced-convection two-phase flow heat exchangers requires a unified treatment involving the three branches of thermal science: thermodynamics, fluid flow, and heat transfer.*

4.7 THE TWO-PHASE MIRROR IMAGE CONCEPT

Although Fig. 4.8 refers specifically to a single horizontal tube operating under conditions typical of refrigeration applications in the *evaporation* mode with *heat addition* when $\Delta x = x_2 - x_1 > 0$, similar flow regimes have been identified in the quasi-mirror-image condensation mode with *heat removal* when $\Delta x = x_2 - x_1 < 0$. The *symmetry* between the condensation and evaporation modes highlighted in Fig. 4.10 is readily understandable for forced-convection tube flow since the hydrodynamic conditions are substantially independent of the direction of heat flow in the *absence* of strong thermal effects such as *nucleate boiling*[8] (in the evaporating mode), as is typically the case in direct-expansion refrigeration evaporators or any thermodynamically efficient forced-convection two-phase flow heat exchangers at $|\bar{T}_w - \bar{T}_{sat}|/\bar{T}_{sat} \ll 1$ when nucleation is fully suppressed.

This *two-phase mirror-image* concept is of more than just academic interest because it makes it possible to *predict* the pressure drop ΔP_{tp}, the effective

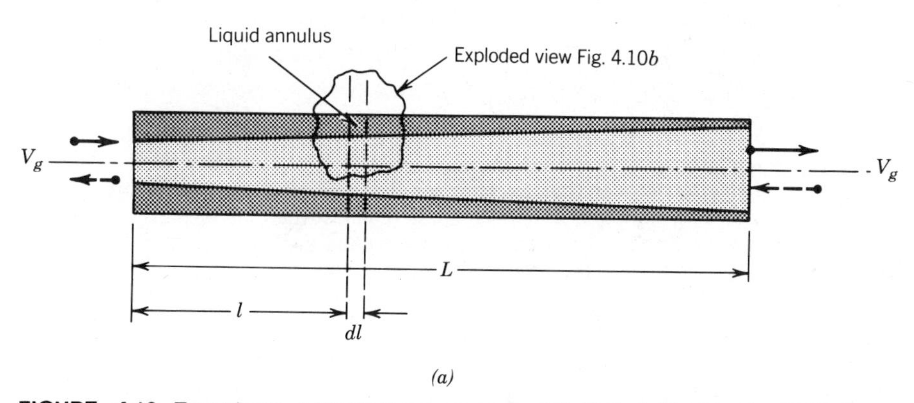

(a)

FIGURE 4.10. Two-phase mirror-image concept. (*a*) Full tube length. *Based on*: $\dot{m}, \bar{x}, \bar{V}_g, \bar{V}_l, \Delta P_{tp}, \bar{P}, \Delta T_{sat}$ and $\bar{T}_{sat}$, identical in evaporating and condensing modes; $\Delta P_{tp}/\bar{P}$ and $|\Delta T_w|/\bar{T}_{sat} \approx 0$ in annular (spray) SSSF regimes; horizontal $(L/D \gg 1)$ tubes and identical local (average) liquid film thickness δ (or $\bar{\delta}$); therefore, $\alpha_{tp} = k_l/\delta$ and $\bar{\alpha}_{tp} = \bar{k}_l/\bar{\delta}$ are identical regardless of heat and fluid flow directions and:

$$|\dot{Q}_e|_{cond} = |\dot{Q}_e|_{evap}$$
$$-(\bar{T}_w - \bar{T}_{sat})_{cond} = (\bar{T}_w - \bar{T}_{sat})_{evap}$$

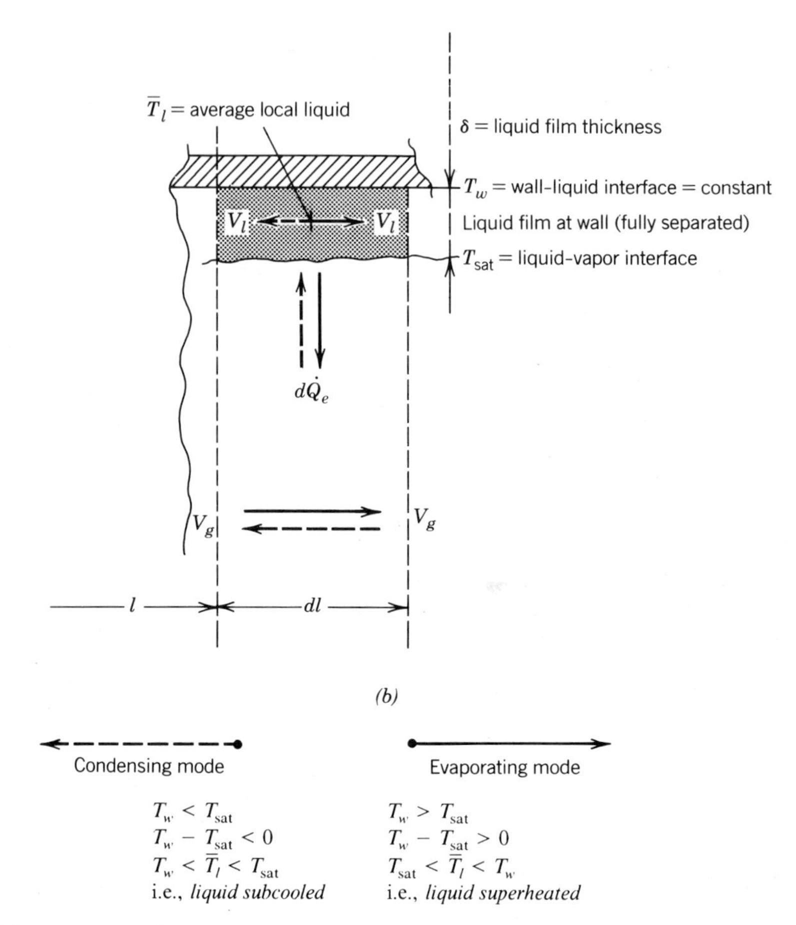

Note: The symmetry between the cooling and heating mode is *maintained* for the *homogeneous* ("fog") *flow limit* of annular-spray regimes when the dispersed fraction D of the liquid flow fraction $1 - x \equiv y$ approaches unity and the liquid film thickness $\delta \gtrsim 0$, therefore, $V_l = V_g$ or $V_g / V_l = 1$ (i.e., zero slip).

friction factor f_{tp}, and the coupled two-phase flow heat transfer coefficient α_{tp} in the forced-convection *condensing mode from empirical data* obtained in the forced-convection *evaporation mode* at the same pressure for a given tube geometry and refrigerant (or vice versa) as verified empirically with internally finned channels [4.22] and demonstrated theoretically [4.50] along the lines highlighted in Chapter 9.

4.8 COMPLEXITY OF TWO-PHASE FLOW PHENOMENA

Recalling our present lack of knowledge on the true nature of turbulence and flow transition phenomena in the simplest case of single-phase Newtonian fluid flows, it is easy to understand why two-phase flows are even more intractable analytically than the simple single-phase turbulent flow regimes. The proliferation of radically different flow regimes alone may explain why no generally applicable simple correlations have yet been proposed, nor are likely to be found, for the effective two-phase friction factor f_{tp} and the related heat transfer coefficient α_{tp} in spite of the vast amount of funding and massive worldwide research efforts on two-phase flow since the launching of the U.S. atoms for peace program in the 1950s.

The state of the art as of 1964 is well summed up by Stein of the Argonne National Laboratory [4.51]:

> One of the most fascinating (or perhaps exasperating) aspects of boiling heat transfer research is the general lack of agreement among investigators in the field... Although there may be current majority opinions as to word pictures of the important physical mechanisms involved, there is almost no agreement as to how they and their interactions could be described mathematically.

Since this quotation is more than 20 years old, one might hope the situation has improved noticeably in the intervening years. Unfortunately, this is not the case, as indicated by the following cursory update on the current state of the art.

4.8.1 State of the Art 1985: Air Conditioning and Refrigeration Applications

The same basic recommendations for forced-convection evaporation in tubes appear in the latest updated 1985 edition of the *ASHRAE Handbook, Fundamentals* [4.52] as in the previous edition [4.23] as indicated by the following excerpts [4.52, Chap. 4]:

> Since the hydrodynamic and heat transfer aspects of two-phase flow are not as well understood as those of single-phase flow, no single set of correlations can be used in all two-phase flow systems to predict pressure drops or heat transfer rate... [This chapter] provides information on the *vast number of correlations* that have been developed to predict heat transfer coefficients and pressure drops in two-phase flow systems... .
>
> *Accurate experimental data defining limits of regimes* and determining effects of the various parameters are *not yet available*. The accuracy of previously proposed

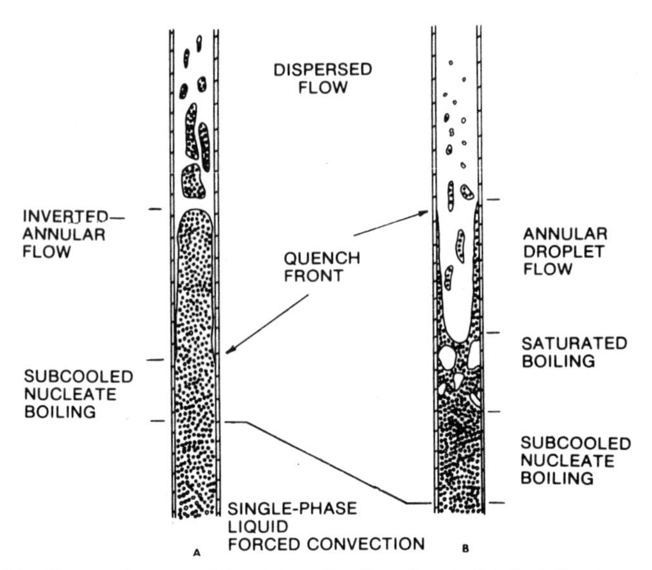

FIGURE 4.11. Flow regimes and heat transfer in reflood: (*a*) fast flooding and (*b*) slow flooding. [From K. Kawaji et al., Reflooding with Steady and Oscillatory Injection, Part 1: Flow Regimes, Void Fraction, and Heat Transfer, *ASME J. Heat Transfer*, **107**, 671 (Aug. 1985). Reproduced with permission.]

correlations to predict the heat transfer coefficient for two-phase flow is not known beyond the range of the test data.

The 1985 ASHRAE update seems to indicate a growing awareness of the almost hopeless analytical complexities associated with two-phase flow phenomena. This impression is further reinforced by the mere fact that a new chapter [4.53] has been allocated exclusively to two-phase flow phenomena which were formerly included in the general heat transfer chapter of the *ASHRAE Handbook, Fundamentals*. It is fair to assume that the purpose of this segregation is to cope more effectively with the growing complexities of two-phase flow phenomena.

4.8.2 State of the Art 1985: Nuclear Applications

More sophisticated instrumentation and computer-aided analytical techniques have now made it possible to reach higher levels of complexity as indicated by Kawaji et al. [4.54], who have identified an important additional type of two-phase flow pattern called the *inverted annular flow* (see Fig. 4.11).

In their rationale to "ressurect" an interest in corresponding states correlations for pool and flow boiling burnout, Sharan et al. [4.55] reassess the currently accepted methods of predicting $q''_{\max}$, the peak boiling heat flux when the flow of the liquid toward the surface is interrupted by a vapor blanket (italics added).

> Thus [the] equation [for q''_{max}], although it is accurate within about $\pm 15\%$, represents a good deal of smoothing and simplification of *complicated phenomena*. If the process of correlation can be sufficiently perfected, it *might reveal* details of the *functional dependence* of q''_{max} that cannot be brought out by the *simplified theories*.

The above rudimentary updates clearly suggest a definite trend toward higher levels of complexity when two-phase flow phenomena are analyzed in the light of conventional mechanistic models. This may explain the renewed interest [4.55–4.61] in the thermodynamic generalizations method based on the extended corresponding states principle (ECSP) proposed at the end of the 1950s for nuclear applications [4.62–4.65].

This relatively new thermodynamic tool had already become quite popular in the refrigeration industry [4.21, 4.66–4.70] since the 1960s because so many different types of refrigerant are used that the economic benefits of such generalizations are much more evident than in the steam power generation industry. The practical application of the ECSP generalization method in two-phase flow, as well as the limiting case of single-phase flow, is addressed in Chapter 6 after the most widely accepted conventional correlations have been reviewed in Chapter 5.

NOTES

1. To be precise a *gas* reverse cycle rather than the *vapor* cycle as discussed by Carnot's first disciple E. Clapeyron in 1834 [4.16]. It is quite remarkable that Sadi Carnot's discovery of the fundamental concepts of refrigeration, the reverse power cycle, *anticipated* the "art" of refrigeration by some 10 years. However, the formalization of Carnot's basic classical thermodynamic concepts into an officially recognized branch of physics by Lord Kelvin and R. Clausius lagged about 20 years behind Perkin's (1834) mechanical refrigeration system [4.14]. In other words, the "official" science of refrigeration lagged behind the art as is usually the case. In actual fact, however, Carnot's theory preceded the new art of mechanical refrigeration.
2. The high heat transfer coefficients between the gas–oil mixture and the hermetic motor make it possible to maintain extremely safe motor winding temperatures even though *discharge* gas is used as a coolant (in contrast to suction gas as in conventional reciprocating gas-cooled hermetic compressors). This reduces the irreversible system losses associated with suction gas-cooled hermetic motors [4.18, 4.19].
3. Any unevaporated liquid droplets that may exist in an apparently slightly superheated gas (nonequilibrium conditions) eventually flash into vapor on contact with the much hotter compressor surfaces. This has two negative effects: first, a reduction of effective compressor volumetric efficiency, and second, a decrease in useful cooling effect per unit mass of refrigerant. As a result, larger compressor displacements are required and the system coefficient of performance can be measurably reduced [4.7].
4. In this context one of the pitfalls of two-phase flow research with surgically *oil-free* systems becomes evident. A key parameter is arbitrarily eliminated and it is not surprising that the results obtained in such *atypical* tests do not always coincide with those obtained in real world direct-expansion evaporators [4.5, 4.6]. It is noteworthy that the real needs of heat transfer engineers are now fully recognized, as evidenced by the abstracts of ASHRAE-sponsored 1985/86 research projects 336-RP, 378-RP, and 469-RP [4.71]. Interested readers will find a wealth of

extremely relevant information on heat transfer with boiling in the keynote address by Professor W. Rohsenow and the papers presented at the ASHRAE 1966 symposium on two-phase flow applications [4.5, 4.28–4.34] which include the extensive recorded discussions generated at this symposium.

5. This illustrates one method of solving the *penetration* problem in large condensers. Alternative solutions used by American manufacturers are highlighted in some of the references cited and in standard texts.
6. The unique correspondence between P_{sat} and T_{sat} is strictly valid under equilibrium conditions when the liquid surfaces are *flat* and *not too far from equilibrium*, that is, in the linear domain of irreversible thermodynamics.
7. Named after Emile Clapeyron who first derived this relationship in his "Memoire sur la Puissance Motrice de la Chaleur" (Motive Power of Heat) [4.16]. A French mining engineer and, like Carnot, a Polytechnicien, Clapeyron derived his equation in 1834, that is, before the official formulation of the first law (conservation) and the second law of thermodynamics on the basis of an infinitesimal Carnot cycle.
8. In typical stable annular (or annular-spray) flow regimes encountered in forced-convection evaporators at $|T_w - T_{sat}|/T_{sat} \ll 1$ and "low" heat fluxes (compared to extremely high heat fluxes in nuclear applications), there is no macroscopic nucleation in the liquid film and the thermal resistance varies linearly with the local liquid film thickness. The same fundamental assumptions are made to derive the well-known Nusselt equation for *natural gravity* condensation [4.72, 4.73], which is essentially also based on the solution of a coupled fluid and heat flow problem to determine analytically the liquid film thickness from which the thermal resistance is computed as in a pure conduction problem. Of course, the *mirror image* makes no sense at the extremely high heat fluxes typically encountered in fossil or nuclear power generation applications [4.74–4.78] since there is no simple linear relationship between the heat flux q'' and $T_w - T_{sat}$ when $q'' \propto (T_w - T_{sat})^a$ with $a \gg 1$. This illustrates again why two-phase flow research results obtained for steam power plant applications cannot be applied blindly to thermodynamically efficient fluid to fluid heat exchangers, or vice versa.

CHAPTER 5

TWO-PHASE FLUID FLOW CORRELATIONS

In this chapter we review recognized semianalytical methods for the prediction of ΔP_{tp} in channels of constant cross section, with special emphasis on forced-convection process heat exchangers.

In accordance with the practice adopted in the technical literature for two-phase flow (see Fig. 5.1) V_g and V_l now represent the average velocities of the gas and liquid phases, respectively, at a given position along the tube axis. We shall occasionally use the overbar in two-phase flow correlations with evaporation (or condensation) to indicate *length average*, for instance, the average axial gas velocity $\overline{V}_g$ to indicate the effective mean velocity of the gas phase varying from a low value $(V_g)_1$ at position 1 to a higher velocity $(V_g)_2$ at exit state 2 in the case of two-phase flow with heat addition, that is, evaporation.

5.1 "NAIVE" ALL VAPOR FLOW MODEL

Before the advent of the more sophisticated nuclear two-phase flow research results, a rudimentary method used in the refrigeration industry [5.1] consisted in simply computing the total two-phase friction pressure drop in an evaporator (or condenser) on the basis of *conventional single-phase correlations* as if the saturated vapor would flow alone through the *whole evaporator* (or condenser) at a constant quality $x = 1$. By making the seemingly extremely conservative assumption of a constant flowing quality $x = 1$ and corresponding constant specific volume $v = xv_g = v_g$ from tube inlet to outlet, $\Delta P_{tp,f}$ could be computed according to Eq. (3.4). As it turned out this "naive" model was not so far off the mark ($\sim$ 30%) in most direct-expansion evaporator applications because the "pessimistic" assumption $v = v_g$ is partly offset by the fact that the *effective* two-phase friction factor f_{tp} is higher than the

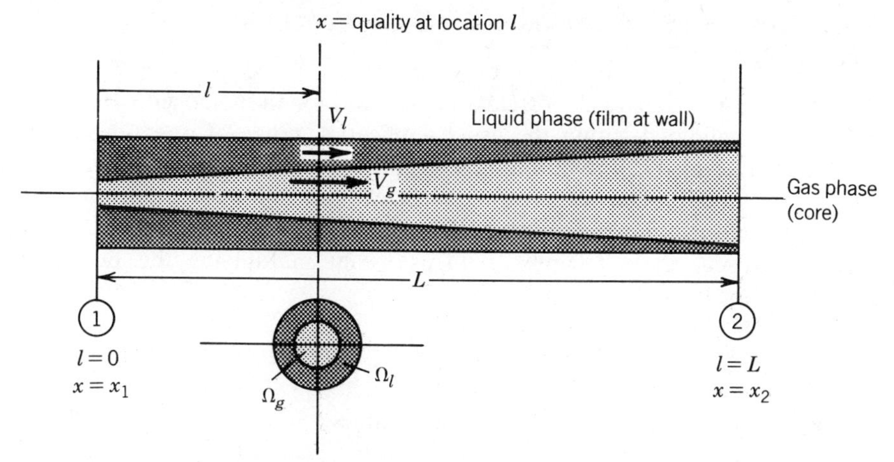

FIGURE 5.1. Two-phase flow nomenclature.

$\dot{m}_g$ = mass flow rate of gas phase = $x\dot{m}$
$\dot{m}_l$ = mass flow rate of liquid phase = $y\dot{m} = (1 - x)\dot{m}$
$\dot{m}$ = mass flow rate of total mixture = $\dot{m}_g + \dot{m}_l$
$x \equiv \dot{m}_g/\dot{m}$ = gas flowing quality
$y \equiv \dot{m}_l/\dot{m}$ = liquid flowing quality = $1 - x$
Ω_g = cross section of tube occupied by gas = $R_g\Omega$
Ω_l = cross section of tube occupied by liquid = $R_l\Omega$
Ω = total cross section of tube = $\Omega_g + \Omega_l = \Omega(R_l + R_g)$
$R_g \equiv \Omega_g/\Omega$ = gas volume fraction in tube
$R_l \equiv \Omega_l/\Omega$ = liquid volume fraction in tube
$R_g + R_l = 1$ or $R_g = 1 - R_l$
$V_g \equiv \dot{m}x/\rho_g R_g \Omega$ = average velocity of gas phase
$V_l \equiv \dot{m}(1 - x)/\rho_l R_l \Omega$ = average velocity of liquid phase
V_g/V_l = velocity or slip ratio = $[x/(1 - x)](\rho_l/\rho_g)(1 - R_g)/R_g$
$V_g/V_l = 1$ = homogeneous or fog or zero-slip flow limit

Note: (1) See also Sec. 5.3.1 for Lockhart and Martinelli's dimensionless parameters and definitions of Reynolds numbers. (2) Picture depicts fully separated annular flow regime in horizontal circular tube (i.e., dispersed liquid fraction $D = 0\%$).

conventional single-phase friction factor f computed in the conventional manner at $x = 1$ due to the presence of the liquid phase flowing at a lower average velocity V_l than the gas phase V_g under certain operating conditions.

5.2 HOMOGENEOUS OR ZERO-SLIP FLOW MODEL ($V_g / V_l = 1$)

Several analytical solutions were derived [4.5, 4.27, 5.2, and references cited therein], based on a simple *homogeneous* flow model, that is, on the assumption that the liquid and vapor phases flow at the same velocity $V_g = V_l$, or a *phase velocity ratio* (also called *slip ratio*) $\equiv V_g/V_l = 1$. This homogeneous flow model is also referred to as the *zero-slip* model in the scientific literature.

5.2.1 Bo Pierre's Correlation for Superheated Vapor at Evaporator Outlet ($x_2 > 1$)

Bo Pierre's semianalytical correlation is based on a homogeneous model empirically validated within the usual application range of direct-expansion refrigeration evaporators with complete evaporation (~ 5–7°C superheat at evaporator outlet). His recommended correlations summarized below have been corrected to comply with the current (American) definition of the friction factor as discussed in Chapter 3. (This meant multiplying the Bo Pierre friction factor by 2).

$$\text{total } friction \text{ pressure drop } \Delta P_{\text{tp},f} = \bar{f}_{\text{tp}}(L/D_h)G^2\bar{v} \tag{5.1a}$$

where $\bar{v} \equiv \bar{x}v_g$ = effective mean specific volume
$\bar{x}$ = effective mean quality = empirical $\varphi_x(D_h, L)$
$\bar{f}_{\text{tp}}$ = mean friction factor = empirical $\varphi_f((\text{Re})_l, K_f$, oil flow fraction)

and

$$(\text{Re})_l \equiv \frac{GD_h}{\mu_l} \equiv \frac{\dot{m}D_h}{\Omega\mu_l} = \text{all liquid Reynolds number} \tag{5.1b}$$

$$K_f \equiv \frac{J(x_2 - x_1)(h_g - h_l)}{Lg} \equiv \frac{J\Delta x h_{lg}}{Lg} = \text{Bo Pierre's boiling number} \tag{5.1c}$$

In the above original (1947) expression of K_f proposed by Bo Pierre, g is the acceleration of gravity and the term J represented the mechanical equivalent of heat in the old technical system of units.

Within his experimental range (see Table 5.1) Bo Pierre could correlate with an excellent degree of accuracy the two empirical terms in Eq. (5.1a): $\bar{x}$ and

TABLE 5.1. Experimental Range and Validity Limits of Bo Pierre's Correlations for $\bar{f}_{\text{tp}}$ and $\bar{x}$

Evaporator tube diameter,[a] D (mm)	12, 18
Evaporator tube length,[a] L (m)	4.08–9.50
Refrigerant mass flow rate, M'_h (kg / h)	15–140
Heat flux,[b] q'' (kcal / m² · h)	1000–26,000
Evaporation temperature (°C)	+10, 0, −10, −15, −20
Inlet condition, x_1	0.10–0.20
Outlet condition (°C)	5–7 super heat
Average vapor quality,[b] $\bar{x}$	0.49–0.81
Oil amount circulated (vol.%)	0–12
Degree of evaporation, ϕ	1.0–0.8

Source: Table II, *ASHRAE Journal*, p. 61, September 1964. Reproduced with permission.
[a]In terms of L/D, a range of $227 < L/D < 792$.
[b]The terms q'' and $\bar{x}$ have been substituted for Bo Pierre's q' and x_m.

$\bar{f}_{tp}$ for his *quasi oil free* tests according to the following equations:

$$\bar{x} = 4.4 D_h^{1/4} L^{-1/2} \tag{5.1d}$$

with D_h and L expressed in meters (m) and

$$\bar{f}_{tp} = 0.0370 K_f^{1/4} \text{Re}^{-1/4} \tag{5.1e}$$

Validity: $(\text{Re})_l K_l^{-1} > 1$, that is, above the critical flow regime transition reported by Bo Pierre at $\text{Re}_l K_f^{-1} = 1$ (5.1f)

Traces of Oil: $< 0.5\%$ by volume in the liquid mixture from the condenser, but *not surgically oil free*;[1] otherwise within the experimental limits shown in Table 5.1 (5.1g)

For oil flow fractions between 6 and 12% by volume in the liquid, corresponding to a slightly lower mass flow fraction, since for the refrigerants tested by Bo Pierre $\rho_l \gtrsim \rho_{oil}$,

$$\bar{f}_{tp} = 0.106 K_f^{1/4} \text{Re}^{-1/4} \tag{5.2a}$$

Validity: 6–12% oil flow fraction by volume in liquid; otherwise the same as in Eq. (5.1a) (5.2b)

The significant negative impact of excessive oil entrainment can be assessed by dividing the friction factor obtained from the above correlations at the same value of $K_f^{1/4} \text{Re}^{-1/4}$:

$$100 \times \frac{\bar{f}_{tp} @ \text{ 6–12\% oil}}{\bar{f}_{tp} @ < 0.5\% \text{ oil}} = 286\% \tag{5.2c}$$

Similar results have been reported [3.4, 3.32] with small hydraulic diameter internally finned channels of the type used in the large screw package chillers described in Figs. 4.4 to 4.6; it is now clear why a maximum allowable oil entrainment target $x_{oil} < 2$–3% was selected for these compact evaporators. The saturation temperature drop penalty at abnormal oil flow fraction of the order of 6–12% would outweigh by far any theoretical gain for the heat transfer coefficient in normal forced-convection evaporators.

Bo Pierre's friction factor correlations were empirically validated under conditions typical of direct-expansion evaporators controlled by a *constant superheat* expansion valve, or $x_2 > 1$, that is, with the *bulk vapor at evaporator outlet superheated*. Under these operating conditions a substantial portion of the evaporator tubes must operate in *dry wall* regimes when f_{tp} and the heat transfer coefficient α_{tp} rapidly converge toward the lower values typical of single-phase superheated vapors in turbulent regimes.

The full experimental range of Bo Pierre's test is summarized in Table 5.1, which is reproduced from the English translation [4.27] of the original (1957) paper in Swedish.

Bo Pierre's analysis essentially reduces the complex two-phase flow phenomena to a conventional single-phase vapor flow correlation with a mean empirically determined specific volume $\bar{v} = \bar{x}(v_g)$ and a corresponding average mixture[2] friction factor $\bar{f}_{tp}$ according to Eq. (5.1e). The effective average quality $\bar{x}$ normally approaches values in the range $0.5 < \bar{x} < 0.8$ and this explains why the naive model $\bar{x} = 1$ discussed in the previous section is not so far off the mark in typical direct-expansion evaporators even though the conventional dry gas friction factor $\bar{f}$ for turbulent flow regimes in ideal smooth tubes is significantly lower than $\bar{f}_{tp}$.

It is important to note that Bo Pierre's correlations are based on tests carried out in an actual refrigeration system [4.27] and conform to the methods adopted by all pragmatic refrigeration engineers.

5.2.2 Correlations for Incomplete Evaporation at Evaporator Outlet ($x_2 < 1$)

The expression for ΔP_{tp} obtained by Soumerai [5.2] is based on incomplete evaporation at the tube outlet in order to operate exclusively in the more effective *wet wall* regimes, with a slight surplus liquid at the tube exit: $y_2 = 1 - x_2 > 0$ in conjunction with a *postevaporator superheater* as shown in Fig. 4.5.

The analytical derivation is also based on a simple homogeneous or zero-slip flow model ($V_g/V_l = 1$) and, additionally, on the assumption that the *average two-phase friction factor* $\bar{f}_{tp}$ *is independent of the total flow rate* $\dot{m}$ and fluid properties for a given channel geometry and wall material. The assumption of a constant $\bar{f}_{tp}$ that is independent of $\dot{m}$ is quite plausible in the light of our discussion of fully rough single-phase flow regimes in Chapter 3, if one visualizes the core of the flowing mixture consisting of the vapor phase with some entrained droplets flowing at $V_g = V_l$ inside an effectively fully rough liquid wall annular boundary. These intuitive postulates were *validated* empirically as shown by Soumerai [4.5, 4.6] in experiments carried out with the *same physical tubes* both in *adiabatic* ($\dot{Q}_e = 0$) and *diabatic flow* ($\dot{Q}_e \neq 0$) with heat *addition* $\Delta x = x_2 - x_1 > 0$ in forced-convection evaporators, as well as heat *removal* $\Delta x = x_2 - x_1 < 0$ for forced-convection condensers [4.22], that is, in the quasi-mirror-image process depicted in Fig. 4.10.

The most significant results in the context of our present discussion are summarized below. For a more complete description of the test setup and validity/accuracy limits of the proposed analytical method, the interested reader is referred to the original references and several discussions in Ref. 5.3.

Generally, the total two-phase pressure drop for a horizontal tube can be expressed as the sum of a frictional component and an acceleration component [4.5, 4.27, 5.2–5.4]:

$$\Delta P_{\mathrm{tp}} = \Delta P_{\mathrm{tp},f} + \Delta P_{\mathrm{tp},a} \tag{5.3}$$

$$\Delta P_{\mathrm{tp}} = \Delta P_{\mathrm{tp},f}\left(1 + \frac{\Delta P_{\mathrm{tp},a}}{\Delta P_{\mathrm{tp},f}}\right) \tag{5.4}$$

For homogeneous flow, the *acceleration component* $\Delta P_{\mathrm{tp},a}$ is

$$\Delta P_{\mathrm{tp},a} = G^2\left[(1 - x_2)(v_l)_2 - (1 - x_1)(v_l)_1 + x_2(v_g)_2 - x_1(v_g)_1\right] \tag{5.5}$$

where $\Delta P_{\mathrm{tp},a}$ = acceleration pressure drop
$G = \dot{m}/\Omega$ total two-phase mass velocity
$1/\rho_l = v_l$ = specific volume of saturated liquid
$1/\rho_g = v_g$ = specific volume of saturated vapor
subscript 1 = channel inlet
subscript 2 = channel outlet

To ensure an acceptable power consumption in typical direct-expansion evaporator design applications, the reduced pressure $P^+ = P/P_{\mathrm{cr}} \ll 1$ and $\Delta P_{\mathrm{tp}}/P \ll 1$; therefore, the compressibility effects can safely be neglected:

$$(v_l)_1 = (v_l)_2 = v_l \tag{5.6}$$

and

$$(v_g)_1 = (v_g)_2 = v_g \tag{5.7}$$

Equation (5.5) reduces to the simpler form:

$$\Delta P_{\mathrm{tp},a} = G^2(\Delta x)v_{lg} = G^2(\Delta x)v_g\left(1 - \frac{v_l}{v_g}\right) \tag{5.8}$$

where

$$\Delta x = x_2 - x_1 > 0$$

$$\textit{Validity:} \quad \frac{\Delta P_{\mathrm{tp}}}{P} \ll 1 \quad \text{and} \quad P^+ = \frac{P}{P_{\mathrm{cr}}} \ll 1$$

Since $v_l/v_g \ll 1$ at $P^+ \ll 1$, this term can also be neglected and Eq. (5.8) reduces to

$$\Delta P_{\mathrm{tp},a} = G^2(\Delta x)v_g = G^2 v_g(x_2 - x_1) > 0 \tag{5.9}$$

It should be noted that the vapor quality change $\Delta x \equiv x_2 - x_1 > 0$ due to *external heat exchange* $\dot{Q}_e$ is always positive with $\dot{Q}_e > 0$ in the accelerating evaporating (fluid heating) mode but negative in the decelerating condensing (fluid cooling) mode when $\Delta x = x_2 - x_1 < 0$. To recognize the difference between the expanding ($\Delta x > 0$) and contracting ($\Delta x < 0$) modes we treat the quality change Δx *and* the *external* rate of heat exchange $\dot{Q}_e$ as *algebraic* quantities but the latent heat of evaporation at constant pressure $h_{lg} \equiv h_g - h_l$ as a *positive* quantity. Thus, the conventional heat balance for a perfectly insulated evaporator or condenser is

$$\dot{Q}_e = \dot{m}(h_2 - h_1) = \dot{m}h_{lg}(x_2 - x_1) \tag{5.10}$$

Solving for $\Delta x = x_2 - x_1$ yields

$$\Delta x = x_2 - x_1 = \frac{\dot{Q}_e}{\dot{m}h_{lg}} \tag{5.11}$$

where h_{lg} is the positive latent heat of evaporation at P = constant.

Therefore, the quality change $\Delta x \neq 0$ can be positive or negative when $\dot{Q}_e \neq 0$, that is, in *diabatic* flow, and $\Delta x = 0$ in quasi-incompressible *adiabatic* flow as indicated below.

$$\Delta x = x_2 - x_1 > 0 \quad \text{in } \textit{evaporating} \text{ mode with } \dot{Q}_e > 0 \tag{5.12a}$$

$$\Delta x = x_2 - x_1 < 0 \quad \text{in } \textit{condensing} \text{ mode with } \dot{Q}_e < 0 \tag{5.12b}$$

$$\Delta x = x_2 - x_1 = 0 \quad \text{or} \quad x_2 = x_1 = \text{constant in } \textit{adiabatic} \text{ flow } \dot{Q}_e = 0 \tag{5.12c}$$

From Eqs. (5.9), (5.11), (5.12a)–(5.12c) it is clear that the acceleration component $\Delta P_{\text{tp},a}$ in Eq. (5.3) and the ratio $\Delta P_{\text{tp},a}/\Delta P_{\text{tp},f}$ in Eq. (5.4) are positive for evaporators, negative for condensers, and negligibly[3] small for adiabatic two-phase flow (but always positive since the "friction heat" is always positive in the latter case).

Thus, at *identical friction pressure drop* $\Delta P_{\text{tp},f}$ the total two-phase pressure drop ΔP_{tp} per Eqs. (5.3) and (5.4) is generally higher in the evaporation than in the condensation process under otherwise identical conditions because the acceleration component $\Delta P_{p,a}$ is positive with heat addition but negative with heat removal. The ratio $\Delta P_{\text{tp}}/\Delta P_{\text{tp},f}$ can be estimated for the evaporating or condensing modes by introducing in Eq. (5.4) the expression of $\Delta P_{\text{tp},a}$ per Eq. (5.9) and $\Delta P_{\text{tp},f}$ computed according to Bo Pierre's correlation [Eq. (5.1a)]. This, together with Eq. (5.12), yields

$$\frac{\Delta P_{\text{tp}}}{\Delta P_{\text{tp},f}} = 1 \pm \frac{|\Delta x| G^2 v_g}{\bar{f}_{\text{tp}}(L/D_h)\bar{x}G^2 v_g} \tag{5.13}$$

or, after canceling the common term $G^2 v_g$ and regrouping,

$$\frac{\Delta P_{\text{tp}}}{\Delta P_{\text{tp},f}} = 1 \pm \frac{1}{\bar{x}/|\Delta x| \bar{f}_{\text{tp}} (L/D_h)} \tag{5.14a}$$

where the signs are applicable as follows: + for evaporators and − for condensers. Since all empirical results [3.4, 4.27 and references cited therein] indicate that $\bar{x} \approx |\Delta x|$ and $\bar{f}$ is typically of the order of 0.04 in fully wet wall regimes, Eq. (5.14a) can be reduced to the following approximate relationship:

$$\frac{\Delta P_{\text{tp}}}{\Delta P_{\text{tp},f}} \approx 1 \pm \frac{25}{L/D_h} \tag{5.14b}$$

In thermodynamically efficient forced-convection evaporators, the aspect ratio L/D_h will typically be in the range of 500–1000 and the term $25/(L/D_h) =$ 0.05–0.025 so that

$$\Delta P_{\text{tp}} \approx \Delta P_{\text{tp},f} \tag{5.14c}$$

within ± 5–10% for $L/D_h \geq 500$–1000.

For the typical spray-annular or annular flow regimes encountered in forced-convection evaporators and condensers, the following general relationships are thus valid for $\Delta P_{\text{tp},f}$ and ΔP_{tp} in the evaporating mode and the *quasi-mirror-image* condensing mode highlighted in Fig. 4.10.

$$(\Delta P_{\text{tp},f})_{\text{evaporating}} \equiv (\Delta P_{\text{tp},f})_{\text{condensing}} \tag{5.15}$$

for a *given* channel configuration, D_h, L/D_h, $\bar{e}/D_h$, and material *identical* fluids, $\dot{m}$, P, and $|\Delta x|_{\text{evaporating}} \equiv |\Delta x|_{\text{condensing}}$, when

$$(x_1)_{\text{evap}} = (x_2)_{\text{cond}} > 0$$

$$(x_2)_{\text{evap}} = (x_1)_{\text{cond}} < 1$$

Therefore

$$(\Delta P_{\text{tp}})_{\text{evap}} > (\Delta P_{\text{tp}})_{\text{cond}}$$

but for practical purposes

$$(\Delta P_{\text{tp}})_{\text{evap}} \simeq (\Delta P_{\text{tp}})_{\text{cond}} \tag{5.16}$$

when $\Delta P_{\text{tp},a}/\Delta P_{\text{tp},f} \ll 1$ or $L/D_h > 500$–1000.

Since in most practical evaporator and condenser design applications $\Delta P_{\text{tp}} \simeq \Delta P_{\text{tp},f}$, we now focus exclusively on the friction component $\Delta P_{\text{tp},f}$

which can be derived quite simply in the case of evaporation [4.4–4.6] on the basis of the conventional single-phase Darcy equation (3.4) for an element of evaporator length dl as follows:

$$dP_{\mathrm{tp},f} = \bar{f}_{\mathrm{tp}} \frac{G^2}{2} v \frac{dl}{D_h} \tag{5.17}$$

where $G = \dot{m}/\Omega$

$v = xv_g + (1 - x)v_l$

In normal direct-expansion evaporators when $x_1 > 0.1$–0.3 and $P^+ \ll 1$, the term $(1 - x)v_l$ is negligibly small compared to xv_g. The total two-phase friction pressure drop for $l = L$ is $\Delta P_{\mathrm{tp},f} = \int_{l=0}^{l=L} dP_{\mathrm{tp},f}$. Moving all the constants including $\bar{f}_{\mathrm{tp}}$ outside the integral sign, we obtain with Eq. (5.17)

$$\Delta P_{\mathrm{tp},f} = \frac{1}{2} \bar{f}_{\mathrm{tp}} \frac{G^2}{D_h} v_g \int_{l=0}^{l=L} x\, dl \tag{5.18}$$

As demonstrated in the original publications [4.5, 5.2], the total integrated pressure drop is independent of the actual distribution of the heat flux along the tube length for a *given* total rate of heat exchange $\dot{Q}_e$ and *prescribed* inlet vapor quality x_1, $\Delta x - x_2 - x_1$ or $x_2 = x_1 + \Delta x$ for a fixed saturation pressure P_2 at the tube outlet. It is therefore permissible to select the simplest case of a constant rate of evaporation $dx/dl = (x_2 - x_1)/L = \text{constant}$ to integrate Eq. (5.18):

$$dl = \frac{L}{x_2 - x_1} dx = (\text{constant})\, dx \tag{5.19}$$

Introducing Eq. (5.19) into (5.18) and integrating between the prescribed limits of x_1 at $l = 0$ and x_2 at $l = L$ yields $\int_{x_1}^{x_2} x\, dx = \frac{1}{2}(x_2^2 - x_1^2)$. Thus, Eq. (5.18) becomes

$$\Delta P_{\mathrm{tp},f} = \tfrac{1}{2} \bar{f}_{\mathrm{tp}} \frac{L}{D_h} G^2 \frac{x_2^2 - x_1^2}{x_2 - x_1} v_g \tag{5.20}$$

Substituting $(x_2 + x_1)(x_2 - x_1)$ for $x_2^2 - x_1^2$ in the above equation and canceling the term $x_2 - x_1$ in the nominator and denominator, we finally arrive at the expressions recommended by Soumerai [4.5, 5.2]:

$$\Delta P_{\mathrm{tp},f} = \bar{f}_{\mathrm{tp}} \frac{L}{D_h} \frac{G^2}{2} (\bar{x} v_g) \equiv (\bar{x} \bar{f}_{\mathrm{tp}}) \frac{L}{D_h} \frac{G^2}{2} v_g \tag{5.21a}$$

where

$$\bar{f}_{tp} = \text{average two-phase friction factor} \tag{5.21b}$$

$$\bar{x}v_g \equiv \bar{v} \equiv 1/\bar{\rho} = \text{effective average specific volume} \tag{5.21c}$$

$$\bar{x}\bar{f}_{tp} = \bar{f}_{eff} = \text{effective average two-phase friction factor based on } v_g \tag{5.21d}$$

$$\bar{x} \equiv \tfrac{1}{2}(x_1 + x_2) \equiv x_1 + \tfrac{1}{2}(x_2 - x_1) \equiv x_1 + \tfrac{1}{2}\frac{\dot{Q}_e}{\dot{m}h_{lg}} \tag{5.21e}$$

with

$$G \equiv \dot{m}/\Omega = \text{total mass velocity of mixture} \tag{5.21f}$$

$$v_g \equiv 1/\rho_g \equiv \text{specific volume of saturated vapor } (x = 1) \tag{5.21g}$$

It is interesting to note that Eq. (5.21a) is essentially the conventional correlation used to compute ΔP_f for a single-phase vapor except for the term $\bar{f}_{eff} = \bar{x}\bar{f}_{tp}$, which is computed with $\bar{x}$ per Eq. (5.21e) and the empirical friction factor $\bar{f}_{tp}$.

The integrated two-phase flow friction pressure drop, or head loss, can be conveniently expressed in dimensionless terms in the same manner as in single-phase fluid flow on the basis of the *concept of effective number of velocity heads*, $(\text{ENVH})_{tp,f}$, with the help of Eq. (5.21a) as follows.

With $G^2\bar{v} = G^2\bar{v}^2/\bar{v}$ and since $G^2\bar{v}^2 \equiv \bar{V}^2$, the term $G^2\bar{x}v_g = G^2\bar{v}$ in Eq. (5.21a) can be replaced by $\bar{V}^2/\bar{v} = \bar{V}^2\bar{\rho}$. Dividing all sides of Eq. (5.21a) by $\frac{1}{2}\bar{\rho}\bar{V}_1^2$ yields

$$\frac{\Delta P_{tp,f}}{\bar{\rho}\,\frac{1}{2}\bar{V}^2} \equiv \frac{\Delta P_{tp,f}\bar{v}}{\frac{1}{2}\bar{V}^2} \equiv (\text{ENVH})_{tp,f} \equiv \bar{f}_{tp}\left(\frac{L}{D_h}\right) \tag{5.22}$$

where $(\text{ENVH})_{tp,f}$ = effective number of velocity heads due to wall friction in two-phase flow, $\bar{V} = G\bar{v} = \dot{m}\bar{v}/\Omega$ = average velocity, and all other terms are as defined for Eq. (5.21a).

The above dimensionless expression in terms of $(\text{ENVH})_{tp,f}$ is particularly useful in forced-convection two-phase flow heat exchanger analysis.

The major conclusions, within the operating limits investigated, [4.5, 5.2] pertaining specifically to two-phase fluid flow aspects are repeated below with a few updated comments shown in square brackets.

1. The *homogeneous* [zero-slip] flow model is a *valid* assumption in annular-dispersed wet wall regimes typical of direct-expansion [or more generally: forced-convection] refrigeration evaporators with the *small hydraulic diameter internally finned channels*, D_h = 1.44 mm, described here [as well as smaller internally finned tubes of similar design with hydraulic diameter down

to $D_h = 1$ mm discussed in an earlier paper [5.5] and unpublished proprietary reports].

2. The *homogeneous* zero-slip flow model seems *applicable* to much *larger diameter bare circular* ($\frac{5}{8}$ *in.* = 14.6 *mm*) *tubes* operating with incomplete evaporation, within the normal range of exit vapor velocities and L/D_h ratios used in practice on the basis of Ashley's earlier (1942) test data [5.6]. In effect, this is an extension of Bo Pierre's homogeneous correlation for complete evaporation in *dry* or *partially dry* wall regimes [Bo Pierre reported [4.27, 5.7] a superheat of the order of 5–7°C above the temperature corresponding to the measured pressure in the *middle* of the evaporator. See also Table 5.1.]

3. The *assumption of fully rough turbulent flow*, $\bar{f}_{\text{tp}} =$ *constant*, identical for refrigerant R-12 and R-22 and independent (or a very weak function) of refrigerant flow rate $\dot{m}$, is *valid for both* the small hydraulic *internally finned* tubes $D_h = 1.44$ mm used in this study and smaller $D_h = 1.1$ mm in earlier references [5.5] and unpublished reports, as well as normal diameter ($\frac{5}{8}$ in. = 14.6 mm) *bare circular copper* tubes used by Ashley [5.6]. The empirical fact that $\bar{f}_{\text{tp}}$ = constant and is independent (or a very weak function) of hydraulic diameter variations of *over one order of magnitude* is also contrary to what would be expected in fully turbulent single-phase flow with conventional Newtonian fluids [see discussion of Moody's chart in Chapter 3]; in this respect, the single-component liquid–vapor mixture (plus traces of oil $< 0.1\%$) exhibits *non-Newtonian characteristics* as suggested by various research workers [5.2, 5.8].

4. The *fully rough* turbulent flow model, $\bar{f}_{\text{tp}}$ = constant, independent of $\dot{m}$ is *compatible* [and fully consistent] with Bo Pierre's and our own empirical correlations for the *coupled heat transfer coefficient* with *incomplete evaporation*; that is, $\bar{\alpha}_{\text{tp}} \propto \dot{m}^{1-n}$ with $1 - n \to 1$ as $n \to 0$, therefore, $\bar{\alpha}_{\text{tp}} \propto (\dot{m})^1$ in the fully rough wall regimes when ΔT_w = constant, independent of $\dot{m}$. (See Section 3.3.2.2. of Ref. 3.32, and Sections 9.7 and 9.8 of Ref. 1.4.) [This is discussed further in Chapter 8 of this volume.]

5. The proposed simple correlations, Eqs. (5.21a)–(5.21g), can be used with sufficient accuracy in direct-expansion or forced-convection evaporator design practice with $x_1 > 0$ and incomplete evaporation in fully wet wall regimes at $x_2 < x_{2,\text{ dry wall}}$ limit, or $x_2 < 0.9$ for normal bare circular copper tubes [5.9, 5.10] and $x_2 < 0.95$–0.98 for the proprietary internal finned tubes addressed in Refs. 4.4–4.6. In particular the *pressure drop with heat exchange* ($\dot{Q}_e \neq 0$) *can be computed like an adiabatic* ($\dot{Q}_e = 0$) *two-phase flow at an actual vapor quality* $x \equiv \bar{x} = \frac{1}{2}(x_1 + x_2)$ per Eq. (5.21e)

It is evident that the last statement, item 5, is also *applicable* in the case of the quasi-*mirror-image forced-convection condensation process* per Eq. (5.15) when

$$(x_1)_{\text{cond}} = (x_2)_{\text{evap}}, (x_2)_{\text{cond}} = (x_1)_{\text{evap}}, \quad \text{and} \quad \bar{x}_{\text{evap}} = \bar{x}_{\text{cond}}$$

5.2.3 Comparison With Other Analytical Models

"Naive" All Gas Flow Model. The integrated friction pressure drop $\Delta P_{tp,f}$ according to the naive model described in Section 5.1 yields the same correlation as Eq. (5.21a) with $\bar{x} = 1$ and $f_{g,x=1}$ computed according to conventional correlations at $\bar{x} = 1$ and $\dot{m}_g = \dot{m}$ instead of $\bar{f}_{eff} = \bar{x}\bar{f}_{tp}$ per Eq. (5.21d). Thus, the naive model should yield the same end result $\Delta P_{tp,f}$ when $f_{g,x=1} = \bar{x}\bar{f}_{tp}$. This condition is satisfied within the estimated $\pm 30\%$ accuracy band of most two-phase flow pressure drop correlations as can be seen from the following rough estimates.

In turbulent flow regimes typically encountered in forced-convention evaporators or condensors, $f_{g,x=1} = 0.025 \pm 30\%$ (see Numerical Analysis 3.2). In contrast, the empirical value of the friction factor $\bar{f}_{tp}$ derived in Refs. 3.4 and 3.32 is substantially higher for *wet wall* regimes (incomplete evaporation) with a constant value $\bar{f}_{tp} = 0.045 \pm 13\%$ for hydraulic diameters ranging from 1.1 to 18 mm. However, with typical inlet and outlet qualities of $x_1 = 0.2$ and $x_2 = 0.95$, respectively, we obtain from Eq. (5.21e) a value $\bar{x} = \frac{1}{2}(0.2 + 0.95) = 0.575$. Thus, the effective $\bar{f}_{eff} = \bar{x}\bar{f}_{tp} = 0.026 \pm 13\%$ falls right in line with $f_{g,x=1} = 0.025 \pm 30\%$ and this explains why the "naive" model has proved quite adequate in many practical refrigeration applications.

Martinelli's Fully Separated Fluid Flow Model. According to the *ASHRAE Handbook, Fundamentals* [5.11] the fully separated *slip-flow* or $V_g/V_l > 1$ Martinelli model [5.12–5.14] "... works reasonably well in predicting pressure drop of boiling refrigerant... although sometimes the homogeneous [4.27] model seems better [3.4]..." (See also the related impact of oil [5.15–5.17].) A plausible explanation for the latter conclusion is that the original (1944) highly idealized *fully separated flow* model proposed by Martinelli fails to recognize the *existence* of an *entrained liquid fraction D* of the *total liquid flowing fraction* $(1 - x)$ and it has since been shown that the entrained liquid fraction has a very significant impact on two-phase flow regimes and pressure drop [5.18–5.20] as might be expected. The homogeneous flow model can thus be visualized as the limiting case of *spray-annular* flow regimes when the dispersed liquid flow fraction D approaches unity or $D(1 - x) \simeq (1 - x) \equiv y$. The fact that the homogeneous flow model is sometimes better than Martinelli's slip-flow model simply means that a substantial fraction D of the total flowing liquid fraction $1 - x \equiv y$ is entrained with the gas phase in typical forced-convection direct-expansion evaporators.

It is also interesting to note that the validity/plausibility of the conclusion $\bar{f}_{tp}$ = constant, independent of flow rate $\dot{m}$, reported in 1966 for conditions typical of forced-convection refrigeration evaporators with incomplete evaporation [4.5] has since (1969/70) been supported by semiempirical correlations recommended for *annular-flow regimes* in applications outside the refrigeration industry with *steam and water* according to the following excerpts [5.11, p. 4.16] italics added:

Wallis's analysis [5.21–5.23] of the flow occurrences is based on interfacial friction between the gas and liquid. The *wavy film corresponds to a conduit of relative roughness* $(\bar{e}/D)$ *about four times the liquid film thickness*. The pressure drop relation... corresponds to the Martinelli-type analysis with: $f_{\text{two-phase}} = (\phi_g)^2 f_g$, where the friction factor f_g (of the gas alone) is taken as 0.02, an appropriate turbulent flow value [the authors refer to Moody's chart shown here as Fig. 3.4]. Although rather simple, Wallis' calculation method is rationally based, and indicates agreement [5.22] with experiments. It can be modified for more detailed consideration of factors such as Reynolds number variation in friction, gas compressibility, and *entrainment* [5.23].

The terms ϕ_g and f_g in the Martinelli type analysis are discussed in the next section.

5.3 FULLY SEPARATED MARTINELLI TYPE SLIP-FLOW MODELS ($V_g / V_l > 1$)

Because of the extreme complexity of two-phase flow phenomena, there exists a very extensive literature associated with it, particularly in connection with the nuclear reactor industry. As of 1963 several thousand references in two-phase flow had been compiled in an "index to the Two-Phase Gas–Liquid Flow Literature" [5.24]. The proliferation of two-phase flow research publications and recommended empirical correlations has continued unabated in the intervening years. To complicate this situation further, the recommended semianalytical correlations have tended to become far more complex with the increased availability of effective low cost electronic computers, more refined instrumentation techniques, and the atypical highly sophisticated demands of the nuclear reactor industry. Most of the two-phase flow models derived for nuclear power applications are based on the more refined *slip-flow* rather than the simpler *zero-slip-flow* models such as those reviewed in the previous sections which have proved quite adequate in many refrigeration and general chemical process applications [5.11].

In view of the enormity of the two-phase literature, we focus exclusively here on the slip-flow model initially proposed by Martinelli and associates [5.12–5.14] for two reasons: first, because this method is still very popular, and second, because the Martinelli model is particularly important in connection with predictions of the *coupled* two-phase flow heat transfer coefficient addressed in Chapter 8.

5.3.1 Two-Component Two-Phase Flow Applications

Lockhart and Martinelli derived an empirical dimensionless correlation for the frictional pressure drop $\Delta P_{\text{tp},f}$ in the isothermal two-phase flow of air and several liquids in horizontal tubes. The flow was substantially incompressible and according to Eq. (5.9), $\Delta P_{\text{tp},a} = 0$, so that the measured ΔP_{tp} was entirely due to friction; that is, $\Delta P_{\text{tp}} = \Delta P_{\text{tp},f}$.

The basic assumptions made by Martinelli and co-workers to derive their analytical model were that the static pressure drop for the liquid phase equals that of the gas phase, regardless of the type of two-phase flow patterns, and that the volume occupied by the liquid plus that occupied by the gas phase under SSSF conditions equals the volume of the pipe. The following key dimensionless parameters were defined:

$$\Phi_l \equiv \left(\frac{(\Delta P/\Delta L)_{\mathrm{tp},f}}{(\Delta P/\Delta L)_{\mathrm{lo}}} \right)^{1/2} \tag{5.23}$$

$$X \equiv \left(\frac{(\Delta P/\Delta L)_{\mathrm{lo}}}{(\Delta P/\Delta L)_{\mathrm{go}}} \right)^{1/2} \tag{5.24}$$

where $(\Delta P/\Delta L)_{\mathrm{tp},f}$ is the *actual* pressure drop gradient of the two-phase mixture due to friction only, whereas $(\Delta P/\Delta L)_{\mathrm{lo}}$ and $(\Delta P/\Delta L)_{\mathrm{go}}$ are *fictitious* single-phase pressure gradients defined as follows:

- $(\Delta P/\Delta L)_{\mathrm{lo}}$ is the pressure gradient that would exist with the liquid phase flowing alone (the subscript lo stands for liquid only).
- Similarly, $(\Delta P/\Delta L)_{\mathrm{go}}$ is the pressure gradient that would exist with the gas phase flowing alone (the subscript go stands for gas only).

Four possible dissipative flow regimes were arbitrarily defined as shown in Table 5.2 by the values of $(\mathrm{Re})_{\mathrm{lo}}$ and $(\mathrm{Re})_{\mathrm{go}}$ for single-phase flow. The four different dissipative flow regimes are identified by the subscripts *vv*, *vt*, *tv*, and *tt*, where the first subscript refers to the flow of liquid alone and the second to the gas alone. Thus, *tt* designates turbulent flow for each phase flowing alone whereas in *vt* regimes the liquid would flow alone in a viscous (or laminar) regime and the gas phase in a turbulent regime.

The superficial (fictitious) Reynolds numbers are computed in a conventional fashion per Eq. (3.32) for circular tubes as shown below and in Table 5.2.

$$\mathrm{Re}_{\mathrm{lo}} \equiv \frac{4\dot{m}_{\mathrm{lo}}}{\pi D_h \mu_l} \tag{5.25}$$

$$\mathrm{Re}_{\mathrm{go}} \equiv \frac{4\dot{m}_{\mathrm{go}}}{\pi D_h \mu_g} \tag{5.26}$$

and a ratio

$$\frac{(\mathrm{Re})_{\mathrm{lo}}}{(\mathrm{Re})_{\mathrm{go}}} \equiv \frac{\dot{m}_{\mathrm{lo}}}{\dot{m}_{\mathrm{go}}} \left(\frac{\mu_g}{\mu_l} \right) \tag{5.27}$$

TABLE 5.2. Four Types of Dissipative Flow Regime According to Martinelli

Type of Dissipative Flow Regime	*tt*	*vt*	*tv*	*vv*
$(\mathrm{Re})_{lo} \equiv \dfrac{4\dot{m}_{lo}}{\pi D \mu_l} = \left(\dfrac{4\dot{m}^a}{\pi D \mu_l}\right)(1-x)$	> 2000	< 1000	> 2000	< 1000
$(\mathrm{Re})_{go} \equiv \dfrac{4\dot{m}_{go}}{\pi D \mu_g} = \left(\dfrac{4\dot{m}}{\pi D \mu_g}\right)x$	> 2000	> 2000	< 1000	< 1000

[a] $\dot{m}$ ≡ total mixture mass flow rate.

From their experimental data Martinelli and co-workers derived empirical universal relationships between the dimensionless parameter Φ_l and X shown as four different curves $\Phi_l(X)$ corresponding to the four dissipative regimes, reproduced here as Fig. 5.2. The two-phase friction pressure drop $\Delta P_{tp,f}$ is then computed with the empirical values of Φ_l of Fig. 5.2 according to

$$\Delta P_{tp,f} = \Delta P_{lo}(\Phi_l)^2 \tag{5.28}$$

where ΔP_{lo} is computed in the conventional manner according to Eq. (3.4) for the liquid phase flowing alone.

It is difficult to make firm quantitative statements on the validity/accuracy limits of Martinelli's method of predicting $\Delta P_{tp,f}$ per Eq. (5.28) on the basis of the square of the empirical parameter Φ_l obtained from Fig. 5.2 because there are many conflicting views on this matter [4.24, 5.25]. However, the average and maximum possible deviations reported by the originators themselves and by independent researchers certainly confirm that the *accuracy is much poorer* than that achieved in *single-phase fluid flow* (see also Chapter 4). The following cautious statements can be ventured:

1. Average and maximum deviations certainly are much greater than conventional single-phase flow correlations.
2. The estimated average deviation is ± 30–40%.
3. The maximum reported deviation is much greater than the average deviation.

The above estimates are based mainly on the following published assessments by several independent researchers (further supported by unpublished proprietrary laboratory test results). According to Stepanoff [5.25],

> Martinelli and his co-workers show that their own test data for the 1 in. pipe fall on a single curve with ±30 percent scatter... Bergelin [5.26], quoting the Jenkins test data [5.27], believes the scatter of points is ±40 percent, and for certain modes of flow it may be as high as 100 percent.

In the above quotation, Stepanoff is referring to the original 1944 [5.12] paper and test results obtained with a 1 in. *glass* tube, that is, an *abnormally*

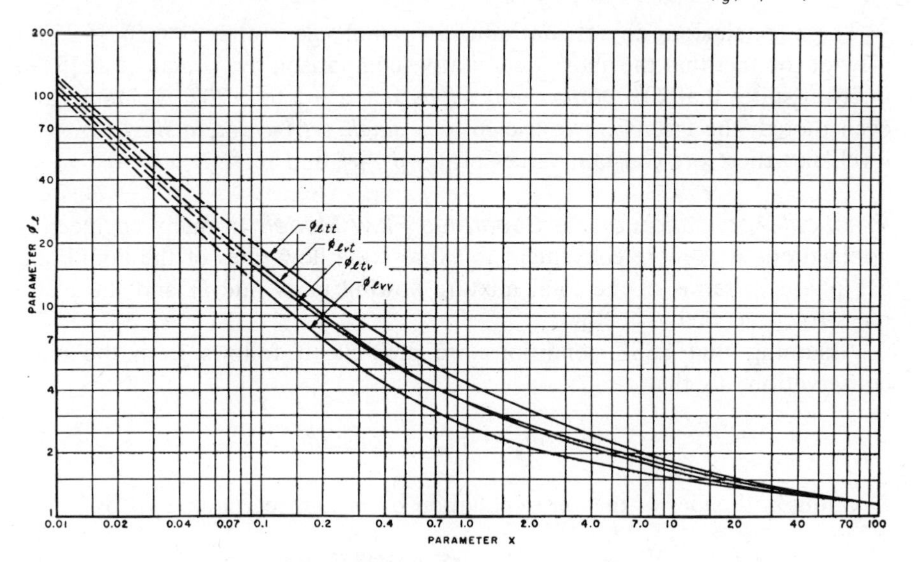

Coordinates of ϕ_l vs Parameter X

X	Turbulent-Turbulent ϕ_l	Turbulent-Viscous ϕ_l	Viscous-Turbulent ϕ_l	Viscous-Viscous ϕ_l	X	Turbulent-Turbulent ϕ_l	Turbulent-Viscous ϕ_l	Viscous-Turbulent ϕ_l	Viscous-Viscous ϕ_l
0.01	128	112	120	105	2.0	3.10	2.62	2.62	2.06
0.02	68.4	58.0	64.0	53.5	4.0	2.38	2.15	2.05	1.76
0.04	38.5	31.0	34.0	28.0	7.0	1.96	1.83	1.73	1.60
0.07	24.4	19.3	20.7	17.0	10	1.75	1.66	1.59	1.50
0.10	18.5	14.5	15.2	12.4	20	1.48	1.44	1.40	1.36
0.2	11.2	8.70	8.90	7.00	40	1.29	1.25	1.25	1.25
0.4	7.05	5.50	5.62	4.25	70	1.17	1.17	1.17	1.17
0.7	5.04	4.07	4.07	3.08	100	1.11	1.11	1.11	1.11
1.0	4.20	3.48	3.48	2.61					

FIGURE 5.2. Relationships between parameter ϕ_l and parameter X. [From *ASHRAE Handbook, Fundamentals*, Chap. 2, Fig. 20, ASHRAE, New York, 1966. Reproduced with permission.]

smooth tube from the standpoint of most practical engineering applications. This may explain the use of a *smooth tube* Blasius exponent $n = 0.20$ to calculate the fictitious single pressure gradients $(\Delta P/\Delta L)_{\text{lo}}$ and $(\Delta P/\Delta L)_{\text{go}}$ and the corresponding empirical dimensionless correlations of Φ_l in terms of X per Eqs. (5.23) and (5.24).

The following excerpts from the *ASHRAE Handbook, Fundamentals* [5.28] assess the accuracy of the Lockhart–Martinelli and Martinelli–Nelson [5.13, 5.14] methods as follows (italics added):

> Hatch and Jacobs [5.29] have concluded that the Martinelli–Nelson method can be used to predict *approximately* the two-phase pressure drop of R-11. For Refrigerant-22 at pressures around *100 psi*, Altman, Norris, and Staub [5.30] have shown that the Martinelli–Nelson method correlated *80 percent* of their pressure drop data within *+5 and −20 percent*.

These conclusions do not contradict Stepanoff's general assessment [5.25] in view of the fact that the above quantitative comparison by Altman et al. [5.30] in effect states that 20% *of their tests fell outside* a +5 to −20% *deviation band* even though the tests were restricted to a *single refrigerant*, *a single pressure* level, and therefore constant values of μ_g, ρ_g, μ_l, and ρ_l.

Practical Application of the Separated Flow Model. In many engineering applications it is more convenient to express the flow rates of the liquid and gas phase in terms of the total mixture flowing mass rate $\dot{m}$ and the phase mass fractions defined below.

Assuming that SSSF conditions are possible, it follows from the mass conservation law that

$$\dot{m} = \dot{m}_{go} + \dot{m}_{lo} = \text{constant} \tag{5.29}$$

Dividing both sides of the above equation by $\dot{m}$ and regrouping yields

$$x + y = 1 \quad \text{or} \quad y = (1 - x) \tag{5.30a}$$

with

$$x \equiv \frac{\dot{m}_{go}}{\dot{m}} = \text{gas mass flowing fraction} \tag{5.30b}$$

$$y \equiv 1 - x \equiv \frac{\dot{m}_{lo}}{\dot{m}} = \text{liquid mass flowing fraction} \tag{5.30c}$$

Substituting $\dot{m}(1 - x)$ for $\dot{m}_{lo}$ and $\dot{m}(x)$ for $\dot{m}_{go}$ in Eqs. (5.25)–(5.27), we obtain the following relationships (also shown in Table 5.2):

$$(\text{Re})_{lo} = \frac{4\dot{m}}{\pi D_h \mu_l}(1 - x) \equiv (\text{Re})_{x=0}(1 - x) \tag{5.31a}$$

where

$$(\text{Re})_{x=0} \equiv \frac{4\dot{m}}{\pi D_h \mu_l} = \text{all liquid limiting case, when } x = 0 \tag{5.31b}$$

and

$$(\text{Re})_{go} = \frac{4\dot{m}}{\pi D_h \mu_g} x \equiv (\text{Re})_{x=1} x \tag{5.31c}$$

where

$$(\text{Re})_{x=1} \equiv \frac{4\dot{m}}{\pi D_h \mu_g} = \text{all gas limiting case, when } x = 1 \tag{5.31d}$$

Dividing Eq. (5.31a) by Eq. (5.32a) yields the ratio

$$\frac{(\mathrm{Re})_{\mathrm{lo}}}{(\mathrm{Re})_{\mathrm{go}}} = \frac{(\mathrm{Re})_{x=0}(1-x)}{(\mathrm{Re})_{x=1}x} = \left(\frac{\mu_g}{\mu_l}\right)\left(\frac{1-x}{x}\right) \tag{5.32}$$

Equation 5.32 clearly shows that the *key ratio* $(\mathrm{Re})_{\mathrm{lo}}/(\mathrm{Re})_{\mathrm{go}}$ is *independent* of the channel geometry, $D_h = 4r_h$, L/D_h, $\bar{e}/D_h$, wall material, and total flow rate $\dot{m}$ and is uniquely defined by the consistency x of the flowing mixture and the viscosity ratio (μ_g/μ_l); that is,

$$(\mathrm{Re})_{\mathrm{lo}}/(\mathrm{Re})_{\mathrm{go}} = \varphi(x, \mu_g/\mu_l \text{ only}) \tag{5.33}$$

Since the viscosity is a function of the types of chemical species—chsp_l and chsp_g—and the (equilibrium) thermodynamic state of the mixture defined by two intensive properties such as T and P, the above ratio is ultimately a function of *five independent variables*:

$$(\mathrm{Re})_{\mathrm{lo}}/(\mathrm{Re})_{\mathrm{go}} = \varphi(x, T, P, \mathrm{chsp}_l, \mathrm{chsp}_g) \tag{5.34}$$

It is important to note that in the case of two-component and single-component fluids without any phase change, x and $y = 1 - x$ *remain constant* from tube inlet to outlet. The situation becomes significantly less complex in single-component two-phase adiabatic flow as highlighted below.

5.3.2 Single-Component Two-Phase Adiabatic Flow: $\dot{Q}_e = 0$

Because there is a one-to-one relationship between T_{sat} and P for a given pure substance, the liquid and gas phase viscosities μ_l and μ_g, respectively, are uniquely defined at a given pressure. Thus, the number of *independent variables* in Eq. (5.34) is reduced from five to three in the most general case, considering all possible pure substances; that is,

$$(\mathrm{Re})_{\mathrm{lo}}/(\mathrm{Re})_{\mathrm{go}} = \varphi(x, P, \text{pure substances}) \quad \textit{valid} \text{ for } \textit{all} \text{ pure substances} \tag{5.35}$$

For any *given pure substance*, such as H_2O in steam power plants, or any one of the many refrigerants listed in Table A.5 of Appendix A, the Re ratio is *uniquely* defined in terms of only *two independent* variables: the *consistency* of the mixture x and the operating *saturation pressure* level P or:

$$(\mathrm{Re})_{\mathrm{lo}}/(\mathrm{Re})_{\mathrm{go}} = \varphi(x, P) \quad \textit{valid} \text{ for a } \textit{given} \text{ pure substance} \tag{5.36}$$

The explicit equations for Martinelli's parameters X_{tt} and X_{vv} can be derived most generally from their definition as follows.

The conventional single-phase pressure drop correlation Eq (3.4) yields for the liquid phase flowing alone at a mass velocity $G_{lo} = G(1 - x)$

$$\left(\frac{\Delta P}{\Delta L}\right)_{lo} = \frac{G^2}{2D_h} f_{lo} \frac{(1-x)^2}{\rho_l} \propto f_{lo} \frac{(1-x)^2}{\rho_l} \tag{5.37}$$

Similarly, for the gas phase flowing alone at a mass velocity $G(x)$,

$$\left(\frac{\Delta P}{\Delta L}\right)_{go} = \frac{G^2}{2D_h} f_{go} \frac{x^2}{\rho_g} \propto f_{go} \frac{x^2}{\rho_g} \tag{5.38}$$

Dividing Eq. (5.37) by (5.38) and taking the square root on both sides yields with Eq. (5.24)

$$X \equiv \left(\frac{(\Delta P/\Delta L)_{lo}}{(\Delta P/\Delta L)_{go}}\right)^{1/2} = \left(\frac{f_{lo}}{f_{go}}\right)^{1/2} \left(\frac{1-x}{x}\right) \left(\frac{\rho_g}{\rho_l}\right)^{1/2} \tag{5.39}$$

The single-phase friction factor f_{lo} can be computed from the general Blasius type correlation for the single-phase friction factor, Eq. (3.44), with μ_l and $G_{lo} = G(1 - x)$ for the liquid phase. The friction factor f_{go} is computed in a similar fashion with $G_{go} = Gx$ and μ_g. Thus, the ratio for f_{lo}/f_{go} in Eq. (5.39) bccomes

$$\frac{f_{lo}}{f_{go}} = \left(\frac{\mu_l}{\mu_g}\right)^n \left(\frac{1-x}{x}\right)^{-n} \tag{5.40}$$

Substituting the above expression for f_{lo}/f_{go} in Eq. (5.39) yields the following most general analytical expression for X_{tt} or X_{vv}:

$$X \equiv \left(\frac{\rho_g}{\rho_l}\right)^{1/2} \left(\frac{\mu_l}{\mu_g}\right)^{n/2} \left(\frac{1-x}{x}\right)^{1-n/2} \tag{5.41a}$$

Validity: SSSF tt and vv regimes with appropriate Blasius exponent n per Eq. (3.44) (5.41b)

Equation (5.41a) reduces for vv flow regimes with $n = 1$ to the usual expression

$$X_{vv} = \left(\frac{\rho_g}{\rho_l}\right)^{0.5} \left(\frac{\mu_l}{\mu_g}\right)^{0.5} \left(\frac{1-x}{x}\right)^{0.5} \tag{5.42a}$$

Validity: vv regimes with Blasius exponent $n = 1$ (5.42b)

For tt flow regimes, using the Blasius exponent $n = 0.2$, on which the empirical $(\Phi_l)_{tt}$ in Fig. 5.2 is based in order to remain consistent with

Martinelli and co-workers we find that Eq. (5.41a) yields the well-known expression for *tt* regimes:

$$X_{tt} = \left(\frac{\rho_g}{\rho_l}\right)^{0.5}\left(\frac{\mu_l}{\mu_g}\right)^{0.1}\left(\frac{1-x}{x}\right)^{0.9} \tag{5.43a}$$

Validity: *tt* regimes with the smooth wall Blasius exponent $n = 0.2$ (5.43b)

With the above explicit equations for X_{vv} and X_{tt}, the two-phase friction pressure drop $\Delta P_{\mathrm{tp},f}$ can be computed on the basis of $(\Delta P)_{\mathrm{lo}}$ according to Eq. (5.28) with the multiplier $(\Phi_l)^2$ computed with the empirical value of Φ_l obtained from Fig. 5.2 at the calculated value of the parameter X_{tt} per Eq. (5.43a) in the case of *tt* regimes. This is the preferred method in nuclear and some refrigeration or process heat exchanger design applications where the bulk of the flowing mixture consists of liquid $(1 - x \gg 0)$ and the vapor flowing fraction $x \ll 1$.

However, using the *liquid phase* pressure drop $(\Delta P)_{\mathrm{lo}}$ as a reference is *inconvenient in conventional evaporator or condenser* applications when $x_{\mathrm{max}} \approx 1$, that is, when the *major part* of the total circuit two-phase pressure drop occurs in the high-vapor-quality region of the channel (i.e., at the outlet section of evaporators, or the inlet section of condensers). For these more typical process heat exchanger applications, it is simpler and generally more accurate to compute $\Delta P_{\mathrm{tp},f}$ on the basis of $(\Delta P)_{\mathrm{go}}$ rather than $(\Delta P)_{\mathrm{lo}}$ as suggested by Soumerai [5.31] with the use of the dimensionless parameter Φ_g defined below:

$$\Phi_g \equiv \left(\frac{(\Delta P/\Delta L)_{\mathrm{tp},f}}{(\Delta P/\Delta L)_{\mathrm{go}}}\right)^{1/2} \tag{5.44}$$

and

$$(\Phi_g)^2 \equiv \frac{(\Delta P/\Delta L)_{\mathrm{tp},f}}{(\Delta P/\Delta L)_{\mathrm{go}}} \tag{5.45}$$

$\Delta P_{\mathrm{tp},f}$ can be calculated for a known value of $(\Phi_g)^2$ according to

$$\Delta P_{\mathrm{tp},f} = (\Delta P)_{\mathrm{go}}(\Phi_g)^2 \tag{5.46}$$

From the definitions of Φ_l and X, per Eqs. (5.23) and (5.24), respectively, it is clear that the product

$$\Phi_l X \equiv \left(\frac{(\Delta P/\Delta L)_{\mathrm{tp},f}}{\cancel{(\Delta P/\Delta L)_{\mathrm{lo}}}}\,\frac{\cancel{(\Delta P/\Delta L)_{\mathrm{lo}}}}{(\Delta P/\Delta L)_{\mathrm{go}}}\right)^{1/2} \tag{5.47}$$

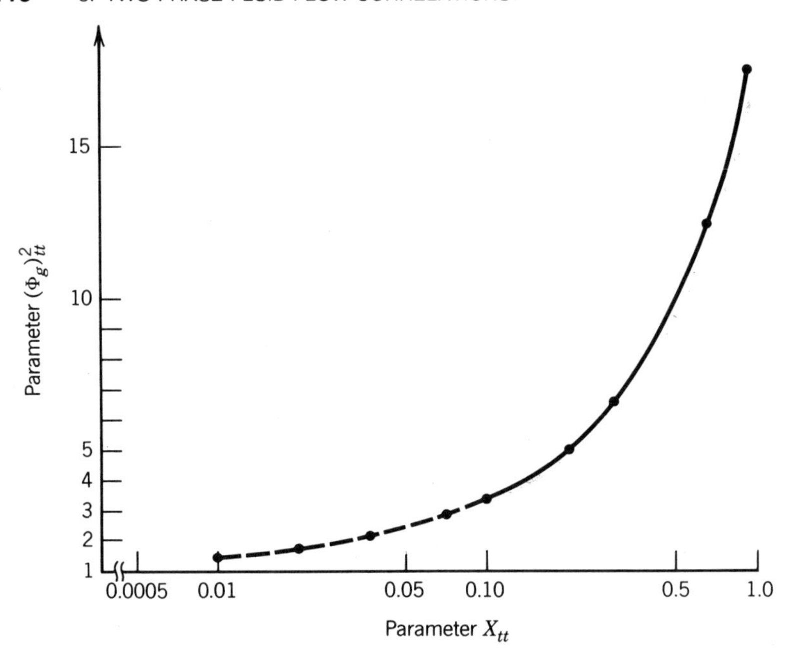

FIGURE 5.3. Relationship between parameters, $(\phi_g)^2_{tt}$ and X_{tt}.

Parameter $(\phi_g)^2_{tt}$ as a function of X_{tt}

X_{tt}	(0.01)	(0.02)	(0.04)	(0.07)	0.10	0.2	0.4	0.7	1.0
$(\phi_l)_{tt}$ from Fig. 5.2	(128)	(68.4)	(38.5)	(24.4)	18.5	11.2	7.05	5.04	4.20
$(\phi_g)^2_{tt} = [X_{tt}(\phi_l)_{tt}]^2$	(1.638)	(1.871)	(2.371)	(2.917)	3.423	5.018	7.952	12.45	17.64

Note: Less accurate values in parentheses were obtained from the *extrapolated* dotted curves $(\phi_l)_{tt}$ in Fig. 5.2.

represents the parameter Φ_g; therefore,

$$\Phi_l X = \Phi_g \tag{5.48a}$$

and

$$(\Phi_l X)^2 = (\Phi_g)^2 \tag{5.48b}$$

Thus, the dimensionless empirical parameter $(\Phi_g)^2$ on the right-hand side of Eq. (5.46) is obtained by squaring the product of $\Phi_l X$, where Φ_l is the empirical parameters shown in Fig. 5.2. The empirical values of $(\Phi_g)^2$ computed in this fashion are shown in graphical and tabular form in Fig. 5.3, for the range of X_{tt} most relevant in normal evaporators or condensers.

With the help of Fig. 5.3 it is a rather simple matter to estimate ΔP_{tp} for adiabatic flow in the usual applications when $\Delta P_{\text{tp}}/P \ll 1$, therefore $\Delta x = 0$

and $x_1 = x_2 = x =$ constant, since according to Eq. (5.9) $\Delta P_{\text{tp},a} = 0$ and

$$\Delta P_{\text{tp}} = \Delta P_{\text{tp},f} = (\Delta P)_{\text{go}}(\Phi_g)^2 \tag{5.49a}$$

Validity: $x =$ constant; $\Delta P_{\text{tp}}/P \ll 1$; adiabatic flow $|\dot{Q}_e| = 0$ (5.49b)

It should be noted that curves and tables similar to those shown in Fig 5.3 could be prepared for $(\Phi_g)^2$ applicable to other dissipative flow regimes such as X_{vv}. We have deliberately confined ourselves here to the turbulent–turbulent dissipative flow regimes (usually corresponding to annular or spray-annular flow patterns) which are typically encountered in actual evaporators and condensers [4.6, 4.21, 4.22, 5.19, 5.28]. The rationale for this fact is that two-phase flow heat exchangers are purposely designed to operate in the more effective stable turbulent–turbulent regimes, basically for the same reasons that single-phase heat exchangers are deliberately circuited for turbulent rather than laminar flow regimes, whenever this is feasible, as discussed in Section 3.12.

5.3.3 Single-Component Two-Phase Flow With Heat Exchange: $\dot{Q}_e \neq 0$

For single-component two-phase flow applications with heat addition, as in evaporators when $\Delta x = x_2 - x_1 > 0$, or heat removal, when $x_2 - x_1 < 0$ in the condensing mode, the empirical multiplier $(\Phi_g)^2$ in Eq. (5.49a) is no longer constant since it is a function of the variable vapor flow fraction x. To calculate $\Delta P_{\text{tp},f}$ it is *necessary to know* the *evolution of the vapor flow fraction x along the channel axial length*, from an inlet value x_1 at $l = 0$ to a specified outlet value x_2 at $l = L$. This requires a knowledge of the external heat flow $d\dot{Q}_e = \dot{m}h_{lg}\,dx$ for an element of length dl at any point along the channel axis l, that is, the local heat flux q''. Substituting $G(\Omega)$ for $\dot{m}$, with $\Omega =$ channel flow cross section, and recalling that $\Omega/a = r_h$, where $a =$ channel internal heat exchanger surface per unit length (or *wetted perimeter* when $\Omega =$ constant), we obtain from $q'' = d\dot{Q}_e/a\,dl$

$$q'' = \left(\frac{G\Omega}{a dl}\right) h_{lg}\,dx = G r_h h_{lg} \frac{dx}{dl} \tag{5.50}$$

Furthermore, the local heat flux must satisfy the basic convection heat rate equation

$$q'' \equiv \alpha_{\text{tp}}(T_w - T_{\text{sat}}) \tag{5.51}$$

where $\alpha_{\text{tp}} =$ two-phase local heat transfer coefficient
$T_w =$ local internal wall temperature
$T_{\text{sat}} =$ local saturation temperature corresponding to local pressure $P = P_{\text{sat}}$

From the conventional heat balance (assuming a perfectly insulated heat exchanger system) we may equate the local heat flux per Eqs. (5.50) and (5.51) and solve for (dx/dl) to obtain

$$\frac{dx}{dl} = \frac{1}{Gr_h(h_{lg})}\alpha_{\text{tp}}(T_w - T_{\text{sat}}) \propto \alpha_{\text{tp}}(T_w - T_{\text{sat}}) \tag{5.52}$$

Thus, the rate of quality change (dx/dl) at any point l along the channel axis is directly proportional to the product of the local two-phase heat transfer coefficient α_{tp} and the local conventional driving force $(T_w - T_{\text{sat}})$ for prescribed G, r_h, and h_{lg}. Since α_{tp} depends on the velocity field (or the local friction factor as discussed in subsequent chapters) and T_{sat} is also a function of $P = P_1 - \Delta P_{\text{tp}}$ at any axial position l, the fluid flow problem cannot generally be solved independently of the coupled heat transfer problem (see also Guideline 9 in Chapter 4).

Martinelli and Nelson, however, have presented an approximate method [5.13] for *forced-circulation boiling of water* which removes the coupling between heat transfer and pressure drop by assuming (as in refrigeration applications [4.4, 4.6, 4.21]) that the vapor quality varies linearly along the tube length. With this assumption $dx/dl = \text{constant} = \Delta x/L$, the local friction pressure drop for an elementary length dl can be computed according to Eq. (5.46) and integrated in a manner similar to the method used in Section 5.2.2 for the homogeneous flow model, noting, however, that the variable $(\Phi_g)^2$ must be kept under the integral sign since it is a function of x. This yields the following most general expression

$$\Delta P_{\text{tp},f} = \overline{M}_g(\Delta P)_{g,x=1} \tag{5.53a}$$

where

$$\overline{M}_g = \text{integrated multiplier} \equiv \frac{1}{|x_2 - x_1|}\int_{x_1}^{x_2}(\Phi_g)^2 x^{2-n}\,dx \tag{5.53b}$$

$(\Delta P)_{g,x=1}$ = single-phase vapor pressure drop computed on the basis of the total mass flow rate $\dot{m}$, as if $x = 1$, from the conventional single-phase correlation, Eq. (3.4) (5.53c)

Validity: Turbulent–turbulent flow regimes with an appropriate value for the generalized Blasius exponent n per Eq. (3.44a) (5.53d)

The above equations are *valid*, *in principle*, for *any* channel *effective wall relative roughness* $(\bar{e}/D)$ with suitable Blasius constants K_n and n. However, because the empirical values of the parameter Φ_l per Fig. 5.2 and corresponding $\Phi_g = X_{tt}\Phi_l$ are a priori valid only with the value $n = 0.2$, it is necessary to use the same numerical value $n = 0.2$ in conjunction with Figs. 5.2 and 5.3.

Thus Eqs. (5.53a) to (5.53b) become

$$\Delta P_{\mathrm{tp},f} = \overline{M}_g(\Delta P)_{g,\,x=1} = \overline{M}_g\left[(f_g)_{x=1}\left(\frac{L}{D_h}\right)\frac{G^2}{\rho_g}\right] \tag{5.54a}$$

where

$$\overline{M}_g \equiv \frac{1}{|x_2 - x_1|}\int_{x_1}^{x_2}(\Phi_g)^2 x^{1.8}\,dx \tag{5.54b}$$

in conjunction with the empirical parameter $(\Phi_g)^2$ in terms of X_{tt} from Fig. 5.3 and $(f_g)_{x=1}$ at $(\mathrm{Re})_{g,\,x=1}$ per Eq. (5.31d), calculated from $(f_g)_{x=1} = 0.184(\mathrm{Re})_{g,\,x=1}^{-0.2}$, or obtained from the smooth tube ($\bar{e}/D_h \cong 0$) curve in Moody's graph (Fig. 3.4).

Validity: Turbulent–turbulent regimes (per Table 5.2), smooth tubes ($\bar{e}/D_h \cong 0$), and $n = 0.2$ (5.54c)

Estimated Accuracy: At best $\pm 30 - 40\%$ (5.54d)

5.3.4 Comparison With Other Analytical Models

The prime purposes of the following numerical analysis is to compare the values of $\Delta P_{\mathrm{tp},f}$ computed on the basis of the Martinelli–Nelson fully separated flow model with those obtained from the simpler homogeneous flow model in a typical heat transfer engineering application.

NUMERICAL ANALYSIS 5.1

Practical comparison of two-phase friction pressure drop prediction methods.

Background

The 1967 edition of the *ASHRAE Handbook, Fundamentals* [5.32] includes a very detailed numerical example to illustrate the application of the Martinelli–Nelson method for the prediction of ΔP_{tp} in direct-expansion evaporator coils. Since the acceleration component $\Delta P_{\mathrm{tp},\,a}$ in $\Delta P_{\mathrm{tp}} = \Delta P_{\mathrm{tp},\,f} + \Delta P_{\mathrm{tp},\,a}$ is calculated on the basis of a homogeneous (zero-slip) flow model per Eq. (5.9), we will ignore the acceleration component $\Delta P_{\mathrm{tp},\,a}$ and focus exclusively on the friction component $\Delta P_{\mathrm{tp},\,f}$ to compare the results obtained on the basis of the fully separated Martinelli model with those calculated from the simpler homogeneous flow models.

In the ASHRAE numerical example, $\Delta P_{\mathrm{tp},\,f}$ is calculated according to the preferred nuclear application methodology using the small *all liquid* single-

TABLE 5.3. Summary of Key Data [5.32, Example 3]

Channel Configuration

$\frac{5}{8}$ in. OD tubing, internal diameter $D = 0.0455$ ft, $\Omega = 0.00162$ ft^2, circuit length $L = 22$ ft, and $L/D = 483.5$.

Specified Design Conditions

Total flow rate $\dot{m} = 7.8$ lb/min; $x_1 = 0.235$, $x_2 = 1.0$, $x_{avg} = \frac{1}{2}(x_1 + x_2) = 0.618$ at 40°F saturation temperature with refrigerant R-12 and corresponding viscosities $\mu_l = 5.95 \times 10^6$, $\mu_g = 2.48 \times 10^{-7}$ slug/ft · s and a dimensionless ratio $\mu_l/\mu_g = 24.0$, densities $\rho_l = 86.4$, $\rho_g = 1.245$ in lb/ft^3 and a dimensionless ratio $\rho_g/\rho_l = 0.01500$.

The constant viscosity and density ratios are used to compute X_{tt} according to Eq. (5.43) which yields

$$X_{tt} = (0.0150)^{0.5}(24)^{0.1}\left(\frac{1-x}{x}\right)^{0.9} = 0.1682\left(\frac{1}{x} - 1\right)^{0.9} \tag{5.55}$$

The assumption of *tt* flow regimes is amply verified at the inlet quality $x_1 = 0.235$. However, at $x \to 1.0$ (and a fortiori $x_2 > 1.0$ with superheated vapor at evaporator outlet), the $(\mathrm{Re})_{lo} \to 0$ so that in the dryer end section of the evaporator the flow regime is actually of the *vt* rather than the *tt* type. This point is ignored along with other "simplifications" made to contend with the complexities of the Martinelli fully separated flow model [5.32, Example 3].

All Liquid and Vapor Pressure Drop

The $(\mathrm{Re})_{l,x=0} = (\dot{m}/\Omega)D/\mu_l = 19{,}000$ yields an all liquid friction factor $f_{l,x=0} = 0.184(19000)^{-0.2} = 0.026$ and corresponding all liquid $(\Delta P)_{l,x=0} = 14.6$ lb/ft^2 $= 0.101$ psi.

Similarly, the $(\mathrm{Re})_{g,x=1} = (\dot{m}/\Omega)D/\mu_g = 461{,}700$ yields an all vapor friction factor $f_{g,x=1} = 0.0136$ and $(\Delta P)_{g,x=1} = 515.5$ lb/ft^2 $= \underline{3.58 \text{ psi}}$.

Two-Phase Pressure Drop

At $x_{avg} = 0.618$ Eq. (5.55) yields $(X_{tt})_{avg} = 0.109$ and a value $(\phi_l)_{avg} = 18$ from Fig. 5.2 for the *tt* regime. The approximate all liquid multiplier computed at $x_{avg} = 0.618$ is $(\overline{M}_l)_{avg} = (\phi_l)^2_{avg}(1 - x_{avg})^{1.8} = 57.4$. Therefore, $\Delta P_{tp,f} = (\overline{M}_l)_{avg}.(\Delta P)_{l,x=0} = 57.4 \times 0.101 = \underline{5.8 \text{ psi}}$

phase pressure drop as a reference, rather than the *all vapor* limiting case as summarized in Table 5.3.

Problem Statement

With the help of the key data summarized in Table 5.3:

(a) Verify that the same numerical value is obtained for $\Delta P_{tp,f}$ computed on the basis of Eq. (5.54a) with $(\overline{M}_g)_{avg} = (\phi_g)^2_{avg}(x_{avg})^{1.8}$.

(b) Compute $\overline{M}_g$ on the basis of the more accurate integral method per Eq. (5.54b) and compare the results with those obtained under (a).

(c) Compare the above results with those obtained on the basis of the simpler homogeneous flow model discussed previously as well the Wallis method recommended in Ref. 5.11.

TABLE 5.4. Calculation of Multiplier $\overline{M}_g$ per Eq. (5.54b)[a]

x	δx_i	x_i	$X_{tt,i}$	$(\phi_g)_i^2$	$\frac{\delta x_i}{x_2 - x_1}(\Phi_g)_i^2(x_i)^{1.8}$
$x_1 = 0.235$					
	0.065	0.2675	0.416	8.2	0.063
0.30					
	0.10	0.35	0.294	6.7	0.132
0.40					
	0.10	0.45	0.201	5.0	0.155
0.50					
	0.10	0.55	0.140	4.1	0.183
0.60					
	0.10	0.65	(0.096)	(3.35)	(0.200)
0.70					
	0.10	0.75	(0.063)	(2.8)	(0.218)
0.80					
	0.10	0.85	(0.035)	(2.25)	(0.220)
0.90					
	0.10	0.95	(0.0119)	(1.65)	(0.197)
$x_2 = 1.00$					

[a] Σ (terms in last column) $\equiv \overline{M}_g = [1/(x_2 - x_1)] \int_{x_1}^{x_2} (\Phi_g)_{tt}^2 x^{1.8}\, dx = 1.368 \cong 1.37$. Less accurate values (in brackets) were obtained from the extrapolated dotted curve $(\Phi_l)_{tt}$ in Fig. 5.2. These more uncertain values account for 61% of $\overline{M}_g$!

Solution

(a) At $(X_{tt})_{\text{avg}} = 0.109$, $(\Phi_l)_{\text{avg}} = 18$, and $x_{\text{avg}} = 0.618$, Eqs. (5.48a), (5.48b) and (5.54a), yield $(\Phi_g)_{\text{avg}}^2 = [(0.109)(18)]^2 = 3.85$, $(\overline{M}_g)_{\text{avg}} = (\Phi_g)_{\text{avg}}^2 (x_{\text{avg}})^{1.8} = 1.619$, and $\Delta P_{\text{tp},f} = \overline{M}_{\text{avg}} (\Delta P)_{g,\,x=1} = 1.619 \times 3.58 = 5.8$. Thus, the value of $\Delta P_{\text{tp},f}$ obtained on the basis of the all gas single-phase pressure drop per Eq. (5.54a) agrees exactly (as it should) with the alternative all liquid flow methodology.

(b) The numerical values of the parameter X_{tt} can be computed as a function of the vapor flowing quality x_i from Eq. (5.43a) to read $(\Phi_g)_i^2$ from Fig. 5.3 and calculate $\overline{M}_g$ more exactly according to Eq. (5.54b) as shown in Table 5.4. This yields a numerical value $\overline{M}_g = 1.37$ which is about 15% lower than $(\overline{M}_g)_{\text{avg}}$ obtained under (a) and this would reduce $\Delta P_{\text{tp},f}$ by the same percentage. However, in view of the typical ± 30–40% uncertainty of the Martinelli calculation method the simpler approximate solution based on x_{avg} is adequate.

(c) With $(\Delta P_{\text{tp},f})_{\text{Martinelli}}$ and $(\Delta P_{\text{tp},f})_{\text{homogeneous}}$ per Eqs. (5.54a)–(5.54c) and (5.21a)–(5.21g), respectively, the following ratio is obtained:

$$\frac{(\Delta P_{\text{tp},f})_{\text{Martinelli}}}{(\Delta P_{\text{tp},f})_{\text{homogeneous}}} = \frac{\overline{M}_g f_{g,\,x=1}}{\bar{x} \bar{f}_{\text{tp}}} \tag{5.56}$$

For the particular case under consideration with $\overline{M}_g = 1.37$ and $f_{g,\,x=1} = 0.0136$, the product $\overline{M}_g f_{g,\,x=1} = 0.0186$. For the homogeneous flow model $\bar{f}_{\text{tp}}$ = constant = 0.045 ± 13% for hydraulic diameters D_h in the range $1 \leq D_h < 15$ mm [3.4, 3.32] and with $\bar{x} = \frac{1}{2}(x_1 + x_2) = 0.618$ the product $\bar{x}\bar{f}_{\text{tp}} = \bar{f}_{\text{tp, effective}} = 0.0278 \pm 13\%$; thus, the following numerical value is obtained from Eq. (5.56):

$$\frac{\left(\Delta P_{\text{tp},f}\right)_{\text{Martinelli}}}{\left(\Delta P_{\text{tp},f}\right)_{\text{homogeneous}}} = \frac{0.0186}{0.0278} = 0.67 = 67\%$$

In this particular application the Martinelli–Nelson method would seem to *underestimate* the actual $\Delta P_{\text{tp},f}$ by about 20–33%, allowing for the ±13% estimated accuracy of $\bar{f}_{\text{tp}}$ for the homogeneous flow model. This deviation, which is within the estimated ±30–40% deviation per Eq. (5.54d), is probably due to the basic assumption of a fully smooth friction factor for $f_{g,\,x=1}$ and various "model simplifications" in the original Martinelli–Nelson method plus a good deal of extrapolation beyond the Martinelli and associates experimental data at $\overline{X}_{tt} \ll 0.1$ (see footnote to Table 5.4).

The above assessment is reinforced by ASHRAE's Handbook more recent recommendation [5.11] to estimate the effective two-phase flow friction factor according to the Wallis method (see Section 5.2.3 in this chapter) from

$$\left(\bar{f}_{\text{tp}}\right)_{\text{eff}} = \overline{M}_g f_g \tag{5.57a}$$

with a typical value

$$f_g = 0.02 = \text{constant} \tag{5.57b}$$

With $\overline{M}_g = 1.37$ and $f_g = 0.02$ Eq. (5.57a) yields

$$\left(\bar{f}_{\text{tp}}\right)_{\text{eff}} = 0.0274$$

which is in excellent agreement with the numerical value $\bar{x}(\bar{f}_{\text{tp}}) = \bar{f}_{\text{tp, eff}} = 0.0278$ based on the simpler homogeneous flow model.

It is noteworthy that even the naive all vapor flow model with $f_g = 0.025 \pm 30\%$ and $\bar{x} = 1$ or $(\bar{f}_{\text{tp}})_{\text{eff}} = 0.025 \pm 30\%$ falls within the normal uncertainty band of recognized two-phase flow pressure drop predictions!

Comments

1. This numerical analysis confirms the advantages (in typical direct-expansion evaporators and many general process two-phase heat exchanger applications) of the simpler homogeneous flow model discussed in the previous section, as well as the estimated deviations of the original Martinelli method of computing $\Delta P_{\text{tp},f}$ according to Eqs. (5.28), (5.49) and (5.54).

2. The significant uncertainties associated with all recognized two-phase flow correlations[4] explain the renewed interest in the thermodynamic generalization methods [5.31] addressed in Chapter 6.

NOTES

1. This way, the important impact of *traces of oil* (such as greater wettability and promotion of fully wet wall rather than the less effective dry wall flow regimes) is still fully considered.
2. For the *homogeneous* or *fog* flow model ($V_l = V_g$), which can be visualized as the limiting case of partially separated two-phase flows when the entrained liquid flow fraction D approaches unity, it is logical to speak of a *thermodynamic mixture* consisting of a gas phase with finely dispersed entrained liquid droplets or fog (see also *continuum hypothesis* in Chapter 7).
3. Actually, $\Delta x > 0$ and Eq. (5.12c) is not generally valid for compressible fluids when $\Delta P_{\text{tp}}/P$ is not negligibly small as in the case of the irreversible adiabatic process (throttling) taking place in the expansion valves of conventional mechanical refrigeration cycles (see Fig. 4.3). However, at the extremely low values $\Delta P_{\text{tp}}/P \ll 1$ typical of efficient two-phase flow heat exchangers, the vapor can be treated as an incompressible fluid (ρ_g = constant, $\Delta x \approx 0$) and Eq. (5.12c) is valid with an adequate level of accuracy.
4. Stepanoff proposes [5.33] an interesting alternative procedure for the prediction of $\Delta P_{\text{tp},f}$ in the case of air–liquid mixtures flowing in horizontal pipes which is based on "the Durand method of pipe friction loss calculation for heterogeneous and non deposit flow of solid water suspensions" [5.34, 5.35].

CHAPTER 6

THERMODYNAMIC GENERALIZATION OF FLUID FLOW DATA

Thermodynamic generalizations based on reduced pressures proposed in the 1960s are reviewed and updated to reflect the latest state of the art. The application of the method is illustrated by numerical analyses and an assessment is made of its value in heat transfer engineering with special emphasis on thermodynamically efficient two-phase forced-convection heat exchangers designed to operate at low temperature differentials. It is shown that this thermodynamic method provides the fluid flow systems engineer and the heat exchanger designer with an additional tool that is simple, effective, and above all more reliable, particularly in evaporator and condenser design practice, than current conventional semiempirical correlations.

6.1 LOW THERMAL LIFT HEAT EXCHANGERS

To identify the basic problem common to all thermodynamically efficient exchangers designed to operate at low fluid-to-fluid temperature differentials (see Chapter 4), we focus on normal comfort cooling and heating refrigeration systems operating at relatively low compression ratios and more specifically on the evaporator in heat pump applications.

The coefficient of performance (COP) of a real heat pump system is significantly lower than that of the ideal reference Carnot cycle operating at the same source and sink temperatures T_L and T_H, respectively:

$$(\text{COP})_C = \frac{1}{\eta_C} = \frac{T_H}{T_H - T_L} = \frac{1}{1 - T_L/T_H} \tag{6.1}$$

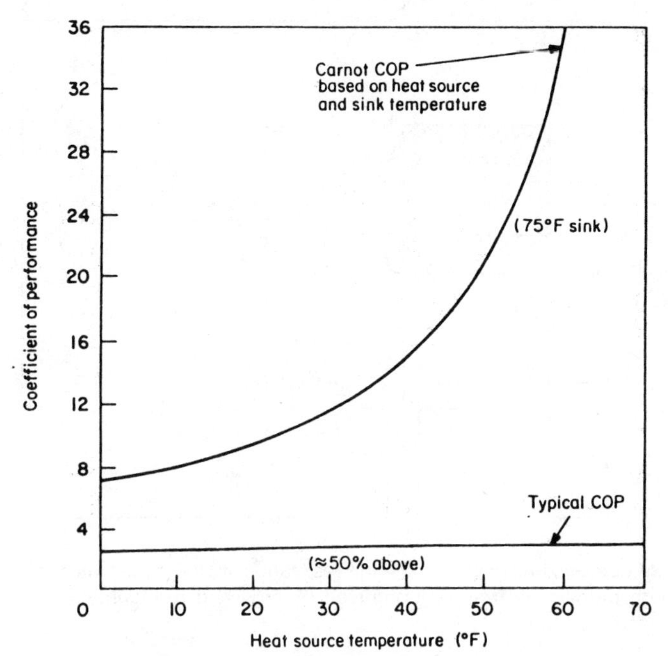

FIGURE 6.1. Variation of COP with temperature. [Adapted with permission from 1983 ASHRAE Equipment Handbook, Chap. 44, Fig. 6, ASHRAE, Atlanta.]

where $\eta_C = 1 - T_L/T_H$ is the Carnot power cycle efficiency. For example, at the standard Air-Conditioning and Refrigeration Institute (ARI) rating point for heating 47°F (8.3°C) outdoor and 70°F (21.1°C) indoor, the ideal $(\text{COP})_C$ computed from Eq. (6.1) is *23*, whereas the actual COP of air-to-air heat pumps ranges from *1.5 to 3.0* according to data published by ARI [6.1, 6.2] for several hundred models having nominal ratings in the cooling mode of 15,500–135,000 Btu/h (4.5–70 kW). An actual COP range of 1.5–3.0 thus corresponds to a *thermodynamic efficiency* = $\text{COP}/(\text{COP})_C$ of only about *6.5–13%*. This means that an ideal Carnot reverse power cycle would yield roughly 7.5–15 times more useful heating capacity than typical ARI air-to-air heat pumps for the same power consumption at the standard rating point!

This rather unsatisfactory performance becomes even worse as the temperature lift $(T_H - T_L)$ is reduced or, more precisely, when $T_L/T_H \rightarrow 1$ and $\eta_C \rightarrow 0$ with a corresponding $(\text{COP})_C \rightarrow \infty$ according to Eq. (6.1) as illustrated at a 75°F (23.9°C) sink temperature in Fig. 6.1. It is clear from Fig. 6.1 that the actual COP increases only slightly with an increasing heat source temperature, whereas the $(\text{COP})_C \rightarrow \infty$, so that when the source temperature approaches the sink temperature, as in mild heating weather, the heat pump thermodynamic efficiency converges toward zero! This undesirable characteristic has a significant negative impact on the all important *seasonal COP* based

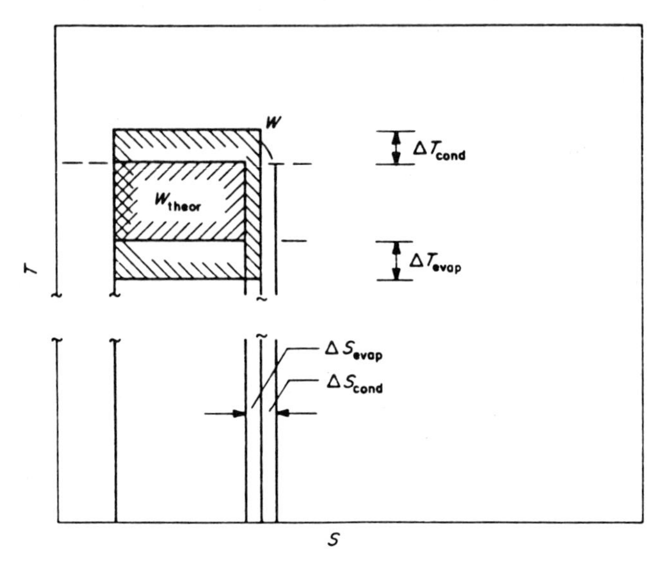

FIGURE 6.2. Losses due to heat transfer resistance in the evaporator and condenser. [Adapted with permission from G. Lorentzen, Energy and Refrigeration, *Int. J. Refrig.* **6** (5/6) 262–273 (1983), Fig. 11.]

on an annual season of operation rather than a single more or less representative standard operating point.

In a recent government-funded study carried out in Switzerland by independent consulting engineers, the seasonal COP of 11 electric-driven heat pumps was accurately measured [6.3]. The results were quite sobering to the energy-conscious Swiss with actual *yearly COP* ranging from a minimum of *1.87* to a maximum of *2.62* against the typical values of 3–4 advertised by local heat pump promoters. The report concludes with the following cautious assessment: "A yearly COP of *up to 3* can be attained *provided* the equipment manufacturer and the Architect-Engineers eliminate current system weaknesses."

An analysis of all irreversible losses in actual single stage low compression ratio refrigeration cycles reveals that the *two main heat exchangers*—the *evaporator* and the *condenser*—account for the *major share of the total system's irreversible losses* at low temperature lifts [4.17, 6.4] as emphasized in Lorentzen's keynote address at the 16th IIR Congress (1983) in Paris. It is evident from Fig. 6.2 that as the outside source temperature approaches the internal sink (room) temperature the theoretical work of compression $W_{theor} \to 0$, whereas the actual work W remains finite and is ultimately due almost entirely to the irreversible losses in the evaporator and condenser, that is, $T_H \Delta s_{evap}$ and $T_H \Delta s_{cond}$, respectively. This means that the ability to predict accurately the thermohydrodynamic performance of evaporators and condensers is the central problem of heat pump optimization studies and this in turn leads to the key question addressed in this chapter: *How can one predict with an adequate level of accuracy the performance of two-phase flow heat exchangers?*

In the light of what was said in Chapter 4, it is evident that the above conclusion is *not* restricted to refrigeration/heat pump systems but applies as well to all heat exchangers designed to operate at small fluid-to-fluid temperature differentials as in conventional steam power plant condensers and a fortiori (in view of their extremely low temperature "lift") to futuristic ocean thermal energy conversion (OTEC) systems.

6.2 TWO-PHASE HEAT EXCHANGER PERFORMANCE PREDICTIONS

For the sake of expediency we consider here the type of two-phase heat exchanger described in Chapter 4, namely, direct-expansion evaporators operating with incomplete evaporation at a vapor quality $x_2 < 1.0$, since they are thermodynamically more efficient (lower average wall to saturated refrigerant temperature differential, $T_w - T_{sat}$) and/or more compact owing to the higher heat transfer coefficients attainable in fully wet wall flow regimes [3.32, 4.4–4.6, 4.18, 4.21, 4.67, 6.5, 6.6] than when the vapor is superheated ($x_2 > 1$) within the evaporator. To ensure dry superheated vapor at the compressor intake, the refrigeration system must be equipped with a *postevaporator* heat exchanger as illustrated in Fig. 4.5, with evaporator exit vapor qualities of the order of 0.95–0.98 < 1.0 for the particular types of compact internally finned channels employed in these package chillers. A somewhat lower maximum exit quality ($x_2 \leq 0.9$) is required with bare circular tube evaporators to avoid the significantly less effective dry wall regimes [5.9].

The optimum thermohydrodynamic design of a direct-expansion evaporator is the result of a compromise between two conflicting goals: On the one hand, it is advantageous to select a high mass velocity G since the *heat transfer coefficient* $\bar{\alpha}_{tp}$ *increases* in direct proportion to G with incomplete evaporation according to Bo Pierre's widely accepted correlations for circular tubes [4.23, 5.9], a result also verified for small hydraulic diameter internally finned tubes [3.32, 4.4].

On the other hand, the two-phase pressure drop ΔP_{tp} increases roughly with the second power of G per Eqs. (5.9) and (5.21a)

$$\Delta P_{tp} \propto \Delta P_{tp,f} \propto \bar{f}_{tp} \left(\frac{L}{D_h} \right) G^2 v_g \bar{x} \propto \bar{f}_{tp} \dot{m}^2 \tag{6.2}$$

for a prescribed hydraulic D_h, $\bar{e}/D_h$, wall material channel aspect ratio (L/D_h), fixed exit pressure P_2, x_1, $x_2 = x_1 + \Delta x$, and mean mixture specific volume $\bar{v} = \bar{x}(v_g)$ since the average two-phase friction factor $\bar{f}_{tp}$ is substantially independent of G in typical wet wall (spray-annular) flow regimes with $x_2 < x_{cr\ dry\ wall} < 1$. Because ΔP_{tp} is associated with a corresponding saturation temperature drop ΔT_{sat}, the inlet saturation temperature $(T_{sat})_1 = (T_{sat})_2 + \Delta T_{sat}$ increases with increasing G and the effective *mean wall to refrigerant temperature difference* $\bar{T}_w - \bar{T}_{sat}$—the mean heat transfer driving force—*decreases*. The lower the maximum $(T_w - T_{sat})_{max}$ at zero pressure drop (i.e., the more thermodynamically efficient the heat exchanger), the more significant the impact of the saturation temperature drop penalty ΔT_{sat}. Thus, the optimum evaporator design is ultimately the result of a trade-off between higher heat transfer coefficients $\bar{\alpha}_{tp}$ at lower $\bar{T}_w - \bar{T}_{sat}$ versus lower $\bar{\alpha}_{tp}$ at higher $\bar{T}_w - \bar{T}_{sat}$.

Consequently, the ability to *predict* ΔP_{tp} *accurately* and to determine with adequate precision the "coupled" ΔT_{sat}, and therefore $\bar{T}_w - \bar{T}_{sat}$, is an *essential* element of any evaporator or condenser design optimization study, particularly in the case of thermodynamically efficient heat exchangers designed for small $\bar{T}_w - \bar{T}_{sat}$ *regardless* of the type of *application* (see also Guideline 9).

This brings us face to face with the following basic dilemma: the best current *semianalytical correlations* proposed to compute the two-phase pressure drop ΔP_{tp} are generally *not adequately accurate* because of the complexity of two-phase flow phenomena, for such rational optimization studies [6.5, 6.6] as shown in Chapters 4 and 5.

The designer is therefore compelled to rely almost exclusively on *ad hoc empirical test data* since he or she cannot afford to optimize the two-phase heat exchanger design on the basis of theoretical calculations for ΔP_{tp} with *deviations* typically over ± 30–40% according to Eq. (5.28). Bear in mind that the deviations of the coupled two-phase flow convection heat transfer coefficient α_{tp} are at best equal to the above estimated values, as previously shown for the single-phase limit of two-phase flow ($y = 1$, $x = 0$ and at $x \geq 1$), a conclusion that is also valid when $0 < \bar{x} < 1$ (as discussed in later chapters).

Fortunately, the *thermodynamic method* of generalization initially proposed for *pool boiling* in the 1960s by researchers[1] of the Leningrad Institute of Physics [4.63, 4.64, 4.66], reassessed and confirmed in the 1980s [4.56–4.58], provides a simple *tool to predict quite accurately* ΔP_{tp} *and the coupled heat transfer phenomena*, for any working fluid and heat exchanger configuration when reliable empirical or theoretical results are available for one reference fluid over a sufficiently wide range of reduced pressures $P^+ \equiv P/P_{cr}$, where P_{cr} is the critical pressure at the characteristic thermodynamic critical point of the particular pure substance under consideration.

It is important to recognize that the thermodynamic generalization method reviewed in the next section is *applicable* to *any* two-phase flow *heat exchangers including condensers regardless* of type of application—evaporation or condensation—and heat exchanger *configuration*. The rationale for this rather sweeping statement is that classical *thermodynamic phenomenological laws* do not depend on a particular physical model and are inherently *independent* of *spatial dimensions* under steady-state steady-flow conditions tacitly assumed in thermal systems component design analyses.

6.3 EXTENDED CORRESPONDING STATES PRINCIPLE GENERALIZATIONS

The scope and prime objectives of the extended corresponding states principle (ECSP) method of generalization[2] can be stated succinctly by quoting the original abstract [4.21]:

> The principle of corresponding states and the extension of the principle to transport properties are briefly reviewed with special emphasis on similar fluids such as halogenated refrigerants, as well as two dissimilar polar substances, water and ammonia. The validity of the theory of thermodynamic similitude [ECSP] is checked from the design engineer's standpoint in a practical and straightforward manner by plotting against reduced pressure published thermo-physical properties, particularly those entering the equations of heat, mass, and momentum transfer. In this paper, the theory of thermodynamic similitude is applied specifically to pressure drop and forced convection heat transfer calculations for single-component, single- and two-phase flow.
>
> The major advantage of the method for the design engineer is that it makes it possible to extend empirical data obtained with one substance in an extremely simple manner to other similar fluids, such as halogenated refrigerants, without additional testing.
>
> There is little doubt that the theory of thermodynamic similitude [ECSP] will find many useful applications since it is, in effect, a *logical extension of nondimensional correlations* used in the field of momentum, heat, and mass transfer. [See Chapter 3 of this volume.]

The validity of the seemingly too simple generalization methods notwithstanding the inherent complexity of two-phase flow phenomena emphasized earlier, becomes more understandable when it is recognized that at identical P^+, G, x_1, $\Delta x = x_2 - x_1$, channel configuration, and wall material, *identical* flow *patterns* or *dissipative flow regimes* will occur for similar fluids—as shown [4.21, 4.67] in the 1960s. The latter paper had the following main objectives:

> It is the purpose of this paper to review the design criteria most widely used in establishing flow patterns as well as those frequently included in pressure drop and heat transfer correlations, and to show how these criteria can be generalized on the basis of thermodynamic similitude [ECSP]. Typical generalized design charts are presented in terms of reduced pressure and flowing mass fraction x for the most widely used thermodynamically similar halogenated refrigerants.

The same basic approach proposed by Borishansky et al. [4.63] and Danilova [4.66] to scale the effect of pressure in nucleate boiling was adopted [4.21, 4.67] for forced-convection tube flow with one significant difference: Instead of limiting its application to specific explicit semiempirical correlations, the thermodynamic dimensionless generalization method was applied in a *less restrictive fashion* to predict the thermohydrodynamic performance of a fluid on the *basis of reduced pressures* from any valid empirical data obtained with another *reference fluid* at the same mass velocity G and same vapor flowing qualities x_1 and $\Delta x = x_2 - x_1$ for any given channel geometry: D_h, L/D_h, wall material, and relative roughness $\bar{e}/D_h$.

The point to stress is that it is *not* necessary with the proposed generalization method [4.21, 4.67] to start from a particular explicit empirical correlation for f_{tp} (or α_{tp}) with its unavoidable validity/accuracy limits but on the contrary *any valid empirical raw data* for one reference fluid for *any* given geometry can be used, even in the absence of accepted correlations. As emphasized in the concluding discussion [4.67], "... this is in fact what makes the thermodynamic similitude such a powerful yet simple tool, particularly in the case of two-phase flow where generally valid correlations are rather scarce." The latter (under)statement is still relevant today [4.23, 4.56–4.61, 5.3, 6.5–6.8] in spite of the major research effort worldwide in two-phase flow in the light of the rudimentary survey of the current state of the art carried out in Chapter 4.

6.3.1 Terminology and Basic Relationships

Let π_i denote any single fluid property, or group of such properties, including transport properties such as $v = 1/\rho$, h, s, μ, k, and $\text{Pr} = c_p\mu/k$, where the subscript i identifies the particular property, or group of properties, under consideration.

A second subscript, either g or l, is used as previously to make a distinction between the gas (or vapor) and the liquid phase.

Thus, the pair of like properties, or group of properties, $(\pi_i)_g$ or $(\pi_i)_l$, and their ratio $(\pi_i)_l/(\pi_i)_g \equiv \pi_i R$ for short, may represent v_g, v_l, and v_l/v_g, respectively, when $\pi_i = v$, or μ_g, μ_l, and μ_l/μ_g, when $\pi_i = \mu$, or $(\text{Pr})_g$, $(\text{Pr})_l$, and $(\text{Pr})_l/(\text{Pr})_g$, when $\pi_i \equiv \text{Pr}$, and so forth.

Within the *existence domain* of the *liquid phase*, that is, from P_t (or T_t) to P_{cr} (or T_{cr}), defining the triple and thermodynamic critical points, respectively, *any pair* of fluid property terms $(\pi_i)_g$, $(\pi_i)_l$, and $\pi_i R \equiv (\pi_i)_l/(\pi_i)_g$ are *uniquely defined* for all pure substances in terms of P (or T) within the two-phase domain of existence limits $P_t < P < P_{cr}$ (or $T_t < T < T_{cr}$) which can be most generally defined in terms of dimensionless reduced pressures as $P_t^+ < P^+ < 1$ (or reduced temperature $T_t^+ < T^+ < 1$).

According to the ECSP, any property π_i, $(\pi_i)_g$, and $(\pi_i)_l$ having a *finite value* at the thermodynamic critical point, which is most generally characterized by the values of P_{cr}, T_{cr} and the compressibility factor Z_{cr} of a given pure substance [4.21], can be expressed as a *reduced*, nondimensional *property* $(\pi_i)^+$, $(\pi_i)_g^+$, and $(\pi_i)_l^+$ according to the following definitions:

$$\pi_i^+ \equiv \frac{\pi_i}{(\pi_i)_{cr}} = \text{reduced property term } \pi_i \tag{6.3a}$$

$$(\pi_i)_g^+ \equiv \frac{(\pi_i)_g}{(\pi_i)_{cr}} = \text{reduced property term } \pi_i \text{ of saturated vapor } (x = 1) \tag{6.3b}$$

$$(\pi_i)_l^+ \equiv \frac{(\pi_i)_l}{(\pi_i)_{cr}} = \text{reduced property term } \pi_i \text{ of saturated liquid } (x = 0) \tag{6.3c}$$

where $(\pi_i)_{cr}$ is the value of the property term π_i at the critical point

Validity: $P_t < P^+ < 1$ (or $T_t < T^+ < 1$), the two-phase liquid vapor phase existence domain of any pure substance (6.3d)

Since the gas and liquid phase ultimately coincide at the critical point, all reduced properties $(\pi_i)_g^+$ and $(\pi_i)_l^+$ must converge toward unity as $P^+ \to 1$; thus the following identity is always satisfied:

$$(\pi_i)_g^+ = (\pi_i)_l^+ = 1 \quad \text{at } P^+ = P/P_{cr} = 1 \text{ and } T^+ = T/T_{cr} = 1 \tag{6.4}$$

It is evident that any property *ratio* $\pi_i R \equiv (\pi_i)_l/(\pi_i)_g$ is identical to the reduce ratio $(\pi_i)_l^+/(\pi_i)_g^+$, since the constant common term $(\pi_i)_{cr}$ cancels out; therefore,

$$\pi_i R \equiv \frac{(\pi_i)_l}{(\pi_i)_g} \equiv \frac{(\pi_i)_l^+}{(\pi_i)_g^+} \equiv \Psi_{\pi_i R}(P^+) \tag{6.5a}$$

where $\Psi_{\pi_i R}(P^+)$ is a *universal* function of the property ratio $(\pi_i)_l/(\pi_i)_g$ in terms of P^+.

Validity: Thermodynamically "perfectly similar" fluids (6.5b)

Since the ECSP generalization method is more accurate[3] and easier to apply in the limiting case of saturated vapors and because the two-phase pressure drop ΔP_{tp} is more conveniently calculated on the basis of the limiting case of all vapor flow as shown in the previous chapter, we first focus on single-phase saturated vapor generalizations.

6.3.2 Single-Phase Vapor Flow Generalizations

From the definition $(\pi_i)_g^+$ per Eq. (6.3b), it is always possible to replace any property term $(\pi_i)_g$ appearing in fluid and heat flow correlations by its value at the critical point multiplied by a universal function $\Psi_{i,g}(P^+)$ of the reduced pressure according to

$$(\pi_i)_g = (\pi_i)_{cr} \Psi_{i,g}(P^+) \tag{6.6a}$$

where, according to the ECSP, $\Psi_{i,g}(P^+)$ is a *universal* function of $(\pi_i)_g^+$ in terms of P^+.

Validity: $P_t^+ < P^+ < 1$ (6.6b)

Deviations: Generally *estimated* to be *less* than 5% for thermodynamically similar fluids and 10–15% for dissimilar fluids. These deviations are quantified more accurately in Numerical Analyses 6.1–6.4. (6.6c)

For our present purpose[4] we follow the same pragmatic approach adopted in Refs. 4.21 and 4.67 and rely entirely on the most recent and accurate (1985) published thermophysical data to test the accuracy of Eq. (6.6a) rather than any theory or model. In other words, we accept the ECSP as a universally valid *phenomenological* law and *assess quantitatively the impact of actual deviations* from this general law on the *end engineering objective* of flow analyses expressed by Guideline 6. The basic method of generalizing empirical or theoretical fluid and heat flow data obtained with a known *reference saturated vapor* to another saturated vapor is outlined below.

With $(\pi_{i,\text{ref}})_g$ denoting a relevant fluid property of an arbitrary reference fluid, we simply compute at the same reduced pressure P^+ the ratio

$$\frac{(\pi_i)_g}{(\pi_{i,\text{ref}})_g} = \frac{(\pi_i)_g^+ (\pi_i)_{cr}}{(\pi_{i,\text{ref}})_g^+ (\pi_{i,\text{ref}})_{cr}} \tag{6.7}$$

Since $(\pi_i)_g^+ = (\pi_{i,\text{ref}})_g^+$ at $P^+ = P_{\text{ref}}^+$ per Eq. (6.6a), the above equation reduces to

$$\frac{(\pi_i)_g}{(\pi_{i,\text{ref}})_g} = \frac{(\pi_i)_{cr}}{(\pi_{i,\text{ref}})_{cr}} \tag{6.8}$$

Momentum and heat and mass transfer correlations being usually expressed in the form of *power laws*, the individual (or group of) property terms π_i most generally appear at some power a_i or as $(\pi_i)^{a_i}$, and Eq. (6.8) can be extended to a more general and useful form to quantify the impact of relevant fluid property terms π_i:

$$\left(\frac{(\pi_i)_g}{(\pi_{i,\text{ref}})_g}\right)^{a_i} = \left(\frac{(\pi_i)_{cr}}{(\pi_{i,\text{ref}})_{cr}}\right)^{a_i} \tag{6.9}$$

where a_i is the empirical exponent of the π_i term in momentum, heat, and mass transfer correlations.

6.3.3 Extension to Saturated Liquids and Single-Component Two-Phase Fluid Flow

Following the same reasoning for the *liquid only* limiting case ($x = 0$) as for the saturated vapor phase ($x = 1$) in Section 6.3.2 yields the same basic relationships except that the suffix g must be replaced by suffix l: that is,

$$(\pi_i)_l = (\pi_i)_{cr} \Psi_{i,l}(P^+) \tag{6.10}$$

where $\Psi_{i,l}$ is a *universal* function of $(\pi_i)_l^+$ in terms of P^+. We proceed in a similar manner to quantify the impact of a particular fluid property, or group of properties, $(\pi_i)_l$:

$$\frac{(\pi_i)_l}{(\pi_{i,\mathrm{ref}})_l} = \frac{(\pi_i)_{cr}}{(\pi_{i,\mathrm{ref}})_{cr}} \tag{6.11a}$$

$$\textit{Validity}: \quad P^+ = P^+_{\mathrm{ref}} \tag{6.11b}$$

$$\textit{Deviation}: \quad \text{Same as Eq. (6.6c)} \tag{6.11c}$$

Note, however, that some of the liquid properties of thermodynamically "dissimilar" substances may deviate more significantly from the ideal ECSP identities than saturated vapors [4.21, 4.67]. Consequently, it is convenient in the case of liquids and particularly in two-phase heat exchanger design applications to generalize on the basis of the easier all vapor case ($x = 1$) along the lines highlighted below.

Any liquid property term $(\pi_i)_l$ can be rewritten, after multiplying and dividing by $(\pi_i)_g$, and recalling that $(\pi_i)_g = (\pi_i)_g^+(\pi_i)_{cr}$, as

$$(\pi_i)_l = \frac{(\pi_i)_l}{(\pi_i)_g}(\pi_i)_g^+(\pi_i)_{cr} = (\pi_i R)(\pi_i)_{cr}(\pi_i)_g^+ \tag{6.12}$$

Thus, the ratio $[(\pi_i)_l/(\pi_{i,\mathrm{ref}})_l]^{a_i}$ reduces with Eqs. (6.12) and (6.6) to

$$\left(\frac{(\pi_i)_l}{(\pi_{i,\mathrm{ref}})_l}\right)^{a_i} = \left(\frac{(\pi_i R)}{(\pi_i R)_{\mathrm{ref}}}\right)^{a_i}\left(\frac{(\pi_i)_{cr}}{(\pi_{i,\mathrm{ref}})_{cr}}\right)^{a_i} \tag{6.13}$$

Equation (6.13) converges, as it should, to the same expression as Eq. (6.9) in the case of similar fluids obeying the ECSP *exactly* when $\pi_i R = (\pi_i R)_{\mathrm{ref}}$. The usefulness[5] of Eq. (6.13) becomes more evident when some of the liquid property terms $(\pi_i)_l$ of dissimilar fluids that deviate more significantly from the exact ECSP identities are considered. The main advantage of Eq. (6.13) is that it makes it possible to quantify the impact of such deviations in a rather simple manner. For instance, when $(\pi_i)_l$ represents the liquid phase viscosity, Eq. (6.13) becomes

$$\left(\frac{\mu_l}{\mu_{l,\text{ref}}}\right)^{a_i} = \left(\frac{\mu_l/\mu_g}{(\mu_l/\mu_g)_{\text{ref}}}\right)^{a_i}\left(\frac{\mu_{\text{cr}}}{\mu_{\text{cr, ref}}}\right)^{a_i} \tag{6.14}$$

Thus, it is possible to assess the impact of $[(\mu_l/\mu_g)]^{a_i} \neq [(\mu_l/\mu_g)_{\text{ref}}]^{a_i}$ on the friction factor when $a_i \equiv n =$ Blasius exponent of Eq. (3.44) or on the key Martinelli–Nelson parameter X_{tt} where the viscosity ratio appears at a power $a_i \equiv n/2$ as shown more concretely in Numerical Analyses 6.1–6.4.

6.4 APPLICATION AND PRACTICAL ACCURACY OF THE ECSP METHOD

To illustrate the *simplicity*, broad *validity*, and remarkable *accuracy* of the ECSP generalization method, we focus here on one concrete practical engineering application: the substitution of the relatively new refrigerant R-502 for R-22. Since its introduction in the 1960s, R-502 has become increasingly popular as a substitute for R-22, particularly in low-temperature refrigeration and heat pump applications, mainly because of the significantly lower (reciprocating) compressor operating temperatures and resulting improved system life expectancy [6.9].

NUMERICAL ANALYSIS 6.1

Generalization of two-phase pressure drop predictions and corresponding ΔT_{sat} penalty.

Problem Statement

(a) Using the most recent thermophysical property data for R-502, verify that the following key fluid flow property generalizations in terms of P^+ conform with ECSP predictions:

(1) The reduced specific volume of the vapor phase ($v_g^+ = v_g/v_{\text{cr}}$) and the specific volume vapor to liquid ratio ($v_g/v_l = \rho_l/\rho_g$).
(2) The molar entropy change, sometimes called the *Trouton entropy*, $Mh_{lg}/T = M(s_g - s_l) = Ms_{lg}$, where M is the *molar* mass.
(3) The vapor viscosity μ_g^+ and the viscosity ratio μ_l/μ_g.

(b) Compute ΔP_{tp} for R-502 on the basis of the ECSP method from accurate $(\Delta P_{\text{tp}})_{\text{R-22}}$ data obtained empirically with the *same* channel geometry and G, therefore identical $\dot{m}$ and same P^+, x_1, Δx, and $\bar{x}$ with the reference fluid R-22.

(c) Assess the maximum possible deviation of the above generalization method in typical turbulent two-phase flow regimes (annular spray) when $\bar{f}_{tp} \propto (\text{Re})_{x=1}^{-n} \propto \mu^n$ with Blasius exponents ranging from $n = 0.20$ for *ideal smooth* to $n = 0$ for *fully rough wet walls* annular two-phase flow regimes [3.32, 4.5].

Solution

(a) At an arbitrary saturation temperature of 35 °F (1.66 °C) we obtain from standard physical property tables (such as those presented in Chapters 16 and 17 of Ref. 4.8) a corresponding saturation pressure $P = 87.54$ psia, or with $P_{cr} = 591$ psia a reduced pressure $P^+ = 87.54/591 = 0.1481$ together with the following values for the key two-phase parameters: $v_g = 0.46955$ and $v_{cr} = 0.02857$ ft^3/lb or a reduced specific volume $v_g^+ = 0.46955/0.02857 =$ *16.44* and a ratio $v_g/v_l =$ *38.57*; $Ms_{lg} = 111.63$ $(0.1264) =$ *14.11* Btu/lb · mol · R (59.08 J/mol · K); $\mu_g = 0.0292$ lb$_m$/ft · h (12.06 Pa · s); $\mu_l/\mu_g =$ 19.03 at $P^+ = 0.1481$. Other points are obtained in the same manner at different values of P^+ for the full range of published R-502 viscosity and specific volume data up to the critical point at $P^+ = T^+ = v^+ = \mu^+ = 1$. The results, summarized in Table 6.1, show a *maximum deviation* of:

(1) substantially less than 5% for v_g^+ and $v_g/v_l \equiv \rho_l/\rho_g$;
(2) less than 1% for the molar entropy, Ms_{lg}; and
(3) somewhat less than 10% and 25% for μ_g^+ and μ_l/μ_g, respectively.

(b) From Eq. (6.2) we obtain

$$(\Delta P_{tp})_{\text{R-502}} = \frac{(v_g)_{\text{R-502}}}{(v_g)_{\text{R-22}}} (\Delta P_{tp})_{\text{R-22}} \tag{6.15}$$

since all the other terms of Eq. (6.2) are identical under the stated conditions.

Equation (6.15) can be further simplified since $v_g = (v_g)^+ v_{cr}$ and $(v_g)^+$ of R-502 $= (v_g)^+$ of R-22 at the same P^+ with an error of $< 5\%$ so that Eq. (6.15) becomes, with $(v_{cr})_{\text{R-502}} = 0.02857$, $(v_{cr})_{\text{R-22}} = 0.030525$ and $(v_{cr})_{\text{R-502}}/(v_{cr})_{\text{R-22}} = 0.02857/0.030525 = 0.936$: $(\Delta P_{tp})_{\text{R-502}} = 0.936(\Delta P_{tp})_{\text{R-22}}$, or, more generally,

$$\Delta P_{tp} = (F_{\Delta P})(\Delta P_{tp})_{ref} \tag{6.16a}$$

where the *constant correction factor* $F_{\Delta P}$ is fully defined by the two fluids' specific volumes at the critical point according to

$$F_{\Delta P} \equiv \frac{v_{cr}}{(v_{cr})_{ref}} \tag{6.16b}$$

TABLE 6.1. Actual Percent Deviations of ECSP Generalizations for R-502 Using R-22 as the Reference Fluid

			1	2	3	4	5	6	7
(a)	R-502 saturation temperature	°F °C	−40 −40	0 −17.8	35 1.67	40 4.4	80 26.7	179.9 82.2	
(b)	Reduced pressure P^+	—	0.03181	0.07461	0.1481	0.16113	0.29767	1.0	
(c)	R-22 saturation temperature at same P^+	°F °C	−23.6 −30.9	18.3 −7.6	54.8 12.7	60.0 15.6	101.5 38.6	204.8 96.0	
	Actual % deviations[a] from ECSP identity at the reduced pressure P^+ line (b) are shown for the key parameters indicated in lines (d)–(j)								Average of columns 1–5 %
(d)	Molar entropy Ms_{lg}	%	0.4	0.5	0.34	0.3	−0.3	0	+0.2
(e)	Reduced specific volume $(v)_g^+$	%	−1.8	−2.5	−2.4	−2.4	−2.3	0	−2.3
(f)	Ratio $v_g/v_l = \rho_l/\rho_g$	%	−1.6	−3.2	−3.1	−3.1	−2.9	0	−2.8
(g)	μ_g and[b] $(\mu)_g^+$	%	−4.3	−3.9	−4.3	−4.5	−5.1	0	−4.4
(h)	Viscosity ratio μ_l/μ_g	%	22.8	15.4	11.2	14.2	1.9	0	+13.1
(i)	$X_{tt,\text{smooth}}$ at $n = 0.2$	%	2.9	3.1	2.8	2.9	1.7	0	+2.7
(j)	$X_{tt,\text{fully rough}}$[c] at $n = 0$	%	0.8	1.7	1.6	1.6	1.5	0	+1.4

[a] $$\% \text{ Deviation} \equiv \frac{(\text{Property of R-502})-(\text{Property of R-22})}{\text{Property of R-22}} \times 100 \quad \text{at the same } P^+.$$

[b] Same % deviations since μ_{cr} of R-502 = 0.074 = μ_{cr} of R-22 per Chapters 16 and 17 of Ref. 4.8.

[c] The deviations on line (j) are equal to one-half of those on line (f) in absolute value and of the opposite sign as seen from Eq. (6.22).

(c) Even if we pessimistically assume for the halogenated refrigerants an inaccuracy band of the order of 10–25% for the *less precise* published *viscosity* data, the impact on the friction factor $\bar{f}_{tp}$ used in Eq. (6.2) for turbulent regimes reduces to 2–5% and 0% for $n = 0.20$ and 0, respectively, which is *negligible* compared to typical two-phase flow deviations of over ± 30–40% per Eq. (5.28).

Comments

1. The above numerical analysis confirms that ΔP_{tp} for R-502 could be *predicted* without any R-502 viscosity data *solely on the basis of the ECSP generalization method* according to Eq. (6.16) from tests with another refrigerant with a deviation of $< 5\%$.

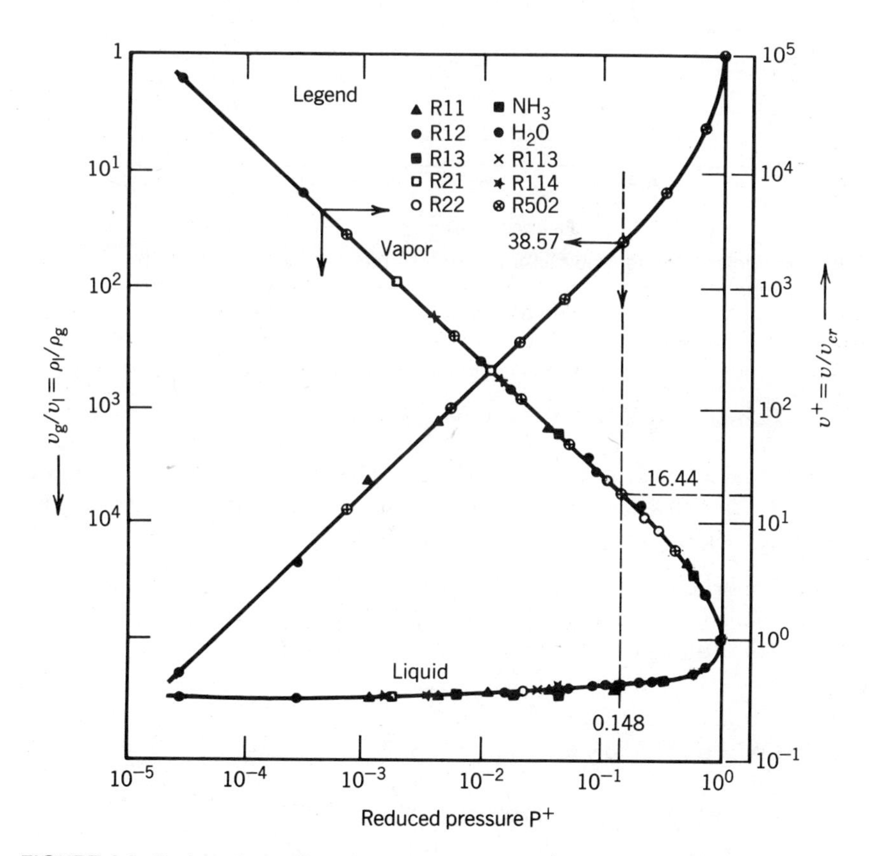

FIGURE 6.3. Reduced specific volume at saturation v^+ and v_g/v_l in terms of P^+. See the actual percent deviations on lines (e) and (f) of Table 6.1. [Updated with permission from H. Soumerai, Application of Thermodynamic Similitude, Part I: Pressure Drop, *ASHRAE J.*, **8**, 78 (June 1966), Fig. 2B.]

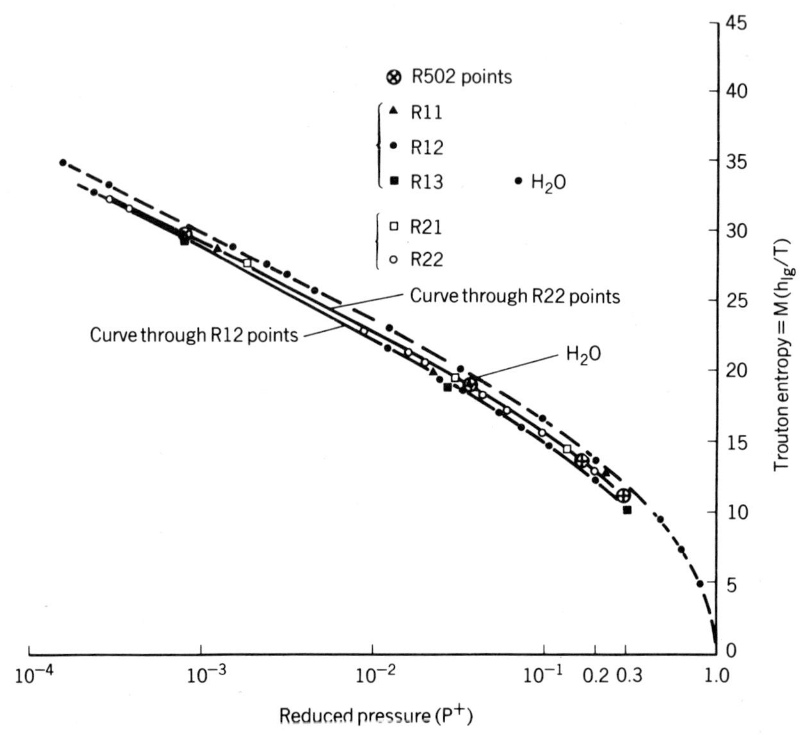

FIGURE 6.4. Molar latent entropy as a function of reduced pressure P^+. The points for R-502 fall within a fraction of 1% on the curve through R-22 points. See actual percent deviation shown on line (d) of Table 6.1. Conversion: J / mol · K = (Btu / lbmol · R)4.184. [Updated with permission from H. Soumerai, Application of Thermodynamic Similitude, Part III: Comments on Applications of Thermodynamic Similitude, ASHRAE J., **8** (Sept. 1966), Fig. A-1.]

2. Universal curves in terms of P^+, updated in the light of the latest published thermophysical data, for the specific volumes, molar entropy, and viscosities are shown as Figs. 6.3, 6.4, and 6.5, respectively.

3. It is important to note that for generalization purposes it is *not* necessary to derive explicit correlations *for any of the* $\Psi_i(P^+)$ functions; the mere knowledge that the $\Psi_i(P^+)$ are indeed *universal functions of* P^+ is sufficient!

NUMERICAL ANALYSIS 6.2

Generalization of raw empirical data.

Problem Statement

The actual values of ΔP_{tp} determined from many reliable tests with R-22 for a given internally finned annular channel configuration are shown in Fig. 6.6. The statistical accuracy of the data, which does *not* depend on any

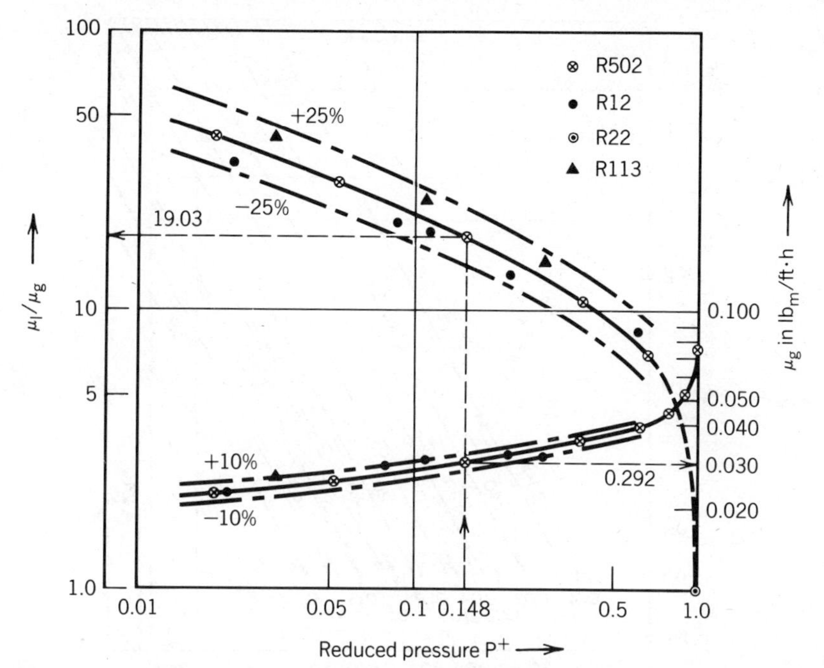

FIGURE 6.5. Viscosity μ_g and μ_l/μ_g in terms of reduced pressure P^+. See the actual percent deviations on lines (g) and (h) of Table 6.1. Conversion factors μPa · s = (lb$_m$/ ft · h)413.38. [Updated with permission from H. Soumerai, Application of Thermodynamic Similitude, Part I: Pressure Drop, *ASHRAE J.*, **8**, (38) (June 1966), Fig. 1.]

particular theory or analytical model, is solely a function of laboratory test accuracy, reproducibility, and manufacturing tolerances. Statistical checks have shown deviations of $\pm 10\%$ with a confidence limit of 95%.

(a) Determine ΔP_{tp} for R-502 from Fig. 6.6 and Eq. (6.15) as well as Eq. (6.16) at $(T_{sat})_2 = 35\,°\text{F}$ (1.66 °C) and for a total flow rate per circuit of $\dot{m} = 171$ lb/h (2.15×10^{-2} kg/s). Compare the results. The flow rate $\dot{m} = 171$ lb/h corresponds to an evaporator circuit cooling capacity of 12,000 Btu/h (3.517 kW) with R-22 at the same exit saturation temperature of 35 °F.

(b) Repeat the above calculation for the *same cooling capacity* as with R-22 and determine the corresponding ΔT_{sat} with R-502.

(c) Compare the above ΔT_{sat} penalty with the values obtained under the same conditions with R-12 and R-22 [4.21, Part I] and assess the accuracy of these calculations.

Solution

(a) At the prescribed $T_{sat} = 35\,°\text{F}$ (1.66 °C) and corresponding $P^+ = 0.1481$, the absolute saturation pressure of the reference refrigerant R-22 is, with $(P_{cr})_{\text{R-22}} = 721.91$ psia, 0.1481(721.91) = 106.91 psia, which corresponds

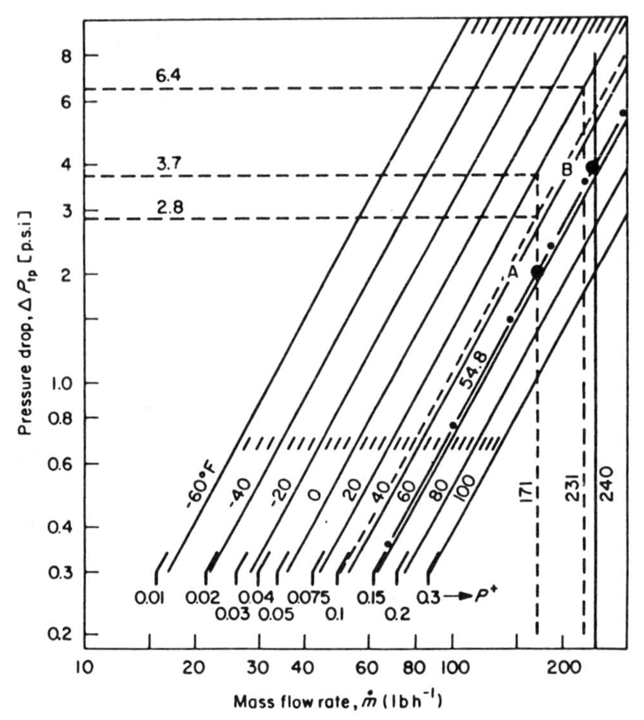

FIGURE 6.6. Experimental pressure drop of R-22 under typical conditions: $x_1 = 0.15$ and $\Delta x = 0.8$. [Updated with permission from H. Soumerai, Application of Thermodynamic Similitude, Part I: Pressure Drop, *ASHRAE J.*, **8**, 78 (June 1966), Fig. 4.]

to a saturation temperature of 54.8°F (12.66°C). From Fig. 6.6 point A, at 54.8°F and $\dot{m} = 171$ lb/h, we read $(\Delta P_{tp})_{R\text{-}22} = 2.05$ psia and obtain from Eq. (6.15) with $(v_g)_{R\text{-}502} = 0.46955$, $(v_g)_{R\text{-}22} = 0.51412$ ft^3/lb or a ratio of $(v_g)_{R\text{-}502}/(v_g)_{R\text{-}22} = 0.9134$ a pressure drop $(\Delta P_{tp})_{R\text{-}502} = 0.9134(2.05) = 1.87$ psia (12.89 kPa).

However, $(\Delta P_{tp})_{R\text{-}502}$ can be computed more simply from Eq. (6.16) with the known critical volumes $(v_{cr})_{R\text{-}502} = 0.02857$ and $(v_{cr})_{R\text{-}22} = 0.030525$: that is, a *constant ratio* $(v_{cr})_{R\text{-}502}/(v_{cr})_{R\text{-}22} = 0.936$ with an insignificant error of $100 \times (0.936 - 0.9134)/0.936 = 2.4\%$. This is well within the estimated $\pm 5\%$ accuracy of the ECSP generalization method.

(b) For the same evaporator capacity, the required flow rate can be computed from $\dot{m}_{R\text{-}502} = \dot{m}_{R\text{-}22}(h_{lg})_{R\text{-}22}/(h_{lg})_{R\text{-}502}$, where h_{lg} is the specific latent heat of evaporation at the prescribed saturation temperature of 35°F (1.66°C) or, from the latest R-22 and R-502 published data $(h_{lg})_{R\text{-}22}/(h_{lg})_{R\text{-}502} = 87.71/62.525 = 1.403$; therefore $\dot{m}_{R\text{-}502} = 171\ (1.403) = 240$. From Fig. 6.6 point B, at $\dot{m} = 240$ and 54.8°F, we now read $(\Delta P_{tp})_{R\text{-}22} = 3.75$. Using Eq. (6.16), $(\Delta P_{tp})_{R\text{-}502} = 3.75(0.936) = 3.51$ psia (24.2

kPa), corresponding to a saturation temperature drop $\Delta T_{sat} = 2.4°F$ (1.33°C), obtained directly from standard refrigerant tables or graphs (or more conveniently from ad hoc saturation temperature versus pressure curves [4.21, Part I]).

(c) The ΔT_{sat} penalty with R-502 can be compared with the results obtained under the same conditions at 12,000 Btu/h (3.517 kW) and exit saturation temperature of 35°F (1.66°C) with R-22 and R-12, [4.21, Part I, Numerical Example 2] as shown below:

R-22	$\Delta T_{sat} = 2°F$ (1.11°C)	$\Delta T_{sat}/(\Delta T_{sat})_{R\text{-}22} = 1.0$
R-502	$\Delta T_{sat} = 2.4°F$ (1.33°C)	$\Delta T_{sat}/(\Delta T_{sat})_{R\text{-}22} = 1.2$
R-12	$\Delta T_{sat} = 6.6°F$ (3.66°C)	$\Delta T_{sat}/(\Delta T_{sat})_{R\text{-}22} = 3.3$

The above results, which are fully in line with those obtained in actual direct-expansion evaporator tests with R-22, R-12, and R-502 [3.4, 4.4, 6.9], explain the significant size and cost reductions achieved in compact direct-expansion chillers with R-22 and R-502 in place of R-12 for a given system COP or, alternatively, the lower $\overline{T}_w - \overline{T}_{sat}$ and consequently *higher COP* attainable with a given evaporator.

From these numerical analyses it is clear that the maximum possible deviation of the ECSP generalization method is well below an unavoidable minimum statistical variation of the reference R-22 data of $\pm 10\%$. Consequently, the predicted R-502 values of ΔP_{tp} and ΔT_{sat} are *substantially as accurate* as the original raw empirical data *obtained with the reference refrigerant* for practical heat exchanger design purposes.

Comments

In actual heat exchanger design applications it is more convenient to compute $\Delta T_{sat,\,R\text{-}502}$ *directly* from the known empirical values of ΔT_{sat} of a reference refrigerant. This could easily be done by recasting the plots of the experimental correlation of ΔP_{tp} shown in Fig. 6.6 as $\Delta T_{sat} = \varphi(\dot{m}, P^+)$, since there is a unique correspondence between ΔT_{sat} and ΔP_{tp} at any prescribed P^+.

NUMERICAL ANALYSIS 6.3

Application of the universal Trouton entropy correlation in terms of P^+.

Problem Statement

(a) Derive from the well-known Clapeyron–Clausius equation (4.12), relating the equilibrium saturation temperature and pressure of any pure fluid, a simple expression of $\Delta T_{\text{sat, R-502}}$ in terms of $\Delta T_{\text{sat, R-22}}$ for identical G, P^+, and channel configuration.

(b) Double check the results obtained in Numerical Analysis 6.2 on the basis of the above expression and assess the accuracy of this simpler generalization method.

Solution

(a) The most general form of Clapeyron's equation expressed on a molar basis in either U.S. technical or SI units is (with M = molar weight, $J = 778$ ft · lb/lb, and $J = 1$ dimensionless in U.S. technical and SI units, respectively)

$$\Delta T_{\text{sat}} M s_{lg} = \Delta P_{\text{sat}} M (v_g - v_l) J = \Delta P_{\text{sat}} M v_g (1 - v_l/v_g) J \qquad (6.17)$$

Solving Eq. (6.17) for ΔT_{sat} and substituting $v_{\text{cr}}(v_g)^+$ for v_g, we obtain for a given $\Delta P_{\text{tp}} = \Delta P_{\text{sat}}$

$$\Delta T_{\text{sat}} = \Delta P_{\text{tp}} M v_{\text{cr}} \left(\frac{v_g^+ (1 - v_l/v_g)}{M s_{lg}} \right) J \qquad (6.18)$$

and a ratio

$$\frac{\Delta T_{\text{sat, R-502}}}{\Delta T_{\text{sat, R-22}}} = \frac{(\Delta P_{\text{tp}} M v_{\text{cr}})_{\text{R-502}}}{(\Delta P_{\text{tp}} M v_{\text{cr}})_{\text{R-22}}} \qquad (6.19)$$

since *all* the terms within the large parentheses on the right-hand side of Eq. (6.18) are identical at the same reduced pressure P^+. Substituting $v_{\text{cr, R-502}}/v_{\text{cr, R-22}}$ per Eq. (6.16) for $(\Delta P_{\text{tp}})_{\text{R-502}}/(\Delta P_{\text{tp}})_{\text{R-22}}$ in Eq. (6.19) yields

$$\frac{\Delta T_{\text{sat, R-502}}}{\Delta T_{\text{sat, R-22}}} = \frac{\left[M (v_{\text{cr}})^2\right]_{\text{R-502}}}{\left[M (v_{\text{cr}})^2\right]_{\text{R-22}}} \qquad (6.20)$$

More generally, by using any arbitrary (similar) fluid as a reference,

$$F_{\Delta T} = \frac{M (v_{\text{cr}})^2}{\left[M (v_{\text{cr}})^2\right]_{\text{ref}}} \qquad (6.21\text{a})$$

where the constant correction factor $F_{\Delta T}$ is fully defined in terms of the *known molecular weights and the specific volumes* at the critical point and

$$\Delta T_{\text{sat}} = (F_{\Delta T})(\Delta T_{\text{sat}})_{\text{ref}} \qquad (6.21\text{b})$$

(b) With the numerical values $M = 111.63$ and $v_{cr} = 0.02857$ for R-502, $M = 86.48$ and $v_{cr} = 0.030525$ for the reference refrigerant R-22 we obtain from Eq. (6.21a) a constant correction factor $F_{\Delta T} = (111.63/86.48)(0.02857/0.030525)^2 = 1.1308$. The value of $\Delta P_{tp} = 3.75$ psia determined in Numerical Analysis 6.2(b) for R-22 at $P^+ = 0.1481$, that is, 106.91 psia, corresponds to a saturation temperature drop $\Delta T_{sat, R\text{-}22}$ of 2.083°F (1.16°C) from Table A-20 in Appendix A. Therefore, with Eq. (6.21b), $\Delta T_{sat, R\text{-}502} = (2.083)(1.1308) = 2.36$°F (1.36°C), which represents a *deviation* of 100 (2.36 − 2.40)/2.40 = *−1.7%* from the value of 2.4°F computed in Numerical Analysis 6.2. This is well within the typical ±10% statistical variations of the R-22 pressure drop raw empirical data and consistent with actual deviations quantified more precisely in Table 6.1 for the specific refrigerant R-502 under consideration.

Comments

1. The constant correction factors $F_{\Delta P}$ and $F_{\Delta T}$ in Eq. (6.16b) and (6.21a), respectively, have also been computed for all the refrigerants discussed in [4.21, 4.67] on the basis of the latest thermophysical property data, using R-22 as the reference refrigerant. The results are summarized for practical applications in Table 6.2. It is evident that the choice of R-22 as the reference refrigerant is arbitrary, and suitable correction factors could be computed according to the same method using any of the other "similar" halogenated refrigerants as a reference according to the specific needs of the design engineer.

2. It should be noted that the generalization method is somewhat less accurate for dissimilar strongly polar molecules such as ammonia (NH_3) and particularly water (H_2O). Nevertheless, even for the most dissimilar substance among those listed in Table 6.2, namely, water, the values of $(v_g)^+$ and v_g/v_l fall right on the universal curves reproduced here as Fig. 6.3 [4.21]. The molar entropy of H_2O lies only 5–12% above the universal curves of the halogenated refrigerants at P^+ of 0.001 and 0.2, respectively, as shown in Fig. 6.4 [4.21, 4.59]. This means that the correction factor $F_{\Delta T}$ for water in Table 6.2 is slightly *overestimated* [in view of Eq. (6.18)] by 5–12% at $P^+ < 0.2$, whereas $F_{\Delta P}$ is theoretically exact, within the presumed accuracy of the published thermophysical data, for all the fluids shown in Table 6.2.

3. The fact that in the case of water the viscosity of the liquid and vapor phase depart more significantly from the universal curves in terms of P^+ (as shown in Ref. 4.21) has no significant impact on ΔP_{tp} and/or ΔT_{sat} in *typical annular-spray flow regimes* in accordance with the recommended correlation proposed in Refs. 4.21, 5.3, 5.22, and 5.23 and reviewed in Chapter 5, when $\bar{f}_{tp}$ = constant and is independent of the Reynolds number, a conclusion further reinforced on the basis of the generalization of the Martinelli–Nelson parameters highlighted in Numerical Analysis 6.4.

TABLE 6.2. Constant Correction Factors[a] $F_{\Delta P}$ and $F_{\Delta T}$ for Various Refrigerants with R-22 as the Reference Fluid

Refrigerant Class	Similar Halogenated Refrigerants									"Dissimilar" Polar Fluids	
Designation R-	11	12	13	21	22	502 Azeotrope R-22 / R-115	113	114	115	717 (NH_3)	718 (H_2O)
$F_{\Delta P}$ per Eq. (6.16b)	0.947	0.940	0.907	1.006	1.0	0.936	0.911	0.901	0.855	2.228	1.641
$F_{\Delta T}$ per Eq. (6.21a)	1.424	1.236	0.995	1.204	1.0	1.1308	1.797	1.604	1.306	0.977	0.561

[a]Computed from 1981 ASHRAE Handbook Fundamentals [4.8] Chapters 16 and 17, using the thermophysical data tables in U.S. technical units as there are some very minor discrepancies (probably well within the unstated accuracy limits of these tables) between the values in U.S. technical and SI units.

NUMERICAL ANALYSIS 6.4

Generalization of the Martinelli-Nelson parameters X_{tt} and X_{vv}.

Problem Statement

The Martinelli–Nelson parameters X_{tt} and X_{vv} can be expressed most generally for single-component two-phase flow according to Eq. (5.41a) derived in Chapter 5. Focusing on the turbulent–turbulent type of flow regimes typically encountered in forced-convection evaporators designed to operate stably at reasonably high velocities and correspondingly high average (or local) two-phase heat transfer coefficients $\bar{\alpha}_{tp}$ (or α_{tp}):

(a) Verify over the full saturation temperature range of single-stage refrigeration/heat pump evaporators that $X_{tt,R\text{-}502} = X_{tt,R\text{-}22}$ or, more generally, that the X_{tt} are identical for all thermodynamically similar pure substances at identical reduced pressure.

(b) Quantify in the case of R-502 the actual percent deviations on the basis of the latest thermophysical property data available and assess the impact of these percent deviations on ΔP_{tp} computed according to the Martinelli–Nelson fully separated two-phase flow model with a smooth tube Blasius exponent $n = 0.20$ and the fully rough limit $n = 0$.

Solution

(a) Comparing the value of X_{tt} for any refrigerant with that of an arbitrary reference refrigerant $X_{tt,\text{ref}}$ at the same quality x, and therefore identical $(1 - x)/x$, we find that Eq. (5.41a) yields, with $\rho_g/\rho_l \equiv v_l/v_g$

$$\frac{X_{tt}}{X_{tt,\text{ref}}} = \left(\frac{(\mu_l/\mu_g)^n}{(\mu_l/\mu_g)^n_{\text{ref}}} \frac{v_l/v_g}{(v_l/v_g)_{\text{ref}}} \right)^{1/2} \tag{6.22}$$

The terms $[(\mu_l/\mu_g)/(\mu_l/\mu_g)_{\text{ref}}]^n$ and $[(v_l/v_g)/(v_l/v_g)_{\text{ref}}]$ being substantially equal to unity at the same reduced pressure P^+ per Figs. 6.3 and 6.5, the square root of their product in Eq. (6.22) is even closer to unity. Thus,

$$\mathrm{X}_{tt} = \mathrm{X}_{tt,\text{ref}} \quad \text{at identical reduced pressure } P^+ \text{ and}$$
$$\text{average (or local) quality } \bar{x} \text{ (or } x) \qquad \text{for } n = 0.2 \rightarrow 0 \qquad (6.23)$$

(b) The actual percent deviations—$100\,(\mathrm{X}_{tt,\text{R-502}} - \mathrm{X}_{tt,\text{R-22}})/\mathrm{X}_{tt,\text{R-22}}$—are shown on lines (i) and (j) in Table 6.1 for a Blasius friction factor exponent $n = 0.2$ as well as the fully rough limiting case $n = 0$. It is evident that such small deviations are insignificant compared to typical deviations of over ± 30–40% for two-phase flow correlations according to the Martinelli fully separated model.

Comments

1. It is clear that the accuracy of the identity expressed by Eq. (6.23) is fully adequate for similar halogenated refrigerants in two-phase flow heat exchanger design applications, both with heat addition in *evaporators* and with heat removal in *condensers*.

2. The generalization method can be extended to the two limiting cases of *single-phase heat exchanger* applications, *without phase changes* when $x \geq 1$ with saturated or slightly superheated vapors (all gas phase) *and* $x \leq 0$ (therefore, $y = 1 - x \geq 1$) for subcooled *liquids*, since the key thermophysical liquid properties at $T = T_{\text{sat}}$ and $P > P_{\text{sat}}$ are, for all practical purposes, independent of the excess pressure: $(P - P_{\text{sat}}) > 0$ at $P^+ \ll 1$.

3. It is noteworthy that Eq. (6.23) is also *valid* for most *dissimilar polar substances* [4.21] such as *water* and *ammonia* within the validity limit of turbulent–turbulent flow regimes in the limiting case of a Blasius exponent $n \rightarrow 0$, equivalent to a fully rough wall flow regime, since the viscosity ratio term $(\mu_l/\mu_g)^{n/2} \rightarrow 1$ when $n \rightarrow 0$ and the effects of viscosity vanish (see Table 6.1). This illustrates very concretely the important point already emphasized for single-phase fluid flow in Chapter 3, namely:

GUIDELINE 10 *The more fully rough the turbulent single- or two-phase flow regimes, the simpler the thermodynamic solutions in contrast to conventional mechanistic analytical models that tend to become more intractable.*

6.5 SUMMARY

The thermodynamic (ECSP) fluid flow generalization method based on reduced pressure is particularly useful in thermally efficient (i.e., $|\bar{T}_w - \bar{T}_{\text{sat}}|/$

$\overline{T}_{sat} \ll 1$) evaporator and condenser design optimization applications since these two-phase heat exchangers account for the major part of the overall system's irreversible losses in low-thermal lift power generation as well as reverse power cycles, and the current conventional analytical methods are generally insufficiently accurate for effective optimization studies. As highlighted in this chapter and in the 1960s [4.21, 4.67], the main advantages of the thermodynamic generalization are the following:

1. The method is simple and permits the prediction of key two-phase flow heat exchanger design parameters with an excellent degree of accuracy when reliable empirical (or analytical) results are available for a single working fluid for any given geometry. Substantially the same accuracy can be expected as that of the original empirical raw data obtained with a reference fluid. The practical and economic significance of this conclusion can be appreciated when one recalls that the *refrigeration and air conditioning industry alone* has standardized the designation of almost *one hundred different refrigerants* (see Tables A.5 and A.6 of Appendix A).

2. When accurate fluid transport properties are *unavailable*, as is frequently the case, this is probably the only practical method conceivable to predict the actual thermohydrodynamic performance of a known system with a new working fluid. This is a particularly useful feature when several alternative refrigerants or working fluids are being considered in thermoeconomic analyses of novel reverse power (refrigeration) and power generation cycles.

3. The ECSP generalization method is a *logical thermodynamic extension* of the *conventional similarity and dimensional analytical techniques* used so effectively in the field of momentum, heat, and mass transfer. Like dimensional analysis, the ECSP generalization technique sheds no light on the underlying physical nature of two-phase flow phenomena. Nevertheless, this theoretical weakness is overshadowed by the *significant practical advantages* of the method as a *useful engineering tool, namely*, its *universality* and *accuracy*.

4. Although the numerical analyses were deliberately limited to direct-expansion evaporators with incomplete evaporation, the methods highlighted here and in Refs. 4.21 and 4.67 are also applicable to *all types of two-phase flow* with heat addition (*evaporators*) or heat removal (*condensers*) and all types of heat exchanger configurations as well as the *limiting case* of *single-phase flow* ($x \leq 0$ or $x \geq 1$) heat exchangers.

The rationale for this rather sweeping conclusion is that the ECSP generalization method is based exclusively on *classical thermodynamic laws* that are *inherently independent of spatial dimensions*[6] under assumed steady-state steady-flow conditions at the heat exchanger inlet and outlet (in/out SSSF). We shall make full use of the basic methodology outlined in this chapter, and the universal functions shown in Figs. 6.3–6.5, particularly the remarkable phenomenological law that the molar entropy is a universal function of reduced pressure, when we address the more complex coupled two-phase flow heat transfer coefficients in subsequent chapters.

NOTES

1. According to Basarow [6.10], Russian physicists have a long tradition in the study of critical phenomena dating back to Mendeleev (1860).
2. In the following excerpts (abstract and introductory paragraphs) from a review paper [6.11] entitled "The Corresponding States Principle, A Review of Current Theory and Practice," presented in 1968 at an Applied Thermodynamics Symposium, the authors highlight succinctly the central and extremely useful role of the CSP and the major contribution made by the Dutch physicist van der Waals in the last century.

   The most powerful tool available for quantitative prediction of the physical properties of pure fluids and mixtures is the CSP.

   The principal application of the corresponding states principle (CSP) is the prediction of unknown properties of many fluids from the known properties of a few. It also makes possible the utilisation of some important developments in statistical mechanics which would otherwise be prohibited by computational difficulties. Historically, the CSP was introduced by van der Waals before 1900, and today all who face the problem of predicting physical properties would agree with the recent statement by Guggenheim [6.12] that the CSP "may safely be regarded as the most useful by-product of the van der Waals equation of state."

3. This is indicated by the following excerpt from [6.13]:

   Dr. Tamami Kusuda (Bureau of Standards, Washington, D.C., member ASHRAE TC 1.3 for heat transfer) pointed out correctly that the principle of corresponding states of transport properties is *fairly well established for the gas phase*, but needs more technical support for the liquid phase, and, along these lines, recommended the following additional references [6.14–6.17] especially [6.17] which includes plots of reduced conductivity, k/k_{cr}, and reduced viscosity, μ/μ_{cr}.

4. The interested reader will find additional information, particularly on thermodynamic similarity criteria in Refs. 4.21 (Part I), 4.60, and references cited therein.
5. Another major practical advantage of Eqs. (6.12) and (6.13) is that useful generalizations in terms of P^+ can be achieved when the values of $(\pi_i)_{cr}$ are not exactly known or for property terms such as h_{lg} and s_{lg} or c_p that either converge to zero or infinity at the critical point.
6. Within the validity limits of the *continuum hypothesis* which is generally satisfied in normal macroscopic thermodynamic systems such as heat exchangers (see Continuum Hypothesis in Chapter 7).

CHAPTER 7

ENGINEERING HEAT TRANSFER AND THERMODYNAMICS

The subjective notion of temperature was known in prehistoric times by humans who already practiced the art of heat transfer when they wrapped themselves in animal skins as protection against a cold environment. After the discovery of fire, "civilized" tribes performed chemical engineering operations involving process heat transfer and thermodynamics to cook their meals, melt metals, and so forth. Thus the art of heat transfer and thermodynamics, in the broadest sense of the word, dates back to prehistoric times.

Before any physical laws could be formulated in the field of thermal sciences, it was necessary to devise accurate methods of measuring and quantifying temperature and a number of eminent physicists, including Galileo Galilei and Isaac Newton, addressed the central challenge of thermal sciences: thermometry.

It is interesting to recall that the precursor of modern thermal sciences—the science of fire—was not the exclusive domain of physicists. Alchemists (this includes Newton), the fathers of modern chemistry, and several physicians made significant contributions to this new branch of physics. Indeed, some of the first liquid-filled thermometers were invented by physicians and the first law of thermodynamics, the conservation law, was officially[1] formulated and roughly demonstrated by a young doctor, Robert Mayer, in 1842 and then verified more accurately by James Prescott Joule in the following year. This early fruitful connection between scientists interested in the physiology of respiration, or life sciences, and physicists provides a historical precedent for the bold interdisciplinary concepts proposed by the pioneer of twentieth century nonequilibrium thermodynamics, Ilya Prigogine [7.1].

7.1 THERMOMETRY (EIGHTEENTH CENTURY)

With the construction, around 1592, of his *thermoscope*, Galileo invented the first thermometer on record [7.2, 7.3] and can thus be considered one of the earliest pioneers of thermal sciences. The thermoscope was an air thermometer relying on the expansion of a permanent gas to measure a temperature change since, in Galileo's words, "it is a recognized true principle that the property of cold is to contract and that of heat to expand." This first thermometer was thus based on laws of thermodynamics, illustrating again the fundamental relationships between the two branches of thermal sciences later called heat transfer and thermodynamics, defined in Chapter 1.

Improved air thermometer designs were later developed [7.4, 7.5] by Amontons (1663–1705) and Lambert (1728–1777). To avoid the complications encountered in the practical use of gas thermometers, many researchers focused their attention on liquid-filled thermometers. In his excellent classical technical thermodynamic treatise, Rudolf Plank [7.2] identifies the following main pioneers: the physician T. Rey (1631) and the Archduke of Toscana, Ferdinand II (1641), who are credited by one of the early pioneers in thermometry, De Reaumur, with the invention of the first alcohol-filled prototypes [7.6]. Practical liquid-filled thermometers were developed and actually produced much later, first by Daniel Gabriel Fahrenheit [7.7] in Amsterdam (1714) and then by Rene Antoine Ferchault De Reaumur [7.6] and Anders Celsius [7.8] in the period extending from 1730 to 1742.

Reasonably accurate *thermometers*, essential tools for heat transfer and thermal power generation engineers and physicists, were thus available about the first half of the eighteenth century. To define a temperature scale, the scientists of the time jointly selected as a *standard reference* two specific thermodynamic states of the most abundant and unique fluid on earth—*water*—namely, its boiling and melting points under normal atmospheric pressure. Unfortunately, they failed to agree on a common temperature scale and *two different systems of units* were born: the degree Celsius (°C) and the degree Fahrenheit (°F) with the corresponding absolute temperature scales—degree Kelvin (K) and Rankine (°R)—which are still used by engineers today in spite of the introduction in the second half of this century of the degree Kelvin as the official SI unit. The melting and the boiling points were arbitrarily set at 0°C and 100°C, respectively, in the *centrigrade scale*, or an *interval of* 100°C *or* K between boiling and melting points, which was arbitrarily divided linearly. In the Fahrenheit scale the following arbitrary values were selected: +32°F for the melting point and 212°F for the boiling point, or an *interval of* 180°F (°R). Thus, a unit interval of 1°C (K) in the Celsius (Kelvin) scale corresponds to a numerical value of 1.8°F (°R) in the Fahrenheit (Rankine) scale.

The range of absolute temperatures encountered in typical technological applications, such as power generation, refrigeration, and chemical process industries, as well as the extreme temperatures achieved under laboratory conditions are highlighted in Fig. 4.1. To show the full range of absolute temperatures on a single graph, a logarithmic scale has been used in Fig. 4.1. This figure also emphasizes the *unattainability of the absolute zero* and the strikingly small range of temperatures in which biological life can be maintained on earth.

In closing, it is worth recalling that measuring the true *temperature of moving fluids* is still a major *challenge* today in spite of modern instrumentation techniques, as emphasized in E. R. G. Eckert's 1984 paper [2.23], which highlights again the intimate relationship between fluid flow, heat flow, and thermodynamics (see note No. 5, Chapter 2).

7.2 HEAT TRANSFER

No mention is made of Isaac Newton's work on thermometry [7.9] in most classical equilibrium thermodynamics treatises. The obvious reason for this omission is that neither the instrument nor the alternative temperature scale proposed by Newton was adopted by his contemporaries. By contrast Newton's name appears in most heat transfer textbooks because he formulated (in 1701) the first heat transfer rate equation on record, now known as *Newton's cooling law*.

7.2.1 The Convection Mode: Newton's Cooling Law

In his efforts to develop a high-temperature thermometer suitable for his research activities in alchemistry, Newton actually carried out transient heat transfer tests "... up to the temperature of coal fire ... " by heating an iron bar above the melting points of various substances (wax and low melting point metals) embedded in it, then measuring the time required for the bar to cool down to various temperature levels. From these experiments Newton proposed in substance the following linear law (using today's terminology and keeping in mind that the first law of thermodynamics was formulated 150 years later): The instantaneous rate of heat loss per unit time of a given body per unit exposed area A, or the heat flux $q'' \equiv \dot{Q}_e/A$, is *directly proportional* to the excess temperature of the body over that of its surroundings, $\Delta T_w = T_w - T$. Expressed in mathematical terms,

$$\dot{Q}_e/A \equiv q'' \propto \Delta T_w = (T_w - T) \tag{7.1}$$

where T_w = solid body surface temperature
T = surrounding fluid temperature

The proportionality expressed by Eq. (7.1) can be reduced to an equality by introducing an *empirical proportionality factor defining the heat transfer coefficient* (sometimes called the *conductance* or *film coefficient*)[2] according to

$$\dot{Q}_e/A \equiv \bar{q}'' \equiv \bar{\alpha}\,\overline{\Delta T}_w = \bar{\alpha}(\bar{T}_w - \bar{T}) \tag{7.2a}$$

Locally,

$$d\dot{Q}_e/dA = q'' = \alpha\,\Delta T_w = \alpha(T_w - T) \tag{7.2b}$$

Equation (7.2a) defines the *average* heat transfer coefficient $\bar{\alpha}$ based on the average wall-to-fluid temperature difference $\overline{\Delta T}_w = \bar{T}_w - \bar{T}$ over the whole surface A, whereas Eq. (7.2b) defines the *local heat* transfer coefficient α based on an elementary surface dA and local temperature difference $\Delta T_w = T_w - T$.

Equations (7.2a) and (7.2b) are the basic rate equations commonly used in heat exchanger design practice today for the convection mode of heat transfer [7.10–7.14]. Thus, the first physical law in the science and the art of heat transfer (according to Guideline 1 in Chapter 1), the earliest heat rate equation, was laid down in 1701 by Isaac Newton, as an *incidental by-product of his unsuccessful attempt* to develop a high-temperature thermometer.

Although the simple linear relationship between a heat flux and a temperature potential expressed by Eq. (7.1) is taken for granted in normal fluid-to-fluid forced-convection heat exchanger design applications,[3] this *first phenomenological heat rate* equation marks a major turning point in the history of physics, consecrated more than a century later when J. B. J. Fourier presented his classical treatise on the analytical theory of heat [7.15]. In the face of his numerous critics at the French Academy of Sciences, Fourier relied heavily on Newton's cooling law as well as later experimental results published by other eighteenth century scientists to defend his axiomatic theory on the propagation of heat.

7.2.2 The Thermal Conduction Mode: A New Discipline in Physics

When Baron Jean Baptiste Joseph Fourier was awarded the prize of the French Academy of Science, in 1812, for his mathematical description of the diffusion of heat in solids, the science of heat transfer was officially recognized as a separate discipline in physics, outside the scope of rational and celestial mechanics, which had been dominant in physical sciences since Newton's time [3.1, 7.16–7.18] particularly among Newton's heirs in France, the leading mathematical physicists at the French Academy of Sciences.

Since the conduction mode of heat transfer normally plays a relatively minor role in steady-state heat exchanger design applications and the subject is treated quite adequately in all introductory heat transfer textbooks, we confine ourselves here to the expression of Fourier's law in the simplest case of *one-dimensional* heat diffusion in *homogeneous and isotropic materials*; that is, the heat flux q'' resulting from thermal conduction is *directly proportional* to the magnitude of the *temperature gradient* dT/dx in the direction of heat diffusion:

$$q'' \propto \left| \frac{dT}{dx} \right| \tag{7.3}$$

By introducing a constant proportionality factor k and bearing in mind that the gradient dT/dx must be negative in the direction of heat flow since heat flows only in the direction of decreasing temperature (in accordance with the second law of thermodynamics), we obtain from Eq. (7.3)

$$q'' \equiv -k\left(\frac{dT}{dx}\right) \tag{7.4}$$

where the empirical constant k, defined by Eq. (7.4), is a transport property of the material known as the *thermal conductivity*. Typical thermal conductivities of solid materials and fluids are listed in Appendix A.

The temperature gradient dT/dx is most generally a function of *position and time* under *non*steady-state conditions, in other words, a *field function*. Thus, according to Eq. (7.4), the heat flux q'' as well as dT/dx are generally functions of spatial and time coordinates. Since surface heat exchangers are designed, tested, and guaranteed for operation under specified inlet/outlet steady-state conditions in actual design practice, we dispense with the complexities of transient heat transfer applications and confine ourselves in this text to *steady-state steady-flow conditions* when the temperature field T and the gradient dT/dx and q'', according to Eq. (7.4), are *independent of time* and functions of spatial position (x, y, z) only. For the special case typical of fluid-to-fluid forced-convection tubular heat exchangers, the temperature distribution in the thin solid wall[4] is for all practical purposes linear and the local temperature gradient on the right-hand side of Eq. (7.4) can be expressed most simply as

$$-\frac{dT}{dx} = \frac{T_H - T_C}{\delta_w} \tag{7.5}$$

This gradient results in a local heat flux

$$q'' = \frac{k}{\delta_w}(T_H - T_C) \tag{7.6}$$

where δ_w is the wall thickness and the subscripts H and C refer to the constant hot and cold temperature boundaries, respectively. The negative sign in Eq. (7.4) can be omitted in Eq. (7.6) since $T_H - T_C$ is always > 0.

7.2.3 The Thermal Radiation Mode (Last Quarter of Nineteenth Century)

The mechanisms of heat transfer by conduction in solids or by convection between a solid boundary and a fluid require the presence of a medium to convey heat from a warmer heat source to a colder receiver. By contrast, radiant heat transfer does not require an intervening medium because heat can be transmitted by radiation across an absolute vacuum. The nature of this radiation was elucidated in Max Planck's pioneering work at the turn of the century (1895–1900) in terms of electromagnetic waves traveling at the velocity of light [7.19].

The basic *rate equation for thermal radiation* from the ideal radiator (the *black body*), discovered *empirically* by Stephan in 1879 and derived *theoretically* by Boltzmann in 1884 on the basis of the *second law of thermodynamics* [7.20, 7.21], states that

$$q''_{\text{rad}} = \sigma T^4 \tag{7.7}$$

where q''_{rad} is the rate of heat transfer by radiation per unit area from one side of the black body of area dA, at an absolute surface temperature of T, and σ is the Stephan–Boltzmann universal constant with a numerical value of $5.670 \times 10^{-8}\,\text{W/m}^2 \cdot \text{K}^4$ (or 0.1713×10^{-8} Btu/h $\cdot$ ft^2 $\cdot$ °R^4 in U.S. technical units). Note that Eq. (7.7) can also be obtained by integrating the well-known blackbody spectral energy distribution derived by Max Planck [7.19, 7.22–7.29].

The blackbody heat rate computed according to Eq. (7.7) represents the upper emissive power of any *real* surface at a prescribed temperature T. The ratio of the emissive power of an actual surface to that of a blackbody is by definition the emissivity ε of the surface and the heat rate equation for nonblack surface becomes

$$q''_{\text{rad}} = \varepsilon \sigma T^4 \tag{7.8}$$

The values of *normal total emissivity* ε_η for several representative surfaces are shown in Table A.9 of Appendix A.

Generally, heat can be transferred from a solid body by convection and thermal radiation concurrently. However, the relative importance of the thermal radiation mode becomes significant only at relatively high absolute temperature because of the low value of the constant σ and since q''_{rad} varies with the fourth power of the absolute temperature according to Eq. (7.8) and is usually insignificant compared to the convection mode q''_{conv} in *normally efficient* fluid-to-fluid heat exchanger applications. For the sake of brevity we shall ignore the radiation mode of heat transfer in the balance of this text and conclude this rudimentary overview of thermal radiation, with a few general remarks that are particularly relevant in a text mainly dedicated to the application of thermodynamic concepts in heat transfer engineering.

In contrast to the conduction and convection mode heat rate equations where the driving forces, $T_H - T_C$ and $T_w - T$ per Eqs. (7.5) and (7.2), respectively, are completely independent of the absolute temperature level, the radiation mode heat rate Eq. (7.7) is expressed in terms of a thermodynamically more meaningful absolute temperature as the main variable at the fourth power and the heat flux q''_{rad} is a nonlinear[5] function of $T_w - T$ over the *full range of absolute temperatures*. Furthermore, several of the fundamental laws of thermal radiation (derived *after* classical equilibrium *thermodynamics* had been firmly *established* as a major branch of physics and the first foundations of statistical equilibrium and nonequilibrium thermodynamics had been laid down in the nineteenth century) are based on second law considerations, for example, Kirchhoff's law and Boltzmann's theoretical derivation of the Stephan–Boltzmann law. The central role of entropy, that is, thermodynamics, in Max Planck's pioneering work on thermal radiation is clearly emphasized in Planck's scientific autobiography [7.30] in which he acknowledges the significant influence of Boltzmann's prior theoretical derivations and probabilistic interpretation of entropy. (see also Note 3 of Chapter 3).

Alfred Kastler stresses the fundamental role of entropy in radiation and modern physics [7.31–7.33] in the opening statement of his presentation at the 1977 International Workshop on Synergetics with the following statement:

> The problem of entropy of radiation was first treated by Max Planck in the years 1895–1900. He considered himself as a disciple of Clausius, the founder of thermodynamics, the man who had introduced the concept of entropy into physics. Planck's aim, when he began his research of blackbody radiation, was to establish an expression for the entropy of this radiation. He was convinced, when he began this work, that thermodynamics, combined with Maxwell's theory of electromagnetism, would give the answer to his question.

We summarize this brief overview of the radiation mode of heat transfer in the form of a major guideline that is particularly relevant in the context of this book in view of the fact that its significance is usually grossly understated in most standard heat transfer textbooks!

GUIDELINE 11 *The first successful application of strictly thermodynamic concepts in the solution of a heat flow rate problem* q''_{rad} *dates back to the fourth quarter of the nineteenth century*!

7.2.4 Comparison of the Conduction and Convection Heat Rate Equations

In spite of their similarity, the heat rate Eqs. (7.2) and (7.4) have a vastly different physical content.

For pure *conduction* heat flow problems in isotropic solids of known constant conductivity, Eq. (7.4) is quite dependable because it correctly *represents the true physical phenomenon*. The temperature field in (isotropic) solids can therefore be predicted with rigorous precision for any given geometry and specified initial and boundary conditions under transient (time-dependent) or steady-state (time independent) conditions, at least in principle, the only difficulty being merely a matter of mathematics. These mathematical difficulties were traditionally resolved in complex situations by approximate graphical finite difference methods and/or analogue model techniques as shown in general heat transfer textbooks [7.34]. With the wide availability of fast low-cost modern electronic computers, these mathematical problems have been substantially eliminated with the aid of computerized finite difference numerical/graphical techniques.

By contrast, the defining Eq. (7.2) is merely a *convenient mathematical device* and *not* the *expression of the complex physical phenomena lumped under the label "convection mode of heat transfer."* Experience has shown that the dimensional heat transfer coefficient $\bar{\alpha}$ (or α) defined by Eq. (7.2a) [or Eq. (7.2b)] depends not only on the units employed [7.35] but also on several physical fluid properties, as well as a large number of additional parameters such as spatial dimensions, flow rates, or velocity of the fluid past the heat exchanger surface, whether or not the fluid is experiencing a phase change (evaporation or condensation), *and frequently* even on the average temperature driving force $\overline{\Delta T_w}$ (or local ΔT_w) appearing in the defining Eqs. (7.2a) and (7.2b). This last qualification simply means that, according to the definition $\bar{q}'' \equiv \bar{\alpha}\overline{\Delta T_w}$ with $\bar{\alpha} = \varphi_\alpha(\overline{\Delta T_w}) \neq$ constant under otherwise fixed conditions, the approximate linear Eq. (7.1) derived from Newton's cooling law must be modified as follows: The *general relationship* between *heat flux* q'' and the conventional local *driving force* ΔT_w for the *convection* mode of heat transfer is represented by

$$q'' = \Delta T_w \alpha = \Delta T_w \varphi_\alpha(\Delta T_w) \propto (\Delta T_w)^a \tag{7.9}$$

when $\alpha = \varphi_\alpha(\Delta T_w) \neq$ constant under *otherwise fixed* conditions with an exponent $a \neq 1$.

Thus, the conventional average (or local) *linear* relationship $\bar{q}'' \propto \overline{\Delta T_w}$ (or $q'' \propto \Delta T_w$) represents the *limiting case* of a more general *nonlinear* law when the exponent $a \to 1$, that is, when $\bar{\alpha}$ (or α) becomes independent of $\overline{\Delta T_w}$ (or ΔT_w) under otherwise fixed conditions.

Table 7.1, based on typical (1954) values listed in McAdams [7.35], provides a broad overview of the range of the average heat transfer coefficient $\bar{\alpha}$, defined by Eq. (7.2a) and obtained in typical technological applications. As shown in the last column of Table 7.1, the average heat transfer coefficient $\bar{\alpha}$ is far from constant and may vary by a factor of 100,000 to 1 (steam dropwise condensation compared to air heating or cooling). The ratios $\bar{\alpha}_{\text{high}}/\bar{\alpha}_{\text{low}}$ shown in Table 7.1 indicate that $\bar{\alpha}$ is also highly variable even within a given class of application. It should be emphasized that the $\bar{\alpha}$ values listed in Table 7.1 are merely intended to show *orders of magnitude* for typical engineering applications. Focusing more specifically on evaporators for the chemical process industries, W. L. Badger, co-author of one of the first chemical engineering[6] textbooks [7.36], states in the chapter entitled "Heat Transfer in Evaporators" [4.42]:

> Changes of ... even greater order of magnitude can be caused by a change in the type of liquid, depth of liquid, diameter of tubes, length of tubes, shape and size

TABLE 7.1. Approximate Range of Values[a] of $\bar{\alpha}$ in Btu / h · ft² · °F

Typical Technological Applications		$\bar{\alpha}_{low}$	$\bar{\alpha}_{high}$	$\frac{\bar{\alpha}_{high}}{\bar{\alpha}_{low}}$	$\frac{\bar{\alpha}_{high}}{\bar{\alpha}_{low} \text{ of air}}$
Without phase change					
	air, heating or cooling	0.2	10	50	50
	steam, superheating	5	20	4	100
	oils, heating or cooling	10	300	30	1,500
	water, heating	50	3,000	60	15,000
With phase changes					
	organic vapors, condensing	200	400	2	2,000
	water, boiling	300	9,000	30	45,000
	steam, film-type condensation	1,000	3,000	3	15,000
	steam, dropwise condensation	5,000	20,000	4	100,000

[a]Based on values in McAdams [7.35]. Multiply by 5.679 to convert to SI units in W / m² · K.

> of the body, and many other factors.... The author has tests of evaporators showing overall heat-transfer coefficients ranging from 4000 to 2 Btu per sq ft per hr per °F.

This is a range of 2000 to 1, noting that the *overall effective* heat transfer includes the effects of the individual heat transfer coefficients on both sides of the heat exchanger surface.

McAdams concludes the introductory chapter of *Heat Transmission* [1.2], the most influential twentieth century textbook among practicing heat transfer engineers in the United States and probably worldwide, with the following statement:

> A large fraction of this text is devoted to a study of the factors which control the coefficient h [$\equiv \alpha$ in this book] and the correlation of these factors so that h may be predicted.

This assessment is still very relevant today, judging by the unabated stream of publications worldwide on convection heat transfer, particularly in the more complex two-phase flow applications highlighted in the preceding chapters. Since the search for the illusive heat transfer coefficients defined mathematically by Eqs. (7.2a) and (7.2b) is still the prime objective of contemporary convective heat transfer research, a fact too frequently overlooked by thermal *systems* engineers [4.44], we paraphrase McAdams' 1954 statement in the following major guideline:

GUIDELINE 12 *The prime objective of applied convective heat transfer research is the study of the factors that control the average (or local) heat* *transfer coefficient* $\bar{\alpha}$ (*or* α) *and the correlation of these factors so that* $\bar{\alpha}$ (*or* α) *may be predicted with an adequate level of accuracy in real world heat exchangers.*

7.3 UNIFIED TREATMENT OF FLUID, HEAT, AND MASS TRANSFER

The pragmatic pioneers of the steam engine revolution in England already knew from practical experience, long before Osborne Reynolds' scientific findings[7] on sinuous flow regimes in 1883 [3.5, 3.7], that the waterside heat transfer coefficient in shell-and-tube steam condensers varies almost in direct proportion to the average water velocity inside tubes in typical turbulent regimes. However, Reynolds [7.37] and his disciple T. E. Stanton [1.11] can certainly be given the credit for formulating the first meaningful mathematical expression for the complex phenomenon of forced convection in the turbulent flow regimes encountered in many heat transfer engineering applications.

7.3.1 Modified Reynolds Analogy Correlations

The more universally valid version of the Reynolds analogy, the well-known Chilton–Colburn correlation, first proposed in 1933 by Colburn [7.38] for the heat transfer coefficient of Newtonian fluids flowing inside long ($L/D_h > 60$) straight circular tubes at *moderate* ΔT_w in stable turbulent regimes, can be expressed most generally as

$$\mathrm{St}(\mathrm{Pr})^{\beta} = \tfrac{1}{2}f_{rh} = \tfrac{1}{8}f \tag{7.10a}$$

where $\mathrm{St} \equiv \alpha/c_pG$ = Stanton number
$\mathrm{Pr} \equiv c_p\mu/k$ = Prandtl number
β = an empirical exponent
$f = 4f_{rh}$ = friction factor based on D_h defined in Chapter 3

Validity/Accuracy: Assessed with suitable empirical values of β in Chapter 8 (7.10b)

The Chilton–Colburn correlation converges, as it should, to the original Reynolds analogy [1.11, 7.37] for Pr = 1 fluids, or more generally when the variation of the empirical exponent β is considered,

$$\mathrm{St} \equiv \frac{\alpha}{c_pG} = \frac{1}{8}f \tag{7.11a}$$

Validity: $(\mathrm{Pr})^{\beta} = 1$ (7.11b)

To simplify the present discussion, since this has no impact on the following variation analysis, we assume $\mathrm{Pr}^{\beta} = 1$ is satisfied so that the original Reynolds analogy is valid "accurately" (the more general case $(\mathrm{Pr})^{\beta} \neq 1$ is addressed in Section 7.10 and more specifically for turbulent regimes in the first part of Chapter 8). Solving Eq. (7.11a) for the heat transfer coefficient yields

$$\alpha = \tfrac{1}{8} c_p G f \propto G f \tag{7.12a}$$

With the universal Blasius type expression of the friction factor f proposed in Chapter 3,

$$f = K_n (\mathrm{Re})^{-n} = K_n \left(\frac{GD}{\mu} \right)^{-n} = K_n \left(\frac{\mu}{GD} \right)^{+n} \tag{7.12b}$$

the heat transfer coefficient will vary with G or $\dot{m}$, for a fixed channel geometry (D_h, $\bar{e}/D_h$, L/D_h) and constant fluid properties [c_p, μ, $(\mathrm{Pr})^{\beta} = 1$], according to

$$\alpha \propto G^{1-n} \propto \dot{m}^{1-n} \tag{7.13}$$

We know from Numerical Analysis 3.2 that the Blasius exponent n for turbulent flow regimes may vary, depending on Reynolds number and $\bar{e}/D_h$, roughly between $0.28 < n < 0$. Thus, the exponent $1 - n$ in Eq. (7.13) can vary between 0.72 and 1; that is,

$$\alpha \propto G^{0.72 \text{ to } 1} \propto \dot{m}^{0.72 \text{ to } 1} \propto \bar{V}^{0.72 \text{ to } 1} \tag{7.14}$$

Equations (7.13) and (7.14) show that the heat transfer coefficient becomes almost directly proportional to $\dot{m}$ or $\bar{V}$ in fully rough turbulent regimes when $n \to 0$, as already reported by Stanton [1.11] in 1897.

The most broadly valid conclusion that may be drawn from the above variation analysis can be stated as follows:

- Under SSSF and otherwise fixed conditions, there is a one to one correspondence between the flow rate $\dot{m}$ and the coefficients α (or $\bar{\alpha}$) and f (or $\bar{f}$). Thus, the single-phase fluid and heat flow phenomena *are coupled* and only one of the two coefficients, α or f, can be treated as an independent variable.
- Stated in analytical terms, there exists under the specified conditions a unique function,

$$\varphi(\alpha / c_p G, f) = 0 \tag{7.15a}$$

that is *valid* for any flow rate $\dot{m}$, under otherwise fixed conditions, provided inlet/output SSSF regimes are possible.

In principle, Eq. (7.15) can be used to predict the heat transfer coefficient from $\mathrm{St} \equiv \alpha/c_pG$ in terms of the friction factor f; that is,

$$\mathrm{St} \equiv \frac{\alpha}{c_pG} = \varphi_{\mathrm{St}}(f) \tag{7.15b}$$

This is the usual application in heat transfer engineering since f is more easily determined than α. Conversely, one can predict f when the variation of α (or the overall thermal performance of the elementary heat exchanger) is known in terms of G or $\dot{m}$.

It should be noted that we have deliberately made no restriction on the type of flow regime in the above validity statement because Eqs. (7.15a) and (7.15b) are *valid* for *any type of flow regime* as demonstrated in Section 7.8.

7.3.2 Boundary Layer Models

It is clear that Eqs. (7.15a) and (7.15b) are valid, as broadly stated, *regardless* of the model one chooses to visualize the physical phenomena expressed by the coefficients α and f. More specifically, the well-known boundary layer analytical models proposed by Prandtl in 1910 [7.39], Taylor in 1916 [7.40], and later twentieth century heat transfer researchers [3.42, 7.41] have their roots in the same fundamental interrelationships between fluid and heat flow phenomena even though their mathematical expressions may look very different from the Chilton–Colburn correlation. This can be shown by considering the following equation recommended by Petukhov [3.42, Eq. (48)] for the heat transfer coefficient in tube flow, which is considered by some researchers [3.43] to be the most sophisticated and accurate correlation available today:

$$\mathrm{St} \equiv \frac{\alpha}{c_pG} \equiv \frac{\mathrm{Nu}}{\mathrm{Re}\,\mathrm{Pr}} = \frac{f/8}{K_1(f) + K_2(\mathrm{Pr})(f/8)} \tag{7.16}$$

where

$$\begin{aligned} f &= (1.82 \log_{10} \mathrm{Re} - 1.64)^{-2} \\ K_1(f) &= 1 + 3.4f \\ K_2(\mathrm{Pr}) &= 11.7 + 1.8\,\mathrm{Pr}^{-1/3} \end{aligned} \tag{7.17}$$

Validity: Fully developed flow in smooth circular tubes, constant fluid properties (or $\Delta T_w \to 0$), $10^4 < \mathrm{Re} < 5 \times 10^6$, and $0.5 < \mathrm{Pr} < 2000$ for both constant heat flux and constant wall temperature boundary condition

Analytical Accuracy: Stated by Petukhov [3.42] as follows.

The disagreement of the predicted Nu [or St] computed on the basis of the expressions of f, $K_1(f)$ and $K_2(\mathrm{Pr})$ [per Eqs. (7.17)] with the analytical [Eq. (7.16)] is within 1% except for the ranges $5 \times 10^5 < \mathrm{Re} < 5 \times 10^6$ and $200 < \mathrm{Pr} < 2000$ where it is 1–2%.

Actual Deviations[8] from Eq. (7.16) are of the order of $\pm 10\%$ as shown in Fig. 3 of Ref. 3.42.

Equations (7.16) and (7.17) can be expressed most generally as

$$\mathrm{St} = \frac{\alpha}{c_p G} = \varphi_{\mathrm{St}}(f, \mathrm{Pr}) \tag{7.18}$$

which for a given fluid at constant temperature (i.e., constant Pr and c_p) reduces to

$$\mathrm{St} = \frac{\alpha}{c_p G} = \varphi_{\mathrm{St}}(f \text{ only}) \tag{7.19}$$

or, as expected, an expression of the same general form as Eq. (7.15b), since the physical law linking the fluid and heat flow phenomena must be independent of the model one chooses to visualize it. For the limiting case of a Pr = 1 fluid, Petukhov's analytical Eq. (7.16) does converge, as it should, to the simple Reynolds analogy,

$$\mathrm{St} = \frac{\alpha}{c_p G} = \frac{\frac{1}{8}f}{1 + 5.1f} \tag{7.20a}$$

or

$$\mathrm{St} \cong \tfrac{1}{8}f \tag{7.20b}$$

within the actual deviation band of $\pm 10\%$ of Eqs. (7.16) with typical friction factor values of the order of 0.03 to less than 0.01 obtained from Petukhov's Eq. (7.17) as previously shown in Table 3.2.

Simply stated, the end engineering objective of *boundary layer* analytical models is to estimate an *effective laminar boundary layer thickness* at the wall, δ_{bl}, to reduce the convective mode of heat transfer to a simple thermal conduction problem by computing the heat flux q'' with Eq. (7.6) according to

$$q'' = \frac{k}{\delta_{\mathrm{bl}}} \Delta T_w \equiv \alpha \, \Delta T_w \tag{7.21}$$

or, from the definition of the heat transfer coefficient,

$$\alpha = \frac{k}{\delta_{\mathrm{bl}}} \tag{7.22}$$

The empirically validated results expressed by Eq. (7.13), previously derived on the basis of the Chilton–Colburn type of Reynolds analogy, can now be visualized in the light of boundary layer theory, by introducing α per Eq. (7.22) into Eq. (7.13) to obtain $k/\delta_{\mathrm{bl}} \propto \dot{m}^{1-n}$ which yields at a prescribed temperature and constant conductivity k

$$\delta_{\mathrm{bl}} \propto \dot{m}^{-(1-n)} = \frac{1}{\dot{m}^{1-n}} \tag{7.23}$$

Equation (7.23) shows that δ_{bl} must decrease with increasing flow rate $\dot{m}$ to account for the variation $\alpha \propto \dot{m}^{1-n}$ per Eq. (7.13). Thus, for a Pr = 1 fluid as the fully rough wall limit is approached and $n \to 0$, Eq. (7.23) converges to $\delta_{\mathrm{bl}} \propto \dot{m}^{-1}$, indicating that the effective boundary film thickness is then inversely proportional to $\dot{m}$, when α varies substantially in direct proportion to the mass flow rate $\dot{m}$, under otherwise fixed conditions.

7.3.3 Rationale for Unified Treatment of Momentum, Heat, and Mass Transfer (UT-MHMT)

The fundamental coupling between momentum and convective heat transfer provides the rationale for a *unified treatment of momentum, heat, and mass transfer* (UT-MHMT) proposed in a number of engineering textbooks [7.42–7.47] published since the 1950s. Rohsenow and Choi [7.42] emphasize this point in their preface:

> Departing from tradition, we have acknowledged only two modes of heat transfer—conduction and radiation. Convection is treated simply as fluid motion. . . .

7.4 VALIDITY LIMITS AND MODELS IN THERMAL SCIENCES

The award winning monograph on the propagation of heat was presented by J. B. J. Fourier to the Institut de France in 1807, that is, 5 years before he was awarded the prize of the Academy of Sciences in 1812. This time lag was due to the fierce opposition of some of the leading mathematicians in the French Academy, who vetoed the publication of the original award winning memoire, thereby forcing Fourier to publish the substance of his work (in a slightly modified book form) almost two decades after its initial presentation. One may rightfully wonder why it took so long for an extremely forceful personality such as Fourier to convince his peers of the validity of his analytical theory of heat diffusion! The answers to this question, thoroughly documented in several books on the works of Fourier [7.48–7.50] can be stated succinctly as follows:

- *Conservatism*. The usual resistance of recognized authorities in any field of human endeavor, including supposedly rational/unemotional scientists, against sudden changes of doctrines (or paradigms) coupled with the very human frailties and susceptibilities of many leading scientists [2.1].
- *Style/Mathematical Elegance*. Lack of absolute mathematical rigor in the eyes of the leading French scientists, particularly the pure mathematician Lagrange.
- *Serious reservations on the validity of the linear heat rate equations* treated as axiomatic truths in Fourier's original monograph.
- Fourier's insistence that the *mathematical equation* of a phenomenological law is sufficient and *requires no physical model.*

Although a more detailed account of Fourier's rebuttals would provide useful insights to heat transfer researchers who have to break new ground (and R & D managers or educators whose mission is to encourage innovation), we confine ourselves here to a short overview of the last two arguments against Fourier's work, since they are more rational and still very relevant today.

7.4.1 Limited Validity of the Linear Heat Rate Equations

In retrospect, it is clear that Fourier lacked adequate empirical data to demonstrate rationally and unequivocally the validity/accuracy limits of his axiomatic laws of heat propagation. He was therefore quite vulnerable to attacks from his opponents, especially in connection with Newton's approximate convective cooling law, Eq. (7.1), described by Fourier in his rebuttals as "a physical law generally recognized by all 'natural philosophers'." This was a rather weak argument that did not convince his critics, particularly the great mathematician Lagrange who showed scant interest in phenomenological laws beyond the scope of Newton's celestial mechanics. This is probably the main reason why the French scientific community had to wait until 1822, and English speaking scientists/engineers until around 1878, to become acquainted with Fourier's masterpiece initially presented at the French Academy of Science in 1807!

In fact, eighteenth century German and French scientists had already questioned, and Fourier's contemporaries, Dulong and Petit, masterfully demonstrated [7.51], the limited validity limits of the linear heat rate equation for pure convection and for the combined convective and radiation modes typical of some high-temperature applications. In essence, they verified empirically that the heat flux q'' varies according to the generally nonlinear law expressed by Eq. (7.9) with an exponent $a \neq 1$. The quantitative assessments by Dulong and Petit of the exponent $a \geq 1$ for single-phase fluids used as coolants (i.e., in the fluid heating mode when $T < T_w$) are in substantial agreement with modern correlations for forced *and* natural convection [7.52]. To simplify the presentation, we do not make any distinction between local and average heat transfer coefficients and we use the symbol α without the overbar.

1. Forced Convection Single-Phase Fluid at Moderate $|\Delta T_w|$

When $\Delta T_w = T_w - T > 0$ as in Newton's cooling test (i.e., fluid heating mode) at moderate ΔT_w, α is independent of ΔT_w and the exponent a in Eq. (7.9) is equal to unity. This holds true in the fluid cooling mode when $T_w < T$ and $\Delta T_w < 0$ at $|\Delta T_w|/T \approx 0$, and the same numerical value α is applicable in the cooling and heating modes; that is,

$$q'' \propto |\Delta T_w|^{a \to 1} \propto |\Delta T_w|^1 \tag{7.24a}$$

Linear Relationship: $a = 1$ (7.24b)

Symmetry: $\alpha_{\text{fluid cooling mode}} = \alpha_{\text{fluid heating mode}}$ (7.24c)

Validity: Pure forced convection in heating and cooling modes when $\Delta T_w/T \approx 0$. (7.24d)

In other words, the smaller the wall-to-fluid temperature differential, or driving force, the more accurate Newton's linear cooling law, $q'' \propto (\Delta T_w)^1$. This is fortunate and quite significant in many engineering applications since *thermodynamically efficient heat exchangers must be designed for* $|\Delta T_w|/T \ll 1$ to minimize irreversible losses. Thus, it is possible to design thermodynamically efficient surface heat exchangers with an adequate accuracy on the basis of the simpler linear law according to the well-known conventional LMTD and/or the alternative effectiveness-NTU methodology highlighted in Section 7.6 and Appendix *B*.

2. Forced Convection Single-Phase Fluid at Large ΔT_w

According to widely accepted modern correlations of α, such as those shown in Table 6, Chapter 3 of Ref. 7.52, the exponent $a \neq 1$ and the symmetry between the fluid heating and cooling modes is broken; that is,

$$q'' \propto |\Delta T_w|^{a \neq 1} \tag{7.25a}$$

Nonlinear Relationship: $a \neq 1$ (7.25b)

Symmetry Broken: $\alpha_{\text{fluid cooling mode}} \neq \alpha_{\text{fluid heating mode}}$ (7.25c)

Validity: Pure forced convection at high ΔT_w (7.25d)

3. Natural Convection Single-Phase Fluids

Recognized correlations for the heat transfer coefficient in the natural convection mode, such as those listed in Table 5, Chapter 3 of Ref. 7.52, show that

$$\alpha \propto |\Delta T_w|^{0.25 \text{ to } 0.33} \tag{7.26a}$$

With $q'' \equiv \alpha|\Delta T_w|$, Eq. (7.26a) results in a nonlinear q'' versus $|\Delta T_w|$ relationship:

$$q'' \propto |\Delta T_w|^a = |\Delta T_w|^{1.25 \text{ to } 1.33} \tag{7.26b}$$

Nonlinear Relationships: $a = 1.25$ to $1.33 > 1$ (7.26c)

Symmetry Broken:
$\alpha_{\text{fluid cooling mode}} \neq \alpha_{\text{fluid heating mode}}$ (7.26d)

Validity: Moderate $|\Delta T_w|$ beyond minimum natural convection threshold (7.26e)

In other words, the relationship between heat flux and driving force is *never exactly linear* in natural convection even at relatively small $|\Delta T_w|$ or $|\Delta T_w|/T \ll 1$. This is understandable since the *convective motion* of macroscopic fluid "packets" is enhanced by the larger density changes associated with larger $|\Delta T_w|$, thus increasing α because of the improved "scrubbing action" at the surface–fluid interface. It is only in the *limiting case of zero bulk convection*, when the stagnant fluid can be treated like a solid medium according to the laws of conduction in solids, that $a = 1$ and the linear law $q'' \propto |\Delta T_w|^1$ applies rigorously.[9]

4. Two-Phase Nucleate Boiling, $T_w > T_{sat}$

In this case the *coolant* changes phase as it absorbs heat from an external heat source and the exponent a may range anywhere from 2 to 5 [4.58, 7.53], when ΔT_w exceeds the minimum differential required to initiate nucleation $(\Delta T)_{\text{i.n.}}$; that is,

$$q'' \propto (\Delta T_w)^a = (\Delta T_w)^{2 \text{ to } 5} \tag{7.27a}$$

Highly Nonlinear Relationship: $a = 2$ to $5 \gg 1$ (7.27b)

Symmetry Broken:
$\alpha_{\text{fluid cooling mode}} \neq \alpha_{\text{fluid heating mode}}$ (7.27c)
(condensation) (evaporation)

Validity: Nucleate boiling beyond a minimum threshold $\Delta T_w > (\Delta T)_{\text{i.n.}}$ = threshold temperature difference to initiate nucleation (7.27d)

It is interesting to note that the nineteenth century issue of *linearity* versus *nonlinearity* is apparently still very much alive [7.54], judging by the major controversies created by two books in the 1940s and 1970s that focus on extremely *nonlinear* applications, that is, high ΔT_w process heat exchangers [7.55] and extremely high heat flux nuclear applications [7.56]. These controversies vividly illustrate the necessity of documenting as fully as feasible the validity/accuracy limits of any recommended design correlation according to Guideline 2 in Chapter 1.

It is important to stress that Fourier, the lucid experimental physicist, was fully aware of the validity/accuracy limits of the convective heat rate equation. As a matter of fact he outlined in his writings [7.48] the types of combined experimental/analytical research program actually carried out by nineteenth and twentieth century researchers to couple[10] the convection and fluid flow phenomena in order to "perfect this theory in the future."

The symmetry in forced convection between the heating and cooling mode at $|\Delta T_w|/T \approx 0$ provides part of the rationale for the mirror-image heat exchanger concept highlighted for forced-convection evaporation/condensation in Section 4.7 and discussed more fully in Chapter 9 with special emphasis on single-phase heat exchangers. Although we focus mainly on thermodynamically efficient heat exchangers in this book, in view of their increasing relevancy since the oil crisis of the 1970s, the basic nonequilibrium thermodynamic methods introduced later for applications within the linear domain can be extended to deal with more complex nonlinear applications as shown in Chapter 9.

7.4.2 Mathematical Expressions of Physical Laws Versus Models

It is instructive to examine the attitude of leading scientists toward the demand of an explanation of physical phenomena, or the discovery of what underlies them. During Fourier's time, the dominant French mathematical physicists firmly believed that the explanation of the nature of heat on the basis of mechanistic models[11] was indispensable to progress in thermal sciences. In this context, Fourier stood prominently *outside the main scientific stream* since his analytical treatment of heat transfer clearly proclaims that *what matters is the equation* of the process regardless of the explanation or the model one wants to adopt to visualize it. In the process, Fourier, the mathematical physicist, also invented several important new mathematical tools (like Newton a century before him) to solve his diffusion problem. As a matter of fact, Fourier is better known outside the heat transfer scientific community for one of his most outstanding (and at the time also controversial) contributions to mathematics: namely, the *Fourier analysis*, which is also one of the most useful tools in the new (1970s) physics of *nonlinear* dissipative dynamic systems [3.6].

By giving up models and laying stress on the equations of phenomena, Fourier started an important scientific trend that bore many fruitful results but also carried the germ, when driven to the extreme, of a less productive activity, best described by quoting Di Francia [1.1]:

> ...an attitude which later rendered physicists critical towards a certain type of mathematical physics... we are referring to the procedure by which, starting from some laws established with *limited precision* one derives by refined and rigorous mathematics, a number of *extremely precise* results, in cases void of

physical interest, which no one is willing or able to verify experimentally! Sometimes one cannot blame physicists for their lack of enthusiasm towards that kind of activity. An activity that can be *misleading* [my italics] and induce us to believe that we know much more about the physical world than we actually do.

It is particularly, gratifying for thermal engineers to recall that while the theorists of the French Academy of Science were arguing for or against *models*, the young French engineer Sadi Carnot published (at his own expense!) the epoch-making memoire *Reflections on the Motive Power of Fire* [1.7], which essentially laid down the foundations of a radically *new scientific method of "thinking physical problems"* namely, classical or phenomenological equilibrium thermodynamics. His most significant scientific breakthrough is particularly remarkable in the context of our present discussion, because he used the *wrong model*, the concept of an indestructable "caloric fluid," to arrive at the *right conclusions*. He demonstrated the essence of the second law of thermodynamics, now known as Carnot's theorem, through the use of a novel system of thinking that German speaking scientists later called "Gedanken Experimente"[12] meaning thought experiments. By defining the well-known *idealized closed cycle* and working with a constant quantity of fluid without any irreversible losses between two isothermal and two adiabatic (isentropic) processes, he showed that his ideal reference cycle would yield the maximum amount of useful mechanical energy per unit of "caloric" input in the power generation mode, or a maximum energy conversion efficiency *regardless* of the *fluid used* in the system. His well-known maximum efficiency relationship (see Chapter 6) is $\eta_C = (T_H - T_L)/T_H$, where T_H and T_L represent the constant-temperature heat *source* and *sink*, respectively, with $T_H > T_L$.

The power of Carnot's thought experiments is vividly demonstrated by the fact that by "conceptually running" his closed cycle in the opposite direction in a *reverse power cycle* mode, when T_H becomes the high-temperature *sink* and T_L the low-temperature *source*, he invented the mechanical refrigeration cycle. By this conceptual exercise he actually *anticipated the art of artificial refrigeration* and can thus be considered the conceptual father of the refrigeration industry (see Chapter 4).

True to what has become almost a tradition in the turbulent evolution of thermodynamics, his epoch-making memoir was simply ignored by the French mathematical physicists and engineers for reasons that are discussed[13] in great detail in several chapters (some in English) of the "Actes de la Table Ronde—Sadi Carnot et l'Essor de la Thermodynamique" held at the Paris Ecole Polytechnique in 1974 on the occasion of the 150th anniversary of Carnot's 1824 memoir [1.8]. His forgotten pioneering work was finally brought back from oblivion 10 years later in a paper entitled "Memoire sur la Puissance Motrice de la Chaleur," published by another graduate of Ecole Polytechnique, the French mining engineer E. Clapeyron [4.16, 7.57]. In this paper, Clapeyron, who adhered more firmly than Carnot to the obsolete caloric model of heat (see Note 1 in this chapter), made several original contributions including the equation that bears his name (reviewed in Section 4.4) and showed the first graphical representation of a finite Carnot cycle in the now famous pressure–volume diagram.[14] Probably the most far-reaching contribution of Clapeyron is the fact that he wrote his paper in a more rigorous engineering style that could easily be understood by thermal engineers beyond the borders of France.

Clapeyron's paper, like Carnot's earlier memoire, raised no interest whatsoever among the French scientific intelligentsia. Fortunately, it had a most profound impact on two eminent engineers/scientists outside France—William W. Thomson (Lord Kelvin) and Rudolph Clausius—with the positive outcome we know: the formalization of a new branch of physics called classical or phenomenological equilibrium thermodynamics.

The impact of Carnot's process of reasoning can best be stated by quoting one of the U.S. representatives [7.58] at the Actes de la Table Ronde meeting in Paris [1.8]:

> "Nothing in the whole range of Natural Philosophy is more remarkable than the establishment of general laws by such a process of reasoning." Those are the words of William Thomson [7.59] used in 1849 to express his wonder at the unexpected power of Sadi Carnot's arguments [4.16]. Carnot had set out to analyze the principles underlying the operation of heat engines; his analysis led him to a completely general theorem providing, in Emile Clapeyron's phrase, "the common link between the phenomena caused by heat in solid bodies, liquids, and gases" [7.60]. The modern reader of Carnot's 1824 memoire, aware of the thermodynamics developed after his death, a thermodynamics that preserves the spirit of Carnot's approach in addition to many of his ideas, can only echo William Thomson's words of praise.

Bearing in mind that the overwhelming majority of thermal engineering breakthroughs achieved to date were based on *phenomenological* laws of classical thermodynamics requiring no physical models, we may conclude this cursory historical overview with the following appropriate engineering-oriented guideline.[15]:

> ***GUIDELINE 13*** *The validity of the mathematical expression of a physical law, or an empirical correlation, and its usefulness in engineering do not depend on an explanation or a model. Although extremely useful as didactic or heuristic tools,* ***physical models*** *are nevertheless* ***not*** *mandatory.*

7.5 THERMOSTATICS AND THERMODYNAMICS

The thermodynamics developed by early nineteenth century pioneers from Carnot, Clausius, Kelvin, Mayer, Joule, and Rankine to Gouy, Stodola, and Planck at the turn of the century is now referred to as *equilibrium thermody-*

namics (ET) or, more appropriately, *thermostatics*[16] to distinguish this classical branch from the new twentieth century *nonequilibrium thermodynamics* (NET) addressing truly dynamic, time-dependent situations. The scope of these two main branches of thermodynamics can best be delineated by their respective axiomatic validity limits, as highlighted below, together with a succinct overview of the two different methodologies successfully used by chemical physicists and thermal engineers to deal with equilibrium and nonequilibrium thermodynamic systems. To keep this overview brief, no specific references are listed for well-known basic concepts addressed in thermodynamic textbooks.[17] Furthermore, we focus here almost exclusively on ideal gases, for the sake of expediency, keeping in mind that the same conclusions can be derived for other real fluids in view of the universal validity of fundamental thermodynamic laws in normal engineering systems.

7.5.1 Continuum Hypothesis

Both branches of thermodynamics ET *and* NET are based on the *continuum hypothesis*, which can be deduced from the definition of the specific mass of a substance. In the case of a gas, the well-known p, v, T relationship of classical ET can be expressed most generally as

$$Pv = Z\left(\frac{\mathscr{R}}{M}\right)T \tag{7.28}$$

where $\mathscr{R}$, M, and Z are the universal gas constant, the "molar weight," and the compressibility factor, respectively. Note that Z accounts for deviations from the perfect gas laws. For a perfect gas, $Z = 1$ and Eq. (7.28) reduces to the well-known perfect gas law which can be solved for $\rho = 1/v$:

$$\text{Specific mass} = \rho = \frac{M}{\mathscr{R}}\left(\frac{P}{T}\right) \tag{7.29a}$$

$$\textit{Validity}: \quad \text{Perfect gas } Z = 1 \tag{7.29b}$$

The average specific mass $\bar{\rho}$ of a substance contained in a finite elementary volume $\Delta\mathscr{V}$ within a larger thermodynamic system of total volume $\mathscr{V}$ delineated by its spatial boundary (as indicated in the closed system depicted in Fig. 7.1) is by definition the ratio of the mass Δm contained in the subsystem of volume $\Delta\mathscr{V}$ or

$$\bar{\rho} \equiv \frac{\Delta m}{\Delta\mathscr{V}} = \text{average specific mass in subsystem} \tag{7.30}$$

The local specific mass ρ at any spatial point, defined by its spatial coordinates (x, y, z) at a given instant of time θ, is by definition the limit of $\bar{\rho}$

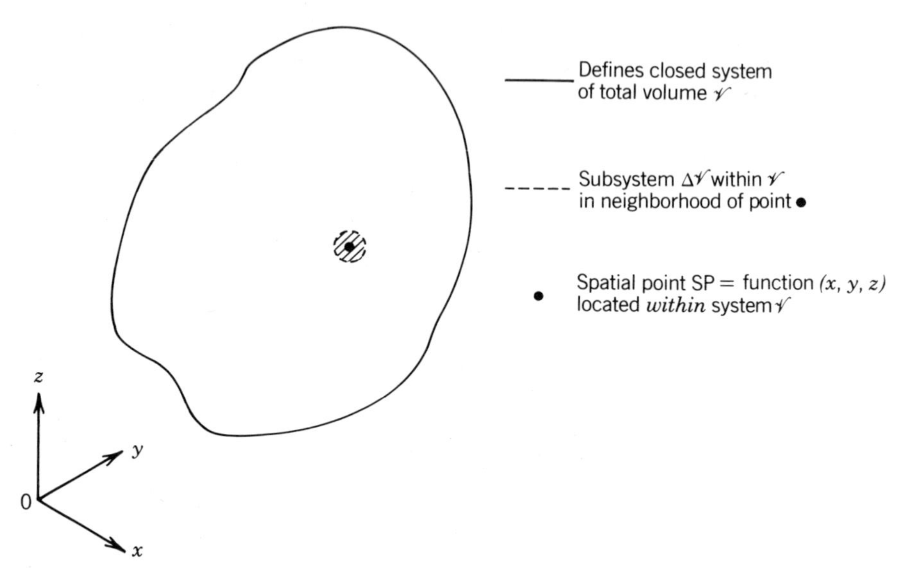

FIGURE 7.1. Closed thermodynamic system.

around point P when $\Delta\mathscr{V} \rightarrow 0$: that is,

$$\rho \equiv \frac{\lim}{\Delta\mathscr{V} \rightarrow 0}\left(\frac{\Delta m}{\Delta\mathscr{V}}\right) \tag{7.31}$$

In other words, the substance is treated like a *continuum*, disregarding the impact of the molecular nature of matter on ρ, which thus becomes a point function of spatial position and, most generally, to include time-dependent nonequilibrium situations, θ, or a field function φ_ρ of spatial coordinates (x, y, z) and time θ:

$$\rho = \varphi_\rho(x, y, z, \theta) \tag{7.32}$$

This simplifying assumption is permissible as long as the smallest spatial dimension of the thermodynamic system (or any defined subsystem therein) is at least several orders of magnitude larger than the mean free path of the molecules. This validity requirement is fortunately amply met by "human scale" components in normal thermal systems such as surface heat exchangers, the main subject of this book. Accordingly, we shall tacitly assume from now on that the continuum hypothesis is satisfied unless otherwise stated.

7.5.2 Continuum Hypothesis in Multiphase Systems

It is important to note, in the context of two-phase flow phenomena addressed in previous chapters, that the continuum hypothesis can be successfully applied to multiphase mixtures provided there is a good dispersion [7.61]. For example, "wet steam" is effectively treated as a continuum in the lower stages of steam turbines where the liquid droplets are finely dispersed throughout the vapor phase. Similarly, because the entrained liquid refrigerant droplets in direct-expansion refrigeration evaporators are dispersed as a "fog" in the gas phase, it is also possible to simplify the highly complex multiphase flow phenomena by treating the resulting *quasihomogeneous mixture* as a continuum (see homogeneous flow model in Chapter 5).

It is essential to keep in mind, that such single-component two-phase mixtures consisting of finely dispersed liquid droplets (fog) in their vapor phase are typically associated with nonequilibrium conditions and thus violate the fundamental assumption of thermodynamic equilibrium addressed in Section 7.5.4! In effect, the liquid droplet's actual temperature $T_{act} \neq T_{sat}$, where T_{sat} is the normal equilibrium saturation temperature (based on flat liquid–vapor interface without surface tension effects) corresponding to the local pressure P, as discussed more fully in advanced thermodynamics and specialized steam turbine texts, or, for the particular case of forced-convection evaporators, in Ref. 3.32.

7.5.3 Phenomenological and Statistical Methodologies

The bitter conflict between the proponents of the phenomenological or macroscopic methodology (the anti-atomists) and the advocates of the statistical microscopic approach (the atomists) largely subsided by the turn of this century and the two alternative methodologies have since become complementary. Both methodological approaches are treated in some modern thermodynamic textbooks, whereas some educators believe the two philosophies do not mix and focus on one methodology. We shall simply acknowledge here that the phenomenological and statistical approaches are alternative methodologies to deal with the same physical reality and that both methods have been applied successfully in equilibrium as well as non-equilibrium thermodynamics.

In this engineering-oriented book, we shall continue to favor the classical or phenomenological approach except when it is useful to appeal to statistical mechanics to provide a plausible explanation of abstract concepts such as entropy (see footnote 3 of Chapter 3). In spite of this pragmatic engineering choice, we benefit fully (though indirectly) from all the breakthroughs achieved on the basis of statistical methods when we use modern thermophysical data. A good case in point is the generalization in terms of reduced pressures of several thermodynamic properties in Chapter 6. Our conclusions were derived from the empirical (i.e., phenomenological) fact that key parameters fall on common universal curves in terms of reduced pressure within stated validity/accuracy limits. However, some of the physical property data used to demonstrate these useful phenomenological relationships were certainly derived with the help of modern statistical thermodynamic concepts (as indicated in Note 2 of Chapter 6).

The pragmatic reasons for favoring the classical phenomenological approach in this book can be summarized as follows:

1. The overwhelming majority of heat transfer engineers have had a fair exposure to the phenomenological approach but none (or a very limited one) to kinetic theory, microscopic thermodynamics, or statistical mechanics.

2. The phenomenological approach is totally adequate and simpler to apply in normal heat exchanger analyses and it is generally more reliable and accurate, as implied by Guideline 13 and stated in the following unbiased assessment (italics added) by Fermi [7.62] who, like most twentieth century physicists, was also an expert in statistical mechanics:

> But the approach in pure [i.e., phenomenological] thermodynamics is different. Here the fundamental laws are assumed as postulates based on experimental evidence, and conclusions are drawn from them without entering into the kinetic mechanism of the phenomena. This procedure has the advantage of being *independent, to a great extent, of the simplifying assumptions that are often made in statistical mechanical considerations. Thus, thermodynamic results are generally highly accurate.*

7.5.4 Fundamental Limitations of Thermostatics

Classical equilibrium thermodynamics (ET), or thermostatics, is based on the axiomatic assumption that the system (or any finite subsystem within it) is in *thermodynamic equilibrium*. Referring to the closed nonreactive system defined by the boundary line in Fig. 7.1, this axiomatic assumption simply means that two intensive state properties, such as T and P, must be independent of time (θ) and spatial location (x, y, z) within the arbitrarily prescribed boundary. Any specific extensive thermodynamic property π_i, such as u, h, and s, is thus uniquely defined by two independent intensive properties, such as T and P, in classical thermodynamics. Consequently, the specific mass ρ [see Eqs. (7.28) and (7.29a)] is also uniquely defined and constant throughout the whole system under consideration. As a corollary, any extensive property π_i expressed on a volumetric basis as $\rho(\pi_i)$, such as $\rho(u)$, $\rho(h)$, and $\rho(s)$, is also constant and independent of time as summarized in the following equations.

$$\rho,\ \pi_i \text{ and } \rho\pi_i = \text{constant at any spatial point } (x, y, z)$$

$$\text{within the system and independent of time } \theta \qquad (7.33)$$

where π_i = any specific (i.e.: per unit mass) extensive property such as internal energy u, enthalpy h, or entropy s

$\rho\pi_i$ = any extensive property per unit volume such as ρu, ρh, and ρs

The equilibrium requirement clearly imposes impossible restrictions on the application of classical equilibrium thermodynamics in heat transfer engineering, since all modes of heat transfer (the heat rate equations) depend on the existence of thermal gradients, that is, on *nonequilibrium* conditions that obviously are *not compatible* with the basic tenets of equilibrium thermodynamics. Furthermore, since thermostatics addresses only end equilibrium states, the key independent *variable time θ never appears* in classical thermodynamics equations.

On the reverse side of the coin, Eq. (7.33) spells out the unique and most powerful feature of ET, namely, its great generality. Fundamental classical thermodynamic relationships are *applicable* to all (normal macroscopic) systems *regardless of substance and physical configuration*. In other words, the same thermodynamic relationships are applicable in the case of a 100 W household refrigerant condenser as for a 1000 MW nuclear steam turbine condenser! The most fundamental advantages and limitations of ET are summarized in the following guideline, with special emphasis on heat transfer engineering applications.

GUIDELINE 14 *(a) Classical thermodynamic relationships are valid for any macroscopic system satisfying the continuum hypothesis regardless of the type of substance, physical size, and configuration. (b) Time-dependent heat transfer problems **cannot** be solved through the **sole** use of time-independent equilibrium thermodynamic concepts.*

One of the main objectives of the elementary heat exchanger analyses carried out in Section 7.6 is to show exactly how pragmatic nineteenth century heat transfer engineers managed to circumvent this major limitation of ET and anticipated, in substance, some of the fundamental concepts of twentieth century nonequilibrium thermodynamics.

7.5.5 Twentieth Century Nonequilibrium Thermodynamics

The end objective of nonequilibrium thermodynamics (NET) can now be defined very broadly in the context of the fundamental limitations of ET, or thermostatics, according to the following guideline.

GUIDELINE 15 *The aim of **nonequilibrium thermodynamics** is to remove the unnatural restraints imposed by the thermodynamic equilibrium hypothesis and other idealizations of thermostatics, thus extending the scope of thermodynamics to address **truly dynamic** situations involving the independent variable time, nonequilibrium conditions, and dissipative (friction) phenomena.*

The mathematical chemical physicist and Nobel laureate, Ilya Prigogine, one of the early pioneers of this extremely broad new branch of physics, also

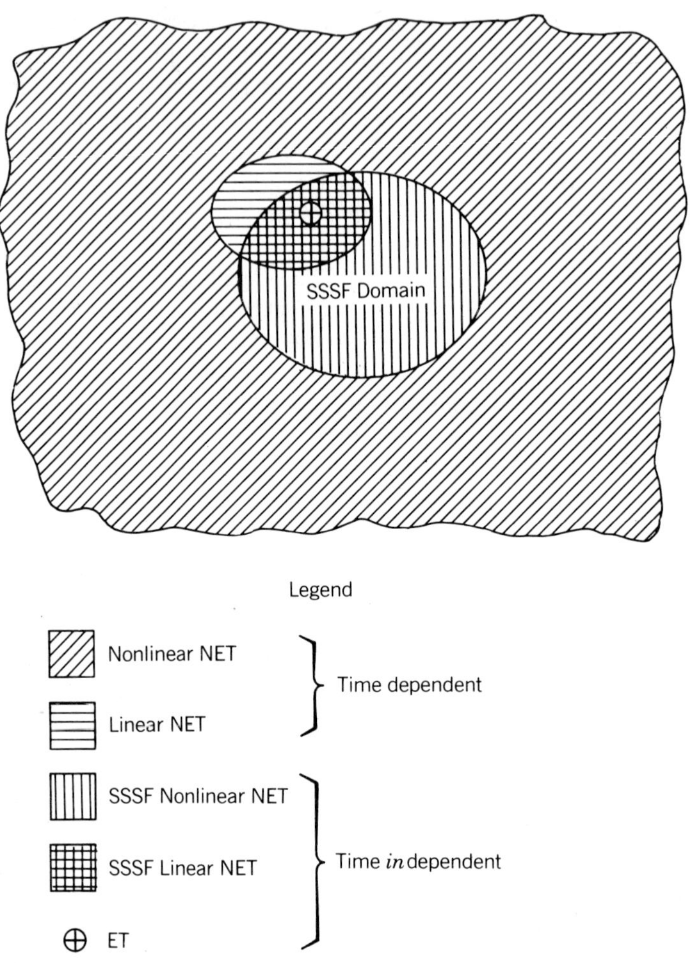

FIGURE 7.2. Schematic domains of nonlinear NET, linear NET, and equilibrium thermodymamics (ET). *Note*: Linear heat flux fluid-to-fluid surface heat exchanger systems fall within the SSSF nonlinear and linear domains indicated for turbulent and laminar flow regimes, respectively.

known under the name of *thermodynamics of irreversible processes*, identifies three major stages in the evolution of thermodynamics. These are summarized below and schematically portrayed in Fig. 7.2, with special emphasis on simple nonreactive (no chemical reactions) systems typically encountered in normal fluid-to-fluid heat exchangers.

First Stage Mid-nineteenth Century: Equilibrium Thermodynamics. This branch of thermodynamics, characterized by the equilibrium hypothesis, can be considered the limiting case of NET when the irreversible entropy production, the fluxes (e.g., the heat flux q''), and the generalized thermodynamic *forces* [corresponding, in the case of convection, to the conventional driving force $\Delta T_w = T_w - T$ per Eq. (7.1)] are all zero.

Second Stage First Half of Twentieth Century: Linear Nonequilibrium Thermodynamics. This branch is characterized by conditions close to equilibrium when the generalized thermodynamic driving forces are weak and the fluxes are linear functions of the forces as in single-phase forced-convection heat exchangers when $|\Delta T_w|/T \approx 0$ per Eq. (7.24).

Third Stage Current State-of-the-Art: Nonlinear Nonequilibrium Thermodynamics. This branch is characterized by nonlinear and more complex relationships between the fluxes and the driving forces under nonequilibrium conditions *far from equilibrium* and frequently associated with the creation of new organized *dissipative structures*, to use the term coined by the leaders of the Brussels School of Irreversible Thermodynamics [2.17, 7.17, 7.18, 7.63–68]. The natural convection *Bénard cells* typically illustrate such new dissipative structures in the *nonlinear* NET domain.

Nonlinear, far from equilibrium phenomena are at the forefront of fundamental research in nonequilibrium thermodynamics as well as modern (macro) physics [2.14]. In the last decades some specialized scientific/technical publications have appeared, such as the *Journal of Non-Equilibrium Thermodynamics*, founded in the 1970s, which addresses particularly nonlinear problems, and papers dealing with complex engineering problems (combustion processes) on the basis of NET theory have occasionally been published in well-known engineering heat transfer journals [7.69].

7.5.6 The Basic Tools of NET

The highly complex and abstract concepts of NET are thoroughly discussed in specialized texts addressed primarily to mathematical/chemical physicists [7.18, 7.67–7.72]. In an attempt to assess the potential usefulness of NET in engineering heat transfer applications, we summarize below, admittedly in a somewhat rudimentary fashion, the fundamental concepts outlined in recent general survey articles [7.64–7.66] and explained more precisely in terms easily understood by the average thermal engineer not familiar with the intricacies of NET by Glansdorff [7.63].

7.5.6.1 Local Equilibrium Theorem. To address nonequilibrium systems the hypothesis of *local equilibrium* within a macroscopic system in a state of nonequilibrium is introduced.

> ***GUIDELINE 16*** *According to the local equilibrium hypothesis, the thermodynamic state of a nonequilibrium system remains locally defined at any spatial point within the system by the same intensive properties as in*

equilibrium thermodynamics and the same classical relationships between state properties (such as the ideal gas law $\rho = M/\mathcal{R}(P/T)$) remain valid locally.

This hypothesis has been validated for macroscopic thermodynamic systems (in the linear and even some nonlinear NE regions) satisfying the continuum hypothesis provided the fluid velocities are small compared to the velocity of sound (i.e., Mach number $\ll 1$). This validity requirement is therefore always satisfied in normal macroscopic heat exchanger systems addressed in this book, because high-velocity fluid flow applications (at Mach numbers near or above unity) are outside the scope of this introductory text.

With the help of the local equilibrium theorem, the nonequilibrium state of a given system can be determined analytically, at least in principle, by summing up the local volumetric terms of $\rho\pi_i$ multiplied by the differential volume $d\mathcal{V}$, that is, $\rho\pi_i\,d\mathcal{V}$ over the full system volume $\mathcal{V}$ to obtain the extensive property term Π_i at an instant θ according to

$$\Pi_i = \int_{\mathcal{V}} \rho\pi_i\,d\mathcal{V} = \Pi_i(\theta) \tag{7.34}$$

Taking as an example the key property of irreversible thermodynamics S, with s, the specific entropy, generally a field function $s = \varphi_s(x, y, z, \theta)$ like $\rho = \varphi_\rho(x, y, z, \theta)$, one obtains from Eq. (7.34) the total system entropy S at a given instant of time θ:

$$S = \int_{\mathcal{V}} \rho s\,d\mathcal{V} = S(\theta) \tag{7.35}$$

Other extensive properties of the complete system in a state of nonequilibrium such as $U(\theta)$ and $H(\theta)$, the total system internal energy and enthalpy, respectively, can be computed at a given instant of time according to Eq. (7.35) simply by substituting u and h for s under the integral sign on the right-hand side of Eq. (7.35).

7.5.6.2 Transfer of Energy and Entropy Production. The distinguishing feature of irreversible thermodynamics is the description of the transfer process for any substance on the basis of the transfer of *entropy* which *uniquely* characterizes the complex interrelated transfer phenomena, as discussed more fully by Luikov and Mikhailov [7.71] who wrote the first advanced textbook fully dedicated to engineering applications of NET concepts.

We shall confine ourselves to those fundamental concepts that are absolutely essential for an assessment of the potential application of NET as a useful tool in normal heat transfer engineering applications, with special emphasis on forced-convection heat exchangers operating under steady-state conditions without chemical reactions.

Using the formalism introduced by the Brussels School of Irreversible thermodynamics, one starts with the balance equation for the total entropy of the complete thermodynamic system $S(\theta)$ and expresses the total differential change dS in a time interval $d\theta$ according to

$$\frac{dS}{d\theta} = \frac{dS_e}{d\theta} + \frac{dS_i}{d\theta} \tag{7.36}$$

where $dS_e/d\theta$ denotes the external contribution due to exchanges with the outside world and $dS_i/d\theta \equiv \mathscr{P}_i$ is the internal entropy production due to all irreversible processes occurring *within* the thermodynamic system. The second law of thermodynamics requires that

$$\mathscr{P}_i \equiv \frac{dS_i}{d\theta} \geq 0 \tag{7.37}$$

At the ultimate equilibrium state of classical equilibrium thermodynamics, the entropy production vanishes and $\mathscr{P}_i \equiv 0$.

The entropy production term $\mathscr{P}_i$, which uniquely characterizes the evolution of the thermodynamic system under nonequilibrium conditions, can be expressed with the help of the local equilibrium hypothesis, using the balance equations for matter, momentum, and energy according to:

$$\mathscr{P}_i \equiv \frac{dS_i}{d\theta} = \int_{\mathscr{V}} d\mathscr{V}\,\sigma[S_i] = \int_{\mathscr{V}} d\mathscr{V} \sum_{\alpha} J_\alpha X_\alpha \tag{7.38}$$

where $\sigma[S_i]$ is the local volumetric (per unit volume) entropy production, which is most generally a function of spatial coordinates (x, y, z) and time θ.

J_α are the flows (or fluxes) of the irreversible processes and X_α the corresponding *generalized thermodynamic forces* which may be either the gradient of transport phenomena (a reformulation of the conventional empirical rate equations) or chemical affinities (in the case of chemical reactions outside the scope of this book).

The equilibrium state in equilibrium thermodynamics (ET) thus represents the limiting case of NET when all the flows (or fluxes) and the generalized forces vanish simultaneously: that is,

$$\mathscr{P}_i = \frac{dS_i}{d\theta} \to 0, \qquad J_\alpha \to 0 \quad \text{and} \quad X_\alpha \to 0 \tag{7.39}$$

Limiting case of NET ≡ equilibrium states of ET

as highlighted in Fig. 7.2.

The current status of the rapidly evolving science of NET can best be assessed by the following quotations from a fairly recent publication by

Glansdorff and Prigogine [7.72]. These excerpts (italics added) are particularly relevant for a factual assessment of the present and future role of NET in heat transfer engineering—the art, rather than the pure science.

> Up to now, there exists a rather complete theory of non-equilibrium thermodynamics only in the *linear range*... corresponding to linear relations between flows and forces...
>
> But what is perhaps the most interesting and important single problem of non-equilibrium thermodynamics concerns the stability theory and this has not yet been studied with sufficient generality: may we always continue to extra-polate results obtained by equilibrium thermodynamics, to far-from-equilibrium conditions? This is clearly not so in a wide class of hydrodynamic problems... For example a horizontal layer of fluid heated from below will for a sufficiently large value of the "adverse" temperature gradient present a convection pattern (the so called Benard instability). *Also* laminar *flow* will for a sufficiently large Reynolds number *become turbulent*.
>
> What we wish is to deduce a stability theory from the thermodynamics of irreversible processes which leads to information, consistent with this kinetic method, but *independent of the detailed normal mode analysis*. But clearly we have then to incorporate in some way into our thermodynamic description the response of the system to fluctuations. In other words *we have to build a generalized thermodynamic theory which includes also a macroscopic theory of fluctuations*.

Focusing here more specifically on the rather narrow (compared to the vast scope of NET) field of forced-convection nonreactive fluid-to-fluid heat exchangers operating stably under steady-state conditions, we may dispense, for now, with the complexities of stability theory and summarize this overview with the following relevant guideline:

> ***GUIDELINE 17*** *A rather complete theory of nonequilibrium thermodynamics is currently available only for the "not far from equilibrium" linear domain when the relationships between* ***all*** *the flows (or fluxes) and forces are linear.*

7.5.7 Heat Transfer Engineering in the Context of NET

The end engineering objective of the thermal system components defined here as a *fluid-to-fluid surface heat exchanger system* (FFSHES) can be stated most generally in thermodynamic terms as follows:

> ***GUIDELINE 18*** *Thermodynamically, the purpose of a* ***fluid-to-fluid surface heat exchanger system*** (FFSHES), *operating under assumed steady-state conditions without chemical reactions, is to cool a hotter fluid stream* $\dot{m}_H$ *from an inlet equilibrium state* $(\mathrm{ES})_{1,H}$ *at* $T_{1,H}$ *to a specified*

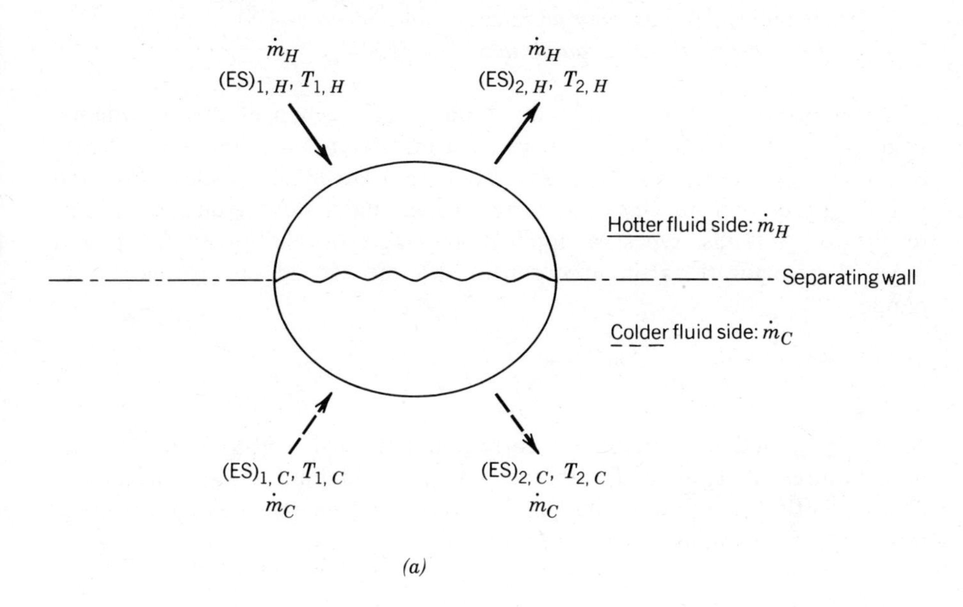

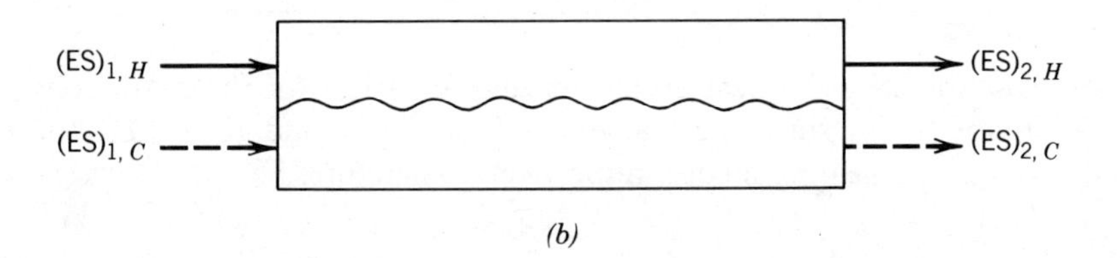

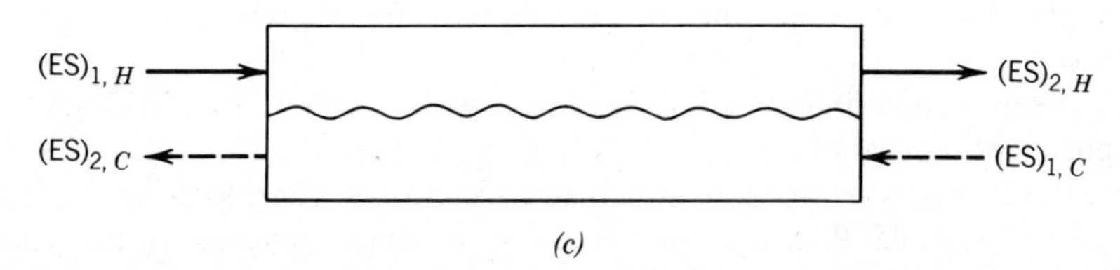

FIGURE 7.3. Schematic thermodynamic diagrams of fluid-to-fluid surface heat exchangers: (*a*) general case, (*b*) pure parallel flow, and (*c*) pure counterflow arrangements. *Note*: ES stands for equilibrium state. First subscripts 1 and 2 refer to inlet and outlet states, respectively; second subscripts *H* and *C* refer to hotter and colder fluids, respectively.

outlet equilibrium state $(ES)_{2,H}$ *by interaction through a separating heat exchanger surface (shown schematically as* ~~ *in Fig.* 7.3) *with a colder fluid stream* $\dot{m}_C$, *heated from an inlet equilibrium state* $(ES)_{1,C}$ *at* $T_{1,C} < T_{1,H}$ *to a prescribed outlet equilibrium state* $(ES)_{2,C}$.

Since all temperatures within the surface heat exchanger thermodynamic system are bounded between the specified maximum and minimum entering equilibrium temperatures $T_{1,H}$ and $T_{1,C}$, respectively, it is clear that the following relationships must always be satisfied under steady-state conditions, regardless of fluids, types of application (single or multiphase flow), and physical configurations, in accordance with the second law of thermodynamics,

$$T_{1,C} < T_C < T_{w,C} < T_{w,H} < T_H < T_{1,H} \tag{7.40}$$

where $T_{w,H}$ and $T_{w,C}$ represent corresponding local wall-to-fluid interface temperatures at T_H and T_C on the hot and cold fluid sides, respectively. Consequently, the following inequalities also apply for the conventional local temperature potentials or *driving forces*:

$$(T_H - T_{w,H}), (T_{w,H} - T_{w,C}), \text{ and } (T_{w,C} - T_C) < \Delta T_{\max} \equiv (T_{1,H} - T_{1,C}) \tag{7.41}$$

at any location with the heat exchanger and a fortiori for the mean conventional driving forces, since, most generally $T_{2,H} < T_{1,H}$ and $T_{2,C} > T_{1,C}$ in the fluid cooling and heating modes, respectively. Therefore,

$$(\bar{T}_H - \bar{T}_{w,H}), (\bar{T}_{w,H} - \bar{T}_{w,C}), (\bar{T}_{w,C} - \bar{T}_C), \text{ and } (\bar{T}_H - \bar{T}_C) \ll \Delta T_{\max} \tag{7.42}$$

where the overbar denotes the effective *average* for the whole heat exchanger surface.

To keep irreversible losses within acceptable limits, FFSHES must be designed for operation at $(\bar{T}_H - \bar{T}_C)/\bar{T}_C \approx 0$. This is evident from simple Carnot cycle analyses or more systematic second law considerations based on the Gouy–Stodola theorem, an extension to open systems of the Carnot principle [4.12, 4.13, 7.73–7.82] and amply verified in actual design practice by typical specifications for the major heat exchangers in power generation and refrigeration applications. As a result, thermodynamically efficient forced-convection heat exchangers typically operate within the validity limits of the conventional linear heat rate equations according to Eq. (7.24). Indeed, the validity of these linear relationships provides the theoretical basis for the well-known conventional LMTD and the equivalent NTU-effectiveness heat exchanger analytical methods highlighted in Section 7.6.

To identify more precisely this increasingly important class of thermodynamically efficient heat exchangers, the following definition is introduced:

GUIDELINE 19 *For the special class of surface heat exchangers defined as* ***linear heat flux heat exchangers*** (LHFHE), *the linear relationships between heat fluxes and driving forces are fully satisfied within the allowable test accuracy band* (*typically* 1–5%) *specified in standard industrial performance test codes. This definition imposes no restriction on hydrodynamic flow regimes as long as specified stability requirements are met.*

With the above definition it is immediately evident that linear heat flux heat exchangers automatically satisfy one of the main validity requirements of linear NET. However, as emphasized in Guideline 17, the linear relationships between *all* flows and driving forces must be satisfied and it is necessary to verify the condition of linearity for the coupled dissipative phenomena associated with real fluid flows. For any given heat exchanger geometry, it was shown in earlier chapters that the isothermal friction pressure drop $\Delta P_f = \rho \, \Delta H_f = \rho \bar{T} \Delta s_f$, and therefore the driving force required to maintain specified steady flow rates $\dot{m}_C$ and $\dot{m}_H$ is most generally proportional to $\dot{m}^{2-n}$ with $n = 1$ and $n = 0.25$ to 0 for stable laminar and turbulent regimes, respectively, or $\Delta P_f \propto \dot{m}^1$ in laminar and $\Delta P_f \propto \dot{m}^{\sim 2}$ in turbulent regimes. It is thus evident that the necessary linear relationship between the hydrodynamic forces and flows is satisfied only in the case of stable laminar flow regimes in *both* fluid circuits. Consequently, all the necessary assumptions of linear NET are fully satisfied only for stable laminar regimes and not for turbulent regimes, even in the case of linear heat flux heat exchangers. The most frequent case of turbulent flow regimes encountered in heat exchanger design practice thus lies outside the axiomatic validity limits of linear NET, according to the current state of nonequilibrium thermodynamic theory. In typical turbulent flow heat exchanger applications, we must consequently either confine ourselves to the conventional semiempirical methods or attempt to remove the severe restriction imposed by the linear relationships between hydrodynamic forces and flows while retaining the other unique and essential features of NET, namely, the central role of all dissipative phenomena. These conclusions are summarized in the following cautiously worded guideline and highlighted in Fig. 7.2.

GUIDELINE 20 *Linear nonequilibrium thermodynamic theory is* (*a*) ***Applicable*** *in the case of linear heat flux heat exchangers, provided they are circuited for operation in stable laminar flow regimes exclusively*; (*b*) ***not applicable*** *even in the case of linear heat flux heat exchangers, when designed to operate in stable turbulent flow regimes*; *and* (*c*) ***potentially applicable*** *in typical turbulent flow linear heat flux heat exchangers provided it proves feasible to eliminate the necessary condition of linearity* *between fluid flows and their driving forces, while retaining all other essential and unique features of nonequilibrium thermodynamics, including the central role of fluid flow dissipation, or friction heat*!

7.6 FIRST LAW ANALYSIS OF HEAT EXCHANGERS

The logarithmic mean temperature difference (LMTD) and the alternative effectiveness versus number of thermal units (ε-NTU) analytical methods are summarized in Appendix B, together with the axiomatic assumptions made to derive these relationships.

The conventional LMTD and NTU equations are derived from the phenomenological conduction and convection heat rate equations in conjunction with the mass and energy conservation laws without any second law thermodynamic considerations. This is in sharp contrast to the twentieth century NET methodologies where entropy production and second law considerations play a central role. To make a clear distinction between these two fundamentally different approaches we refer to the conventional LMTD and NTU methodology as the *first law* methods.

One of the prime objectives of the following analyses is to determine exactly how pragmatic heat transfer engineers, who originally derived these first law analytical methods, managed to circumvent the inherent limitations of equilibrium thermodynamics and in the process anticipated some fundamental concepts of twentieth century nonequilibrium thermodynamics.

Since the laws of classical thermodynamics used to derive the conventional methods are independent of the physical configuration (see Guideline 14), keeping in mind that the heat transfer coefficients are assumed to be known constants, it is permissible and expedient to focus on the simplest single-tube heat exchanger system, without loss of generality. These theoretical and numerical analyses are handled as concrete case studies to illustrate the major points we want to stress, while at the same time addressing typical problems faced by heat exchanger engineers in actual heat exchanger design practice.

We consider the archetype single-channel fluid-to-fluid surface heat exchanger system (FFSHES) depicted in Fig. 7.4*a*, in which a quasi-ideal gas (such as air at moderate pressures) is heated by pure saturated steam condensing on the outside wall of the channel at a constant saturation temperature T_{sat} corresponding to the measured absolute pressure P_{sat} and set ourselves the following main objectives.

1. Derive the conventional NTU as well as LMTD equations and compare the simple arithmetic mean temperature difference AMTD with the LMTD, that is, quantify the ratio AMTD/LMTD.
2. Analyze the experimental/analytical methods adopted by heat transfer researchers to determine reliable empirical values of the heat transfer

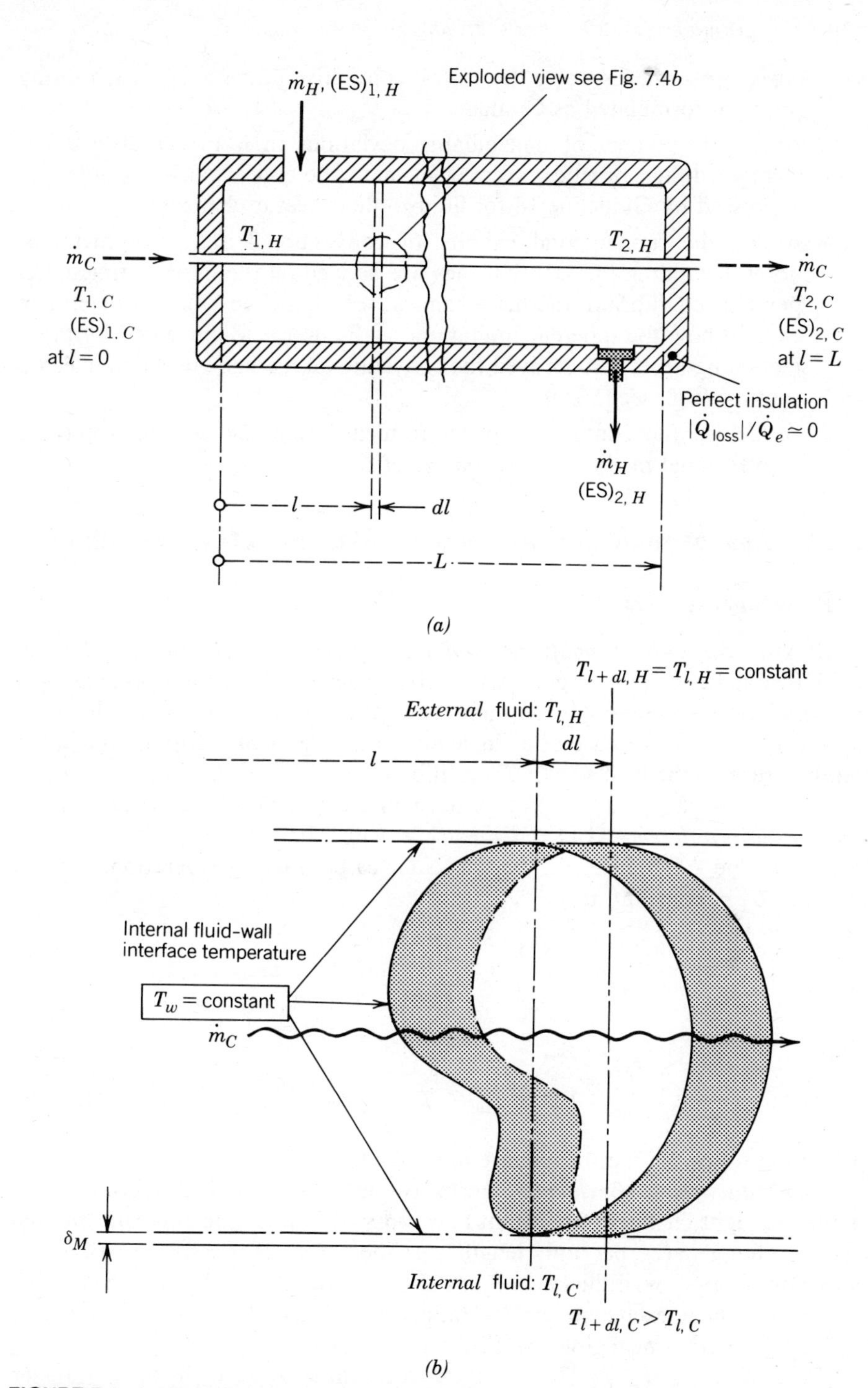

FIGURE 7.4. Archetype single-channel heat exchanger for constant external fluid temperature $T_{1,H} = T_{2,H} = T_{sat}$ with internal fluid in the heating mode, $T_{1,C} < T_{sat}$: (*a*) overall configuration and (*b*) exploded view of infinitesimal heat exchanger "slice" of constant straight cross section Ω_i and otherwise arbitrary geometry.

coefficients, that is, solve the central convective heat transfer engineering problem formulated as Guideline 12.

3. Assess the impact of unavoidable deviations in $\bar{\alpha}$ in actual heat exchanger design practice, keeping in mind the end engineering objective expressed by Guideline 18 for fluid-to-fluid heat exchangers.
4. Assess the fundamental validity of conventional analytical methods, notwithstanding the fact that they are derived on the basis of time-independent equilibrium thermodynamics (see Guideline 14), and determine exactly how the inherent limitations of ET were circumvented by pragmatic engineers who derived the conventional methods in the nineteenth century (e.g., see [1.11]).
5. Introduce the concept of mean residence time $\overline{\Delta\theta}$ and interpret the conventional method in the context of $\overline{\Delta\theta}$.

7.6.1 Assessment of Conventional NTU and LMTD Equations

Problem Statement

Making the same *simplifying assumptions as* in the case of conventional analytical methods (see Appendix B), derive a general dimensionless analytical expression to compute the required heat exchanger surface (i.e., the total *length of travel* in the case of a single-circuit heat exchanger) to approach the temperature of the heat source T_{sat} within a specified *smaller terminal temperature difference* SD $\equiv T_{sat} - T_{2,C}$ or normalized in terms of the *larger terminal temperature difference* LD $\equiv T_{sat} - T_{1,C}$ with the *approach factor* AF $\equiv$ SD/LD. The AF reduces, in the particular case under consideration, $T_{1,H} = T_{2,H} = T_{sat} =$ constant, to

$$\mathrm{AF} \equiv \frac{\mathrm{SD}}{\mathrm{LD}} \equiv \frac{\mathrm{SD}}{\Delta T_{max}} \equiv \frac{T_{sat} - T_{2,C}}{T_{sat} - T_{1,C}} \tag{7.43a}$$

with

$$\mathrm{LD} \equiv T_{sat} - T_{1,C} \equiv \Delta T_{max} \tag{7.43b}$$

when the following parameters are prescribed:

Elementary Heat Exchanger. Perfectly clean (zero fouling thermal resistance) straight channel of constant flow cross section Ω_i and constant internal heat exchange area per unit length a_i. Therefore, $D_{h,i} = 4r_{h,i} = 4\Omega_i/a_i =$ constant. Copper wall thickness $\delta_M \ll D_{h,i}$, or $\delta_M/D_{h,i} \cong 0$; thus, the outside to internal heat exchange surface ratio (A_o/A_i) can safely be estimated as $A_o/A_i = 1$ with a negligible (and conservative) error.

Heat Source. $T_{1,H} = T_{2,H} = T_{sat} =$ constant $> T_{1,C}$ with heat transfer coefficients falling within the range indicated in Table 7.1 for steam condensers or $\bar{\alpha}_H \approx 2000 \times 5.7$ W/m$^2 \cdot$ K.

Heat Sink. Mass flow rate $\dot{m}_C$, initial equilibrium state $T_{1,C}$ at $P_{1,C}$ and constant specific heat $(c_p)_C$ with a constant value $\bar{\alpha}_C$ falling within the typical range shown in Table 7.1 for air, or

$$\bar{\alpha}_C \approx 5 \times 5.7 \text{ W/m}^2 \cdot \text{K and } \bar{\alpha}_C/\bar{\alpha}_H \approx 5/2000 = 0.0025 = \tfrac{1}{4}\%!$$

Solution

Referring to the infinitesimal heat exchanger "slice" of thickness dl and volume $\Omega_i dl$, shown schematically in Fig. 7.3b, we obtain from the mass and energy conservation laws and the conventional heat rate equations

$$d\dot{Q}_e = (\dot{m}c_p)_C\, dT_C = (T_{\text{sat}} - T_C)U_i a_i\, dl \tag{7.44}$$

on the basis of the usual assumptions: the kinetic and potential energy terms are negligible, the fluid properties (c_p, ρ, k, μ) are constant, the heat exchanger is perfectly insulated ($\dot{Q}_{\text{loss}} = 0$), and there is no axial heat flow. For $T_H = T_{\text{sat}} = \text{constant} > T_C$ (i.e., gas heating mode) as well as fixed values of $(\dot{m}c_p)_C$ and $U_i a_i$. Eq. (7.44) can be integrated after separating the variables and keeping the constant terms outside the integral signs as follows:

$$\int_{T_{1,C}}^{T_{2,C}} \frac{dT_C}{T_{\text{sat}} - T_C} = \frac{U_i a_i}{(\dot{m}c_p)_C} \int_{l=0}^{l=L} dl \tag{7.45}$$

This yields

$$-\ln(T_{\text{sat}} - T_{2,C}) + \ln(T_{\text{sat}} - T_{1,C}) = \ln\left(\frac{T_{\text{sat}} - T_{1,C}}{T_{\text{sat}} - T_{2,C}}\right) = \frac{U_i a_i L}{(\dot{m}c_p)_C} \tag{7.46}$$

where $T_{2,C}$ is the specified terminal gas equilibrium temperature and L is the required length of travel with a corresponding single-circuit heat exchanger surface $A_i = a_i L$. With the definitions of the "normalized" smaller temperature difference SD/LD = AF per Eq. (7.43a), the above equation reduces to

$$\ln\left(\frac{1}{\text{AF}}\right) = -\ln(\text{AF}) = \text{NTU} \tag{7.47}$$

or

$$\text{AF} \equiv \frac{T_{\text{sat}} - T_{2,C}}{T_{\text{sat}} - T_{1,C}} \equiv \frac{\text{SD}}{\text{LD}} = e^{-\text{NTU}} \tag{7.48a}$$

$$\frac{1}{\text{AF}} \equiv \frac{T_{\text{sat}} - T_{1,C}}{T_{\text{sat}} - T_{2,C}} \equiv \frac{\text{LD}}{\text{SD}} = e^{+\text{NTU}} \tag{7.48b}$$

where, by definition,

$$\frac{U_i a_i L}{(\dot{m}c_p)_C} \equiv \frac{U_i A_i}{(\dot{m}c_p)_C} \equiv \text{NTU} \equiv \text{number of thermal units} \qquad (7.49)$$

With a constant heat source temperature $T_H = T_{\text{sat}}$, it is permissible to write

$$T_{2,C} - T_{1,C} = (T_{\text{sat}} - T_{1,C}) - (T_{\text{sat}} - T_{2,C}) \equiv \text{LD} - \text{SD} \qquad (7.50)$$

The total rate of heat exchange absorbed by the colder fluid from an inlet to an outlet, equilibrium states 1 and 2, respectively, is $\dot{Q}_e = (\dot{m}c_p)_C(T_{2,C} - T_{1,C})$, which can be expressed with AF = SD/LD as

$$\dot{Q}_e = (\dot{m}c_p)_C \text{LD}\left(1 - \frac{\text{SD}}{\text{LD}}\right) = (\dot{m}c_p)_C \text{LD}(1 - \text{AF}) \qquad (7.51)$$

Since $\dot{Q}_e$ is directly proportional to LD $= T_{\text{sat}} - T_{1,C} = \Delta T_{\text{max}}$ according to Eq. (7.51), it is permissible and convenient to normalize $\dot{Q}_e$ in terms of LD; that is,

$$\frac{\dot{Q}_e}{\text{LD}} = (\dot{m}c_p)_C(1 - \text{AF}) \qquad (7.52)$$

The total heat exchanger capacity expressed on a "per degree LD" basis by Eq. (7.52) is particularly useful, for instance, in practical condenser design and performance analyses (see Appendix C), as well as for the more fundamental second law considerations addressed in subsequent chapters.

The overall mean temperature difference ΔT_{om} defined by the identity

$$\dot{Q}_e \equiv U_i A_i \Delta T_{\text{om}} \qquad (7.53)$$

can be determined analytically by equating Eqs. (7.51) and (7.53) and solving for ΔT_{om} to obtain, with Eq. (7.47) and the definition of NTU per Eq. (7.49),

$$\overline{\Delta T}_{\text{om}} = \text{LMTD} = \text{LD}\frac{1 - \text{AF}}{\ln(1/\text{AF})} \qquad (7.54\text{a})$$

Equation (7.54a), of course, is a special limiting case (when the temperature of one of the two fluids remains constant) of the conventional and more general expression valid for pure counterflow and parallel flow single-phase heat exchangers (see Appendix B):

$$\text{LMTD} = \frac{\text{LD} - \text{SD}}{\ln(\text{LD}/\text{SD})} = \frac{\text{LD}(1 - \text{SD}/\text{LD})}{\ln(\text{LD}/\text{SD})} \qquad (7.54\text{b})$$

Validity: Pure parallel flow *and* counterflow

Since the LMTD is directly proportional to LD per Eq. (7.54), it is again permissible, and frequently convenient, to *normalize the logarithmic mean temperature difference in terms of the larger terminal temperature difference*, LD, according to the following fully dimensionless logarithmic expressions:

$$\frac{\text{LMTD}}{\text{LD}} = \frac{1 - \text{AF}}{\ln(1/\text{AF})} = \frac{1 - \text{AF}}{-\ln(\text{AF})} \tag{7.55}$$

The *exact* LMTD can be compared to the *approximate* arithmetic mean temperature difference (AMTD) which is by definition

$$\text{AMTD} \equiv \tfrac{1}{2}(\text{LD} + \text{SD}) \tag{7.56}$$

or, in terms of AF = SD/LD and LD,

$$\text{AMTD} = \tfrac{1}{2}\text{LD}(1 + \text{AF}) \tag{7.57}$$

which can be also normalized in terms of LD as

$$\frac{\text{AMTD}}{\text{LD}} = \tfrac{1}{2}(1 + \text{AF}) \tag{7.58}$$

Dividing Eq. (7.58) by Eq. (7.55) yields the following ratio:

$$\frac{\text{AMTD}}{\text{LMTD}} = \frac{(1 + \text{AF})\ln(1/\text{AF})}{2(1 - \text{AF})} \geq 1 \tag{7.59}$$

which is thus uniquely defined in terms of the normalized smaller temperature difference AF = SD/LD. The ratios AMTD/LMTD computed from Eq. (7.59) are summarized in Table 7.2 for values of AF extending over the full possible range $0 \leq \text{AF} \leq 1$.

Column 4 of Table 7.2 clearly shows that although the AMTD always overestimates the logarithmic mean temperature difference, the percentage deviations (column 5) become insignificant at $\text{AF} \geq 0.5$ compared to the typical heat transfer coefficient uncertainties (tests, manufacturing tolerances, and theoretical model inaccuracies) encountered in heat transfer engineering applications, which are typically of the order of ± 10–20% even in the simplest cases of single-phase heat exchangers. For $\text{AF} \geq 0.5$, it is therefore permissible to set LMTD = AMTD and the same effective mean temperature difference is thus obtained on the basis of the two most extreme arrangements of heat exchanger design practice: pure parallel flow and pure counterflow.

Since pure parallel and counterflow are the most extreme heat exchanger flow arrangements conceivable, one can immediately conclude that the effective mean temperature difference for the intermediate *cross flow* arrangement must also converge toward the arithmetic mean AMTD, as can easily be

TABLE 7.2. Comparison of AMTD and LMTD at Different Approach Factors $1 \leq$ AF ≤ 0

AF	$\frac{\text{LMTD}}{\text{LD}}$	$\frac{\text{AMTD}}{\text{LD}}$	$\frac{\text{AMTD}}{\text{LMTD}}$	$100\left(\frac{\text{AMTD} - \text{LMTD}}{\text{LMTD}}\right)$ in %
1	1	1	1	0
0.9	0.949	0.95	1.001	+0.1
0.8	0.896	0.90	1.004	+0.4
0.7	0.841	0.85	1.015	+1.5
0.6	0.783	0.80	1.022	+2.2
0.5[a]	0.721	0.75	1.040	+4.0
0.4	0.655	0.70	1.069	+6.9
0.3	0.581	0.65	1.118	+11.8
0.2	0.497	0.60	1.207	+20.7
0.1	0.3909	0.5500	1.407	+40.7
0.05	0.3170	0.5250	1.656	+65.6
0.01	0.2149	0.5050	2.349	+134.9
0.005	0.1878	0.5025	2.676	+167.6
0.001	0.1446	0.5005	3.461	+246.1
$\rightarrow 0$	$\rightarrow 0$	$\rightarrow 0.5$	$\rightarrow \infty$	$\rightarrow \infty$

[a]AMTD = LMTD with an error of less than 4% both in pure parallel flow *and* pure counterflow heat exchangers, for $0.5 \leq$ AF ≤ 1.

verified by referring to Appendix B (since $\Delta T_{lm} \equiv F \Delta T_{lm, CF}$ and $F \rightarrow 1$ as shown in Figures 11.10–11.13 in Appendix B).

7.6.2 Empirical Determination of the Heat Transfer Coefficient

Problem Statement

Using the above analytical results, determine the actual average internal heat transfer coefficients $(\bar{\alpha}_i)_C$ from empirical values of $T_{2,C}$ obtained under SSSF conditions with a given channel geometry at representative values of $L/D_{h,i} > 60$ (when the inlet effects become negligible for typical turbulent flow regimes) in laboratory or field tests carried out at different flow rates $\dot{m}_C$ within the intended application range of the parameters $T_H = T_{\text{sat}}$ and $T_{1,C}$.

Solution

To solve this central convective heat transfer engineering problem (see GL 12), it is first necessary to compute the empirical overall heat transfer coefficient $\bar{U}_i$. In the case under consideration, the actual $\bar{U}_i$ based on the total internal surface $A_i = a_i L$ can be computed from the experimental results with a known total internal surface A_i at different flow rate $\dot{m}_C$, measured $T_{2,C}$, $T_{1,C}$, and corresponding heat exchanger capacity $\dot{Q}_e$ computed primarily from the energy conservation law according to $\dot{Q}_e = (\dot{m}c_p)_C(T_{2,C} - T_{1,C})$. In reliable laboratory or field tests, $\dot{Q}_e$ is double checked by a secondary heat balance on the other fluid side of the surface heat exchanger. In the case under consideration, this double check requires that the following relationship be satisfied:

$$\dot{Q}_e = (\dot{m}c_p)_C(T_{2,C} - T_{1,C}) = \dot{m}_H(h_{1,H} - h_{2,H}) \tag{7.60a}$$

within a specified test accuracy typically of the order of ± 1–5% or

$$\text{Allowable maximum deviation typically } \pm 1\text{–}5\% \tag{7.60b}$$

where $h_{1,H}$ and $h_{2,H}$ are the enthalpy of the steam at inlet and condensate at outlet, respectively.

At any known value of $\dot{Q}_e$, measured LD $= T_{\text{sat}} - T_{1,C}$, and SD $= T_{\text{sat}} - T_{2,C}$, the overall heat transfer coefficient $\overline{U}_i$ can be computed most generally for parallel or counterflow from Eqs. (7.53) and (7.54) with a known value of A_i according to

$$\overline{U}_i = \frac{\dot{Q}_e}{A_i(\text{LMTD})} = \frac{\dot{Q}_e}{A_i}\frac{\ln(\text{LD}/\text{SD})}{\text{LD} - \text{SD}} \tag{7.61}$$

when both fluids undergo a temperature change. For $T_H = T_{\text{sat}} =$ constant, the above equation, with $\dot{Q}_e = (\dot{m}c_p)_C(\text{LD} - \text{SD})$, reduces to

$$\overline{U}_i = \frac{(\dot{m}c_p)_C}{A_i}\ln\left(\frac{\text{LD}}{\text{SD}}\right) = \frac{(\dot{m}c_p)_C}{A_i}\ln\left(\frac{1}{\text{AF}}\right) \tag{7.62}$$

To determine the internal heat transfer coefficient $\bar{\alpha}_i$, it is generally necessary to break down the empirical overall thermal resistance obtained from Eq. (7.62) into its individual components by appealing to the well known relationship

$$\frac{1}{\overline{U}_i A_i} = \frac{1}{\bar{\alpha}_i A_i} + \frac{1}{(k_M/\delta_M)\overline{A}_M} + \frac{1}{\bar{\alpha}_o A_o} \tag{7.63}$$

where the subscripts i and o refer to the internal and outside channel surfaces, respectively, and $\overline{A}_M$ represents the effective average surface for the radial conduction heat flow through the separating metal wall of thickness δ_M and conductivity k_M. Equation (7.63) simply states that the overall thermal resistance is the sum of the individual resistance. Note that this relationship is also derived from the first law of thermodynamics.

In the case under consideration, which is typical of "bare" tube heat exchangers, with $A_o/A_i \cong 1$ and (since $A_i < \overline{A}_M < A_o$) $\overline{A}_M/A_i \cong 1$, it is permissible to eliminate the terms A_i, A_o, and $\overline{A}_m$ in Eq. (7.63), which reduces to the simpler expression

$$\frac{1}{\overline{U}_i} = \frac{1}{\overline{\alpha}_i} + \frac{1}{k_M/\delta_M} + \frac{1}{\overline{\alpha}_o} = \frac{1}{\overline{\alpha}_i}\left(1 + \frac{\overline{\alpha}_i}{k_M/\delta_M} + \frac{\overline{\alpha}_i}{\overline{\alpha}_o}\right) \tag{7.64}$$

With $k_M = 401\ \mathrm{W/m\cdot K}$ for copper and δ_M of the order of 1 mm$(= 10^{-3}\ \mathrm{m})$, the second term in the large parentheses on the right-hand side of Eq. (7.64) $\overline{\alpha}_i/(k_M/\delta_M) < 5(5.7)/[(401)(10)^3] \approx 7 \times 10)^{-5}$ so that with $\overline{\alpha}_i/\overline{\alpha}_o \approx 0.0025 = 0.25\%$ the numerical value of the large parentheses differs from unity by less than 1%, which is negligibly small compared to unity, particularly in the context of normal test deviations of 1–5% for $\dot{Q}_e$ and typical uncertainties of the order of ± 5–20% for the best correlations of $\overline{\alpha}_i$ (even in the simplest case of single-phase flow heat exchangers). It is therefore permissible, in line with Guidelines 2 and 7, to set

$$\overline{\alpha}_i = \overline{U}_i = \frac{(\dot{m}c_p)_C}{a_i L}\ln\left(\frac{1}{\mathrm{AF}}\right) \quad \text{within an acceptable accuracy of better than 1\%} \tag{7.65}$$

Thus, the overall thermal resistance is concentrated fully on the gas side, which is said to be "controlling" in this case, and the temperature of the wall–gas interface T_w substantially coincides with the constant temperature $T_{1,H} = T_{2,H} = T_{\mathrm{sat}}$ or

$$T_{\mathrm{sat}} = T_w = \text{wall–fluid interface temperature} = \text{constant} \tag{7.66}$$

In other words, this analysis has focused on the classical T_w = constant boundary case of convective heat transfer addressed in most introductory textbooks. The main advantage of this classical single-channel test setup is that the empirical values of $\overline{\alpha}_i$ thus obtained are generally more accurate than those computed from empirical $\overline{U}_i$ values with the help of Eq. (7.64) when $\overline{\alpha}_i/\overline{\alpha}_o \neq 0$. This explains why condensing steam is favored by many researchers to determine empirically the actual heat transfer coefficients in channels of different geometries.

An alternative test setup to determine $\overline{\alpha}_i$ for the T_w = constant boundary condition, without using steam, in a *mirror-image heat exchanger system* (MIHES), is outlined in Chapter 9. The extremely useful single-phase mirror-image concept can also be extended to two-phase forced-convection heat exchangers in evaporation and condensation [7.83].

Several semiempirical methods of breaking down the overall heat transfer coefficient $\overline{U}_i$ in its main components, when $\overline{\alpha}_i A_i \approx \overline{\alpha}_o A_o$, as is normally the case in properly balanced heat exchangers, are described in standard heat transfer textbooks. One of these methods, known as the *Wilson-plot technique* [7.84], is particularly useful in direct-expansion refrigeration evaporators, when the internal and external thermal resistances are properly balanced ($\bar{\alpha}_i a_i L \approx \bar{\alpha}_o a_o L$) as is typically the case with halogenated refrigerants [3.32, 4.4, 5.5, 5.6,].

7.6.3 Impact of Heat Transfer Coefficient Deviations

Problem Statement

Practical experience indicates that the empirical results obtained according to Eq. (7.65) can be correlated, in the case under consideration, exclusively in terms of $\dot{m}_C$ for a given channel hydraulic diameter $D_{h,i}$ with an accuracy of ± 10–20%. Determine the impact of such deviations in $\bar{\alpha}_i$ on the actual approach factor AF and corresponding heat exchanger capacity $\dot{Q}_e = (\dot{m}c_p)_C(T_{2,C} - T_{1,C}) = (\dot{m}c_p)_C\text{LD}(1 - \text{AF})$ at different nominal design values of AF.

Solution

Let the subscripts act and nom refer to the actual and nominal values of $\bar{\alpha}_i$, that is, $(\bar{\alpha}_i)_{\text{act}} = (\bar{\alpha}_i)_{\text{nom}} \pm 10$–$20\%$. Since under otherwise prescribed conditions the exponent NTU is directly proportional to $\bar{\alpha}_i = \overline{U}_i$ per Eq. (7.49), we obtain from Eqs. (7.48) and (7.51) a ratio

$$\frac{(\dot{Q}_e)_{\text{act}}}{(\dot{Q}_e)_{\text{nom}}} = \frac{1 - e^{-(\text{NTU})_{\text{act}}}}{1 - e^{-(\text{NTU})_{\text{nom}}}} = \frac{1 - (\text{AF})_{\text{act}}}{1 - (\text{AF})_{\text{nom}}} \tag{7.67}$$

It is clear from Eq. (7.67) that the impact of a given % variation in $\bar{\alpha}_i$ will depend strongly on the nominal design value $(\text{AF})_{\text{nom}}$. The capacity ratios $(\dot{Q}_e)_{\text{act}}/(\dot{Q}_e)_{\text{nom}}$ computed from Eq. (7.67) are shown in terms of $(\text{AF})_{\text{nom}}$ in Table 7.3 for a $\pm 10\%$ and $\pm 20\%$ variation in $\bar{\alpha}_i$.

As could be expected, the end effect of a given % deviation in $\bar{\alpha}_i$ becomes less and less significant as AF $\rightarrow$ 0 and conversely more pronounced as AF $\rightarrow$ 1 when $(\dot{Q}_e)_{\text{act}}/(\dot{Q}_e)_{\text{nom}} \rightarrow (\bar{\alpha}_i)_{\text{act}}/(\bar{\alpha}_i)_{\text{nom}}$. In other words, the more

TABLE 7.3. Impact on $(\dot{Q}_e)_{\text{act}}/(\dot{Q}_e)_{\text{nom}}$ of $\pm 10 - 20\%$ Deviations in Heat Transfer Coefficients

Nominal Approach Factor		1.0	0.75	0.5	0.25	0.1	0.01
(1)	at 1.1 U_{nom}	110%	108.5	106.7	104.3	102.3	100.3
(2)	at 0.9 U_{nom}	90%	91.2	92.8	95.0	97.1	99.4
(3)	Total spread	20%	17.3	13.9	9.3	5.2	0.9
(4)	at 1.2 U_{nom}	120%	116.8	112.9	108.1	104.1	100.6
(5)	at 0.8 U_{nom}	80%	82.2	85.1	89.3	93.5	98.5
(6)	Total spread	40%	34.6	27.8	18.8	10.6	2.1

effective the heat exchanger design, that is, the lower the nominal approach factor $(AF)_{nom}$, the less sensitive the heat exchanger becomes to deviations in $\bar{\alpha}_i$ (or more generally $\bar{U}_i$). Thus,"efficient" heat exchangers designed for a very close approach factor AF are generally much less sensitive to normal variations in $\bar{\alpha}_i$ (or $\bar{U}_i$) typically encountered in convective heat exchanger design practice.

Conversely, Eq. (7.67) indicates that the empirical values of $\bar{U}_i$ computed from Eq. (7.65) and the resulting values of $\bar{\alpha}_i$ (and most generally also $\bar{\alpha}_o$) become increasingly inaccurate as the approach factor $AF = SD/LD \rightarrow 0$ because of unavoidable inaccuracies associated with all temperature measurements. This is a particularly serious problem when the larger terminal temperature difference LD is small, as is typically the case in thermodynamically efficient refrigeration applications as well as the main convective fluid-to-fluid heat exchangers in the power generation industry, such as feedwater heaters and condensers. Since $\bar{U}_i \propto \ln(1/AF) = \ln(LD/SD)$ according to Eq. (7.65), it is evident that at typical LD values of less than $10\,°C/K$ with approach factors $AF < 0.1$, and therefore smaller terminal temperature differences of less than $1\,°C/K$, that the calculated heat transfer coefficients are extremely sensitive to minute temperature measurement inaccuracies.

The possible impact of temperature measurement inaccuracies of $\pm 0.5\,°C$ and $\pm 0.1\,°C$ on the empirical heat transfer coefficient determined from tests on the basis of Eq. (7.65) is quantified in Table 7.4 for actual values $(LD)_{act}$ and $(AF)_{act}$ shown on lines (1), (2), and (3), respectively. On the optimistic assumption that $\dot{Q}_e$ is measured with perfect accuracy, the ratio $\bar{U}_{meas}/\bar{U}_{actual}$ can be expressed with Eq. (7.65) according to

$$\frac{U_{meas}}{U_{act}} = \frac{\ln(LD/SD)_{meas}}{\ln(LD/SD)_{act}} \tag{7.68}$$

TABLE 7.4. Impact of Temperature Measurement Deviations of ±0.5 and ±0.1° C

(1) $(LD)_{act}$	°C	10	10	10	10	10	10	10
(2) $(SD)_{act}$	°C	10	7.5	5	2.5	1.0	0.5	0.1
(3) $(AF)_{act} = (SO)_{act}/(LD)_{act}$		1	0.75	0.5	0.25	0.1	0.05	0.01
$(LD)_{meas}$ and $(SD)_{meas}$ deviate by ±0.5 °C								
(4)[b] $(U_{meas}/U_{act})_{min}$	%	95.0	94.5	93.3	90.4	84.5	78.5	62.2
(5)[c] $(U_{meas}/U_{act})_{max}$	%	105.0	106.0	107.8	112.4	127.9	∞	a
(6) Total spread	%	10.0	11.5	14.5	22.0	43.9	∞	a
$(LD)_{meas}$ and $(SD)_{meas}$ deviate by ±0.1 °C								
(7)[d] $(U_{meas}/U_{act})_{min}$	%	99.0	98.9	98.6	97.9	96.3	94.2	85.2
(8)[e] $(U_{meas}/U_{act})_{max}$	%	101.0	101.2	101.5	102.2	104.1	107.1	∞
(9) Total spread	%	2.0	2.3	2.9	4.3	7.8	12.9	∞

[a]SD < 0!; [b]0.5 °C high; [c]0.5° C low; [d]0.1 °C high; [e]0.1 °C low.

to obtain the numerical values shown in % on line (4) computed on the assumption that the LD and SD are both 0.5 °C high, whereas the ratios on line (5) are based on the assumption that LD and SD are underestimated by 0.5 °C. The results shown in lines (7) and (8) of Table 7.4 were obtained in a similar fashion on the assumption of more exact temperature measurements with deviations of only ±0.1 °C.

As Table 7.4 clearly shows, experiments primarily intended for heat transfer coefficient correlations should be run as far as possible at relatively high approach factors of the order of AF > 0.5 to improve the accuracy of the resulting data, particularly in extremely low LD applications.[18] When the tests are carried out at AF > 0.5, the mean temperature differences may be computed simply on the basis of the AMTD rather than the LMTD (see Table 7.2), as is indeed the case in most convective heat transfer research studies reported in the technical literature.

In view of the central significance of the convective heat transfer coefficients in heat transfer engineering (see Guideline 12), the above conclusions, which are quite general and not restricted to the single-channel heat exchanger system analyzed in this elementary case study, are summarized in the form of a guideline.

GUIDELINE 21 *The smaller the approach factor,* AF = SD/LD, *the less sensitive the heat exchanger to unavoidable deviations in heat transfer coefficients. Conversely, experiments primarily intended for heat transfer coefficient correlations should be carried out at relatively large* AF, *if possible at* AF > 0.5, *to improve the accuracy of the resulting empirical heat transfer coefficients.*

At large approach factors, AF ≥ 0.5, *it is permissible to compute the effective average temperature difference on the basis of the arithmetic mean temperature difference* (AMTD), *noting that the same results are then obtained for the two most extreme flow arrangements: pure parallel flow and pure counterflow and the theoretical differences between the various flow arrangements become insignificant.*

7.6.4 Fundamental Validity of Conventional Methods

Problem Statement

Keeping in mind that the above conventional analytical methods were derived on the basis of nineteenth century classical equilibrium thermodynamic laws, assess their fundamental validity (notwithstanding the limitations of ET stated as Guideline 14) focusing on the exact physical meaning of the local flowing fluid bulk, or *mixing cup* temperature T_C and the related definition of the heat transfer coefficient, and examine the definition of T_C in the context of the twentieth century local equilibrium theorem (Guideline 16).

Solution

From a close examination of the analytical steps followed in deriving Eq. (7.47), it is clear by referring to Eq. (7.44) and Fig. 7.4*b* that T_C at any axial position l is a fictive uniform local temperature that disregards the actual nonuniform temperature profile (i.e., nonequilibrium conditions) prevailing within the infinitesimal system "slice" of volume $\Omega_i\, dl$. This fictive uniform temperature, usually called the local *bulk* or *mixing cup* temperature, is in fact the ultimate equilibrium thermodynamic temperature that would be measured locally at the axial position l if the infinitesimal thermodynamic system ($\Omega_i\, dl$) could be instantaneously insulated from the heat source $T_w \neq T_C$. Through the device of this fictive local equilibrium temperature, it is possible to circumvent the inherent limitation of equilibrium thermodynamics, solely on the premise that the convective heat rate equation is valid, as stated by Eq. (7.2), with a constant phenomenological heat transfer coefficient defined on the basis of the same fictive local equilibrium temperature, as experience has amply verified. Thus, with the help of a fictive (or potential) local equilibrium temperature T_C, it is possible to integrate Eq. (7.44) according to Eq. (7.45) from a specified inlet equilibrium state $(\mathrm{ES})_1$ at $l = 0$ to specified outlet equilibrium state $(\mathrm{ES})_2$ at $l = L$, without any concern whatsoever about the true local temperature and velocity profile within the (constant) channel cross section Ω_i at any axial location.

The conventional analytical methods, being essentially based on the application of classical equilibrium thermodynamics, in conjunction with empirically determined (assumed) constant heat transfer coefficients, are a priori valid (like all classical thermodynamic laws) for any system configuration, any substance, and any flow regime. This explains why the LMTD or NTU analytical methods can be applied successfully both for laminar and turbulent stable flow regimes, notwithstanding the fact that the true velocity and temperature profiles within a given section geometry are radically different for these two extreme single-phase flow regimes.

In summary, the complex nonequilibrium convective heat transfer phenomena are "lumped" in a single empirical factor, the illusive heat transfer coefficient, defined on the basis of a fictive (or potential) local equilibrium temperature.

It is interesting to note in this context, that the coupled fluid flow phenomena are similarly treated in actual engineering design practice on the basis of a fictive velocity $\bar{V}$, which is based on an idealized frictionless uniform (Bernoulli) velocity profile. To verify this, it is only necessary to recall that the central fluid flow problems (as stated in Guideline 6 of Chapter 3) can be solved for turbulent and laminar regimes with Eq. (3.4), where the velocity head term $\bar{V}^2/2$ is based on a fictive uniform velocity $\bar{V}$ in conjunction with a semianalytical friction factor calculated on the basis of a Reynolds number, which is also defined on the basis of the same fictive velocity $\bar{V}$. Thus, the true velocity profile in a given cross section is ignored in recognized pressure drop (or head loss) calculations which are thus fundamentally based on an ideal fictive uniform velocity $\overline{V}$ that would exist in the absence of friction dissipation, that is, a Bernoulli velocity (see Chapter 2) rather than the true nonuniform axial velocities in the channel cross section.

The ultimate effects of wall friction dissipation are ultimately lumped in a single term called the *friction factor*, just as the ultimate effects of the nonequilibrium temperatures are lumped in the convective heat transfer coefficient to compute a forced-convection fluid-to-fluid heat exchanger. In this respect, the conventional fluid and heat flow analytical methods are fully consistent.

As strange as this may seem at first, the above conclusions are *not* incompatible with the various *boundary layer* models for the prediction of the heat transfer coefficients in fully developed turbulent channel flow highlighted previously. Although the boundary layer analyses focus on the true (fully developed) nonuniform velocity and temperature profiles within the channel cross section, the fact remains that their end engineering objective is the determination of the illusive heat transfer coefficient, along the lines presented in Section 7.3.2, which is then applied in the conventional heat exchanger design methods exactly as shown above, that is, on the basis of fictive uniform fluid velocities and temperatures!

In the context of the new theory of irreversible thermodynamics highlighted previously, it can be seen that the nineteenth century heat transfer engineers [1.11], who first proposed the conventional methodology, essentially anticipated the twentieth century concept of local equilibrium in the infinitesimal subsystem $(\Omega_i dl)$ within a macroscopic system $(\Omega_i L)$ in a steady state of nonequilibrium, although the actual temperature variations within the channel cross section at any axial position l are ignored for pragmatic reasons. In the conventional analytical procedure the full length heat exchanger system of volume $\Omega_i L$, total length L, and constant cross section Ω_i is in effect broken down in infinitesimal heat exchanger "slices" $(\Omega_i dl)$ stacked in series at "local" equilibrium temperature $T_C \neq T_w$ for the fluid heating mode under consideration $T_w > T_C$ per Fig. 7.4 (but the method is also valid for the symmetrical fluid cooling mode when $T_H < T_w$). Since the exponent NTU defined by Eq. (7.49) is directly proportional to $A_i = a_i L \propto L$, the required length of travel can be determined by solving Eq. (7.49) for L according to

$$L = \left(\frac{(\dot{m}c_p)_C}{a_i \overline{U}_i} \right) \text{NTU} \equiv (K_1)\text{NTU} \propto \text{NTU} \tag{7.69a}$$

where

$$\frac{(\dot{m}c_p)_C}{a_i \overline{U}_i} = K_1 = \text{constant} \tag{7.69b}$$

when

$$\dot{m}_C, (c_p)_C, a_i, \text{ and } \overline{U}_i \text{ are fixed} \tag{7.69c}$$

It is noteworthy that the constant term K_1 defined by Eq. (7.69b) has the dimension of a length and physically represents the actual length of travel required for an NTU = 1 heat exchanger. According to Eq. (7.69a), this is actually the length of travel $L_{1/e}$ for an approach factor AF $= e^{-1} = 1/e = 0.3679 \simeq 0.37$ when NTU = 1; therefore,

$$K_1 \equiv L_{1/e} = \frac{(\dot{m}c_p)_C}{a_i \bar{U}_i} = \text{length of travel for AF} = 1/e \text{ at NTU} = 1 \quad (7.69d)$$

with a corresponding constant factor CF = 1 − AF $\simeq$ *0.63*. When the constant K_1 is known, the required length of travel for any specified design AF (and corresponding CF = 1 − AF) can readily be computed from Eq. (7.69) or, vice versa, the approach factor AF can be predicted for different lengths of travel L. Assuming, as an illustration, that the length $L_{1/2}$ required for an approach factor AF $= \frac{1}{2}$ has been determined purely empirically (i.e., without any model or theory), it is easy to compute AF, and therefore CF = 1 − AF, at any other length of travel $L \neq L_{1/2}$ for K_1 = constant since NTU $\propto L$ and AF = exp(−NTU) according to

$$\text{AF} = \left(\tfrac{1}{2}\right)^{l_{\text{red}}} = \frac{1}{(2)^{l_{\text{red}}}} \quad (7.70a)$$

where

$$l_{\text{red}} \equiv \frac{L}{L_{1/2}} = \text{reduced length of travel in terms of } L_{1/2} \quad (7.70b)$$

with

$$L_{1/2} = \text{length of travel for AF} = \tfrac{1}{2} \quad (7.70c)$$

Note that the subscript 1/2 indicates the arbitrary reference AF $= \frac{1}{2}$ used in this particular example.

The above equations simply show that by stacking two identical elementary heat exchangers of length $L_{1/2}$ in series, the resulting total AF is reduced from $\frac{1}{2}$ to $(\frac{1}{2})^2 = \frac{1}{4}$. This conclusion is logical since for a given $\Delta T_{\max} = T_w - T_{1,C} = (\Delta T_{\max})_{\text{No. 1}}$, the fluid must leave heat exchanger No. 1 at a thermodynamic temperature approaching T_w within $\frac{1}{2}$ $(\Delta T_{\max})_{\text{No. 1}}$ (since AF $= \frac{1}{2}$) which is at the same time the inlet temperature of heat exchanger No. 2. Thus, $(\Delta T_{\max})_{\text{No. 2}} = \frac{1}{2}$ $(\Delta T_{\max})_{\text{No. 1}}$ for T_w = constant and, with an AF $= \frac{1}{2}$, the fluid will leave heat exchanger No. 2 at $(\frac{1}{2})(\frac{1}{2})\Delta T_{\max} = \frac{1}{4}\Delta T_{\max}$ as indicated by Eq. (7.70a) with $l_{\text{red}} = 2$. By following the same reasoning, it is easy to see that for a total length of travel $L = 3\ L_{1/2}$ or $l_{\text{red}} = 3$, which is equivalent to three identical $L_{1/2}$ heat exchangers stacked in series, that

$AF = (\frac{1}{2})^3 = (\frac{1}{2})^{l_{red}}$ and so forth, in agreement with Eq. (7.70a). An alternative interpretation in terms of mean residence time is described in Section 7.6.5 and some practical implications of this asymptotic approach to the maximum end equilibrium temperature $T_2 = T_w$ when AF $\to 0$ are discussed in Chapter 8 and interpreted in the context of nonequilibrium thermodynamics in Chapter 9.

7.6.5 Mean Residence-Time Concept

Problem Statement

Consider the well-known chemical engineering concept of mean residence time $(\overline{\Delta\theta})$ and reinterpret the results previously obtained in the light of $\overline{\Delta\theta}$.

Solution

For steady flow regimes the extremely useful chemical engineering concept of mean residence time $(\overline{\Delta\theta})$ can be expressed most generally [7.85], when the cross section Ω_i is *not* constant, according to

$$\text{mean residence time} = \overline{\Delta\theta} = \frac{\text{total mass of fluid in system}}{\dot{m}} \tag{7.71}$$

For the elementary heat exchanger system depicted by Fig. 7.4, Eq. (7.71) reduces to the simpler form

$$\overline{\Delta\theta} = \frac{\Omega_i L\rho}{\dot{m}} = \frac{L}{(\dot{m}/\Omega_i)(1/\rho)} \tag{7.72}$$

since Ω_i = constant and the fluid flowing inside the channel is treated as an incompressible (ρ = constant) fluid in the derivation of the conventional analytical method. By definition $G = \dot{m}/\Omega_i$ and $G/\rho = \overline{V}$; therefore, Eq. (7.72) can be written in the simpler (almost) self-evident form:

$$\overline{\Delta\theta} = \frac{L}{\overline{V}} \tag{7.73}$$

for a prescribed channel, fluid, and mass flow rate, Ω_i, ρ, and $\dot{m}$ are constant. Therefore, the average velocity $\overline{V}$ is also constant and independent of L and the mean residence time $\overline{\Delta\theta}$ is directly proportional to the length of travel L according to Eq. (7.73): that is,

$$\overline{\Delta\theta} \propto L \quad \text{for } \overline{V} = \text{constant at prescribed } \Omega_i,\ L,\ \rho, \text{ and } \dot{m} \tag{7.74}$$

Thus, for a prescribed $\overline{V}$, the length of travel L is equivalent (up to constant) to a time dimension.[19] As a result, the reduced length defined by Eqs. (7.70b) and (7.70c) can alternatively be visualized with Eq. (7.73) as reduced mean residence time:

$$(\overline{\Delta\theta})_{\text{red}} \equiv \frac{\overline{\Delta\theta}}{(\overline{\Delta\theta})_{1/2}} = \text{reduced mean residence time} \tag{7.75a}$$

where

$$(\overline{\Delta\theta})_{1/2} \equiv \frac{L_{1/2}}{\overline{V}} = \text{mean residence time for AF} = \text{CF} = \tfrac{1}{2} \tag{7.75b}$$

Since with Eq. (7.73)

$$(\overline{\Delta\theta})_{\text{red}} \equiv \frac{\overline{\Delta\theta}}{(\overline{\Delta\theta})_{1/2}} \equiv \frac{L}{L_{1/2}} \tag{7.76}$$

it is permissible to expand the relationship between AF and $l_{\text{red}} = L/L_{1/2}$ expressed by Eq. (7.70a) as follows

$$\text{AF} = \left(\tfrac{1}{2}\right)^{(\overline{\Delta\theta})_{\text{red}}} = \left(\tfrac{1}{2}\right)^{l_{\text{red}}} \tag{7.77}$$

The numerical values obtained from Eq. (7.77) in terms of either $\overline{\Delta\theta}/(\overline{\Delta\theta})_{1/2}$ or $L/L_{1/2}$ are summarized in Table 7.5 to emphasize that the "hidden variable", time, reappears, as it should in a time-dependent heat transfer problem, even though the conventional analytical heat exchanger design methods are (partly) derived on the basis of classical, *time-independent* thermodynamics.[20]

It is now possible to reinterpret Eqs. (7.70a)–(7.70c) in the light of $(\overline{\Delta\theta})_{\text{red}}$ instead of the reduced length l_{red} as follows. For a given channel geometry Ω_i, $L = L_{1/2}$ prescribed T_w and T_1, there can exist under in/out SSSF conditions only one mass flow rate $\dot{m}_{1/2}$ to meet a specified approach factor AF $= \frac{1}{2}$ since otherwise a stable operation would *not* be conceivable. To this unique $\dot{m}_{1/2}$ corresponds a unique velocity $\overline{V}_{1/2}$ and with Eq. (7.73) a unique mean

TABLE 7.5. Approach Factor AF as a Function[a] of Reduced Lengths of Travel $L/L_{1/2}$ or Equivalent Reduced Mean Residence Time $\overline{\Delta\theta}/\overline{\Delta\theta})_{1/2}$

$L/L_{1/2}$	=	1	2	3	4	5	z
$\overline{\Delta\theta}/(\overline{\Delta\theta})_{1/2}$	=	1	2	3	4	5	z
AF = $(T_w - T_2/(T_w - T_1)$	=	$(\frac{1}{2})^1$	$(\frac{1}{2})^2 = \frac{1}{4}$	$(\frac{1}{2})^3 = \frac{1}{8}$	$(\frac{1}{2})^4 = \frac{1}{16}$	$(\frac{1}{2})^5 = \frac{1}{25}$	$(\frac{1}{2})^z$

[a]Computed from Eqs. (7.76) and (7.77) where the subscript 1/2 refers to the length of travel and mean residence time required at AF $= \frac{1}{2}$.

residence time $(\overline{\Delta\theta})_{1/2}$. By stacking two, three, and so on identical elementary heat exchangers in series, the total mean residence time is doubled, tripled, and so on, and the resulting total system approach factor (AF) varies exponentially with $(\overline{\Delta\theta})_{red}$ in accordance with Eq. (7.77).

In line with chemical engineering practice [7.86] it is generally more convenient to use an NTU = 1 heat exchanger as a reference with a corresponding AF = $1/e$ per Eq. (7.48). By defining a *relative length* of travel in terms of $L_{1/e}$ as $L_{rel} \equiv L/L_{1/e}$ and a corresponding relative mean residence time $(\overline{\Delta\theta})_{rel} = \overline{\Delta\theta}/(\overline{\Delta\theta})_{1/e}$, where $(\overline{\Delta\theta})_{1/e} = L_{1/e}/\overline{V}$, the following fully dimensionless relationships are obtained:

$$\frac{T_w - T_2}{T_w - T_1} = \text{AF} = e^{-L_{rel}} = e^{-(\overline{\Delta\theta})_{rel}} \tag{7.78a}$$

with

$$L_{rel} \equiv \frac{L}{L_{1/e}} = \text{relative length of travel} \tag{7.78b}$$

$$\overline{\Delta\theta}_{rel} \equiv \frac{\overline{\Delta\theta}}{(\overline{\Delta\theta})_{1/e}} = \text{relative mean residence time} \tag{7.78c}$$

where $L_{1/e}$ = reference length of travel for AF = $1/e$ at NTU = 1
$(\overline{\Delta\theta})_{1/e}$ = reference mean residence time for AF = $1/e$ at NTU = 1

Note that the reference length $L_{1/e}$ and $(\overline{\Delta\theta})_{1/e}$ are fully defined for a prescribed channel geometry, flow rate, and fluid, when $\overline{U}_i$ and, for a given a_i, the product $a_i\overline{U}_i$ are known ($a_i\overline{U}_i$ being the overall heat transfer coefficient per unit axial length) from Eqs. (7.48) and (7.49) which yield, with $\Omega_i/a_i = r_{h,i} = \frac{1}{4}D_{h,i}$ per Eq. (3.32)

$$L_{1/e} = \frac{(\dot{m}c_p)_C}{a_i\overline{U}_i} = \frac{\overline{V}\rho c_p r_{h,i}}{\overline{U}_i} \tag{7.78d}$$

$$(\overline{\Delta\theta})_{1/e} \equiv \frac{L_{1/e}}{\overline{V}} = \frac{\rho c_p r_{h,i}}{\overline{U}_i} \tag{7.78e}$$

7.6.6 General Conclusions

1. The main conclusions derived from the above variation analyses are applicable to other fluid-to-fluid surface heat exchanger configurations provided the simplifying assumptions made to derive the conventional analytical methods (see Appendix B) are satisfied. The reason for this rather sweeping

generalization is that the conventional methods are essentially based on universally valid classical thermodynamic laws in conjunction with (assumed) constant phenomenological heat transfer coefficients.

2. Because the actual empirical heat transfer coefficients are determined empirically in actual practice from the conventional LMTD and/or NTU equations, the methodology is *self-consistent* in the sense that any measurable effects of deviations from the simplifying assumptions are automatically *lumped* in the phenomenological heat transfer coefficient obtained from tests. This explains why the conventional methods can be applied successfully in real world heat exchangers even when one or several of the simplifying assumptions are clearly not fulfilled. However, it is important to keep in mind the validity limits of such heat transfer correlations as further discussed in Chapter 9.

3. Whenever the simplifying assumptions used to derive the conventional methods are so grossly violated that no meaningful correlations of the resulting empirical heat transfer coefficients are feasible, it is logical to question the practical usefulness of the conventional methods and to modify them to fit the particular situation, if this is possible, or else to seek radically new solutions.

4. The methodology by which pragmatic heat transfer engineers managed to circumvent the fundamental limitations of equilibrium thermodynamics are summarized in the form of a guideline because of its fundamental significance and, more importantly, its practical usefulness in more complex situations such as two-phase flow heat exchanger applications.

GUIDELINE 22 *The conventional heat exchanger analytical methods are based on fictive local uniform equilibrium temperatures, rather than real local nonequilibrium temperatures, in conjunction with constant empirical heat transfer coefficients defined on the basis of the same fictive local uniform equilibrium temperatures. The effects of nonequilibrium conditions are lumped in the empirical heat transfer coefficient so defined.*

Similarly, the accepted analytical solutions for the central engineering fluid flow problems stated in Guideline 6 are based on fictive uniform (Bernoulli) velocities that would exist in the absence of friction rather than real local axial velocities. The effects of nonuniform velocities, primitively caused by wall friction dissipation, are lumped in phenomenological friction factors computed from Reynolds numbers that are also based on the same fictive uniform velocities.

It is clearer in the context of the above guideline why the search for the illusive heat transfer coefficient and the coupled friction factor constitutes the prime objective (Guideline 12) of convective heat transfer research studies dedicated to heat exchanger design applications. In principle, the thermodynamic aspects are fully resolved by the conventional LMTD or NTU methods. Consequently, the RD & E efforts in convection focus almost exclusively on the search of the coupled average coefficients $\bar{\alpha}$ and $\bar{f}$ (or local α and f).

7.7 UNIFIED *THERMODYNAMIC* TREATMENT

From what has been said so far, it is clear that the nineteenth century segmentation of thermal engineering sciences into three tightly separated disciplines has become rather archaic and, what is worse, counterproductive. Recognizing this fact, we consider a fully unified *thermodynamic* treatment of momentum, heat, and mass transfer (UTT-MHMT) which is justified both on theoretical, and more importantly in technological applications, on practical grounds.

7.7.1 Theoretical Grounds for a Unified *Thermodynamic* Treatment

Since some typical forced-convection fluid-to-fluid heat exchanger design applications fall within the validity limits of the linear branch of NET theory now widely accepted by physicists, it is permissible and possible (in principle) to use the modern tools of *irreversible thermodynamics* to solve some typical heat transfer engineering problems, keeping in mind that second law considerations were successfully applied already in the nineteenth century to help formulate the thermal radiation heat rate equations (see Guideline 11).

Closer to home, we have seen in Section 7.3 that several well-known heat transfer textbooks published since the 1950s focus on the fundamental coupling between heat, mass, and fluid flow phenomena to solve convective heat and mass transfer engineering problems more effectively. This relatively new methodology, now widely accepted in heat transfer engineering, is known as a unified treatment of momentum, heat, and mass transfer or UT-MHMT. Recalling that the flow of real fluids is always associated with an internal irreversible positive entropy change and since entropy *is* thermodynamics (see Guideline 5(b) in Chapter 1), we may go one step further than the conventional UT-MHMT methodology by considering a more complete interdisciplinary unified approach, including the youngest branch of thermal sciences: thermodynamics. This automatically leads to a unified thermodynamic treatment of momentum, heat, and mass transfer or UTT-MHMT.

Because we are primarily interested here in the application of thermal sciences as useful engineering tools in heat exchanger design practice (rather than purely theoretical considerations), it is appropriate to recapitulate a few of the demonstrated practical advantages and, in complex two-phase flow heat exchanger applications, the necessity of a fully interdisciplinary approach.

7.7.2 Practical Grounds for a Unified *Thermodynamic* Treatment

Some of the practical thermodynamic tools previously identified are listed below.

- Conventional LMTD and NTU-ε Analytical Methods

  The very popular LMTD or NTU-ε heat exchanger design methodology is essentially based on the application of equilibrium thermodynamics tools, mainly the first law of classical thermodynamics in conjunction with the phenomenological linear heat rate equations.

- Corresponding States Principle Generalizations

  The extremely powerful method of generalizing fluid flow data in terms of reduced pressures, described in Chapter 6, is based on the extended corresponding states principle (ECSP), that is, a classical thermodynamic tool. The same thermodynamic approach is applied in Chapter 8 to generalize in a very simple fashion, and with an excellent degree of accuracy, highly complex two-phase flow heat transfer phenomena in real world evaporators and condensers, either from semianalytical correlations or directly from raw empirical data.

  The most significant advantage of this generalization method is its universal validity. Like all classical thermodynamic concepts the ECSP generalization is a universal tool, a priori applicable to simple internal as well as the most complex external flows of single-phase and multiphase fluids, with and without heat addition or removal, regardless of chemical species and system geometry.

- Equilibrium Saturation Temperature Drop Generalizations on the Basis of Molar Entropy

  The Clapeyron–Clausius equation being derived from an infinitesimal ideal Carnot cycle (i.e., the essence of the second law of classical thermodynamics) as shown in Chapter 4, the method of predicting the two-phase flow saturation temperature drop ΔT_{sat} associated with ΔP_{tp}, described in Chapter 6, is thus fundamentally based on the application of classical thermodynamic second law considerations.

  The extremely useful phenomenological law indicating that the molar entropy is a universal function of reduced pressure provides a more fundamental rationale for the far-reaching generalizations of complex two-phase flow phenomena in nucleate and forced-convection evaporation (and in the quasi-mirror-image process of condensation) addressed in Chapters 8 and 9.

- Thermodynamic Coupling of Fluid and Heat Flow in Forced-Convection Evaporators and Condensers

  As emphasized by Guideline 9 in Chapter 4, a unified treatment of heat and fluid phenomena is *indispensable* in normally efficient two-phase forced-convection heat exchangers with heat addition as in evaporators, or removal as in condensers, since the conventional heat transfer driving forces are inextricably coupled with the integrated two-phase friction head loss, that is, total internal dissipation. Because the internal friction dissipation is always associated with a positive *entropy production*, we have in effect anticipated in Chapter 4 one of the major concepts of nonequilibrium thermodynamics emphasized in this chapter, namely, the central role of all related dissipative phenomena, including friction dissipation, which is axiomatically ruled out of consideration in conventional methods based on the simplifying assumption that $\dot{Q}_f = \dot{m}\,\Delta H_f = \dot{m}\overline{T}\Delta s_f = 0$.

7.7.3 Friction Dissipation: The Hidden Variable

The arguments in favor of a fully unified thermodynamic treatment are certainly more convincing in two-phase flow applications since it is mandatory to recognize the thermodynamic coupling expressed by Guideline 9 in order to design thermally efficient evaporators or condensers. In fact, the *unconventional* thermodynamic approach initially proposed in [3.32, 4.4, 4.5] imposed itself for the simple reason that the conventional LMTD or NTU methodology was not applicable in forced-convection thermodynamically efficient evaporators. The situation became particularly intolerable in the case of advanced design enhanced surface compact refrigeration evaporators (in the 1960s) because the application of the conventional methods proved to be not only inaccurate (as one would expect since some of the key simplifying assumptions are grossly violated) but worse, *misleading*. The conventional methodology led to conclusions diametrically opposed to empirical results obtained in extensive laboratory and field tests, as shown in the above-mentioned references. This made it mandatory to discard the conventional (first law) methods and derive adequate semianalytical solutions reflecting more faithfully the actual conditions encountered in real world compact evaporators and condensers. This search led to several ad hoc solutions based on second law thermodynamic considerations focusing on the coupled total friction dissipation.

Although the fluid and heat flow coupling is far less obvious in single-phase than in two-phase flow applications, one is led to suspect the existence of a similar coupling between the total internal friction dissipation $\Delta H_f = \overline{T}\Delta s_f$ or $\dot{Q}_f = \dot{m}\,\Delta H_f = \dot{m}\overline{T}\Delta s_f$ and the total external rate of heat exchange $\dot{Q}_e$ in single-phase flow heat exchangers, in view of the universal validity of classical thermodynamic laws.[21]

One of the purposes of the forced-convection heat exchanger variation analyses presented in Section 7.8 is to demonstrate the existence of the fundamental coupling between $\dot{Q}_f$ and $\dot{Q}_e$ in single-phase flow heat exchangers and derive some useful general conclusions from this primitive phenomenological law.

In Part I of these variation analyses we appeal exclusively to classical thermodynamic considerations to derive very general functional relationships, *solely* on the premise that prescribed inlet/outlet steady-state steady-flow (in/out SSSF) conditions are satisfied, noting that these relationships are valid for stable turbulent as well as laminar flow regimes. From this primitive coupling between $\dot{Q}_f = \dot{m}\,\Delta H_f$ and $\dot{Q}_e = \dot{m}c_p(T_2 - T_1)$, the existence of the coupling between heat transfer coefficients and friction factors is derived for stable turbulent as well as laminar regimes, *without any* a priori theory or model regarding the exact nature of these coupled phenomena.

Explicit universal relationships are then derived in Section 7.9 for the two most extreme stable single-phase flow regimes conceivable: streamline and fully rough turbulent flows.

In Section 7.10 a generalized Chilton–Colburn type of Reynolds analogy is formulated, which is applicable to turbulent *and* laminar regimes.

7.8 VARIATION ANALYSIS, PART I: THERMODYNAMIC CONSIDERATIONS

We again take advantage of the universal validity of classical thermodynamic laws by focusing on the archetype elementary single-channel gas heater depicted in Fig. 7.4 and assume the heat exchanger designer has to determine the total surface required to meet the following customer specifications. For the sake of simplicity we have dropped the subscript C for the cold fluid (*fluid heating* mode as shown in Fig. 7.4, bearing in mind that the same relationships would also be valid for a $T_H > T_w$ in the *fluid cooling* mode):

- *Fixed design temperatures* $T_H = T_{\text{sat}}, T_1, T_2$, therefore a constant heat exchanger approach factor AF $= (T_{\text{sat}} - T_2)/\Delta T_{\max}$ and, in view of the definition of the *contact factor* $\equiv$ CF $= (T_2 - T_1)/(T_{\text{sat}} - T_1) = 1 - \text{AF} =$ constant at various specified design flow rates $\dot{m}$.
- The single-channel heat exchanger (and by extension multicircuit tubular heat exchangers, as shown in Chapter 8) must meet some stringent *quantitatively defined stability requirements* for any of the prescribed fixed design flow rates. These flow rates may vary, depending on application, by a factor of 10 to 1.

Using the minimum design flow rate $\dot{m}_{\min}$ as a convenient reference, any other fixed design flow rate can be expressed in dimensionless form by introducing a reduced flow rate $\dot{m}_{\text{red}}$ defined as follows:

$$\frac{\dot{m}}{\dot{m}_{\min}} \equiv \dot{m}_{\text{red}} \tag{7.79}$$

and the full range of possible design flow rates becomes

$$1 \le \dot{m}_{\text{red}} \le 10 \tag{7.80}$$

with the nondimensional flow rate $\dot{m}_{\mathrm{red}}$ selected as the key independent design variable.

To meet this archetype heat exchanger design problem accurately, reliably, and economically, the designer first plans the "paper study" outlined below:

- Agree with the customer on realistic measurable criteria to ensure that the stringent in/out SSSF specifications can be verified in normal laboratory and/or field tests and incorporate them as an integral part of an approved standard performance test code. In formulating the standard test code, the designer must keep in mind that the end user is not at all interested in the internal subtleties of heat transfer and fluid flow phenomena, as the user's only concern is to ensure stable operation at the specified inlet and outlet equilibrium states and flow rate (Guideline 18 reflects this end user oriented philosophy).
- Choose the largest and smallest practical tube diameters, $(D_{h,i})_{\mathrm{largest}}$ and $(D_{h,i})_{\mathrm{smallest}}$, from standard heat exchanger tubing available from factory stock (better cost and availability). The suitable application range of each standard tube diameter is to be based on Reynolds numbers in the range

$$10{,}000 \leq \mathrm{Re} \leq 100{,}000 \quad \text{in } \textit{turbulent} \text{ regimes for the full design range } 1 \leq \dot{m}_{\mathrm{red}} \leq 10 \tag{7.81}$$

or, in the case of highly viscous fluids, within the stable laminar range

$$200 \leq \mathrm{Re} \leq 2000 \quad \text{in } \textit{laminar} \text{ regimes for the full design range } 1 \leq \dot{m}_{\mathrm{red}} \leq 10 \tag{7.82}$$

in order to avoid uncertain transitional flow regimes and thus safely meet the stringent in/out SSSF specifications.
- Carry out conceptually (i.e., *thought experiment*) a minimum test program anticipating more complete future customer requirements and incorporating essential fluid flow constraints such as total wall friction dissipation $\Delta P_f/\rho$ and associated total pumping power $\dot{m}(\Delta P_f/\rho) \equiv \dot{Q}_f$ = total *friction heat* dissipation rate.[22]

7.8.1 Inlet / Outlet SSSF Standard

Problem Statement

Define measurable in/out SSSF criteria that impose the least restrictions on the heat exchanger design.

Solution

There is no need to "reinvent the wheel" since all recognized industrial test codes (e.g., ASME and ASHRAE) for components (including fluid-to-fluid surface heat exchangers), normally designed for operation under globally defined steady-state conditions, provide adequate quantitative criteria to verify that stable conditions are met during suitably long test intervals. The main points regarding these standard in/out SSSF specifications that need to be emphasized here are the following:

- The standard industrial codes are less restrictive than the eighteenth century hydrodynamic definition of SSSF conditions derived from well-behaved streamline flow theory (see Chapter 2). In contrast to the classical hydrodynamic SSSF definition, standard thermal component in/out SSSF specifications spell out acceptable maximum fluctuations at *inlet and outlet only*, without imposing any restrictions within the component itself.
- Standard in/out specifications *permit fluctuations* (time dependency) *within* the surface heat exchanger as long as the specified inlet/outlet criteria are met. This realistic feature is useful in turbulent flow regimes which are not truly time independent even on a macroscopic scale (see Chapter 3). It is a *necessary* feature in complex two-phase flow heat exchangers, since significant macroscopic fluctuations are very likely at some locations within the heat exchanger (e.g., for some of the two-phase flow patterns shown in Figs. 4.8 and 4.11) even though they may be sufficiently dampened at the heat exchanger outlet to comply with the standard outlet stability requirements. As a matter of fact, one of the special semiempirical solutions devised to cope with the complexities of direct-expansion compact refrigeration evaporators described in Refs. 3.32 and 4.4 is based on an average time integrated wall temperature $\tilde{T}_w$ to allow for such unavoidable (macroscopic) fluctuations within the evaporator.
- The classical hydrodynamic SSSF definition thus represents a rather narrow limiting case within the more general and significantly less restrictive thermal component test code specifications for SSSF conditions at inlet and outlet only. The two definitions are fully equivalent only for the (unusual) well-behaved *streamline* flow regimes.
- Since natural and engineering fluid flow phenomena are more frequently chaotic (see Chapters 3–6) and thus do *not* comply rigorously with the eighteenth century hydrodynamic SSSF definition, we focus, from now on, exclusively on the standard in/out SSSF definition.

7.8.2 Primitive Coupling Between $\dot{Q}_f$ and $\dot{Q}_e$ for Turbulent Flows

Problem Statement

Conceptually, run the performance tests at $1 \leq \dot{m}_{\text{red}} \leq 10$ with the largest $D_{h,i}$ first, to establish the existence in stable turbulent flow regimes of a one-to-one relationship between $\dot{Q}_f = \dot{m}(\Delta P_f/\rho)$ and $\dot{Q}_e = \dot{m}c_p(T_2 - T_1)$ for

TABLE 7.6. Single-Phase Heat Exchanger Variation Analysis — Part I: Thermodynamic Considerations

(1)	Re	10,000	Re	100,000
(2)	$\dot{m}$	$\dot{m}_{min}$	$\dot{m}_{red}(\dot{m}_{min})$	$10\dot{m}_{min}$
(3)	$\dot{m}_{red} = \dot{m}/\dot{m}_{min}$	1	$\dot{m}_{red}$	10
(4)	L	L_1	$L_{\dot{m},red}$	$(L)_{10}$
(5)	ΔH_f	$(\Delta H_f)_1$	$(\Delta H_f)_{\dot{m},red}$	$(\Delta H_f)_{10}$
(6)	$\dot{Q}_e$	$1(\dot{Q}_e)_{min}$	$\dot{m}_{red}(\dot{Q}_e)_{min}$	$10(\dot{Q}_e)_{min}$
(7)	$\dot{Q}_f$	$\dot{m}_{min}(\Delta H_f)_1$	$\dot{m}_{red}\dot{m}_{min}(\Delta H_f)_{\dot{m},red}$	$10\dot{m}_{min}(\Delta H_f)_{10}$
(8)	$\bar{\alpha}$	$\bar{\alpha}_1$	$\bar{\alpha}_{\dot{m},red}$	$\bar{\alpha}_{10}$
(9)	$\bar{f}$	$\bar{f}_1$	$\bar{f}_{\dot{m},red}$	$\bar{f}_{10}$
(10)	Re	200	Re	2000

the full design range $1 \leq \dot{m}_{red} \leq 10$. From this primitive relationship derive, as a necessary corollary, the existence of a general Reynolds type of relationship between the conventional empirical heat transfer coefficient and friction factor. This theoretical analysis is to be based exclusively on first law thermodynamic considerations, the assumptions that in/out SSSF requirements are satisfied, and the empirical knowledge that real fluid flows are always associated with a dissipation $\Delta H_f > 0$.

Solution

The key analytical results obtained for stable turbulent flow regimes are summarized in Table 7.6 and highlighted below (by referring to the relevant columns and/or line numbers in the table).

For the selected largest channel, $D_{h,i}$, Ω_i, and a_i are known constants as well as μ and ρ for a given fluid (gas in this case) at the fixed specified temperatures T_1, T_2, and $T_{sat} = T_w$. Thus, the actual flow rate $\dot{m}_{min}$ corresponding to $(\text{Re})_{min} = 10{,}000$ can be computed from the definition of the Reynolds number, Eq. (3.3). Since $\text{Re} \propto \dot{m}$ under the prescribed design condition, $\dot{m} \propto \dot{m}_{min}$ as shown on line (2) for the full design range $10{,}000 \leq \text{Re} \leq 100{,}000$ and corresponding design range of the reduced flow rates $1 \leq \dot{m}_{red} \leq 10$ on line (3).

We now conceptually carry out a test at the maximum design flow rate $\dot{m}_{max} = 10 \times \dot{m}_{min}$, or $\dot{m}_{red} = 10$, with an extremely long tube $(L/D \gg 1)$ at the specified temperatures T_1 and $T_H = T_{sat} = T_w$ and find that $(T_2)_{act} > (T_2)_{design\ target}$, indicating that the initial length of travel so selected is excessively long. We therefore reduce the tube length and repeat the tests until ultimately, by a trial and error procedure, the design objective $(T_2)_{act} = (T_2)_{design}$ is achieved, within the accuracy limits allowed in the approved test code. At this point, the measured elementary heat exchanger tube length meeting the design targets at $\dot{m}_{red} = 10$ and the corresponding unique empirical value of the total pressure drop or head loss are recorded. These purely empirical values L_{10} and $(\Delta P/\rho)_{10} = (\Delta H)_{10}$ are entered under the column $\dot{m}_{\text{red}} = 10$ on lines (4) and (5), respectively. Note that the subscript refers to the reduced design flow rate $\dot{m}_{\text{red}} = 10$ corresponding to $\dot{m}_{\text{max}}$.

The same trial and error procedure is repeated at lower $\dot{m}_{\text{red}}$ values within the design range $1 \le \dot{m}_{\text{red}} \le 10$ and the purely empirical results thus obtained are tabulated in the appropriate column $\dot{m}_{\text{red}}$ on lines (4) and (5).

For the fixed design values of T_1, T_2 the first law of classical thermodynamics yields at $\dot{m}_{\text{min}}$, $(\dot{Q}_e)_{\text{min}} = \dot{m}_{\text{min}} c_p (T_2 - T_1)$. Since under the prescribed design conditions the term $c_p(T_2 - T_1)$ is constant, it is evident that the total heat exchange rate $\dot{Q}_e$ at any value of $\dot{m}$ is directly proportional to $\dot{m}$; therefore,

$$\dot{Q}_e = \dot{m}_{\text{red}} \left(\dot{Q}_e \right)_{\text{min}} \tag{7.83}$$

as indicated on line (6).

By definition, the total *friction heat* or dissipation per unit time at any flow rate $\dot{m}$ is

$$\dot{Q}_f = \dot{m}\, \Delta H_f \tag{7.84}$$

or, with $\dot{m} = \dot{m}_{\text{min}} \dot{m}_{\text{red}}$ and the corresponding empirical value $(\Delta H_f)_{\dot{m}_{\text{red}}}$, we obtain the following expression:

$$\left(\dot{Q}_f \right)_{\dot{m}} = \dot{m}_{\text{red}} \dot{m}_{\text{min}} \left(\Delta H_f \right)_{\dot{m}_{\text{red}}} \tag{7.85}$$

shown on line (7) of Table 7.6.

As a corollary of the in/out SSSF condition, there must exist a unique correspondence between $\dot{m}$ and L and, consequently, a unique one-to-one relationship between any of the parameters shown on lines (4)–(7) at the prescribed design conditions (otherwise SSSF conditions would *not* be satisfied). Thus, we have established the existence of "primitive" relationships of the following most general form:

$$\dot{Q}_e = \varphi_{\dot{Q}_e}\left(\dot{m}_{\text{red}} \right) \tag{7.86}$$

$$\dot{Q}_f = \varphi_{\dot{Q}_f}\left(\dot{m}_{\text{red}} \right) \tag{7.87}$$

Consequently,

$$\dot{Q}_f = \varphi_{\dot{Q}_f}\left(\dot{Q}_e \right) \tag{7.88}$$

or, dividing Eq. (7.87) by Eq. (7.86),

$$\frac{\dot{Q}_f}{\dot{Q}_e} = \varphi_{\dot{Q}_f/\dot{Q}_e}\left(\dot{m}_{\text{red}} \right) \tag{7.89}$$

These relationships, valid within the stated design specifications, that is, a fixed approach factor AF and in/out SSSF conditions, are *primitive in the sense that they are derived directly from measurements* on the basis of the conservation laws, without making use of any derived concepts such as friction factors or heat transfer coefficients at this point.

From the definition of the conventional heat transfer coefficient, it is clear that a single empirical value $\bar{\alpha}_{\dot{m}_{\text{red}}}$ can be computed from the definition $\bar{\alpha}$ for any value of the independent variable $\dot{m}_{\text{red}}$ with the corresponding unique values of L and $\dot{Q}_e$, since the effective mean temperature driving force is a known constant under the stated design conditions. The value of $(\bar{\alpha})_{\dot{m}_{\text{red}}}$ thus obtained at any $\dot{m}_{\text{red}}$ is indicated on line (8) of Table 7.6.

A unique value $f_{\dot{m}_{\text{red}}}$ can also be computed at any $\dot{m}_{\text{red}}$ from the conventional fluid flow pressure drop, or head loss, correlations with the measured values of $(\Delta H_f)_{\dot{m}_{\text{red}}}$ and $L_{\dot{m}_{\text{red}}}$ and therefore $(L/D_{h,i})_{\dot{m}_{\text{red}}}$ as indicated on line (9) of Table 7.6.

Following the same reasoning as in the case of the primitive coupling between $\dot{Q}_f$ and $\dot{Q}_e$, one immediately arrives at a similar coupling between the derived design parameter $\bar{\alpha}$ and $\bar{f}$, in the form

$$\bar{\alpha} = \varphi_{\bar{\alpha}}(\bar{f}) \tag{7.90}$$

which is, of course, in full agreement with the well-known semiempirical relationships for turbulent regimes reviewed in Section 7.3.

7.8.3 Primitive Coupling Between $\dot{Q}_f$ and $\dot{Q}_e$ for Laminar Flows

Problem Statement

Repeat the previous analysis for stable laminar, instead of turbulent, flow regimes.

Solution

Following the same reasoning for stable laminar flow regimes at $200 < \text{Re} < 2000$ and corresponding $1 \leq \dot{m}_{\text{red}} \leq 10$ leads to exactly the same general conclusions as in the case of stable turbulent flow regimes. It is only necessary to substitute in Table 7.6 the Reynolds numbers shown on line (10) for those on line (1). Lines (2)–(9) remain valid exactly as stated for stable laminar flow regimes with the corresponding Re values shown on line (10).

Of course, the absolute values of $\dot{m}_{\text{min, lam}}$ will be generally significantly lower at $(\text{Re})_{\text{min}} = 200$ for laminar flow than at $\dot{m}_{\text{min, turb}}$ and $(\text{Re})_{\text{min}} = 10{,}000$ for a given channel geometry and fluid. For a prescribed channel geometry, identical design specifications, with $\dot{m}_{\text{min, lam}} = (200/10{,}000)\dot{m}_{\text{min, turb}} = \frac{1}{50}\dot{m}_{\text{min, turb}}$, and since $\dot{m} \propto \text{Re} \propto \dot{Q}_e$ for a prescribed $T_2 - T_1$, the actual total rate of heat exchange $\dot{Q}_e$ is correspondingly smaller.

Consequently, all the conclusions derived from Table 7.6 for turbulent regimes remain valid as broadly stated for stable laminar flow regimes, including Eqs. (7.83)–(7.90).

Comments

1. Although some of the results presented here are rather well known, particularly the coupling between $\bar{\alpha}$ and $\bar{f}$ in turbulent regimes, the point to stress is that they have been derived solely on the premise that in/out SSSF conditions are possible without making use of any theoretical models or explanation of the detailed nature of the coupled fluid and heat flow phenomena. In this sense, the approach followed is fully in line with classical thermodynamic methodologies.

2. Because it is not possible to quantify the types of relationship postulated by Eqs. (7.86)–(7.90) solely on the basis of classical thermodynamic laws (see Guideline 14(a)), we must now appeal to the phenomenological heat and flow rate equations as highlighted in Part II of this variation analysis.

7.9 VARIATION ANALYSIS, PART II: EXPLICIT CORRELATIONS OF $\dot{Q}_f / \dot{Q}_e$

General Background. The prime objective is to derive an explicit equation for the general relationship between $\dot{Q}_f$ and $\dot{Q}_e$ postulated by Eqs. (7.88) and (7.89) on the basis of widely accepted semiempirical correlations for $\bar{\alpha} = \alpha_\infty$ and $\bar{f} = f_\infty$ in long tubes ($L/D \gg 1$) or fully developed tube flow regimes. Since this does not affect the most general conclusions derived from this analysis, we simplify this exercise by assuming a Pr = 1 fluid and circular tubes, and focus first on the two most extreme single-phase flow regimes conceivable: fully rough wall turbulent flow and streamline flow.

7.9.1 Pr = 1 Fluids, Fully Rough Turbulent Flows

Problem Statement

Quantify the results obtained in Part I for the fully rough wall limit of stable turbulent flow regimes, selecting first the largest standard heat exchanger circular tube of internal diameter $(D_{h,i})_{\text{largest}}$.

Solution

For a prescribed approach factor AF, Eq. (7.47) yields the required $(\text{NTU})_{\text{design}} = \ln(1/\text{AF})$ which is therefore fixed. The required length of travel is obtained by solving $(\text{NTU})_{\text{design}} = a_i L \bar{U}_i / \dot{m} c_p$ per Eq. (7.49) for L, or with $\bar{U}_i = \bar{\alpha}_i$ in this case:

$$L = \frac{\dot{m} c_p}{a_i \bar{\alpha}_i} (\text{NTU})_{\text{design}} \tag{7.91}$$

As shown in Section 7.3, the general Chilton–Colburn correlation converges toward the simple Reynolds analogy for a Pr = 1 fluid; therefore, $\bar{\alpha}_i$ can be expressed with Eq. (7.10a) according to

$$\bar{\alpha}_i = \tfrac{1}{8} f_{\text{fr}} c_p G = \tfrac{1}{8} \frac{f_{\text{fr}} c_p \dot{m}}{\Omega_i} \tag{7.92}$$

Introducing the above expression for $\bar{\alpha}_i$ in Eq. (7.91), we obtain, with $\Omega_i / a_i \equiv r_{h,i} \equiv \frac{1}{4} D_i$ for circular tubes:

$$L = (\text{NTU})_{\text{design}} \frac{\dot{m} c_p \Omega_i}{\frac{1}{8} a_i f_{\text{fr}} \dot{m} c_p} = D_i (\text{NTU})_{\text{design}} \left(\frac{2}{f_{\text{fr}}} \right) \tag{7.93a}$$

or

$$\frac{L}{D_i} = (\text{NTU})_{\text{design}} \left(\frac{2}{f_{\text{fr}}} \right) \tag{7.93b}$$

Since $(\text{NTU})_{\text{design}}$ and D_i are fixed and f_{fr} = constant in fully rough wall turbulent regimes, it is clear from Eq. (7.93a) that

$$L_{\dot{m}} = L_{\dot{m}_{\min}} = \text{constant and independent of } \dot{m} \tag{7.94a}$$

or

$$l_{\text{red}} \equiv \frac{L_{\dot{m}}}{L_{\dot{m}_{\min}}} = \frac{(L/D_i)_{\dot{m}}}{(L/D_i)_{\dot{m}_{\min}}} = 1 = \text{constant} \tag{7.94b}$$

as shown on line (4) on the left-hand side of Table 7.7

The mean residence time $\overline{\Delta\theta} = L/\bar{V}$ per Eq. (7.73) is inversely proportional to $\dot{m}$ and therefore proportional to $1/\dot{m}$ since $\bar{V} \propto \dot{m}$ and L = constant according to Eq. (7.94). Thus, the reduced mean residence time

$$(\overline{\Delta\theta})_{\text{red}} \equiv \frac{(\overline{\Delta\theta})_{\dot{m}}}{(\overline{\Delta\theta})_{\dot{m}_{\min}}} = \frac{1}{\dot{m}_{\text{red}}} \tag{7.95}$$

as shown on line (5).

According to the conventional method of predicting ΔH_f, Eq. (3.4), the total head loss is proportional to $\bar{V}^2 \propto G^2 \propto \dot{m}^2 \propto \dot{m}_{\text{red}}^2$ since f_{fr} is constant and independent of Re and flow rate $\dot{m}$ or $\dot{m}_{\text{red}}$ under the prescribed conditions for the limiting case fully rough turbulent regimes. Thus, $(\Delta H_f)_{\text{red}} \propto \dot{m}_{\text{red}}^2$ as shown on line (6).

The variation of $(\dot{Q}_f)_{\text{red}} = (\dot{Q}_f)/(\dot{Q}_f)_{\dot{m}_{\min}}$ and the ratio $(\dot{Q}_f/\dot{Q}_e)_{\text{red}} \equiv (\dot{Q}_f/\dot{Q}_e)_{\dot{m}}/(\dot{Q}_f/\dot{Q}_e)_{\dot{m}_{\min}}$ shown respectively on lines (7) and (8) follow immediately from the definition of $\dot{Q}_f = \dot{m}\,\Delta H_f$. It is also evident that the ratio $(\dot{Q}_f/\dot{Q}_e)_{\text{red}}$ = line (7)/line (3) = $(\Delta H_f)_{\text{red}}$ on line (6), as expected.

TABLE 7.7. Single-Phase Heat Exchanger Variation Analysis—Part II: Explicit Correlations

Flow Regime	Stable Fully Rough Turbulent Flow					Stable Laminar Flow				
(1) Re	10,000	···	Re	···	100,000	200	···	Re	···	2000
(2) $\dot{m}_{red}$	1	···	$\dot{m}_{red}$	···	10	1	···	$\dot{m}_{red}$	···	10
(3) $(\dot{Q}_e)_{red}$	1	···	$\dot{m}_{red}$	···	10	1	···	$\dot{m}_{red}$	···	10
(4) l_{red}	1	···	1	···	1	1	···	$\dot{m}_{red}$	···	10
(5) $(\overline{\Delta\theta})_{red}$	1	···	$1/\dot{m}_{red}$	···	$1/10$	1	···	1	···	1
(6) $(\Delta H_f)_{red}$	1	···	$\dot{m}^2_{red}$	···	$(10)^2$	1	···	$\dot{m}^2_{red}$	···	$(10)^2$
(7) $(\dot{Q}_f)_{red}$	1	···	$\dot{m}^3_{red}$	···	$(10)^3$	1	···	$\dot{m}^3_{red}$	···	$(10)^3$
(8) $(\dot{Q}_f/\dot{Q}_e)_{red}$	1	···	$\dot{m}^2_{red}$	···	$(10)^2$	1	···	$\dot{m}^2_{red}$	···	$(10)^2$
(9) $(\bar{\alpha}/\bar{f})_{red}$	1	···	$\dot{m}_{red}$	···	10	1	···	$\dot{m}_{red}$	···	10
(10) $\bar{\alpha}_{red}$	1	···	$\dot{m}_{red}$	···	10	1	···	1	···	1
(11) $\bar{f}_{red}$	1	···	1	···	1	1	···	$1/\dot{m}_{red}$	···	$1/10$

The key data summarized on lines (9) to (11) reflect the empirical correlations of $\bar{\alpha}$ and $\bar{f}$ used in the derivation of the explicit results shown on lines (4)–(8).

7.9.2 Laminar Flows

Problem Statement

Repeat the previous analysis for stable *laminar* flow regimes.

Solution

The fully developed laminar tube flow heat transfer coefficient $\alpha_\infty = \bar{\alpha}$ for $L/D \gg 1$ can be expressed according to the following well-known dimensionless equation for the constant wall temperature boundary condition[23] under consideration here, that is $T_H = T_{sat} = T_w =$ constant:

$$(\mathrm{Nu})_\infty \equiv \overline{\mathrm{Nu}} \equiv \frac{\bar{\alpha}_i D_{h,i}}{k} = \text{constant} \tag{7.96}$$

showing that under otherwise fixed conditions $\bar{\alpha}_i$ is constant and independent of the flow rate $\dot{m}$. Therefore, $(\bar{\alpha}_i)_{red} = 1$ as shown on line (10) on the right-hand side of Table 7.7. Consequently, for a prescribed approach factor and corresponding fixed $(\mathrm{NTU})_{design}$, it is readily apparent from Eq. (7.91) that the required heat exchanger length of travel must vary in direct proportion to $\dot{m}$ or $\dot{m}_{red}$ as indicated on line (4) in the laminar section of Table 7.7.

Since according to the Hagen–Poiseuille equation (see Chapter 3) $(\Delta P)_f$ and $(\Delta P_f/\rho) = \Delta H_f$ are directly proportional to the product $\dot{m}L$ under otherwise fixed conditions, it follows immediately since $L \propto \dot{m}$ at a prescribed approach factor [per line (4)] that $(\Delta H_f)_{\text{red}} = m_{\text{red}}^2$ exactly as in the case of fully rough turbulent regimes [although the absolute values of $(\Delta H_f)_{\dot{m}_{\text{min, lam}}}$ and $(\Delta H_f)_{\dot{m}_{\text{min, turb}}}$ are radically different] as shown on line (6).

From the alternative equation for the prediction of the total pressure drop, or head loss, in stable laminar regimes at $L/D \gg 1$, Eq. (3.7), it is evident that $f_{\text{lam}} \propto \dot{m}^{-1} \propto m_{\text{red}}^{-1}$; therefore, the term $\bar{f}_{\text{red}} = \bar{f}/\bar{f}_{\dot{m}_{\text{min}}} = 1/\dot{m}_{\text{red}}$ as shown on line (11).

Following exactly the same reasoning as in the case of fully rough turbulent regimes yields the results shown on the remaining lines on the right-hand side of Table 7.7.

7.9.3 Common Relationship $\dot{Q}_f / \dot{Q}_e$ for Turbulent and Laminar Flows

Problem Statement

Compare the results obtained in Sections 7.9.1 and 7.9.2 and identify a common quantitative relationship between $\dot{Q}_f$ and $\dot{Q}_e$ for fully rough turbulent and streamline flows, notwithstanding the fact that we are dealing with the two most extreme single-phase flow regimes conceivable (most chaotic and most orderly).

Solution

A line by line comparison of the results shown on the fully rough turbulent and laminar sides of Table 7.7 brings to light the following significant points.

- Because the fluid and convective heat flow phenomena are coupled, it is not at all surprising that the end results shown on lines (4), (5), (10), and (11) are radically different since we are dealing with the two most extreme and radically different flow regimes conceivable in (incompressible) single-phase fluid flows: the most chaotic pattern on the left-hand side and the most orderly structure on the right-hand side of Table 7.7.
- In contrast, it is quite startling and indeed most remarkable that the results shown on *lines* (6)–(9) *are identical* for any arbitrarily selected constant approach factor *notwithstanding the radically different characteristics of these two most extreme stable flow regimes.*

7.9.4 Broad Validity of Second Power Law $\dot{Q}_f / \dot{Q}_e \propto \dot{m}_{\text{red}}^2$

Problem Statement

Reassess all the previous results and conclusions with the smallest standard circular tube of internal diameter $(D_{h,i})_{\text{smallest}}$.

Solution

The results summarized in Table 7.7 remain exactly valid as stated for the smallest standard heat exchanger tube of internal diameter $(D_{h,i})_{\text{smallest}}$ [even though the absolute values of the reference $\dot{m}_{\text{min}}$ at $(\text{Re})_{\text{min}}$ are of course different] for the following reasons:

In the case of fully rough turbulent flow regimes considered so far, if we assume the same effective relative wall roughness $(\bar{e}/D_{h,i})_{\text{smallest}} = (\bar{e}/D_{h,i})_{\text{largest}}$, exactly the same constant values of f_{fr} are obtained at the same Re for the two different diameters as shown in Chapter 3 (Moody's charts, Colebrook's equations, and full similarity).

In the more typical case of heat exchanger design practice, the effective mean wall roughness $\bar{e}$ remains practically constant, as discussed in Chapter 3; therefore $\bar{e}/(D_{h,i})_{\text{smallest}} > \bar{e}/(D_{h,i})_{\text{largest}}$ and it is evident (see Moody's chart, Fig. 3.4) that f_{fr} for the smallest $D_{h,i}$ will also be constant if the largest tube of the family falls within the fully rough zone, $f_{\text{fr}} = \text{constant}$, for $10{,}000 \le \text{Re} \le 100{,}000$ as assumed in this variation analysis. Consequently, all the results obtained with the largest tube diameter remain exactly valid as stated, even though the absolute values of f_{fr} and the coupled $\bar{\alpha}$ will generally be different (larger!) for the smallest tube diameter selected.

In the case of laminar flow regimes, both the friction factor $\bar{f}$ and the coupled heat transfer coefficient $\bar{\alpha}$ are uniquely defined according to Eqs. (3.7) and (7.96), respectively, in terms of Re only under the stated design conditions (the effects of the "natural" wall roughness being effectively suppressed in streamline flow). Consequently, the right-hand side of Table 7.7 for laminar flow remains valid exactly as stated for any tube diameters.

Summary

The results shown in Table 7.7 and the conclusions derived from this analysis remain valid for the full range of standard heat exchanger tube diameters, and it is permissible to express these relationships between $\dot{Q}_f$ and $\dot{Q}_e$ most generally as follows:

$$\left(\dot{Q}_f\right)_{\text{red}} = \dot{m}^3_{\text{red}} \tag{7.97a}$$

$$\left(\Delta H_f\right)_{\text{red}} = \left(\dot{Q}_f/\dot{Q}_e\right)_{\text{red}} = \dot{m}^2_{\text{red}} \tag{7.97b}$$

Validity: In/out SSSF conditions; prescribed constant approach factor AF; laminar or fully rough turbulent flow regimes; straight circular tubes; and normal heat exchanger tubing and pipe diameters (7.97c) where the reduced parameters are defined as

$$\begin{aligned}
\left(\dot{Q}_e\right)_{\text{red}} &\equiv \dot{Q}_e/\dot{Q}_{e\dot{m}_{\min}} \\
\left(\dot{Q}_f\right)_{\text{red}} &\equiv \dot{Q}_f/\dot{Q}_{f\dot{m}_{\min}} \\
\left(\Delta H_f\right)_{\text{red}} &\equiv \Delta H_f/\Delta H_{f\dot{m}_{\min}} \\
\left(\frac{\dot{Q}_f}{\dot{Q}_e}\right)_{\text{red}} &\equiv \frac{\dot{Q}_f/\dot{Q}_e}{\dot{Q}_f/\dot{Q}_{e\dot{m}_{\min}}}
\end{aligned} \tag{7.97d}$$

with the subscript $\dot{m}_{\min}$ referring to the lowest design flow rate selected for stable laminar or turbulent regimes.

Comments

1. The phenomenological laws expressed by Eqs. (7.97a)–(7.97d) display typical characteristics of classical thermodynamics laws summarized in Guideline 14(a), namely, their general validity and independence of detailed mechanisms (laminar and turbulent) or substances (as long as Pr and all physical properties are constant) or channel configuration. This strongly confirms the thermodynamic implications of the results of this variation analysis inferred by the fact that friction dissipation is always associated with a positive increase of the key variable of classical or NE thermodynamics: entropy.

2. On the premise that the primitive phenomenological laws expressed by Eqs. (7.97a) and (7.97b) must be fulfilled, we derive in Part III some generalized relationships between the Stanton number $\overline{\text{St}} = \bar{\alpha}/c_pG$, the friction factor $\bar{f} = 4\bar{f}_{rh}$, and $\Pr \equiv c_p\mu/k$, which are applicable in stable laminar as well as turbulent flow regimes. These generalized relationships are confirmed by recognized solutions for laminar Hagen–Poiseuille channel flows and turbulent flows inside channels of any *natural* wall roughness as shown in Part III (for $\Pr = 1$ fluids in the case of turbulent flows). The extension of the proposed modified Chilton–Colburn types of Reynolds analogies for turbulent flow applications in channels of arbitrary natural wall roughness with $\Pr \neq 1$ fluid is addressed in Chapter 8.

3. It is noteworthy that the results derived from this variation analysis by focusing on total friction dissipation could not be derived on the basis of conventional (first law) analytical methods since they are based on the axiomatic assumption of zero friction dissipation: $\dot{Q}_f = \dot{m}\,\Delta H_f = 0$ or, more precisely,

$$\frac{\dot{Q}_f}{\dot{Q}_e} \equiv \frac{\dot{m}\,\Delta H_f}{\dot{m}c_p|T_2 - T_1|} \equiv \frac{\Delta H_f}{c_p|T_2 - T_1|} \equiv \frac{\Delta T_f}{|T_2 - T_1|} = 0 \tag{7.98a}$$

where

$$\Delta T_f \equiv \frac{\Delta H_f}{c_p} = \textit{apparent}\text{ temperature rise due to friction dissipation} \tag{7.98b}$$

The term *apparent* has been deliberately selected to emphasize the fact that ΔT_f does not always represent a *real* temperature rise as, for example, in the case of ideal gases (because in an adiabatic throttling process the enthalpy must remain constant, and T = constant with c_p = constant, for ideal gases; therefore, $(\Delta T_f)_{act} = 0$).

Although Eq. (7.98a) is satisfied within a totally acceptable error of the order of 1% in many *normal temperature* applications, it ceases to be valid at high Mach numbers $\gtrless 1$ (i.e., at near and supersonic flow velocities [7.87]) and even at normally low fluid velocities in many typical refrigeration, cryogenic, or thermodynamically efficient heat exchangers when ΔT_f can be of the same order of magnitude as $|T_2 - T_1|$ or, $\Delta T_f/|T_2 - T_1| \approx 1$. It is therefore not altogether surprising that the rather *unconventional* analytical approach described here originated from the refrigeration industry where typically $\Delta T_f/|T_2 - T_1| = \dot{Q}_f/\dot{Q}_e \neq 0$. By focusing on $\dot{Q}_f = \dot{m}\,\Delta H_f$, and therefore on $\Delta H_f = \bar{T}\Delta s_f$, we in effect acknowledge the central significance of friction dissipation in convective heat exchangers and thus establish a practical bridge with modern (twentieth century) concepts of nonequilibrium thermodynamics [see Guideline 20(c)] as further discussed in Chapter 9. By contrast, all first law heat exchanger analytical methods *negate* the central significance of friction dissipation by axiomatically setting $\dot{Q}_f = 0$ to simplify the mathematical solution of the complex Navier–Stokes equations (particularly in the case of turbulent flows).

7.10 VARIATION ANALYSIS, PART III: GENERALIZED CHILTON–COLBURN CORRELATION FOR LAMINAR AND TURBULENT REGIMES

General Background and Objectives. We now capitalize on the results obtained in Part II of this variation analysis to formulate, on the basis of empirically validated correlations, a modified Chilton–Colburn type of Reynolds analogy. Some useful generalizations of laminar Hagen–Poiseuille flow correlations are derived on the basis of the proposed more universal form of the Reynolds analogy. This leads to a quantitative assessment of the practical advantages of turbulent flow regimes in forced-convection heat exchangers in Section 7.11.

Since we confine ourselves here to fully developed flows (or $L/D_h \gg 1$) when $(\mathrm{Nu})_\infty \equiv \overline{\mathrm{Nu}} \equiv \mathrm{Nu}$ or $\alpha_\infty \equiv \bar{\alpha} \equiv \alpha$ and $f_\infty \equiv \bar{f} \equiv f$, we simplify the writing by omitting the subscript ∞ and the overbar for α and f and all related dimensionless numbers.

7.10.1 Laminar Flows in Circular Tubes

Problem Statement

Derive a simple relationship $\mathrm{St} = \varphi_{\mathrm{St}}(f; \mathrm{Pr})$ for stable fully developed laminar Hagen–Poiseuille flows in circular tubes.

Solution

For fully developed Hagen–Poiseuille flows in circular tubes $\mathrm{Nu} = \alpha D_h/k$ has a constant value determined analytically (on the basis of the simplifying assumption that all physical properties are constant) as shown in standard textbooks [3.20, 3.31, 7.88] according to

$$(\mathrm{Nu})_q = \text{constant} = 4.364 \equiv C_{\mathrm{Nu},q} \quad \text{for constant heat flux } q'' \tag{7.99a}$$

$$(\mathrm{Nu})_T = \text{constant} = 3.66 \equiv C_{\mathrm{Nu},T} \quad \text{for constant wall temperature } T_w \tag{7.99b}$$

where C_{Nu} represents the constant numerical value of Nu and the subscripts q and T allow for a distinction between the two classical wall boundary conditions indicated in Eqs. (7.99a) and (7.99b). Since many actual design applications fall between these two boundary conditions and the theoretical ratio $C_{\mathrm{Nu},q}/C_{\mathrm{Nu},T} = 4.364/3.66 = 1.19$ indicates a percentage spread of less than 20% between these two extreme wall boundary cases, we introduce an arithmetic mean constant $(\mathrm{Nu})_{\mathrm{am}} \equiv C_{\mathrm{Nu,am}}$ with the following constant value according to Eqs. (7.99a) and (7.99b):

$$C_{\mathrm{Nu,am}} = (\mathrm{Nu})_{\mathrm{am}} = \tfrac{1}{2}\left[(\mathrm{Nu})_q + (\mathrm{Nu})_T\right] = 4.00 \tag{7.99c}$$

Validity: Constant q'' *and* T boundary
Maximum Theoretical Deviation: Less than $\pm 10\%$

Keeping in mind the simplifying assumptions (isothermal conditions and constant fluid properties) made to derive theoretically the constants shown as Eqs. (7.99a) and (7.99b), and recalling the extreme sensitivity of laminar flows to nonisothermal conditions and inlet effects (see Section 3.11) and the uncertainties associated with transport properties such as k and μ (see also Chapter 6, and appendix on "Thermophysical Properties" in Ref. 3.3, pp. 461–462), it is evident that a theoretical error of less than $\pm 10\%$ is not significant in most practical heat exchanger design applications.

To reduce the conventional Nu correlation to the more useful Reynolds analogy type of correlation, in terms of the Stanton number St, we take advantage of the identity

$$\mathrm{Nu} \equiv \mathrm{St}\,\mathrm{Pr}\,\mathrm{Re} = \left(\frac{\alpha}{\cancel{c_p}\cancel{G}}\right)\left(\frac{\cancel{c_p}\cancel{\mu}}{k}\right)\left(\frac{\cancel{G}D_h}{\cancel{\mu}}\right) \equiv \frac{\alpha D_h}{k} \tag{7.100}$$

We then substitute the product St Pr Re for Nu in Eqs. (7.99a), (7.99b), and (7.99c) and solve for St to obtain

$$\mathrm{St} = C_{\mathrm{Nu}}\mathrm{Re}^{-1}\mathrm{Pr}^{-1} \tag{7.101}$$

Recalling the generalized Blasius equation for f introduced in Chapter 3,

$$4f_{rh} = f = K_n \mathrm{Re}^{-n} \tag{7.102}$$

with

$$n = 1 \quad \text{for } \textit{all} \text{ fully developed laminar flow regimes} \tag{7.103}$$

and

$$K_n = 64 \quad \text{for circular tube} \tag{7.104}$$

we can derive the laminar flow Reynolds type analogy by multiplying and dividing the right-hand side of Eq. (7.101) by $K_n = 64$, thus obtaining

$$\text{St} = \frac{\alpha}{c_p G} = C_{\text{St}} f \, \text{Pr}^{-1} = (4C_{\text{St}}) f_{rh} \text{Pr}^{-1} \tag{7.105a}$$

where

$$C_{\text{St}} \equiv \frac{C_{\text{Nu}}}{K_n} = \text{constant} \tag{7.105b}$$

For the circular tube case under consideration, the constants $C_{\text{St},q}$, $C_{\text{St},T}$, and $C_{\text{St,am}}$ corresponding to the Nusselt constants $C_{\text{Nu},q}$, $C_{\text{Nu},T}$, and $C_{\text{Nu,am}}$ per Eqs. (7.99a), (7.99b), and (7.99c), respectively, have the following numerical values:

$$C_{\text{St},q} = \text{constant} = 6.82 \times 10^{-2} \quad \text{for } q'' = \text{constant} \tag{7.106a}$$

$$C_{\text{St},T} = \text{constant} = 5.72 \times 10^{-2} \quad \text{for } T = \text{constant} \tag{7.106b}$$

$$C_{\text{St,am}} = \text{constant} = 6.27 \times 10^{-2} \simeq 1/16$$
$$\text{for constant } q'' \text{ or } T \text{ with a theoretical deviation} < 10\% \tag{7.106c}$$

It is important to note that the constant C_{St} is defined in Eq. (7.105b) on the basis of $f = 4f_{rh}$ per Eq. (7.102). Consequently, the above numerical values need to be multiplied by a factor of 4 when used in conjunction with the alternative hydraulic radius-based friction factor f_{rh} as indicated by the right-hand side of Eq. (7.105a).

7.10.2 Laminar Flows in Noncircular Channels

Problem Statement

Extend the results obtained with circular tubes to typical noncircular straight channels in fully developed laminar Hagen–Poiseuille flows.

Solution

The most general form of the modified Reynolds analogy for laminar flows per Eqs. (7.105a) and (7.105b) is applicable to noncircular channels with appropriate values of the constant K_n and C_{St} obtained from the profusion of

TABLE 7.8. Effect of Cross-Sectional Shape on f and Nu in Fully Developed Laminar Duct Flow

	Cross-Section Geometry	$f\,\mathrm{Re}_{D_h}$ [a]	$(\pi D_h^2/4)/A_{\mathrm{duct}}$	$\mathrm{Nu} = hD_h/k$ [b] Uniform q''	Uniform T_0
(1)	Equilateral triangle (60°)	13.3	0.605	3	2.35
(2)	Square	14.2	0.785	3.63	2.89
(3)	Circle	16	1	4.364	3.66
(4)	Rectangle $4a \times a$	18.3	1.26	5.35	4.65
(5)	Parallel plates	24	1.57	8.235	7.54
(6)	Parallel plates, One side insulated	24	1.57	5.385	4.86
(7)	Average Nu, Lines (1) to (6)			4.69	4.325
(8)	Maximum / minimum deviation (in %)			+75.6 −36.0	+74.3 −45.6
(9)	Average Nu, lines (2), (3), and (4)			4.448	3.733
(10)	Maximum / minimum deviation (in %)			+20.2 −32.6	+24.6 −22.5
(11)	Ratio [line (7) / line (9)] × 100 (in %)			105.4	115.9

Source: Adapted from Table 3.2, A. Bejan, *Convective Heat Transfer*, Wiley, New York, 1984, p. 78. Reproduced with permission.

[a] According to the definition used in the present book the term $f\,\mathrm{Re}_{D_h}$ should read $f_{f_h}\mathrm{Re} = \frac{1}{4}K_n$ per Eq. (3.7).

[b] The conventional U.S. symbol h is used by Bejan [3.3] in lieu of the European symbol α selected in the present text to avoid confusion (see Chapter 9) with the symbol h also used for enthalpy in the U.S. technical literature.

analytical results published in the heat transfer literature for Hagen–Poiseuille laminar flows. Table 7.8 shows a useful compilation of friction factor constants ($\frac{1}{4}K_n$) and corresponding constants $C_{\mathrm{Nu},q}$, $C_{\mathrm{Nu},T}$ for laminar fully developed flow in some common duct geometries.

The constants C_{St} in the generalized laminar Eq. (7.105a), computed from Eq. (7.105b) on the basis of the data in Table 7.8, are summarized in Table 7.9.

TABLE 7.9. Moderate Effect of Cross-Section Geometry on Stanton Constants C_{St} in Fully Developed Channel Flow

	Cross-Section Geometry	Laminar Flow $(C_{St})_l$ [a]			Laminar to Turbulent Ratio [b]
		$C_{St,q}$	$C_{St,T}$	$C_{St,am}$	$(C_{St,am})_l/(C_{St})_t$
(1)	Triangle, 60°	0.0564	0.044 [c]	0.0503 [c]	0.4024 [c]
(2)	Square	0.0646	0.0514	0.0580	0.4640
(3)	Circle	0.0682	0.0572	0.0627	0.5015
(4)	Rectangle, $4a \times a$	0.0731	0.0635	0.0683	0.5464
(5)	Parallel plates	0.0858 [d]	0.0785 [d]	0.0820 [d]	0.6560 [d]
(6)	Parallel plates, one side insulated	0.0561 [c]	0.0506	0.0534	0.4269
(7)	Average C_{St}, lines (1)–(6)	0.0674	0.0576	$0.0625 = \frac{1}{16}$	$0.4997 = \frac{1}{2}$
(8)	Maximum / minimum deviation (in %)	+27.3 −16.7	+36.4 −23.2	+31.3 −19.5	+31.3 −19.5
(9)	Average C_{St}, lines (2), (3), and (4)	0.0686	0.0574	0.0630	$0.5040 = \frac{1}{2}$
(10)	Maximum / minimum deviation (in %)	+6.5 −5.8	+10.7 −10.4	+8.4 −7.9	+8.4 −7.9
(11)	Ratio [line (7) / line (9)] × 100 (in %)	98.3	100.3	99.2	99.1 ≃ 100%

[a] Constants computed from Eq. (7.105b) with f, and Nu data in Table 7.8.

[b] Based on $(C_{St})_t = \frac{1}{8}$ for turbulent flow in straight circular tubes which is valid (~ 10%) for normal noncircular constant cross-section geometries (see Section 3.11). Therefore, the ratio $(C_{St,am})_l/(C_{St})_t = (C_{St,am})_l/(\frac{1}{8}) = 8(C_{St,am})_l$.

[c] Lowest values in column.

[d] Highest values in column.

Comments

1. It is clear that all the conclusions previously derived for laminar flows in circular straight tubes, and Eqs. (7.105a) and (7.105b) in particular, remain valid for fully developed laminar flows in noncircular constant cross-section channels as long as C_{Nu} and therefore C_{St} per Eq. (7.105b) are constant (i.e., independent of $\dot{m}$) and the linear relationship between $\dot{m}$ and $\Delta P_f/\rho = \Delta H_f$ characterizing laminar flows is satisfied, that is, $f_{lam} \propto (\mathrm{Re})^{-1} \propto \dot{m}^{-1}$.

2. It is evident from a cursory comparison of Tables 7.8 and 7.9 that the constants C_{St} used in conjunction with the laminar modified Reynolds analogy Eq. (7.105a) are much more uniform for different duct geometries than the corresponding Nusselt constants. The percentage variations of C_{St} shown on lines (8), (10) and (11) in Table 7.9, which are significantly smaller than the corresponding percentage variations of C_{Nu} on lines (8), (10) and (11) in Table 7.8, demonstrate this conclusion rather vividly.

7.10.3 Pr = 1 Fluids, Turbulent Flows in Channels of Arbitrary Natural Wall Roughness

Problem Statement

Verify the conclusions derived for fully rough turbulent regimes in the case of constant cross-section straight channels (circular *and* noncircular) with any of the *natural* relative wall roughnesses $\bar{e}/D_{h,i}$ indicated on Moody's chart (Fig. 3.4) for Pr = 1 fluids.

Solution

For Pr = 1 the Chilton–Colburn correlation, Eq. (7.10), reduces to the original Reynolds analogy; therefore, under the conditions prescribed for this variation analysis,

$$\alpha_i = \tfrac{1}{2} f_{rh} c_p G = \tfrac{1}{8} f c_p G \propto f \dot{m} \tag{7.107}$$

Validity: Any $\bar{e}/D_{h,i}$ in Moody's chart for Pr = 1 fluids

Note that Eq. (7.107) is also applicable for normal fully enclosed Ω_i = constant noncircular straight channels as indicated in Refs. 7.89 and 7.90 and more specifically for the coupled fluid flow friction factor in Section 3.10 of the present book. On the basis of the above equation and the generalized Blasius friction factor correlation,

$$4 f_{rh} = f = K_n (\mathrm{Re})^{-n} \propto \dot{m}^{-n} \tag{7.108}$$

valid, as highlighted in Chapter 3 (see also Table 3.3), for any natural friction factor curve in Moody's chart with a pair of suitable constants K_n and n, the ratio $(\alpha_i)_{\dot{m}}/(\alpha_i)_{\dot{m}_{\min}}$ can be expressed as

$$(\alpha_i)_{\mathrm{red}} \equiv \frac{(\alpha_i)_{\dot{m}}}{(\alpha_i)_{\dot{m}_{\min}}} = \left(\frac{\dot{m}}{\dot{m}_{\min}}\right)^{1-n} = \dot{m}_{\mathrm{red}}^{1-n} \tag{7.109}$$

under the conditions specified for these variation analyses. Furthermore, for a fixed (arbitrary) approach factor and corresponding $(\mathrm{NTU})_{\mathrm{design}}$ it follows from $(\mathrm{NTU})_{\mathrm{design},\,\dot{m}_{\min}} = (\mathrm{NTU})_{\mathrm{design},\,\dot{m}}$ that

$$\left(\frac{a_i L \alpha_i}{m c_p}\right)_{\dot{m}} \equiv \left(\frac{a_i L \alpha_i}{m c_p}\right)_{\dot{m}_{\min}} \tag{7.110}$$

Solving the above equation for the ratio $l_{\mathrm{red}} = L/L_{\dot{m}_{\min}}$ yields with Eq. (7.109)

$$l_{\mathrm{red}} = \frac{L_{\dot{m}}}{L_{\dot{m}_{\min}}} = \frac{(\alpha_i)_{\dot{m}_{\min}}}{(\alpha_i)_{\dot{m}}} \frac{\dot{m}}{\dot{m}_{\min}} = \frac{1}{\dot{m}_{\mathrm{red}}^{1-n}} \dot{m}_{\mathrm{red}} = \dot{m}_{\mathrm{red}}^{n} \tag{7.111}$$

which indicates that l_{red} generally increases slightly with increasing flow rate $\dot{m}_{\mathrm{red}}$ at typical values of the exponent n in the range $0 < n < 0.32$ as shown in Chapter 3. For the limiting case of fully rough wall regimes when $n \to 0$, $l_{\mathrm{red}} \to 1 =$ constant in agreement with the results shown on line (4) of Table 7.7 for fully rough wall turbulent regimes.

We now compare the total wall friction head $(\Delta H_f)_{\dot{m}}$ computed from the conventional relation, Eq. (3.4), at $\dot{m}$ and corresponding $L_{\dot{m}}$, to the head loss $(\Delta H_f)_{\dot{m}_{\min}}$ at $\dot{m}_{\min}$ and corresponding $L_{\dot{m}_{\min}}$ according to

$$(\Delta H_f)_{\mathrm{red}} \equiv \frac{(\Delta H_f)_{\dot{m}}}{(\Delta H_f)_{\dot{m}_{\min}}} \equiv \frac{(fL\dot{m}^2)_{\dot{m}}}{(fL\dot{m}^2)_{\dot{m}_{\min}}} \tag{7.112}$$

which yields with $f \propto \dot{m}^{-n}$ and Eq. (7.111)

$$\frac{(\Delta H_f)_{\dot{m}}}{(\Delta H_f)_{\dot{m}_{\min}}} = \left(\frac{\dot{m}}{\dot{m}_{\min}}\right)^{-n} \left(\frac{\dot{m}}{\dot{m}_{\min}}\right)^{+n} \left(\frac{\dot{m}}{\dot{m}_{\min}}\right)^{2} \tag{7.113}$$

Since the first two terms on the right-hand side of Eq. (7.113) cancel out, we again obtain the second power relationship demonstrated previously for the limiting case of fully rough wall turbulent regimes

$$(\Delta H_f)_{\mathrm{red}} \equiv \frac{(\Delta H_f)_{\dot{m}}}{(\Delta H_f)_{\dot{m}_{\min}}} = \left(\frac{\dot{m}}{\dot{m}_{\min}}\right)^{2} = \dot{m}_{\mathrm{red}}^{2} \tag{7.114}$$

as shown on line (6) of Table 7.7.

Comments

The phenomenological laws expressed by Eqs. (7.97a)–(7.97d) and demonstrated previously for all stable (Hagen–Poiseuille) laminar flows (including $\mathrm{Pr} \neq 1$) and for stable fully rough turbulent flows at $\mathrm{Pr} = 1$ have now been validated for turbulent flows in channels of any arbitrary *natural* relative wall roughness $\bar{e}/D_{h,i}$ for $\mathrm{Pr} = 1$ fluids.

7.10.4 Pr ≠ 1 Fluids, Turbulent Flows in Channels of Arbitrary Natural Wall Roughness

Problem Statement

Verify empirically the modified form of the Chilton–Colburn correlation suitable for $\mathrm{Pr} \neq 1$ fluids and stable turbulent flows in ideal smooth ($\bar{e}/D_{h,i} \simeq 0$) as well as nonsmooth channels with the same natural relative wall roughness $\bar{e}/D_{h,i} \neq 0$ indicated in Moody's chart (Fig. 3.4).

Solution

Many accepted dimensionless heat transfer coefficient correlations (originally derived from dimensional analysis considerations) are expressed in terms of Nusselt numbers $\mathrm{Nu} = \varphi_{\mathrm{Nu}}(\mathrm{Re}, \mathrm{Pr})$, usually in the form of the product of the powers of Re and Pr according to

$$\mathrm{Nu} = C_{\mathrm{Nu}}(\mathrm{Re})^{a}\mathrm{Pr}^{b} \tag{7.115}$$

where C_{Nu}, a, and b are empirical constants valid within (usually) stated limits. To reduce Nu correlations of this type to the fully equivalent dimensionless correlations in terms of a Stanton number St, we substitute the product St Pr Re for Nu in Eq. (7.115) and solve for St to obtain

$$\mathrm{St} = C_{\mathrm{Nu}}(\mathrm{Re})^{a-1}(\mathrm{Pr})^{b-1} = C_{\mathrm{Nu}}(\mathrm{Re})^{-(1-a)}(\mathrm{Pr})^{-(1-b)} \tag{7.116}$$

where $1 - a$ and $1 - b$ are > 0 since a and b are ≤ 1.

The widely accepted correlation recommended by McAdams (his Eq. (9.10a) in Ref. 1.2) for turbulent regimes in smooth circular tubes and constant cross-section noncircular channels is a concrete example of the Nu equation expressed most generally by Eq. (7.115) with $C_{\mathrm{Nu}} = 0.023$, $a = 0.8$, and $b = 0.4$, yielding with Eq. (7.116) an equivalent correlation in terms of St with constant exponents $1 - a = 0.2$ and $1 - b = 0.6$. These results are summarized below, together with the very precisely stated validity/accuracy limits that made McAdams' classical textbook so popular among practicing heat transfer engineers:

$$\mathrm{St\,Pr\,Re} = \mathrm{Nu} = 0.023(\mathrm{Re})^{0.8}(\mathrm{Pr})^{0.4} \tag{7.117a}$$

which is equivalent to

$$\mathrm{St} = 0.023(\mathrm{Re})^{-0.2}(\mathrm{Pr})^{-0.6} \tag{7.117b}$$

Validity: $10^4 \leq \mathrm{Re} \leq 1.2 \times 10^5$; $0.7 \leq \mathrm{Pr} \leq 120$; long straight smooth circular tubes (and fully enclosed normal constant cross-section channels); all fluid properties based on *bulk* temperature
Maximum *Deviation*: $\pm 40\%$ (probable $\pm 20\%$). (7.117c)

With the generalized Blasius friction factor correlation $f = K_n(\mathrm{Re})^{-n}$, Eq. (3.44), and the pair of constants $K_n = 0.184$ and $n = 0.2$ for smooth tubes per

Eq. (3.47) obtained from McAdams' recommended correlation for $f_{rh} = \frac{1}{4}f$, Eq. (7.117b) reduces after multiplying and dividing by $K_n = 0.184$ to

$$\mathrm{St} = \frac{0.023}{0.184}\left[0.184(\mathrm{Re})^{-0.2}\right](\mathrm{Pr})^{-0.6} = \tfrac{1}{8}f(\mathrm{Pr})^{-0.6} \tag{7.118}$$

Comparing the above equation with the generalization Chilton–Colburn equation (7.10) introduced in Section 7.3,

$$\mathrm{St} = \tfrac{1}{8}f(\mathrm{Pr})^{-\beta} = \tfrac{1}{2}f_{r_h}(\mathrm{Pr})^{-\beta} \tag{7.119}$$

it is evident that Eq. (7.118) coincides with the Chilton–Colburn correlation except that the Pr exponent, $\beta = 0.6$, has a slightly lower numerical value than the average Chilton–Colburn exponent $\beta = \frac{2}{3} = 0.666$.

The fact that the *quasistandard* Chilton–Colburn exponent $\beta = \frac{2}{3}$ is slightly higher than the exponent $\beta = 0.6$ obtained from McAdams' popular Nu correlation, Eq. (7.117a), is not too surprising if one keeps in mind the unavoidable uncertainties associated with all friction factor correlations and the coupled heat transfer coefficient, as previously emphasized in Chapter 3.

One should also remember that the approximate empirical value $\beta = \frac{2}{3}$, proposed by Chilton–Colburn in the 1930s [7.38], is by no means an exact universal constant: It simply represents a reasonable compromise, based on the then existing empirical data, which has proved sufficiently accurate in many normal process heat exchanger design applications.

More recent correlations yield slightly different values for the constant exponent β within normal practical design limits in terms of Pr and Re as

TABLE 7.10. Constants C_{Nu}, $n = 1 - a$, and $\beta = 1 - b$ in Eq. (7.116)

Source	C_{Nu}	n	β	Pr	Re × 10^4	Wall Roughness	ΔT_w
McAdams' Eq. (9.10a) in Ref. 1.2, our Eq. (7.117)	0.023	0.2	0.6	0.7–120	1–12	Smooth copper drawn tubing	Moderate
Grober et al.'s Eq. (8.7.15) in Ref. 7.95	0.024	0.2	0.5	0.7 – 1	[a]	[a]	[a]
	0.024	0.2	0.6	1–14			
	0.024	0.2	0.7	> 14			
Hausen's Eq. (8.10.3) in Ref. 7.95	0.024	0.214	0.55	0.7–10	[a]	[a]	[a]
Petukhov's Eq. (48) in Ref. 3.42, our Eq. (7.16)	[b]	0.21–0.08	0.7–0.4	0.5–2000	1–500	Smooth	≈ 0

[a] Not listed in Grassmann [7.95] but presumably similar to McAdams' normal design limits.

[b] Quantified in Chapter 8 along with β in terms of Re or n for different typical design values of Pr.

shown in Table 7.10. Note that the exponents $n \equiv 1 - a$ and β listed in Table 7.10 were determined from the recommended Nu types of correlation per Eq. (7.115), according to Eq. (7.116). It is clear that β is more generally a weak function of Pr in typical turbulent flow applications.

The essential point to note, however, is that β can be treated as a constant within a rather wide range of Pr numbers. When one accepts as an empirical fact that $(\mathrm{Pr})^{\beta}$ = constant, it is easy to see that all the conclusions derived in the previous analyses remain valid as stated for $\mathrm{Pr} \neq 1$. The reasons for this generalization is that the constant $(\mathrm{Pr})^{\beta}$ term cancels out in all the mathematical operations made to derive Eqs. (7.97a)–(7.97d)

Therefore,

$$\text{Eqs. (7.97a)–(7.97d) are valid for } \mathrm{Pr} \neq 1 \tag{7.120a}$$

provided

$$\mathrm{St} = \tfrac{1}{2} f_{rh} (\mathrm{Pr})^{\beta} = \tfrac{1}{8} f (\mathrm{Pr})^{\beta} \tag{7.120b}$$

with

$$\beta = \text{constant within practical design ranges of } (\mathrm{Re})_{\min} \leq \mathrm{Re} \leq (\mathrm{Re})_{\max} \text{ and } (\mathrm{Pr})_{\min} \leq \mathrm{Pr} \leq (\mathrm{Pr})_{\max} \tag{7.120c}$$

To determine the empirical exponent β in the generalized Chilton–Colburn type of Reynolds analogy, we are guided by the following clues:

1. β must be a weak function of Pr according to Table 7.10 for turbulent flows in smooth tubes.
2. Considering the whole field of single-phase laminar and turbulent flow regimes, it is evident that β must also be a function of Re (and presumably $\bar{e}/D_{h,i}$ for nonsmooth channels), since experimental evidence indicates that β decreases significantly from $\beta = 1$ for laminar flows when the Blasius exponent n in Eq. (3.44) is equal to unity, down to $\beta \approx 0.5$ for turbulent flows in ideal smooth tubes when the Blasius exponent $n \approx 0.2$, and presumably lower values for rougher channels.

Recalling that the Blasius exponent n is a measure of the slope of the friction factor curve $f = \varphi_f(\mathrm{Re}, \bar{e}/D_{h,i})$ in Moody's chart, as shown in Chapter 3, one is instinctively led to the plausible conclusion that β should decrease further with decreasing values of $n \to 0$ since the effects of viscosity and conductivity (i.e., $\mathrm{Pr} = c_p \mu / k$ for c_p = constant) become less and less significant as the *fully rough* wall regimes are approached. This intuitive notion is further reinforced when the following excerpt from Petukhov's [3.42, p. 520] in-depth assessment of the most reliable correlations for turbulent tube flows is interpreted in the context of the proposed generalized Chilton–Colburn correlation, Eq. (7.10). In this excerpt we have substituted our symbols a and

b per Eq. (7.115) for Petukhov's exponents m and n to avoid confusion with the terminology used in the present text.

> Comparing Eq. (53) [our Eq. (7.115)]... it is easy to see that with Eq. (53) at constant c, a, and b it is impossible to describe to a reasonable accuracy the change of Nu number with Re and Pr over a wide range of these parameters. A direct comparison of Eq. (53) with experimental data leads to the same conclusion. Allen and Eckert... have shown that for Re $\sim 1.3 \times 10^4 - 11 \times 10^4$ and Pr $= 8$, Eq. (53) (when $c = 0.023$, $a = 0.8$, and $b = 0.4$) produces an error of up to 20%. An equation of the type (53) can be used for Nu $=$ Nu(Re, Pr) only assuming that, c, a, and b are functions of Re and Pr. For Re $\sim 10^4 - 5 \times 10^6$ and Pr (or Sc) $\sim 0.5 - 10^5$, a changes from 0.79 to 0.92, while b varies from 0.33 to 0.6.

Translated in terms of the generalized Chilton–Colburn correlation with $n \equiv 1 - a$ and $\beta \equiv 1 - b$ per Eq. (7.116), the above conclusions simply mean that the exponents n and β must vary from 0.21 down to 0.08 and from 0.67 down to 0.4, respectively, as highlighted in Table 7.10. Keep in mind that the extremely wide range of parameters Re, Pr (or Schmidt number Sc) analyzed by Petukhov exceed by far the normal practical range of efficient process heat exchanger design applications, in turbulent regimes (which is understandable because aerospace applications are emphasized in Petukhov's article).

Comments

The empirical determination of the exponents β for normal fluid-to-fluid heat exchanger applications, addressed in the first part of Chapter 8, is based on Petukhov's most exact correlation, Eq. (7.16), together with the results derived from these variation analyses.

7.10.5 General Conclusions

1. It is important to reiterate that the variation analyses described in Sections 7.8–7.10 are based on the following two major assumptions:
   (a) Inlet/outlet steady-state steady-flow conditions.
   (b) The design approach factor AF – SD/LD is fixed.

These two basic assumptions thus define the two *sufficient and necessary* conditions for the rigorous validity of the results obtained from these variation analyses.

2. It is essential to note that when the approach factor AF = SD/LD *and* the larger temperature difference LD are fixed, the irreversibilities due to the finite thermodynamic temperature differences between the heat source (T_H) and the heat sink (T_C) are also fixed. This means that the entropy production and irreversible losses associated with these prescribed finite temperature differences are *fixed* or *preordained* under the conditions specified for these variation analyses. Thus, the heat exchanger design engineer whose duty it is to meet these prescribed design conditions (typically specified by the thermal

system project engineer) has no control on these thermodynamic losses. The heat exchanger designer, however, can influence the system thermodynamic losses associated with friction dissipation, such as $\dot{m}\,(\Delta H_f)_{\text{tot}}$ in the present case, if the maximum $(\Delta H_f)_{\text{tot}} = (\Delta P_f)_{\text{tot}}/\rho$ is not specified by the system engineer.

Whenever the system engineer also specifies the maximum allowable $(\Delta H_f)_{\text{tot, max}}$ to the heat exchanger designer, as is normally done (or should be done), then all irreversible losses (or rates of entropy production) are preordained, and the heat exchanger designer no longer has any control over *second law losses*. In performing his mission, which is simply to meet the specified targets economically, the experienced heat exchanger designer will endeavor to utilize, as effectively as possible, the available $(\Delta H_f)_{\text{tot, max}}$ allocated to the heat exchanger by the system designer, knowing that in typical turbulent flow regimes smaller, lower-cost heat exchanger surfaces are attainable by reducing individual friction losses, as shown in Chapter 8.

3. The phenomenological relationships equations (7.97a)–(7.97d), verified on the basis of accepted analytical and/or semiempirical correlations for the friction factor f and heat transfer coefficient α, stating that

$$\frac{\dot{Q}_f}{|\dot{Q}_e|} \equiv \frac{\dot{m}\,\Delta H_f}{\dot{m}c_p|T_2 - T_1|} \equiv \frac{\Delta H_f}{c_p|T_2 - T_1|} \equiv \frac{\Delta T_f}{|T_2 - T_1|} = \left(\frac{\dot{m}}{\dot{m}_{\min}}\right)^2 \tag{7.121}$$

have significant practical and theoretical implications discussed more fully in Chapters 8 and 9, respectively.

The fundamental thermodynamic implications become immediately evident when one recalls that ΔH_f can always be expressed as $\overline{T}\Delta s_f$. By substituting $\overline{T}\Delta s_f$ for ΔH_f in Eq. (7.121), one obtains (as shown later) the following relationship:

$$\frac{\dot{Q}_f}{|\dot{Q}_e|} = \frac{\overline{T}\Delta s_f}{c_p|T_2 - T_1|} = \frac{\Delta s_f}{c_p|T_2 - T_1|/\overline{T}} \equiv \frac{\Delta s_f}{|\Delta s_e|} = \left(\frac{\dot{m}}{\dot{m}_{\min}}\right)^2 \tag{7.122a}$$

with

$$|\Delta s_e| \equiv \frac{c_p|T_2 - T_1|}{\overline{T}} = \text{entropy change due to} \left|\dot{Q}_e\right| \tag{7.122b}$$

valid for

$$\left|\frac{T_2 - T_1}{\overline{T}}\right| \ll 1 \tag{7.122c}$$

4. A profusion of theoretically accurate analytical correlations have been published in the heat transfer literature [7.91] for fully developed laminar flows inside many typical channel configurations. For this reason, and since the more effective turbulent flows are more typical in thermodynamically efficient heat exchangers, we emphasize mainly stable turbulent single-phase flow regimes and the corresponding two-phase flow regimes [Martinelli *tt* dissipa-

tive model and/or annular-spray and homogeneous (fog) flow patterns] after the following quantitative assessment of the fundamental advantages of single-phase turbulent flow regimes.

7.11 ADVANTAGES OF TURBULENT FLOW REGIMES

The major advantage of turbulent flow regimes in normal heat exchanger design practice is quantified by comparing the total heat exchanger surface required at a prescribed approach factor for turbulent and laminar fully developed channel flows and the corresponding pumping costs $\dot{m}\,\Delta H_f$, neglecting for the time being individual head losses, which can become significant in multipass heat exchangers, as highlighted in Chapter 8.

We focus first on straight single-pass circular tubes and then extend the analysis to include typical noncircular fully enclosed channels of constant cross section. Furthermore, because any natural wall roughness $\bar{e}/D_h > 0$ would evidently further enhance the advantages of turbulent flow regimes, we consider only ideal smooth wall channels ($\bar{e}/D_h = 0$) in this *worst case* analysis.

For such an order of magnitude analysis, it is sufficiently accurate to use the average Chilton–Colburn Pr exponent $\beta = \frac{2}{3}$ for turbulent flows in smooth tubes and an average laminar constant $C_{\mathrm{St,\,am}} = \frac{1}{16}$ per eq. (7.106c). Dividing Eq. (7.119) with $C_{\mathrm{St}} = \frac{1}{8}$ by Eq. (7.105a) with $C_{\mathrm{St,\,am}} = \frac{1}{16}$ yields the ratio $(\mathrm{St})_t/\mathrm{St}_l = \alpha_t/\alpha_l$

$$\frac{\alpha_t}{\alpha_l} = \frac{(\mathrm{St})_t}{(\mathrm{St})_l} = \frac{1/8}{1/16}\,\frac{\mathrm{Pr}^{-2/3}}{\mathrm{Pr}^{-1}} = 2(\mathrm{Pr})^{1/3} \quad \text{valid for } \dot{m}_l = \dot{m}_t \qquad (7.123)$$

where the subscripts t and l refer to the stable turbulent and laminar flow regimes, respectively.

We now consider the theoretical (imaginary) regime transition point where $f_t \equiv f_l$ and compute the corresponding theoretical transition Reynolds number $(\mathrm{Re})_{\mathrm{cr,\,theor}}$ by equating the generalized Blasius correlation for f, Eq. (3.44), with $K_n = 64$, $n = 1.0$ for laminar, and $K_n = 0.184$, $n = 0.2$ for turbulent flows; that is,

$$64\,(\mathrm{Re})^{-1} = 0.184(\mathrm{Re})^{-0.2} \qquad (7.124a)$$

which yields with $(\mathrm{Re})^{0.8} = 64/0.184$ a numerical value

$$\mathrm{Re} = (\mathrm{Re})_{\mathrm{cr,\,theor}} \simeq 1500 \qquad (7.124b)$$

for the theoretical critical transition from laminar to turbulent flow regimes in smooth circular tubes. The value of $(\mathrm{Re})_{\mathrm{cr,\,theor}} = 1500$ is obviously hypothetical since we know from experience (Chapter 3) that the real flow regime is certainly laminar at $\mathrm{Re} < 2100$. Nevertheless, nothing prevents us from using the flow rate $\dot{m}_{\mathrm{cr,\,theor}}$ corresponding to $(\mathrm{Re})_{\mathrm{cr,\,theor}} = 1500$ as an arbitrary reference flow rate $(\mathrm{Re})_{\mathrm{min},\,t}$ instead of the arbitrary value of 10,000 used in the previous variation analysis for turbulent flow regimes. We therefore repeat

TABLE 7.11. Variation Analysis:[a] Turbulent Versus Laminar Flow Regimes in Straight Circular Smooth Wall Tubes

(1)	$\dot{m}_{red}$	1	1.5	2	3	4	5	10
(2)	Re	1500	2250	3000	4500	6000	7500	15,000
(3)	$l_{rel,l}$	1	1.5	2	3	4	(5)	(10)
(4)	$l_{red,t}$	(1)	(1.08)	1.14	1.25	1.32	1.38	1.58
	$Pr = \frac{1}{8}$		When constant in Eq. (7.125): $2(Pr)^{1/3} = 1$					
(5)	L_t / L_l	(1)	(0.72)	0.57	0.42	0.33	(0.28)	(0.16)
(6)	$(\Delta H_f)_t / (\Delta H_f)_l$		⟵ $1.0 = 1 / 2(Pr)^{1/3}$ ⟶					
	$Pr = 1$		When constant in Eq. (7.125): $2(Pr)^{1/3} = 2$					
(7)	L_t / L_l	($\frac{1}{2}$)	(0.36)	0.29	0.21	0.16	(0.14)	(0.08)
(8)	$(\Delta H_f)_t / (\Delta H_f)_l$		⟵ $0.50 = \frac{1}{2} = 1 / 2(Pr)^{1/3}$ ⟶					
	$Pr = 8$		When constant in Eq. (7.125): $2(Pr)^{1/3} = 4$					
(9)	L_t / L_l	($\frac{1}{4}$)	(0.18)	0.14	0.10	0.08	(0.07)	(0.04)
(10)	$(\Delta H_f)_t / (\Delta H_f)_l$		⟵ $0.25 = \frac{1}{4} = 1 / 2(Pr)^{1/3}$ ⟶					

[a] Based on prescribed constant approach factor $(AF)_l = (AF)_t = (T_w - T_2)/(T_w - T_1)$, where subscripts t and l refer to turbulent and laminar, respectively, with:

- $L_t / L_l = A_t / A_l$ = total heat exchange surface ratio in terms of $\dot{m}_{red}$ per Eq. (7.125) with a Blasius exponent $n = 0.2$ for turbulent flow in smooth tubes.
- $\dot{m}_{red} = 1$ at $(Re)_{cr,\ theor} = 1500$ when $f_t = f_l = 0.043$.
- Parentheses indicate that one of two regimes is highly improbable.

the variation analysis exactly as previously, except for a new arbitrary reference $Re = (Re)_{cr,\ theor} = 1500$ and corresponding minimum (fictive) flow rate $\dot{m}_{min,\ turb}$ at $(Re)_{cr,\ theor}$, for turbulent flow regimes up to a maximum $(Re)_{max} =$ 15,000 and corresponding $\dot{m}_{red} = 1$–10 as shown on lines (1) and (2) of Table 7.11.

From the previous variation analyses, it is evident that $l_{red,\ l} \propto \dot{m}_{red}$ as shown on line (3) and $l_{rel,\ t} \propto \dot{m}^n_{red}$ with the values shown on line (4) for an approximate constant Blasius exponent $n = 0.2$ assumed for this rough comparison. At $\dot{m}_{red} = 1$ for $Re = (Re)_{cr,\ theor}$, $f_t = f_l = 0.043$; therefore, with Eq. (7.123) and since $L \propto 1/\alpha_i$ for a prescribed AF and $\dot{m}$,

$$\frac{L_t}{L_l} = \frac{1}{2(Pr)^{1/3}} \quad \text{at } \dot{m} = \dot{m}_{min} \tag{7.125a}$$

which yields the numerical values 1, $\frac{1}{2}$, and $\frac{1}{4}$ at $Pr = \frac{1}{8}$, 1, and 8, respectively; this is shown on lines (5), (7), and (9) in the column headed $\dot{m}_{red} = 1$ of Table 7.11.

Since $l_{red} \propto \dot{m}_{red}$ in laminar flows, whereas $l_{red,\ t} \propto \dot{m}^n_{red}$ in turbulent flows, we obtain the ratio

$$\frac{L_t}{L_l} \equiv \frac{A_t}{A_l} = \frac{(\dot{m}/\dot{m}_{min})^{-(1-n)}}{2(Pr)^{1/3}} \quad \text{for } 1 \le \frac{\dot{m}}{\dot{m}_{min}} \le 10 \tag{7.125b}$$

and the numerical values shown on lines (5), (7), and (9) at $Pr = \frac{1}{8}$, 1, and 8, respectively. Note that the values shown in parentheses in Table 7.11 are highly improbable (since experience has shown that either one of the two flow regimes is highly improbable at the corresponding Re).

Focusing our attention on lines (5), (7), and (9) and on the gray region $3000 \le \text{Re} \le 6000$ where both laminar and turbulent flows are conceivable, the economical advantages in terms of first cost and heat exchanger compactness in turbulent regimes become quite evident. Even for an extremely low $\text{Pr} = \frac{1}{8}$, the surface ratio $A_t/A_l = L_t/L_l$ shown on line (5) is on the order of 40% and the ratio decreases further down to $\approx$ 20% and $\approx$ 10% at Pr values of 1 and 8, respectively.

Since most fluids encountered in normal exchanger applications have $\text{Pr} > 0.5$, as shown in Appendix A (excluding atypical liquid metals with $\text{Pr} \ll 1$), it is clear that turbulent regimes are significantly more economical than laminar flows even if we were to disregard pumping costs.

Furthermore, the operating pumping costs (and pumps first costs!) which are proportional to $\dot{m}\,\Delta H_f$ are also significantly lower at $\text{Pr} \ge \frac{1}{8}$ in turbulent regimes as indicated by the ratios $(\Delta H_f)_t/(\Delta H_f)_l$ shown on lines (6), (8), and (10) for Pr values of $\frac{1}{8}$, 1, and 8, respectively.

The constant ratio $(\Delta H_f)_t/(\Delta H_f)_l = 1$ shown on line (6) at $\text{Pr} = \frac{1}{8}$ can readily be explained as follows. At $\dot{m}_{\text{red}} = 1$, with $\text{Re} = (\text{Re})_{\text{cr, theor}} = 1500$ and $f_t = f_l$, it is evident that $(\Delta H_f)_t = (\Delta H_f)_l$ or $(\Delta H_f)_t/(\Delta H_f)_l = 1$ since $L_t = L_l$ per line (5). We also know from the previous variation analyses that $\Delta H_f \propto \dot{m}^2_{\text{red}}$ for *both* laminar and turbulent flow regimes; therefore, the ratio $(\Delta H_f)_t/(\Delta H_f)_l = \text{constant} = 1$ for $1 \le \dot{m}_{\text{red}} \le 10$.

Similarly, for $\text{Pr} = 1$ at $\dot{m}_{\text{red}} = 1$ and $f_t = f_l$, the ratio $(\Delta H_f)_t/(\Delta H_f)_l = \frac{1}{2}$ since $L_t/L_l = \frac{1}{2}$ and this ratio will remain constant for $\dot{m}_{\text{red}} > 1$. Following the same reasoning, $(\Delta H_f)_t/(\Delta H_f)_l = \frac{1}{4} = \text{constant}$ for $\text{Pr} = 8$, or most generally

$$\frac{(\Delta H_f)_t}{(\Delta H_f)_l} = \frac{1}{2(\text{Pr})^{1/3}} = \text{constant} \quad \text{for } 1 \le \frac{\dot{m}}{\dot{m}_{\text{min}}} \le 10 \qquad (7.126)$$

Comments

1. Although rudimentary, this analysis demonstrates the significant economical advantages of turbulent flow regimes in normal heat exchanger applications for $\text{Pr} > 0.5$ fluids. This explains why practicing heat transfer engineers always strive to achieve turbulent rather than laminar flow regimes whenever feasible, through proper circuiting in multicircuit heat exchangers and/or by a judicious choice of channel hydraulic diameters.

2. The above general conclusions derived on the basis of circular tubes clearly remain substantially valid for the typical noncircular channels in view of what was said in Section 7.10.2 and the results shown in Table 7.9.

3. The higher the fluid Pr numbers the greater the potential advantages of turbulent flow regimes as is evident according to Eq. (7.125) and (7.126) and

the numerical results summarized in Table 7.11 at Pr numbers ranging from $\frac{1}{8}$ to 8.

4. With typical Pr numbers (see Appendix A) ranging from $\Pr \approx 1$ for gases and vapors at moderate pressures (well below the critical thermodynamic point) and $\Pr = 13$ to 0.9 for water (the favorite reference liquid in thermal sciences and engineering), it is usually possible to operate in the more favorable turbulent regimes at acceptable friction heads ΔH_f and corresponding total wall friction dissipation rate $\dot{Q}_f = \dot{m}\,\Delta H_f$. The situation is quite different for higher Pr fluids.

5. It is easy to show that it is not possible to operate in stable turbulent flow regimes, as in the case of water (see Chapter 3), with such a highly viscous and high Pr fluid as engine oil with normal heat exchanger tube diameters as outlined below.

For a given design approach factor $(\mathrm{AF})_{\mathrm{des}}$ and corresponding $(\mathrm{NTU})_{\mathrm{des}}$ the required length of travel, per Eq. (7.69) for the simple T_w = constant boundary condition, can be computed with $U_i = \alpha_i$, from

$$\frac{L}{(\mathrm{NTU})_{\mathrm{des}}} = \frac{\dot{m}c_p}{a_i\alpha_i} = \left(\frac{\Omega_i}{a_i}\right)\frac{c_pG}{\alpha_i} \equiv r_{h,i}\left(\frac{c_pG}{\alpha_i}\right) \tag{7.127}$$

Substituting α_i per Eq. (7.10) in the above equation yields

$$\frac{L}{(\mathrm{NTU})_{\mathrm{des}}} = \frac{r_{h,i}c_pG}{\frac{1}{8}fc_pG(\Pr)^{-\beta}} = r_{h,i}\left(\frac{8}{f}\right)(\Pr)^{\beta} \tag{7.128}$$

The length of travel required in turbulent regimes for oil and water at a given $(\mathrm{NTU})_{\mathrm{des}}$, same channel geometry (i.e., same $D_{h,i}$, a_i, and Ω_i), and same Re can be compared by computing the ratio $L_{\mathrm{oil}}/L_{\mathrm{water}}$ obtained from Eq. (7.128) with $(r_h)_{\mathrm{oil}} = (r_h)_{\mathrm{water}}$ and $f_{\mathrm{oil}} = f_{\mathrm{water}}$ at $(\mathrm{Re})_{\mathrm{oil}} = (\mathrm{Re})_{\mathrm{water}}$:

$$\frac{L_{\mathrm{oil}}}{L_{\mathrm{water}}} = \left(\frac{\Pr_{\mathrm{oil}}}{(\Pr)_{\mathrm{water}}}\right)^{\beta} \tag{7.129}$$

keeping in mind that the friction factor is nearly constant in turbulent flow regimes with an average value $f \simeq 0.025 \pm 30\%$ for $10{,}000 < \mathrm{Re} < 120{,}000$ in the case of ideal smooth tubes, as shown in Chapter 3.

For the same Re the following conditions must be satisfied:

$$\left(\frac{GD_{h,i}}{\mu}\right)_{\mathrm{oil}} = \left(\frac{GD_{h,i}}{\mu}\right)_{\mathrm{water}} \tag{7.130}$$

Thus, for identical $D_{h,i}$ and Ω_i,

$$\frac{G_{\mathrm{oil}}}{G_{\mathrm{water}}} = \frac{\dot{m}_{\mathrm{oil}}}{\dot{m}_{\mathrm{water}}} = \frac{\mu_{\mathrm{oil}}}{\mu_{\mathrm{water}}} \tag{7.131}$$

and the average velocity ratio, with $\overline{V} \equiv G/\rho$, according to

$$\frac{\overline{V}_{\text{oil}}}{\overline{V}_{\text{water}}} = \left(\frac{\mu_{\text{oil}}}{\mu_{\text{water}}}\right)\left(\frac{\rho_{\text{water}}}{\rho_{\text{oil}}}\right) = \frac{\nu_{\text{oil}}}{\nu_{\text{water}}} \tag{7.132}$$

results in a ratio $(\Delta H_f)_{\text{oil}}/(\Delta H_f)_{\text{water}}$,

$$\frac{(\Delta H_f)_{\text{oil}}}{(\Delta H_f)_{\text{water}}} = \frac{[f(L/D_{h,i})\overline{V}^2/2]_{\text{oil}}}{[f(L/D_{h,i})\overline{V}^2/2]_{\text{water}}} = \left(\frac{L_{\text{oil}}}{L_{\text{water}}}\right)\left(\frac{\overline{V}_{\text{oil}}}{\overline{V}_{\text{water}}}\right)^2 \tag{7.133}$$

for identical f and $D_{h,i}$.

Thus, with $\overline{V} = G/\rho$ and Eqs. (7.129)–(7.132), the above equation reduces to

$$\frac{(\Delta H_f)_{\text{oil}}}{(\Delta H_f)_{\text{water}}} = \left(\frac{(\text{Pr})_{\text{oil}}}{(\text{Pr})_{\text{water}}}\right)^{\beta}\left(\frac{\mu_{\text{oil}}}{\mu_{\text{water}}}\right)^2\left(\frac{\rho_{\text{water}}}{\rho_{\text{oil}}}\right)^2 = \left(\frac{(\text{Pr})_{\text{oil}}}{(\text{Pr})_{\text{water}}}\right)^{\beta}\left(\frac{\nu_{\text{oil}}}{\nu_{\text{water}}}\right)^2 \tag{7.134}$$

or, alternatively, with $\Delta H_f = (\Delta P_f)v = (\Delta P_f)/\rho$

$$\frac{(\Delta P_f)_{\text{oil}}}{(\Delta P_f)_{\text{water}}} = \left[\frac{(\text{Pr})_{\text{oil}}}{(\text{Pr})_{\text{water}}}\right]^{\beta}\left(\frac{\mu_{\text{oil}}}{\mu_{\text{water}}}\right)^2\left(\frac{\rho_{\text{water}}}{\rho_{\text{oil}}}\right)^1 \tag{7.135}$$

Finally, since the corresponding total friction dissipation $\dot{Q}_f = \dot{m}\,\Delta H_f$, one obtains with $\dot{m}_{\text{oil}}/\dot{m}_{\text{water}} = \mu_{\text{oil}}/\mu_{\text{water}}$ the ratio $(\dot{Q}_f)_{\text{oil}}/(\dot{Q}_f)_{\text{water}}$:

$$\frac{(\dot{Q}_f)_{\text{oil}}}{(\dot{Q}_f)_{\text{water}}} = \frac{\dot{m}_{\text{oil}}}{\dot{m}_{\text{water}}}\frac{(\Delta H_f)_{\text{oil}}}{(\Delta H_f)_{\text{water}}} = \left(\frac{(\text{Pr})_{\text{oil}}}{(\text{Pr})_{\text{water}}}\right)^{\beta}\left(\frac{\mu_{\text{oil}}}{\mu_{\text{water}}}\right)^3\left(\frac{\rho_{\text{water}}}{\rho_{\text{oil}}}\right)^2 \tag{7.136}$$

The above equations yield, with $\beta \cong \frac{2}{3}$, the ratios shown on lines (3)–(7) of Table 7.12 for $L_{\text{oil}}/L_{\text{water}}$, $\dot{m}_{\text{oil}}/\dot{m}_{\text{water}}$, $\overline{V}_{\text{oil}}/\overline{V}_{\text{water}}$, $(\Delta H_f)_{\text{oil}}/(\Delta H_f)_{\text{water}}$, and $(\dot{Q}_f)_{\text{oil}}/(\dot{Q}_f)_{\text{water}}$.

Considering that turbulent flow regimes with water result in friction heads and velocities, which are just about the right order of magnitude in normally efficient heat exchangers, we see from Table 7.12 that turbulent flow regimes, with such high viscosity and high Pr fluids as engine oil, are not conceivable in normal heat exchanger design practice, on practical and theoretical grounds. On *practical* grounds, the *operating* pumping costs and pump *first costs* are prohibitive [see lines (6) and (7) of Table 7.12], as is the tube erosion at such extreme velocities [see line (5)], when we keep in mind that the usual maximum allowable velocity limit for water is about 8–10 ft/s (or 3 m/s)! On *theoretical* grounds, $\dot{Q}_f = 0$—the fundamental assumption of conventional heat exchanger analysis—would be grossly violated at friction *dissipation rates several orders of magnitude higher* than that with water.

The above analysis provides a rationale for the reasonable validity limit stated by McAdams for his recommended Eq. (7.117), namely, $0.7 < \text{Pr} < 120$,

TABLE 7.12. Turbulent Regimes: High Pr Engine Oil Versus Water

(1) Temperature		273 K	273 K[a]	273 K[a]	370 K	370 K	370 K
(2) Fluid property terms		$(FP)_{oil}$	$(FP)_{water}$	$\frac{(FP)_{oil}}{(FP)_{water}}$	$(FP)_{oil}$	$(FP)_{water}$	$\frac{(FP)_{oil}}{(FP)_{water}}$
μ in $N \cdot s/m^2$		3.85	1.75×10^{-3}	2200	1.86×10^{-2}	2.89×10^{-4}	64.4
ρ in kg/m^3		891.1	1000	0.899	841.8	960.6	0.876
ν in m^2/s		4.28×10^{-3}	1.75×10^{-6}	2446	2.2×10^{-5}	3.0×10^{-7}	73.4
Pr		47,000	12.99	3618.2	300	1.8	166.7
(3) L_{oil}/L_{water}	Eq. (7.129)			235.7			30.3
(4) $\dot{m}_{oil}/\dot{m}_{water}$	Eq. (7.131)			2200			64.4
(5) $\overline{V}_{oil}/\overline{V}_{water}$	Eq. (7.132)			2446			73.4
(6) $(\Delta H_f)_{oil}/(\Delta H_f)_{water}$	Eq. (7.134)			1.41×10^9			1.63×10^5
(7) $(\dot{Q}_f)_{oil}/(\dot{Q}_f)_{water}$	Eq. (7.136)			3.1×10^{12}			1.05×10^7

[a]At normal freezing point 273.15 K for water.

with $10\,000 \leq Re \leq 120\,000$. Extremely high Pr up to 2000 along with Re values as high as 5,000,000 are totally unrealistic in normally efficient fluid-to-fluid surface heat exchanger applications.[24]

7.12 HEAT TRANSFER ENGINEERING: THE ART

In contrast to mathematical physicists engaged in fundamental fluid, heat, and mass transfer research, engineers must face all significant problems in non-idealized equipment (see "Inlet Effects" in Chapter 3) and contend with such imponderable, yet vital, factors as the impact of surface fouling (an aging process)[25] and unavoidable manufacturing tolerances, as well as test inaccuracies in real world heat exchanger design practice. The impact of some of these rather imponderable factors (in the sense that they cannot be assessed precisely except by statistical methods) are approximately quantified in this section to show their overriding importance in actual heat exchanger design practice and confirm what practicing heat transfer engineers know from experience, namely, that it is futile and counterproductive to strive for *theoretically perfectly accurate* heat transfer coefficient correlations.

7.12.1 Impact of Fouling Resistance in Water-Cooled Condensers

The significant effect of the *scale deposit* or *fouling factor* on the water side of shell-and-tube condensers is the central theme of the paper reproduced here as Appendix C [7.92]. Note that for the remainder of this discussion all figures and tables without the prefix 7 (for Chapter 7), refer to this appendix.

Clearly, the analytical procedure described in this paper is identical to the modern NTU-effectiveness methodology reviewed previously, in spite of the use of a different terminology, which was then quite popular in the refrigeration and chemical process industries. It is only necessary to substitute the "new" terms NTU and ε for the "old" symbols X and CF, respectively, to recognize this.

As shown in Fig. 6 ((Appendix C) the performance for a fixed design water scaling factor $1/\alpha_s$ of a rather large family of standardized condensers (see tabulation in Fig. 6) can be mapped on a single graph by plotting the specific heat rejection per degree LD per circuit in terms of total circuit length of travel, using the water flow rate per circuit as a parameter. These performance curves were computed, as indicated, on the basis of a unique overall heat transfer coefficient curve obtained empirically with clean new condensers (all using the same externally finned tube labeled type C), which proved adequately accurate for the full range of anticipated water velocities and operating temperatures. The corresponding curve for $U_{0.0005}$ in terms of water velocity, including the minimum design *scale deposit factor* $1/\alpha_s = 0.0005$ (in U.S. technical units) for *clean water* applications, is shown in Fig. 3. Because of the major impact of $1/\alpha_s$ and the wide range of scale deposit factors, depending on local water quality, the paper focuses on suitable effective length correction factors (as shown in Fig. 8) to predict the heat rejection at different scale deposit factors or, vice versa, to estimate the actual fouling factor from field test results. It should be noted that according to the terminology used in the present book the U value per linear foot of the tube used in Appendix C is equivalent to $U_i a_i = \text{Btu/h} \cdot \text{F} \cdot \text{ft}$ in U.S. technical units, or $\text{W/m} \cdot \text{K}$ in SI units.

Extensive laboratory and field tests with clean new $(1/\alpha_s \cong 0)$ condensers showed average to maximum deviations from the average U_{clean} correlations of ± 10 and 20%, respectively.

Problem Statement

(a) Using the relevant data in Appendix C, quantity the impact of ± 10 and 20% U_{clean} deviations on the actual performance of any unit in this standardized line of condensers under the following extreme design conditions: $\text{CF} = 1 - \text{AF} = 0.5$ and 0.95 at fouling factors $1/\alpha_s = 0.0005$ to 0.05 (in the U.S. technical units shown in Appendix C) for a typical water velocity = 7 ft/s (2.13 m/s).

(b) Assess the above results in the light of the inherent uncertainties associated with accurate fouling factor predictions.

Solution

(a) Figure 7 shows that the standard externally finned tube C used in this series of condensers has an average clean $(1/\alpha_s = 0)$ overall heat transfer coefficient per unit length of tube $(U_{\text{avg}})_{\text{clean}} = 145\ \text{Btu/h} \cdot {}^\circ\text{F} \cdot \text{ft})$ at 7 ft/s water velocity. With an internal surface per unit length $a_i = 0.178\ \text{ft}^2/\text{ft}$, or

$1/a_i = 5.62$ ft/ft^2, the overall thermal resistance at any fouling resistance factor $(1/\alpha_s)$ is

$$\frac{1}{U_{1/\alpha_s}} = \frac{1}{U_{\text{clean}}} + \frac{1}{\alpha_s a_i} = \frac{1}{U_{\text{clean}}} + 5.62\left(\frac{1}{\alpha_s}\right) \tag{7.137}$$

For a prescribed scale deposit factor $1/\alpha_s = 0.0005$, Eq. (7.137) yields, with $(U_{\text{avg}})_{\text{clean}} = 145$, $1/U_{\text{avg}})_{1/\alpha_s} = (1/145) + (0.0005)\ 5.62 = 0.009705$, or an average overall heat transfer coefficient per unit length of $(U_{\text{avg}})_{1/\alpha_s} = 1/0.009705 = 103.0$. When the actual value U_{act} of the clean tube is 20% lower than the average clean design value, $(U_{\text{avg}})_{\text{clean}}$ or $(U_{\text{act}})_{\text{clean}} = 0.8\ (145)$, we obtain from Eq. (7.137) $(U_{\text{act}})_{1/\alpha_s} = 87.5$ at the same $1/\alpha_s = 0.0005$ and thus a ratio

$$\left(\frac{(\text{NTU})_{\text{act}}}{(\text{NTU})_{\text{avg}}}\right)_{0.0005} = \left(\frac{U_{\text{act}}}{U_{\text{avg}}}\right)_{0.0005} = 0.849 = 85\% \tag{7.138}$$

which is higher than the 80% ratio $(U_{\text{act}})_{\text{clean}}/(U_{\text{avg}})_{\text{clean}}$ due to the effect of the additional *fixed* scale deposit resistance term in Eq. (7.137).

The ultimate impact of a -20% deviation in $(U_{\text{avg}})_{\text{clean}}$ on the condenser capacity will depend, of course, on the specified design approach factor $(\text{AF})_{\text{des}}$. For a given $(\text{AF})_{\text{des}}$, prescribed water flow rate and LD, $(\dot{Q}_e)_{\text{des}}$ is proportional to $\text{CF} = 1 - (\text{AF})_{\text{des}}$, and the required length of travel is determined from $(\text{NTU})_{\text{des}} = -\ln(\text{AF})_{\text{des}} = \ln(1/\text{AF})_{\text{des}}$. Since NTU is directly proportional to U, we can write

$$(\text{NTU})_{\text{act}} = \left[\frac{\left(U_{1/\alpha_s}\right)_{\text{act}}}{\left(U_{1/\alpha_s}\right)_{\text{avg}}}\right](\text{NTU})_{\text{des}} \tag{7.139}$$

where the term in large square brackets is obtained from Eq. (7.138). The actual $(\text{AF})_{\text{act}}$ can now be computed at $(\text{NTU})_{\text{act}}$ from $(\text{AF})_{\text{act}} = \exp[-(\text{NTU})_{\text{act}}]$ and, since $(\dot{Q}_e)_{\text{act}}$ is proportional to $(\text{CF})_{\text{act}} = 1 - (\text{AF})_{\text{act}}$, we can assess the ultimate impact of deviations in $(U_{\text{avg}})_{\text{clean}}$ from

$$\frac{\left(\dot{Q}_e\right)_{\text{act}}}{\left(\dot{Q}_e\right)_{\text{des}}} = \frac{1 - (\text{AF})_{\text{act}}}{1 - (\text{AF})_{\text{des}}} \tag{7.140}$$

For example, at $(\text{AF})_{\text{des}} = 0.5$, and therefore $(\text{NTU})_{\text{des}} = -\ln(\text{AF})_{\text{des}} = -\ln(0.5) = 0.693$, we obtain from Eq. (7.139) $(\text{NTU})_{\text{act}} = 0.589$ and a corresponding $(\text{AF})_{\text{act}} = \exp(-0.589) = 0.555$, yielding with Eq. (7.140) a ratio $(\dot{Q}_e)_{\text{act}}/(\dot{Q}_e)_{\text{des}} = 0.445/0.5 = 0.89 = 89\%$. The results summarized in Table 7.13 were obtained in a similar fashion.

(b) In view of the great uncertainties associated with an exact prediction of the fouling factor $1/\alpha_s$ and its very significant impact in typical water-cooled condenser applications (as shown in Appendix C), it is clear from the percentage deviations in $\dot{Q}_e$ shown in Table 7.13 that deviations of the order of ± 10–20% in $(U_{\text{avg}})_{\text{clean}}$ are quite acceptable in this (and in most) heat

TABLE 7.13. Actual to Design Condenser Capacity Ratio $(\dot{Q}_e)_{act}/(\dot{Q}_e)_{des}$ at Extreme Design Approach and Fouling Factors for a Typical Water Velocity of 7.0 ft / s (2.13 m / s)

(1)	Design approach factor SD / LD	—	0.50	0.50	0.05	0.05
(2)	Design fouling factor[a] $(1/\alpha_s)$	$\frac{h \cdot {}^\circ F \cdot ft^2}{Btu}$	0.0005	0.0050	0.0005	0.0050
(3)	With $(U_{act}/U_{avg})_{clean} = 0.9$	%	94.8	98.5	98.7	99.7
(4)	With $(U_{act}/U_{avg})_{clean} = 1.1$	%	104.6	101.3	101.0	100.3
(5)	Total spread	%	9.8	2.8	2.3	0.6
(6)	With $(U_{act}/U_{avg})_{clean} = 0.8$	%	89.0	96.7	97.0	99.2
(7)	With $(U_{act}/U_{avg})_{clean} = 1.2$	%	108.9	102.4	101.8	100.5
(8)	Total spread	%	19.9	5.7	4.8	1.3

[a]Numerical values of $1/\alpha_s = 0.0005$ and 0.0050 are equivalent to $\alpha_s = 2000$ and 200 Btu / h · ° F · ft² or 11,360 and 1136 W / m²·K, respectively.

exchanger design applications. This is quite fortunate because most of the widely accepted real world heat transfer correlations, as opposed to extremely idealized theoretically exact correlations (based on ideal smooth walls, surgically clean tubes, and fully developed flows), are accurate within an uncertainty band typically of the order of ± 10–20%. As an additional safety against imponderable factors such as $1/\alpha_s$, among others, experienced system and/or heat exchanger design engineers frequently specify deliberately close approach factors (AF $\rightarrow$ 0) in critical applications, in order to make the heat exchanger less sensitive to unforeseen contingencies, as shown in Table 7.13 at an approach factor SD/LD = 0.05 and previously highlighted in Table 7.3.

7.12.2 Impact of Manufacturing Tolerances

The very significant variations in f and the coupled α due to unavoidable wall roughness variations $\Delta e/\bar{e}$ in typical turbulent flow regimes have been quantified in Chapter 3.

Even if one were to disregard statistical variations in e/D and optimistically assume perfectly clean ideal smooth tubes, one must keep in mind that heat exchangers are normally made of tubular and sheet metal components, produced to relatively loose manufacturing tolerances (in contrast to thermal machines such as engines and compressors). United States tube manufacturers, for example, typically specify the outside diameter and a nominal wall thickness. For standard condenser tubes, the normal allowable tube-wall thickness tolerance of $\pm 10\%$ yields inside diameter variations as shown in Table 7.14 (the smallest and the largest standard nominal outside diameters, each in the lightest and heaviest nominal wall available, are highlighted in this Table). The *effect of these variations can be significant*, particularly in heavy wall, small diameter tubes, as shown in the following sensitivity analyses.

TABLE 7.14. Sensitivity Analysis: Impact of ±10% Wall Thickness Variation

Nominal Dimensions			Actual Variations (%)						
Outside	Wall Thickness	Inside	Inside	ΔH_f at Prescribed $\dot{m}$[a]			$\dot{m}$ at Prescribed ΔH_f[b]		
Diameter		Diameter	Diameter	Laminar	Turbulent		Laminar	Turbulent	
				$n = 1$	$n = 0.2$	$n = 0$	$n = 1$	$n = 0.2$	$n = 0$
$\frac{1}{2}$ in	Heaviest: 0.109 in.	0.282 in.	±7.7	∓30.8	∓37.0	∓38.5	±30.8	±20.6	±19.3
(1.27×10^{-2} m	Lightest: 0.035 in.	0.430 in.	±1.6	∓6.4	∓7.7	∓8.0	±6.4	±4.3	±4.0
2 in.	Heaviest: 0.134 in.	1.732 in.	±1.55	∓6.2	∓7.4	∓7.8	±6.2	±4.1	±3.9
(5.08×10^{-2} m)	Lightest: 0.065 in.	1.870 in.	±0.75	∓3.0	∓3.6	∓3.8	±3.0	±2.0	±1.9

[a]Calculated from Eq. (7.147b).
[b]Calculated from Eq. (7.149b).

7.12.2.1 Prescribed Flow Rate. The variation of the total friction head loss, ΔH_f, with a variation $\pm\Delta D$ from a nominal value of the internal tube diameter D, when $\dot{m}$ is prescribed and all other parameters are fixed, can be estimated from Eq. (3.4) and the general Blasius equation for f, Eq. (3.44), with $\text{Re} = (4\dot{m}/\pi D\mu)$ per Eq. (3.32) as follows:

$$\Delta H_f = f\left(\frac{L}{D}\right)\frac{\overline{V}^2}{2} = K_1 f D^{-5}\dot{m}^2 \tag{7.141}$$

with K_1 = constant for constant values of L, ρ, and μ, since $\overline{V} \propto \dot{m}/\Omega \propto \dot{m}/D^2$.

Introducing $f = K_n(\text{Re})^{-n} = K_n\ (4\dot{m}/\pi D\mu)^{-n}$ in Eq. (7.141), with $\mu =$ constant, one obtains

$$\Delta H_f = K_2 \dot{m}^{2-n} D^{-(5-n)} \tag{7.142}$$

where K_2 is a new constant term.

For a prescribed value $\dot{m}$ = constant, Eq. (7.142) reduces to

$$\Delta H_f = K_3 D^{-(5-n)} \tag{7.143}$$

where K_3 = constant, or expressed in logarithmic form,

$$\ln(\Delta H_f) = \ln(K_3) + \ln(D^{-(5-n)}) = \ln(K_3) - (5-n)(\ln D) \tag{7.144}$$

Equation (7.144) is of the general form

$$Y = \text{constant} - (5-n)X \tag{7.145}$$

where $Y \equiv \ln(y)$ with $y \equiv \Delta H_f$, $X \equiv \ln(x)$, x = independent variable = D, and $5 - n$ = a constant term, independent of x.

Differentiating both sides of Eq. (7.145) and recalling that $d(\ln y) = dy/y$ and $d(\ln x) = dx/x$ yields

$$\frac{d(\Delta H_f)}{\Delta H_f} = 0 - (5-n)\frac{dD}{D} \tag{7.146}$$

Therefore, with small relative diameter changes $\delta D/D \ll 1$,

$$\frac{\delta(\Delta H_f)}{\Delta H_f} = -(5-n)\frac{\delta D}{D} \tag{7.147a}$$

or

$$\%\text{ variation in }(\Delta H_f) = -(5-n)(\%\text{ variation in }D) \tag{7.147b}$$

for

$$\dot{m} = \text{constant} \tag{7.147c}$$

valid with $n = 1$ for laminar flow, therefore, $-(5-n) = -4$ and $n = 0.2$ to 0 for turbulent regimes in *smooth* to *fully rough* wall tubes with corresponding values $-(5-n) = -4.8$ to -5.0.

The percentage variations in ΔH_f at prescribed $\dot{m}$ shown in Table 7.14 were computed from Eq. (7.147b) with $-(5-n) = -4$, -4.8, and -5.0 for laminar, smooth wall, and fully rough wall turbulent flows, respectively.

7.12.2.2 Prescribed Friction Head. Equation (7.142) yields, for $\Delta H_f =$ constant,

$$\dot{m}^{2-n} = \frac{\Delta H_f}{K_2} D^{5-n} \equiv K_4 D^{5-n} \tag{7.148}$$

Following the same reasoning as in the previous sensitivity analysis, we obtain

$$d[\ln(\dot{m})] = +\left(\frac{5-n}{2-n}\right) d[\ln(D)] \tag{7.149a}$$

or

$$\%\text{ variation in }\dot{m} = \left(\frac{5-n}{2-n}\right)(\%\text{ variation in }D) \tag{7.149b}$$

for

$$\Delta H_f = \text{constant} \tag{7.149c}$$

with $(5-n)/(2-n) = 4$ for laminar flow at $n = 1$, decreasing to 2.66 and 2.50 for turbulent flow in smooth ($n = 0.2$) and fully rough ($n = 0$) tubes, respectively. The percentage variations in $\dot{m}$ for a prescribed ΔH_f shown in Table 7.14 were calculated accordingly.

In view of the fundamental coupling between fluid and convective heat flow phenomena (particularly in the more favorable turbulent regimes when $\bar{\alpha} \propto \bar{V}^{0.8 \text{ to } 1.0}$), it is clear that the impact of variations in D on the capacity $\dot{Q}_e$ of a simple elementary heat exchanger will be just as significant as the percentage variation in $\dot{m}$ shown in Table 7.14. Furthermore, since most published heat transfer data for internal flows are based on standard commercial tubes (rather than hypothetical tubes specially produced with zero manufacturing tolerances to *exact* nominal dimensions) and the results correlated on the basis of these nominal internal diameters, it is evident that the maximum $\pm 40\%$ (and probable $\pm 20\%$) deviations stated in Eq. (7.117c) for the very popular correlation recommended in 1954 McAdams, Eq. (7.117a), are due in large part to tube wall thickness manufacturing tolerances in addition to statistical variations from the average relative wall roughness $\bar{e}/D$, cleanliness, and possible inlet effects in empirical results obtained from tests carried out with abnormally short tubes (see also Chapter 3).

7.12.3 Rational Choice of Heat Transfer Coefficient Correlations

If we consider the additional test uncertainties associated with *single-sample experiments* [7.93], as well as the approximate nature of key thermophysical properties such as k, μ, and Pr [7.94], it is not at all surprising that individual empirical heat transfer coefficients obtained from *nominally identical* tubes by different researchers in different laboratories may differ by as much as 20% as reported by Petukhov [3.42]. It is mainly on the basis of pragmatic considerations of this kind that professional heat transfer engineers (who have learned to contend with such unavoidable statistical variations and uncertainties) continue to favor the adequately accurate, statistically well documented, simpler "old" correlations proposed by McAdams [1.2], Kern [1.3], and later disciples, over more recent and more sophisticated computer-aided theoretical solutions departing significantly from conditions encountered in real world heat exchangers.

Although they are evident to practicing heat transfer engineers, the above conclusions are summarized in the form of an engineering guideline, mainly because they tend to be overlooked by pure mathematical physicists engaged in heat transfer research.

GUIDELINE 23 *It is counterproductive to strive for accuracies of the order of* 1–2% *for heat transfer coefficients and friction factor correlations, in view of unavoidable manufacturing tolerances, test inaccuracies, the approximate nature of some key transport properties, and such uncertain effects as fouling resistance.*

Accuracies of the order of 10–20% *are adequate and more realistic in real world forced-convection heat exchangers. This explains why professional heat exchanger designers (and most engineering-oriented heat transfer textbooks) continue to favor some of the simpler adequately validated "older" correlations recommended by pragmatic authors, such as McAdams* [1.2] *and later disciples, over recent, more sophisticated theoretical correlations based on highly idealized theoretical models.*

NOTES

1. Sadi Carnot (1796–1832) had already formulated the essence of the first law and estimated the *heat equivalent* of mechanical energy as close as R. Mayer according to his posthumously published private papers [7.96].
2. The symbol α used in the European technical literature has been selected in lieu of h to avoid confusion with the standard American symbol h for the enthalpy.
3. This is generally a useful approximation that becomes very accurate in thermodynamically efficient forced-convection heat exchangers as discussed later.
4. The assumption of a linear temperature distribution in a cylindrical tube wall is sufficiently accurate when $\delta_w/D \ll 1$, a condition typically satisfied in normal (bare tubular) heat exchangers without extended surface.
5. Even though q''_{rad} can be conveniently expressed according to $q''_{rad} = \alpha_{rad} (T_w - T) \propto \Delta T_w$ over a relatively narrow range of absolute temperatures T_w and T, where α_{rad} = equivalent heat transfer coefficient due to thermal radiation. This is the method used in standard heat transfer textbooks, such as Table 7.2, p. 179 of Ref. 1.2, which shows values of $\alpha_{conv} + \alpha_{rad}$ to compute the heat loss from horizontal pipes in a room at 80° F due to natural convection *and* thermal radiation.
6. In their preface, the authors refer to the first treatise in this broad interdisciplinary branch of engineering, *Principles of Chemical Engineering* by Walker, Lewis, and McAdams, published in 1923 as part of the McGraw-Hill chemical engineering series. This series was later to include McAdams' classic *Heat Transmission* [1.2]. Significant contributions to the science and art of momentum, heat, and mass transfer resulted from the creation in the United States of the new branch of engineering called *chemical engineering* in the first quarter of this century. The preface of *Elements of Chemical Engineering* [7.36] is written by Arthur D. Little, who is credited by Badger and McCabe with the formalization in December 1915 of the concept of chemical engineering unit operations presented in a report to the Corporation of the Massachusetts Institute of Technology.

7. Confirming the general rule that the *art* usually precedes the *science*.
8. Actual deviations of the order of $\pm 10\%$ are not surprising in the light of what was said in Chapter 3.
9. The *Benard instability*, essentially a natural convection phenomenon, is a classical example used by the Brussels School of non-equilibrium thermodynamics [7.97, 7.98] to illustrate the new types of structure (or spontaneous self-organization) possible "far from equilibrium" when the relationships between *generalized forces* and *fluxes* are no longer linear. Benard patterns are also discussed in modern engineering texts dealing with natural convection [7.99].
10. Since Fourier had outlined in his writings the basic methods of coupling the velocity and temperature fields to predict the convective heat rate equations, it is not too surprising to find the name of his first supporter and closest disciple in France, Navier, associated with the most general equations (the Navier–Stokes equations) of convective heat and fluid flow phenomena [7.100]. What is more remarkable is the indirect connection with Stokes in the United Kingdom through his close association with William Thomson (Lord Kelvin) [7.101]. Thomson, who read French fluently, "made the acquaintance—a notable event in his career—of Fourier's Theorie Analytique de la Chaleur; reading it through in a fortnight" [1.13].
11. This philosophy ultimately led to the modern branches of thermodynamics now known as statistical equilibrium and/or nonequilibrium thermodynamics.
12. According to Kangro [7.102], the term "Gedanken Experiment" was proposed by Oersted in 1830 [Oersted, *Isis*, **8**, 854–857 (1831)]. Oersted, and later W. Wien, formulated the validity requirements for such fictive experiments.
13. The work of Sadi Carnot was apparently ignored by his contemporaries in France mainly for two *contradictory* reasons [1.8]: (1) it seemed to address engineering problems *too pragmatic* and mundane for the dominant mathematical physicists at the French Academy of Sciences (including Fourier [7.50] who by this time had become a respected member of the establishment) and (2) it was considered *too theoretical*, abstract, and rather irrelevant by pragmatic French thermal engineers more interested in quickly catching up with the superior British technology in steam engines (short-range "nuts and bolts" objectives).
14. Sadi Carnot shows only an infinitesimal cycle in his memoir: where $T_H - T_C = 1°\,C$, and $(T_H - T_L)/T_H \approx 0$ or $T_H/T_L \approx 1$. By contrast, Clapeyron addressed high absolute temperature ratios (T_H/T_L) and corresponding high absolute pressure ratios as shown in modern engineering thermodynamic textbooks.
15. There are other important precedents to confirm this guideline. A good case in point is the most fundamental law of classical mechanics—Newton's law of universal gravitation—which has certainly proved quite useful in physical sciences and engineering even though we still do not know today the true nature of gravitation! It is worth recalling in this connection that Newton himself declared that he had not been able to discover the cause nor did he want to frame any hypothesis on the true nature of gravitation, according to his historical statement "hypotheses non fingo" [7.103, 7.104]. However, some physicists believe it is fundamentally more elegant to propose a theoretical model first to anticipate physical phenomena that are verified subsequently [7.105].

16. According to Kangro [1.8, footnote on p. 229], the term "Thermodynamik" appeared in the German (language) scientific literature in 1837 in the work of Andreas Baumgartner, entitled "Anfangsgruende der Naturlehre nach Ihrem Gegenwaertigen Zustande mit Ruecksicht auf Mathematische Begrundung," in Vienna in 1837. It should be pointed out that Baumgartner applies the term "Thermodynamik" to *thermal radiation* and *conduction phenomena* whereas the *steam engine* is addressed under the heading "*Thermostatik*."
17. Most engineering thermodynamic textbooks are based on the classical (or phenomenological) methods [7.106–7.126] rather than the statistical approach [7.127–7.130], although some textbooks address both methodologies [7.131, 7.132].
18. Typical of refrigeration and particularly ultralow or cryogenic applications.
19. Just as the original definition of the fundamental dimension "time" in physics can be associated to a length dimension (i.e., the trajectory of the earth around the sun).
20. See Guidelines 14 and 15 in this chapter.
21. Within the validity limits of the continuum hypothesis.
22. This assumes that individual friction losses are negligibly small compared to ΔP_f due to wall friction. The more general case, when individual friction losses are significant, is addressed in Chapter 8.
23. Nonisothermal conditions can have a significant impact in laminar flow regimes, as shown in Chapters 3 and 8.
24. Keep in mind that Petukhov's test data emphasize aerospace applications.
25. True steady-state conditions are not possible in this context!

CHAPTER 8

PRACTICAL APPLICATIONS

Making full use of the tools reviewed previously, we focus first on turbulent single- and two-phase (single-component) channel flows, then extend the corresponding states generalization methods introduced in Chapter 6 to solve complex heat transfer design problems with, and without, phase changes in heat exchangers of any arbitrary geometry, including external as well as internal flows.

8.1 SINGLE-PHASE TURBULENT FLOWS IN STRAIGHT CHANNELS

In view of the direct coupling between the convective heat transfer coefficient and the friction factor, similar nonisothermal and inlet effects can be expected for $\bar{\alpha}$ as those highlighted in Chapter 3 for the friction factor $\bar{f}$.

8.1.1 Impact of Nonisothermal and Inlet Effects

In normal heat exchanger design practice the fully developed heat transfer coefficient α_∞ or the corresponding Stanton and Nusselt numbers St_∞ and Nu_∞, per Eq. (7.10a), or (7.16), are corrected by (semi)empirical multipliers to allow for nonisothermal and inlet effects as follows:

$$\frac{\bar{\alpha}}{\alpha_\infty} = \frac{\overline{\mathrm{St}}}{\mathrm{St}_\infty} = \frac{\overline{\mathrm{Nu}}}{\mathrm{Nu}_\infty} = M_{\Delta T} M_{L/D} \tag{8.1a}$$

or

$$\bar{\alpha} = \alpha_\infty M_{\Delta T} M_{L/D} \tag{8.1b}$$

where the correction factors $M_{\Delta T}$ and $M_{L/D}$ account for nonisothermal conditions and inlet effects, respectively. The effective average heat transfer coefficient $\bar{\alpha}$ for a total channel length L and corresponding $\overline{\mathrm{St}}$ or $\overline{\mathrm{Nu}}$ identified by the overbar converge toward α_∞ and corresponding St_∞ or Nu_∞ when $\overline{\Delta T}_w = \bar{T}_w - \bar{T}_b \to 0$ and $L/D \to \infty$; that is,

$$\bar{\alpha} = \alpha_\infty \quad \text{or} \quad \overline{\mathrm{St}} = \mathrm{St}_\infty \quad \text{or} \quad \overline{\mathrm{Nu}} = \mathrm{Nu}_\infty \tag{8.2a}$$

for

$$M_{\Delta T} = M_{L/D} = 1 \tag{8.2b}$$

when

$$\overline{\Delta T}_w = \bar{T}_w - \bar{T}_b \to 0 \quad \text{and} \quad \frac{L}{D} \to \infty \tag{8.2c}$$

where:

$$\bar{T}_b = \text{average fluid bulk temperature} \tag{8.2d}$$

The multipliers $M_{\Delta T}$ and $M_{L/D}$ are calculated according to the following generally accepted semiempirical correlations [1.2, 1.3, 7.29]:

$$M_{\Delta T} = \left(\frac{\mu_b}{\mu_w}\right)^{0.14} \tag{8.3a}$$

where

$$\mu_b = \text{viscosity at average bulk temperature } \bar{T}_b \tag{8.3b}$$

$$\mu_w = \text{viscosity at average (or constant) wall temperature } \bar{T}_w \text{ (or } T_w) \tag{8.3c}$$

Note that $M_{\Delta T}$ can be expressed with sufficient accuracy for common gases with $\mu \propto T^a$ (see Section 3.11) according to

$$M_{\Delta T} = \left(\frac{\bar{T}_b}{\bar{T}_w}\right)^{0.14a} \tag{8.4a}$$

or

$$M_{\Delta T} = \left(\frac{\bar{T}_b}{\bar{T}_w}\right)^{0.0952} \simeq \left(\frac{\bar{T}_b}{\bar{T}_w}\right)^{0.10} \tag{8.4b}$$

Validity: Common gases with $\mu \propto T^a$ and $a \simeq 0.68$ (8.4c) and

$$M_{L/D} = 1 + \frac{1}{(L/D)^{0.7}} \tag{8.5a}$$

Validity: Square-edged entries (8.5b)

according to McAdams [1.2], whereas an insignificantly lower exponent $\frac{2}{3}$, in lieu of 0.70, is recommended in a more recent (1982) engineering heat transfer text [7.29] published by the German society of engineers (VDI).

It should be noted that Eqs. (8.3)–(8.5) were validated on the basis of tests carried out with *normally smooth* heat exchanger tubes, therefore:

Validity: Equations (8.3)–(8.5) are valid a priori for *smooth tubes only* in conjunction with Eqs. (7.10)–(7.20) (8.6)

Since the fluid and convective heat flow phenomena are closely coupled, it is logical to expect, on the basis of what was said in Chapter 3 (see in particular Tables 3.5 and 3.6), that the viscosity exponent will tend to *decrease* below 0.14 and the L/D exponent *increase* above 0.7 for *rougher* tube walls, when $\bar{e}/D \gg 0$.

The multipliers $M_{\Delta T}$ and $M_{L/D}$ computed from Eqs. (8.3a) and (8.5a) are shown in Tables 8.1 and 8.2, respectively. Keeping in mind typical deviations of the order of ± 10–20% for the best correlations of α_∞, we see that the

TABLE 8.1. Temperature Correction Factor $M_{\Delta T}$ for Turbulent Flow in Smooth Tubes per Eq. (8.3a)

μ_b / μ_w	$M_{\Delta T}$	Change[a] (%)
0.1	0.724	−27.6
0.25	0.824	−17.6
0.50	0.908	−9.2
0.75	0.960	−4.0
0.80	0.969	−3.1
0.85	0.978	−2.2
0.90	0.985	−1.5
0.95	0.993	−0.7
1.00	1.00	0
1.05	1.007	+0.7
1.11	1.015	+1.5
1.18	1.023	+2.3
1.25	1.031	+3.1
1.33	1.041	+4.1
2.0	1.102	+10.2
4.0	1.214	+21.4
10.0	1.380	+38.0

[a] Change (%) = $100(M_{\Delta T} - 1)$.

TABLE 8.2. Inlet Effect Correction Factor $M_{L/D}$ for Turbulent Flow in Smooth Tubes per Eq. (8.5a)

L/D	$M_{L/D}$	Change[a] (%)
1	2.00	+100.0
2	1.62	+62.0
5	1.32	+32.0
10	1.20	+20.0
20	1.123	+12.3
50	1.065	+6.5
100	1.040	+4.0
200	1.025	+2.5
500	1.013	+1.3
1000	1.008	+0.8
2000	1.005	+0.5

[a]Change (%) = $100(M_{L/D} - 1)$.

correction factors $M_{\Delta T}$ and $M_{L/D}$ can usually be neglected except when the ratio μ_b/μ_w deviates significantly from unity and $L/D < 60$ in the case of ideal smooth tubes. One of the objectives of Numerical Analyses 8.1 and 8.2 is to determine exactly under what conditions it is permissible to neglect the correction factors $M_{\Delta T}$ and $M_{L/D}$, that is, to set $\bar{\alpha} = \alpha_\infty = \alpha$ and thus omit the distinction between the various heat transfer coefficient definitions, without incurring any measurable loss of accuracy.

NUMERICAL ANALYSIS 8.1

Thermodynamically efficient heat exchangers.

Problem Statement

Quantify the impact of nonisothermal conditions at small relative temperature differences $\overline{\Delta T}_w/\overline{T}_b \ll 1$, typical of thermodynamically efficient heat exchangers, for oil, water, and common gases at $\overline{T}_b = 310$ K and compare the results.

Solution

The numerical results obtained from $M_{\Delta T}$, computed according to Eqs. (8.3a) and (8.4b) and summarized in Table 8.3, can be highlighted by referring to the columns for oil, water, and common gases as follows:

- At an extremely low $\overline{\Delta T}_w$ of ± 1 K, or $\overline{\Delta T}_w/\overline{T}_b = \pm \frac{1}{3}\%$, $M_{\Delta T} \simeq 1.0$ even in the case of an extremely viscous liquid such as engine oil at 310 K. The % changes in heat transfer coefficients due to $\overline{\Delta T}_w$ shown on lines (6) and

TABLE 8.3. Percentage Change in $\bar{\alpha}$ due to $\overline{\Delta T_w}$ for Oil, Water, and Common Gases[a]

	$\bar{T}_b$ (K)	$\bar{T}_w$ (K)	$\overline{\Delta T_w} = \bar{T}_w - \bar{T}_b$ (K)	$100\left(\frac{\overline{\Delta T_w}}{\bar{T}_b}\right)$%	$\frac{\bar{T}_b}{\bar{T}_w}$	Oil (%)	Water (%)	Gases (%)
	(a)	(b)	(c)	(d)	(e)	(f)	(g)	(h)
(1)	310	280	−30	−9.7	1.107	−26.0	−9.5	+1.0
(2)	310	290	−20	−6.5	1.069	—	−6.0	+0.6
(3)	310	295	−15	−4.8	1.051	—	−4.4	+0.5
(4)	310	300	−10	−3.3	1.033	−8.8	−2.9	+0.3
(5)	310	305	−5	−1.6	1.016	−5.2	−1.4	+0.16
(6)	310	309	−1	−0.3	1.003	≃ −1	−0.3	+0.03
(7)	310	310	0	0	1	0	0	0
(8)	310	311	+1	+0.3	0.997	≃ +1	+0.3	−0.03
(9)	310	315	+5	+1.6	0.984	+3.6	+1.4	−0.15
(10)	310	320	+10	+3.3	0.969	+8.5	+2.6	−0.3
(11)	310	325	+15	+4.8	0.954	—	+3.9	−0.5
(12)	310	330	+20	+6.5	0.939	—	+5.0	−0.6
(13)	310	340	+30	+9.7	0.911	+24.0	+7.3	−0.9

[a]Percentage change in $\bar{\alpha}$ computed with Eq. (8.3a) and μ from Tables A.1 and A.2 in Appendix A for water and engine oil, respectively, and from Eq. (8.4b) for common gases. *Note:* At $\overline{\Delta T_w}$ falling inside the limits delineated by the dashed line (---), the definition of linear heat flux heat exchangers (per GL 19) is satisfied with an accuracy of 5% for oil and water and 1% for gases.

(8) are negligibly small compared to typical deviations of the order of ± 10–20% for α_∞ at $\overline{\Delta T_w} \simeq 0$ as expected.

- At $\overline{\Delta T_w} = \pm 10$ K, or $\overline{\Delta T_w}/\bar{T}_b \simeq \pm 3\%$, the % changes shown on lines (4) and (10) are already measurable for oil (−8.8% and +8.5% in the fluid cooling and heating modes, respectively), whereas they are still negligibly small for liquid water and common gases with percentage changes of −2.9/+ 2.6% and +0.3/− 0.3%, respectively.
- At $\overline{\Delta T_w} \equiv \pm 30$ K, or $\overline{\Delta T_w}/\bar{T}_b \simeq \pm 10\%$, the percentage changes shown on lines (1) and (13) are quite significant for oil (−26/+ 24%), moderate for water (−9.5/+ 7.3%), and still negligibly small for gases (+1/− 1%).

Comments

1. It is interesting to note that the results summarized in Table 8.3, which are strictly valid for ideal smooth tubes only, practically duplicate those shown for smooth tubes (at a Blasius exponent $n = 0.25$) in Table 3.6 of Chapter 3. This is not altogether surprising in view of the close coupling between fluid and convective heat flow phenomena and further reinforces the intuitive notion that the impact of nonisothermal conditions ($\overline{\Delta T_w} \neq 0$) must become less pronounced in rougher pipes ($\bar{e}/D \gg 0$) and may ultimately become insignificant at the fully rough wall limit, when $n \to 0$ (see numerical results at $n = 0.15$ and $n = 0$ in Table 3.6), for normal Pr number working fluids used in power generation and reverse power (refrigeration/heat pump) cycles.

2. As the main fluid-to-fluid convective heat exchangers used in power generation and refrigeration/heat pump cycles are usually designed for operation in the more favorable stable turbulent regimes with normal gases and/or liquids (rather than highly viscous and high Pr fluids such as oil) at $\overline{\Delta T_w}/\overline{T}_b \ll 1$, in order to reduce thermodynamic losses, it is quite understandable, in the light of Table 8.3, why recommended heat transfer coefficient correlations for such applications are based on $M_{\Delta T} = 1$. As a matter of fact, the more thermodynamically efficient the heat exchanger, that is, the closer the ratio $\overline{\Delta T_w}/\overline{T}_b \to 0$, the closer $M_{\Delta T} \to 1$, since at the limit $\overline{\Delta T_w}/\overline{T}_b \equiv 0$ the correction factor $M_{\Delta T}$ is exactly equal to unity! Thus, linear heat flux (LHF) heat exchanger systems can now be defined more precisely as the class of heat exchangers satisfying the condition $M_{\Delta T} = 1$ (typically within a 1–5% accuracy per Guideline 19) on both the hot and cold fluid sides, or on the controlling side of the heat exchanger. When $\alpha_i a_i \simeq U_i a_i$, the dashed line shown in Table 8.3 delineates the validity limits of the LHF heat exchanger assumption.

NUMERICAL ANALYSIS 8.2

Normally effective heat exchangers.

Background

Although the following definition of normally effective heat exchangers is somewhat arbitrary, it does reflect actual heat exchanger design practice in many critical industrial applications and makes it possible to simplify the analysis of normally effective heat exchangers without sacrificing accuracy.

> ***GUIDELINE 24*** *Any heat exchanger designed for an effectiveness* $\varepsilon \geq 80\%$, *that is, an approach factor* AF $\leq 20\%$, *belongs by definition to the class of normally effective fluid-to-fluid heat exchangers.*

Problem Statement

Determine the minimum aspect ratios L/D and corresponding lengths of travel L of normally effective heat exchangers[1] with smooth drawn tubes in nominal outside diameters of 0.5–2 in. and internal diameters varying between $10^{-2} \leq D_i \leq 5 \times 10^{-2}$ m, within a practical stable turbulent flow range of $10{,}000 \leq \text{Re} \leq 120{,}000$ for the following:

- Common gases and vapors, such as air or hydrogen, with a substantially constant Pr = 0.7 ≈ 1.0.
- Typical liquids within Pr numbers in the range 1 < Pr < 15, or a representative average value of Pr = 8.

- Highly viscous liquids with large Pr numbers up to 120, the upper validity limit of McAdams' recommended equations for turbulent tube flows, per Eq. (7.117c).

For this "worst case" assessment of the impact of inlet effects, the archetype constant wall temperature boundary condition single-tube heat exchanger is considered (see Section 7.6).

Solution

The required length of travel can be computed from Eqs. (7.47) and (7.49) at an NTU $\geq \ln(1/\text{AF}) = \ln(1/0.2) = 1.609$ with AF ≤ 0.2, or

$$L \geq 1.609\left(\frac{\dot{m}c_p}{a_i}\right)\frac{1}{U_i} \tag{8.7}$$

Since most generally $\alpha_i > U_i$ according to Eq. (7.64), we underestimate the actual length of travel by using α_i instead of U_i in Eq. (8.7); therefore,

$$L > 1.609\frac{\dot{m}c_p}{a_i}\left(\frac{1}{\alpha_i}\right) \tag{8.8}$$

Substituting $\Omega_i G$ for $\dot{m}$ and $\frac{1}{8}fGc_p(\text{Pr})^{\beta}$ for α_i (according to the Chilton–Colburn correlation) in the above equation and recalling that $\Omega_i/a_i \equiv r_{h,i} \equiv \frac{1}{4}D_i$ for circular channels, we obtain

$$L > 3.22(1/f)D_i(\text{Pr})^{\beta} \tag{8.9}$$

or in dimensionless form

$$(L/D_i)/(\text{Pr})^{\beta} > 3.22(1/f) \tag{8.10}$$

Computing the inverse of the friction factor f, from McAdams' recommended Blasius correlation, Eq. (3.47), with $K_n = 0.184$ and $n = 0.2$, yields the numerical values shown on line (2) of Table 8.4 for extreme values of Reynolds numbers typically encountered in normal heat exchanger design practice of Re = 10,000 and 120,000.

For this order of magnitude analysis, it is sufficiently accurate to use the approximate constant Chilton–Colburn exponent $\beta = \frac{2}{3}$ to compute the values of L, L/D_i per Eqs. (8.9) and (8.10) and the corresponding correction factor $M_{L/D}$ calculated from Eq. (8.5) shown in Table 8.4 at Pr numbers of 0.7, 8, and 120.

- It is clear from Eqs. (8.9) and (8.10) as well as the numerical data summarized in Table 8.4 that the inlet correction factors $M_{L/D}$ are negligibly small compared to typical deviations of the order of ± 10–20%

TABLE 8.4. Impact of Inlet Effects for Ideal Smooth (0.5–2 in.) Tubes in Normally Effective Heat Exchangers

(1)	Re	=		10,000		to	120,000	
(2)	(L/D_i) $/(\mathrm{Pr})^{\beta}$	>		110.4		to	181.5	
(3)	Pr	=	0.7	8	120[a]	0.7	8	120[a]
(4)	L/D_i	>	87	442	2677	143	726	4402
(5)	$M_{L/D}$[b]	<	1.044	1.014	1.004	1.031	1.010	1.003
(6)	L (in m)[c] for $D_i = 0.01$ m	>	0.87	4.42	26.77	1.43	7.26	44.02
(7)	L (in m)[c] for $D_i = 0.05$ m	>	4.35	22.1	133.9	7.15	36.3	220.1

[a]Extreme friction head loss at high Pr and Re numbers (see Section 7.11).
[b]By setting $M_{L/D} = 1$ in practical design applications, one errs *insignificantly* on the safe side since $\bar{\alpha}$ is theoretically $> \alpha_{\infty}$.
[c]Multiply by 3.2808 to convert to feet.

in normally effective tubular heat exchangers, even in the worst case of common gases with $\mathrm{Pr} \approx 1$.

- Since L/D_i is directly proportional to $(\mathrm{Pr})^{\beta}$, it is evident that the inlet correction factor of single-pass heat exchangers becomes rapidly insignificant as Pr increases substantially beyond unity. In fact, the length of travel becomes so large at high Pr numbers [see lines (6) and (7) of Table 8.4] that it is necessary to resort to a multipass arrangement to keep the heat exchanger envelope within acceptable limits, particularly in the case of large nominal tube diameters.

Comments

When the internal fluids' thermal resistance is not "controlling," that is, $\alpha_i a_i \approx \alpha_o a_o$, it is usually economical to balance the thermal resistance on both sides of surface heat exchangers so that $\alpha_i a_i \simeq \alpha_o a_o$. The overall thermal resistance is then $1/U_i a_i \simeq 2/\alpha_i a_i$, that is, twice as large as the value $1/\alpha_i a_i$ used in the above calculations, even if one neglects the additional thermal resistances due to wall conduction and scale deposits, that is, assuming perfectly clean surfaces on both the hot and cold sides of the heat exchanger. This means that the L/D_i ratios shown in Table 8.4 would have to be at least doubled for *well-balanced* heat exchangers and this explains why the above worst case analysis generally underestimates the length of travel required in many practical applications. Therefore, the inlet effects are typically even less significant than indicated by Table 8.4, thus further reinforcing the general conclusion that it is permissible for practical design purposes to set $\bar{\alpha}_i = (\alpha_i)_{\infty} \equiv \alpha_i$ in the case of normally effective surface heat exchangers without any compromise in accuracy.

8.1.2 Modified Chilton–Colburn Correlation

We focus in this section exclusively on normally effective and thermodynamically efficient heat exchangers when $M_{\Delta T}$ and $M_{L/D} = 1$ to determine the Prandtl exponent β in Eq. (7.10a) along the lines outlined in Chapter 7 (see in particular Section 7.10.4). Since with $M_{\Delta T} = M_{L/D} = 1$ the heat transfer coefficients $\bar{\alpha} \equiv \alpha_{\infty} \equiv \alpha$, we simplify the writing and dispense with the overbar and subscript ∞, using the same nomenclature as in Eqs. (7.10)–(7.16) for the balance of this discussion.

To determine as accurately as feasible the empirical exponent in Eq. (7.10a), we use what modern experts[2] consider to be the most advanced and accurate heat transfer coefficient correlation, namely, Petukhov's *exact* Eq. (48) (in Ref. 3.42) introduced in Chapter 7 as Eqs. (7.16) and (7.17), as follows.

Petukhov's Eqs. (7.16) and (7.17) reduce to the form

$$\mathrm{St} = \frac{f/8}{\phi_{\mathrm{Ptk}}(\mathrm{Re}, \mathrm{Pr})} \tag{8.11}$$

where the function ϕ_{Ptk}, the denominator on the right-hand side of Eq. (8.11), is a unique function of Pr and f according to Eq. (7.16) and therefore fully defined in terms of Pr and Re only, according to Eq. (7.17).

The modified Chilton–Colburn correlation per Eq. (7.10a) can be expressed most generally as

$$\mathrm{St} = \frac{f/8}{(\mathrm{Pr})^{\beta}} \tag{8.12}$$

Equation (8.12) will yield exactly the same St number as Petukhov's recommended correlation when the right-hand sides of Eqs. (8.11) and (8.12) are identical; that is,

$$\frac{f/8}{(\mathrm{Pr})^{\beta}} \equiv \frac{f/8}{\phi_{\mathrm{Ptk}}(\mathrm{Re}, \mathrm{Pr})} \tag{8.13}$$

therefore

$$(\mathrm{Pr})^{\beta} \equiv \phi_{\mathrm{Ptk}}(\mathrm{Re}, \mathrm{Pr}) \tag{8.14}$$

Taking the logarithm on both sides of Eq. (8.14) yields

$$\beta[\ln(\mathrm{Pr})] = \ln[\phi_{\mathrm{Ptk}}(\mathrm{Re}, \mathrm{Pr})] \tag{8.15}$$

and solving for β yields

$$\beta = \frac{\ln[\phi_{\mathrm{Ptk}}(\mathrm{Re}, \mathrm{Pr})]}{\ln(\mathrm{Pr})} \tag{8.16}$$

TABLE 8.5. Determination of Prandtl Exponent β in the Modified Chilton–Colburn Correlation

			Reynolds Number[a]							
			10,000	5×10^4	10^5	120,000	5×10^5	10^6	5×10^6	(10^7)[b]
Eq. (8.18d):	$100f$		3.1437	2.0930	1.7969	1.7294	1.3115	1.1612	0.8981	(0.8116)
Eq. (8.18b):	$K_1(f)$		1.1069	1.0712	1.0611	1.0588	1.0446	1.0395	1.0305	(1.0276)
Eq. (8.18e):	$f^{1/2}$		0.1773	0.1447	0.1340	0.1315	0.1145	0.1077	0.0948	(0.9009)
			Blasius exponent n							
			0.281	0.229	0.212	0.208	0.181	0.171	0.150	(0.143)
Eq. (8.18c)	$K_3(\mathrm{Pr})$ at	Pr	**Exponent $\beta = \beta(n, \mathrm{Pr})$ for $0.5 \leq \mathrm{Pr} \leq 120$ from Eq. (8.16)**							
	−1.826	0.50	0.353	0.310	0.293	0.289	0.259	0.247	0.222	(0.213)
	−0.9433	0.66	0.253	0.248	0.242	0.240	0.225	0.218	0.202	(0.195)
	−0.666	0.80[c]	0.050	0.114	0.128	0.129	0.144	0.147	0.149	(0.148)
	+2.723	2	0.669	0.551	0.512	0.503	0.440	0.414	0.366	(0.348)
	+4.938	3	0.623	0.528	0.495	0.487	0.433	0.411	0.368	(0.352)
	+10.31	6	0.601	0.525	0.498	0.492	0.446	0.427	0.389	(0.375)
	+18.69	12	0.598	0.535	0.512	0.506	0.466	0.449	0.415	(0.410)
	+99.18	120	0.612	0.571	0.556	0.553	0.526	0.514	0.490	(0.480)

[a] Usual design range $10{,}000 < \mathrm{Re} < 120{,}000$.
[b] Values in parentheses at $\mathrm{Re} = 10^7$ are extrapolated beyond validity limit $\mathrm{Re} = 5 \times 10^6$ of Petukhov's correlations.
[c] Empirical exponent β becomes less accurate at $\mathrm{Pr} \to 1$ but impact on $(\mathrm{Pr})^{\beta}$ or $(\mathrm{Pr})^{\beta/2}$ is insignificant.

The Prandtl exponent β can be computed from Eq. (8.16) for any value of the parameter Pr in terms of Re with, according to Eqs. (7.16) and (7.17),

$$\phi_{\mathrm{Ptk}}(\mathrm{Re}, \mathrm{Pr}) \equiv K_1(f) + K_2(\mathrm{Pr})\left[(\mathrm{Pr})^{2/3} - 1)\right]\left(\tfrac{1}{8}\right)^{1/2} f^{1/2} \tag{8.17}$$

or

$$\phi_{\mathrm{Ptk}}(\mathrm{Re}, \mathrm{Pr}) \equiv K_1(f) + K_3(\mathrm{Pr}) f^{1/2} \tag{8.18a}$$

and

$$K_1(f) \equiv 1 + 3.4f \tag{8.18b}$$

$$K_3(\mathrm{Pr}) \equiv \left[4.1366 + 0.6364(\mathrm{Pr})^{-3}\right]\left[(\mathrm{Pr})^{2/3} - 1)\right] \tag{8.18c}$$

$$f = (1.82 \log_{10}\mathrm{Re} - 1.64)^{-2} \tag{8.18d}$$

$$f^{1/2} = (1.82 \log_{10}\mathrm{Re} - 1.64)^{-1} \tag{8.18e}$$

as highlighted in Table 8.5 for a reasonable range of normal fluid[3] Prandtl numbers in the range $0.5 \leq \mathrm{Pr} \leq 120$ over the full validity range of Re numbers, $10{,}000 \leq \mathrm{Re} \leq 5{,}000{,}000$, according to Eq. (7.17). Keep in mind that these results are a priori valid for *smooth wall* heat exchanger tubes *only*. Less "exact" values of β obtained by extrapolating Petukhov's recommended correlation up to $\mathrm{Re} = 10^7$ are shown in parentheses.

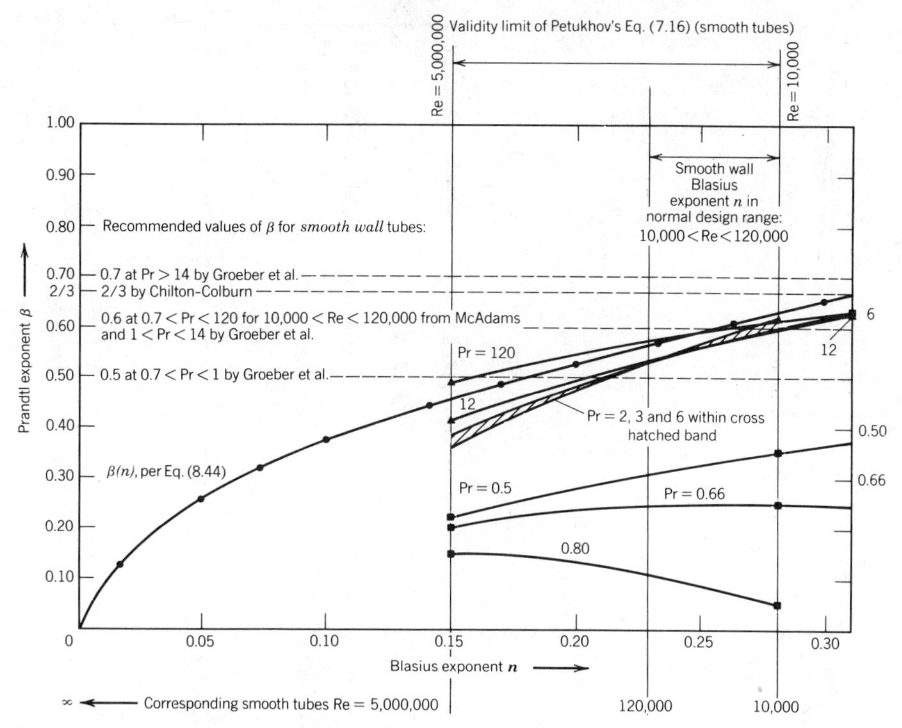

FIGURE 8.1. Prandtl exponent β in terms of Blasius exponent n for normal turbulent range $0 < n < 0.3$.

The Prandtl exponents β have been broken down into two fundamentally different regions in Table 8.5. First, for common gases and vapors with $0.5 \leq \text{Pr} \leq 1$, and second, for normal liquids (i.e., excluding liquid metals) with $\text{Pr} \geq 1$. For normal liquids the impact of the parameter Pr is assessed more closely in the usual range $1 < \text{Pr} < 12$ which is more typical of efficient fluid-to-fluid heat exchanger applications in turbulent regimes than $\text{Pr} \gg 12$ in view of what was said previously in Section 7.11 (see Table 7.12).

The empirical Blasius exponent n shown in Table 8.5 was calculated according to Eq. (3.55) with the corresponding values of f obtained from Petukhov's Eq. (8.18d), exactly in the same manner as in Chapter 3, Numerical Analysis 3.2.

To obtain a broad overview of the variation of the Prandtl exponent β in the modified Chilton–Colburn correlation, Eq. (8.12), the numerical results summarized in Table 8.5 are also shown graphically in Figs. 8.1 and 8.2 as curves $\beta = \beta(n, \text{Pr})$ on normal linear graph paper, using the Pr number as an independent constant parameter. Figure 8.1 focuses on turbulent regimes with Blasius exponent n in the range $0 < n < 0.32$, whereas Fig. 8.2 covers the full range of Blasius exponent numbers including the limiting laminar case when

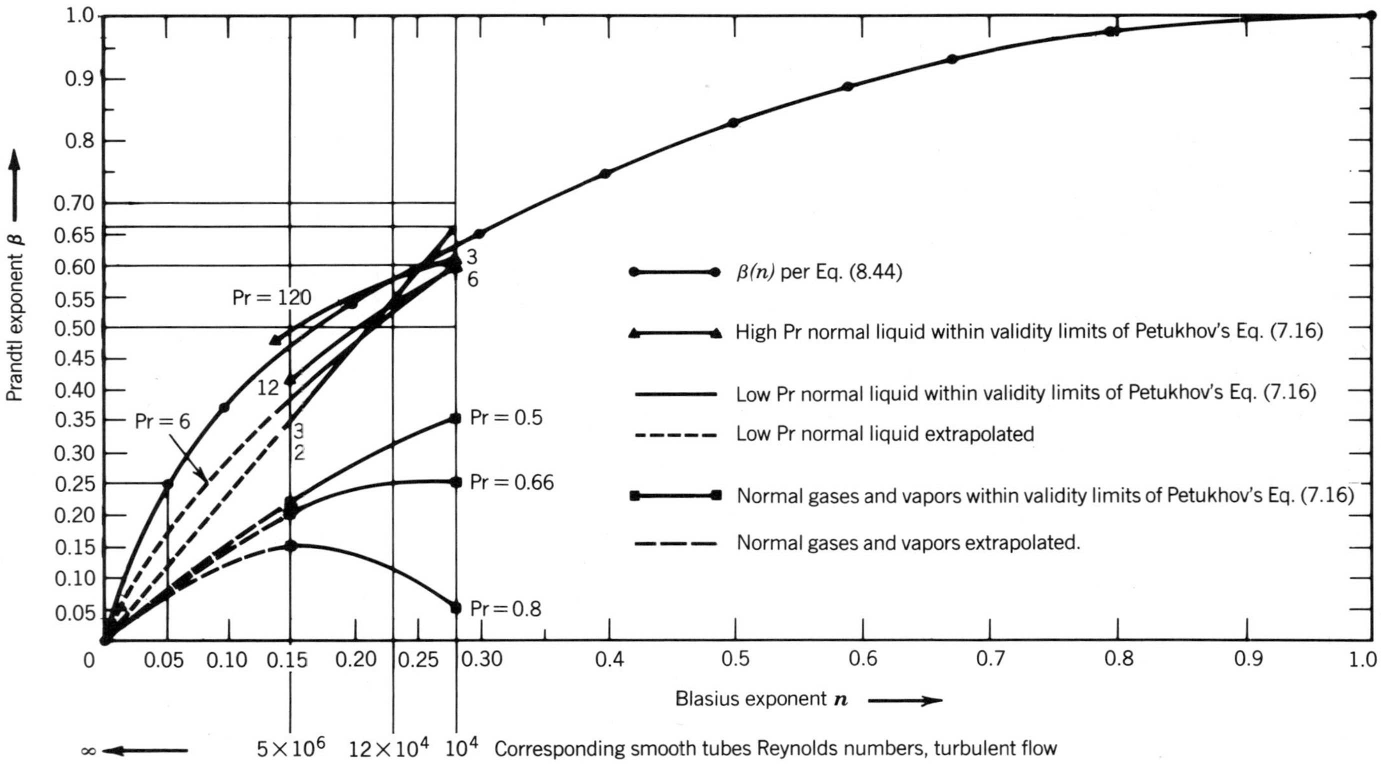

FIGURE 8.2. Prandtl exponent β in terms of Blasius exponent n for full range $0 < n < 1.0$.

$n = 1$ and $\beta = 1$ as well as the fully rough turbulent limit when $n \to 0$ and $\beta \to 0$.

8.1.2.1 *Preliminary Assessment of Prandtl Exponent* β. For a bird's eye view of the factors influencing the empirical exponent β we refer mainly to Fig. 8.2.

General Trends. The plots of $\beta(n)$ clearly show the following:

- β is a function of the Blasius exponent n and generally *decreases* with a *decreasing* Blasius exponent n, as expected when the flow regimes become more chaotic (or turbulent) even in the case of smooth wall tubes at $\mathrm{Re} \gg 120{,}000$ when $n \to 0$.
- For common ($1 < \mathrm{Pr} < 10$) liquids the Pr number has a relatively minor impact on β in the normal design range $10{,}000 < \mathrm{Re} < 120{,}000$ for *smooth wall* tubes. However, significantly different numerical values are obtained for β in the case of common gases at $0.5 < \mathrm{Pr} < 1$ from Petukhov's most accurate correlation for α. Consequently, it is necessary to discuss separately the two main Pr number groups: $\mathrm{Pr} > 1$ and $\mathrm{Pr} < 1$.

Common Liquids: Pr > 1. Since according to the modified Chilton–Colburn correlation $\alpha \propto (1/\mathrm{Pr})^{\beta}$, it is evident that one tends to *underestimate* α at $\mathrm{Pr} > 1$ when the exponent β is *overestimated*. In other words, it is *prudent for practical design applications to select an average constant exponent β on the high side.*

Focusing mainly on the normal design range $10{,}000 < \mathrm{Re} < 120{,}000$, it is possible to assess the empirical values for β recommended by various researchers per Table 7.10 and highlighted in Figs. 8.1 and 8.2 as follows:

- A constant $\beta = \frac{2}{3}$ according to the standard Chilton–Colburn correlation is a slightly conservative but sensible approximation in many tubular heat exchanger applications, particularly in the case of high Pr liquids always associated with high μ and therefore operating at relatively low Re and corresponding high Blasius exponents $n \simeq 0.3$.
- A constant exponent $\beta = 0.6$ according to Eq. (7.118) derived from McAdams' recommended correlation, Eq. (7.117), is "on average" probably more accurate than the Chilton–Colburn exponent $\beta = \frac{2}{3}$ but may underestimate the impact of high Pr numbers at low Re numbers typically encountered with more viscous high Pr liquids.
- The exponents $\beta = 0.6$ for $1 \leq \mathrm{Pr} \leq 14$ and $\beta = 0.7$ for $\mathrm{Pr} > 14$ recommended by Groeber et al. represent a reasonable compromise between the slightly conservative Chilton–Colburn $\beta = \frac{2}{3}$ and the potentially optimistic $\beta = 0.60$ from McAdams' Eq. (7.117). It should be noted that the reason for the increase in β with increasing Pr is mainly due to the fact

that the more viscous high Pr liquids will have to operate at lower Re numbers (and corresponding higher exponent n) at the relatively low pressure drops allowed in normally efficient fluid-to-fluid heat exchangers.

- For common liquids with constant Pr numbers in the range $1 < \mathrm{Pr} < 10$, β decreases with decreasing values of the Blasius exponent n. This infers that the smooth wall empirical exponent $\beta_{\bar{e}/D_i=0}$ shown in Table 8.5 and Figs. 8.1 and 8.2 is generally higher than the actual exponent $\beta_{\bar{e}/D_i \gg 0}$ for "rougher" pipes having a *natural* relative wall roughness $\bar{e}/D_i \gg 0$ as summarized in the following cautious statements:

$$\beta_{\bar{e}/D_i \gg 0} < \beta_{\bar{e}/D_i \simeq 0} \tag{8.19a}$$

and

$$\alpha_{\bar{e}/D_i \gg 0} > \alpha_{\bar{e}/D_i \simeq 0} \tag{8.19b}$$

for $\mathrm{Pr} > 1$ liquids at

$$(\mathrm{Re})_{\bar{e}/D_i \gg 0} \equiv (\mathrm{Re})_{\bar{e}/D_i \simeq 0} \tag{8.19c}$$

when the Blasius exponent

$$n_{\bar{e}/D_i \gg 0} < n_{\bar{e}/D_i = 0} \tag{8.19d}$$

Common Gases and Vapors: $0.5 < \mathrm{Pr} < 1$. Because $\alpha \propto (1/\mathrm{Pr})^{\beta}$ and $(1/\mathrm{Pr}) > 1$ when $\mathrm{Pr} < 1$, it is clear that α *increases* with *increasing* values of β. In contrast to typical liquids, α is overestimated when one overestimates β in the case of $\mathrm{Pr} < 1$ fluids. In other words, it is *prudent for design purposes to select low values of the empirical exponent* β in the case of common gases.

If we accept that Petukhov's recommended heat transfer correlations for turbulent tube flows inside circular tubes are the most advanced and accurate equations available to date [3.43], then it is clear that all the generally accepted Prandtl exponents in the range $0.5 \le \beta \le \frac{2}{3}$ slightly overestimate the advantages of lower Pr numbers at $0.5 < \mathrm{Pr} < 1$. It should be emphasized, however, that at the typical values $\mathrm{Pr} \approx 1$, the impact of an error in the exponent β is not too significant and would probably remain undetected in most *single-sample* experiments [7.93].

By referring to Figs. 8.1 and 8.2 it can be seen that the exponent β for a given constant $\mathrm{Pr} \lessapprox 1$ does decrease and seems to converge, as it theoretically should, toward $\beta \to 0$ when $n \to 0$ at the fully rough wall limit of turbulent regimes. This confirms what was stated in Guideline 10, namely, the more chaotic the flow regime, the simpler the solution of the coupled heat and fluid flow problem (and what was postulated[4] in Ref. 4.21).

8.1.3 Integrated Modified Reynolds Analogy

The archetype constant wall temperature single-channel heat exchanger can now be solved most generally on the basis of Eq. (7.128), which yields for a specified design approach factor $(\mathrm{AF})_{\mathrm{des}}$ and corresponding prescribed $(\mathrm{NTU})_{\mathrm{des}}$ with $f_{rh} = \frac{1}{4}f$ and $D_h = 4r_h$

$$L = 2(\mathrm{Pr})^{\beta}(\mathrm{NTU})_{\mathrm{des}}\left(\frac{D_h}{f}\right) \tag{8.20}$$

or, in fully dimensionless terms,

$$\frac{L}{D_h} = 2(\mathrm{Pr})^{\beta}(\mathrm{NTU})_{\mathrm{des}}\left(\frac{1}{f}\right) \tag{8.21}$$

Noting that the above equations can also be written

$$f\left(\frac{L}{D_h}\right) = 2(\mathrm{Pr})^{\beta}(\mathrm{NTU})_{\mathrm{des}} \tag{8.22}$$

and recalling that the left-hand side of Eq. (8.22) represents the dimensionless wall friction head loss $(\mathrm{NVH})_f$ per Eq. (3.5), we can expand Eq. (8.22) with the definition of $(\mathrm{NVH})_f$ according to

$$\frac{\Delta H_f}{\overline{V}^2/2} \equiv \frac{\Delta P_f}{\rho(\overline{V}^2/2)} \equiv (\mathrm{NVH})_f \equiv f\left(\frac{L}{D_h}\right) \equiv (\mathrm{Pr})^{\beta}2(\mathrm{NTU})_{\mathrm{des}} \tag{8.23}$$

Substituting $\ln(1/\mathrm{AF})_{\mathrm{des}}$ for $(\mathrm{NTU})_{\mathrm{des}}$ per Eq. (7.47) in the above equation, we finally obtain

$$(\mathrm{NVH})_f = (\mathrm{Pr})^{\beta}2\ln\left(\frac{1}{\mathrm{AF}}\right)_{\mathrm{des}} = (\mathrm{Pr})^{\beta}\ln\left(\frac{1}{\mathrm{AF}}\right)^2_{\mathrm{des}} \tag{8.24}$$

indicating that $(\mathrm{NVH})_f$ is uniquely defined in terms of $(\mathrm{AF})_{\mathrm{des}}$ for a given fluid Prandtl number. Thus, the required dimensionless friction head $(\mathrm{NVH})_f$ is directly proportional to $(\mathrm{Pr})^{\beta}$ at a prescribed $(\mathrm{AF})_{\mathrm{des}}$ and corresponding $(\mathrm{CF})_{\mathrm{des}} = 1 - (\mathrm{AF})_{\mathrm{des}}$; that is,

$$(\mathrm{NVH})_f \propto (\mathrm{Pr})^{\beta} \tag{8.25}$$

For $\mathrm{Pr} = 1$ fluids or the limiting fully rough wall case when $\beta \to 0$ as the Blasius exponent $n \to 0$, Eq. (8.24) reduces to

$$(\mathrm{NVH})_f = 2\ln\left(\frac{1}{\mathrm{AF}}\right) = \ln\left(\frac{1}{\mathrm{AF}}\right)^2 \tag{8.26a}$$

valid for

$$\text{Pr} = 1 \quad \text{or the limiting case} \quad (\text{Pr})^{\beta} \to 0 \tag{8.26b}$$

8.1.3.1 Discussion. It is most remarkable that the hydraulic diameter D_h does *not* appear in Eqs. (8.24)–(8.26). Thus, for any arbitrarily selected average design velocity $\overline{V}$ [i.e., within the empirical validity limits of the modified Reynolds analogy in turbulent regimes, as stated for instance in Eq. (7.117c)], $(\text{NVH})_f$ is uniquely defined in terms of $(\text{AF})_{\text{des}}$ for a given fluid with fixed known physical properties (such as Pr, μ, and ρ) at the prescribed average bulk temperature. Therefore, according to the definition of $(\text{NVH})_f$ shown on the left-hand side of Eq. (8.23), the same total wall friction head ΔH_f must be dissipated regardless of channel geometry D_h. In effect, the total integrated wall friction head ΔH_f is preordained when $\overline{V}$ and $(\text{AF})_{\text{des}}$ are specified and the heat transfer problem reduces to a straightforward fluid flow problem, that is, finding the required length of travel L or aspect ratio L/D_h to dissipate the prescribed $(\text{NVH})_f$ according to Eq. (8.24).

Since at a specified average bulk temperature there is a one-to-one correspondence between $\overline{V}$ and ΔH_f for a given fluid (i.e., at fixed values of Pr, ρ, and μ) and a prescribed $(\text{AF})_{\text{des}}$ according to Eqs. (8.23) and (8.24), it follows logically that only one unique in/out SSSF velocity is possible for a prescribed total wall friction head $(\Delta H_f)_{\text{des}}$ regardless of hydraulic diameter D_h. This conclusion is fully consistent with the results of the variation analyses obtained in Chapter 7 on the premise that the necessary inlet/outlet steady-state steady-flow condition is respected. Once this unique velocity $\overline{V}$ has been determined, the length of travel L, or the dimensionless aspect ratio L/D_h, can readily be computed according to conventional fluid flow correlations. Thus, Eqs. (8.23) and (8.24) fulfill the basic intent of the modified Reynolds analogy [8.1] on a length integrated basis!

The unique in/out SSSF velocity at prescribed $(\Delta H_f)_{\text{des}}$ *and* $(\text{AF})_{\text{des}}$ can readily be determined by computing $(\text{NVH})_f$ from Eq. (8.24) and solving the left-hand side of Eq. (8.23) for $\overline{V}^2$ according to

$$\overline{V}^2 = \frac{2(\Delta H_f)_{\text{des}}}{(\text{NVH})_f} \tag{8.27}$$

Therefore,

$$\overline{V} = \frac{\left[2(\Delta H_f)_{\text{des}}\right]^{1/2}}{(\text{NVH}_f)^{1/2}} \tag{8.28}$$

and, with Eq. (8.24),

$$(\text{NVH}_f)^{1/2} = (\text{Pr})^{\beta/2}\left[2\ln\left(\frac{1}{\text{AF}}\right)\right]^{1/2} = (\text{Pr})^{\beta/2}\left[\ln\left(\frac{1}{\text{AF}}\right)^2\right]^{1/2} \tag{8.29}$$

The numerator on the right-hand side of Eq. (8.28) has the dimension of a velocity [since $(\mathrm{NVH})_f$ is dimensionless] and represents the potential inviscid nozzle velocity obtained at a total head $\Delta H_f = \Delta P_f/\rho = g\,\Delta z_f$ per Eq. (2.7) and Numerical Analysis 2.1. It is convenient to introduce the virtual velocity $\overline{V}_{\mathrm{virt}}$ defined by

$$\overline{V}_{\mathrm{virt}} = (2\,\Delta H_f)^{1/2} \equiv \left(\frac{2\,\Delta P_f}{\rho}\right)^{1/2} = (2g\,\Delta z_f)^{1/2} \tag{8.30}$$

Using Eq. (8.28) and the fact that $\dot{m} = \Omega G \equiv \Omega\rho\overline{V} \propto \overline{V}$, we obtain

$$\frac{\overline{V}}{\overline{V}_{\mathrm{virt}}} = \frac{G}{G_{\mathrm{virt}}} = \frac{\dot{m}}{\dot{m}_{\mathrm{virt}}} = \frac{1}{(\mathrm{NVH}_f)^{1/2}} \tag{8.31}$$

or, after substituting the right-hand side of Eq. (8.29) for $(\mathrm{NVH}_f)^{1/2}$ in Eq. (8.31),

$$\frac{\overline{V}}{\overline{V}_{\mathrm{virt}}} = \frac{G}{G_{\mathrm{virt}}} = \frac{\dot{m}}{\dot{m}_{\mathrm{virt}}} = \frac{1}{(\mathrm{Pr})^{\beta/2}[2\ln(1/\mathrm{AF})]^{1/2}} \tag{8.32}$$

For $\mathrm{Pr} = 1$ or when $(\mathrm{Pr})^{\beta/2} \to 1$ for $\beta/2 \to 0$ at finite values of Pr not significantly different from unity, Eq. (8.32) reduces to the simpler form

$$\frac{\overline{V}}{\overline{V}_{\mathrm{virt}}} = \frac{G}{G_{\mathrm{virt}}} = \frac{\dot{m}}{\dot{m}_{\mathrm{virt}}} = \frac{1}{[2\ln(1/\mathrm{AF})]^{1/2}} \tag{8.33a}$$

valid for

$$\mathrm{Pr} = 1 \quad \text{or} \quad \beta/2 \to 0 \quad \text{with } \mathrm{Pr} \approx 1 \tag{8.33b}$$

The numerical values of the velocity ratio $\overline{V}/\overline{V}_{\mathrm{virt}}$, obtained from Eq. (8.33) at approach factors AF ranging from 0.5 down to 0.001 for extremely effective heat exchangers with a $\mathrm{CF} = 1 - \mathrm{AF} = 0.999 = 99.9\%$, are shown in Table 8.6.

For fluids with $(\mathrm{Pr})^{\beta/2} \neq 1$ the velocity ratios $\overline{V}/\overline{V}_{\mathrm{virt}}$ at the same prescribed $(\mathrm{AF})_{\mathrm{des}}$ and $(\Delta H_f)_{\mathrm{des}}$ as those obtained from Eq. (8.33a) (and Table 8.6) for $(\mathrm{Pr})^{\beta/2} = 1$ must be multiplied by a velocity correction factor (VCF)

TABLE 8.6. Velocity Ratio $\overline{V}/\overline{V}_{\mathrm{virt}}$ in Terms of AF for $(\mathrm{Pr})^{\beta} = 1$ per Eq. (8.33)

AF	0.5	0.4	$e^{-1} = 0.368$	0.3	0.2[a]	0.1[a]	0.01[a]	0.001[a]
$\overline{V}/\overline{V}_{\mathrm{virt}}$	0.841	0.739	$2^{-1/2} = 0.707$	0.644	0.557	0.466	0.333	0.269
CF = 1 − AF	0.5	0.6	0.632	0.7	0.8	0.9	0.99	0.999

[a]Normally effective heat exchangers according to GL 24.

TABLE 8.7. Velocity Correction Factor (VCF) per Eq. (8.34)[a]

(1)	Prandtl number	0.5	0.7	0.9	1	1.5	3	6	120
(2)	VCF at maximum β	1.129	1.064	1.019	1	0.868	0.681	0.534	0.187
(3)	VCF at minimum β	1.017	1.009	1.003	1	0.904	0.760	0.639	0.302
(4)	$\frac{1}{2}$ [line (2) + line (3)]	1.073	1.036	1.011	1	0.886	0.720	0.587	0.245
(5)	Line (2) / line (4)	1.052	1.027	1.008	1	0.980	0.945	0.911	0.765

[a]Smooth wall tubes within usual design range 10,000 < Re < 120,000 and corresponding empirical maximum / minimum Prandtl exponents shown in Table 8.5 (i.e., $\beta_{max} \simeq 0.35$, $\beta_{min} \simeq 0.05$ for Pr < 1 and $\beta_{max} \simeq 0.7$, $\beta_{min} \simeq 0.5$ for Pr > 1).

for $(\mathrm{Pr})^{\beta/2} \neq 1$ obtained by dividing Eq. (8.32) by Eq. (8.33a); that is,

$$\mathrm{VCF} \equiv \frac{\overline{V}_{(\mathrm{Pr})^{\beta/2} \neq 1}}{\overline{V}_{(\mathrm{Pr})^{\beta/2} = 1}} = \frac{1}{(\mathrm{Pr})^{\beta/2}} \tag{8.34}$$

The values of the velocity correction factor computed from Eq. (8.34) at the lower and higher values of the experimental Prandtl exponent β shown in Table 8.5 for normal liquids (Pr > 1) and gases (0.5 < Pr < 1) in stable turbulent regimes at $10{,}000 \leq \mathrm{Re} \leq 120{,}000$ are summarized in Table 8.7. The validity limits of Eq. (8.34) and the VCF shown in Table 8.7 can be stated as follows:

Validity of Eq. (8.34):

$$\text{Prescribed } \overline{T}_b,\ (\mathrm{AF})_{\mathrm{des}}, \text{ and } (\Delta H_f)_{\mathrm{des}} \tag{8.35a}$$

$$\text{Constant fluid properties: } \mu,\ \rho, \text{ and Pr at } \overline{T}_b \text{ and } \overline{P}_b \tag{8.35b}$$

$$\text{Stable turbulent flows for } 10{,}000 < \mathrm{Re} < 120{,}000 \tag{8.35c}$$

$$\text{In ideal smooth tubes and channels: } \bar{e}/D_h \simeq 0 \tag{8.35d}$$

Constant exponent β in the ranges

$$0.49 \leq \beta \leq 0.67 \quad \text{for liquids with } 2 \leq \mathrm{Pr} \leq 120 \tag{8.35e}$$

$$0.05 \leq \beta \leq 0.35 \quad \text{for gases with } 0.5 \leq \mathrm{Pr} \leq 0.8 \tag{8.35f}$$

On the basis of the preliminary conclusions derived in Section 8.1.2.1, it is permissible to "amend" the above relationships for real world rougher pipes and channels [noting that Eqs. (8.35a)–(8.35c) remain valid as stated] as follows:

$$\text{Tubes and constant cross-section channels with a natural wall roughness } \bar{e}/D_h \gg 0 \tag{8.36a}$$

$$\beta < 0.49\text{–}0.67 \quad \text{for liquids with } 2 \leq \mathrm{Pr} \leq 120 \tag{8.36b}$$

$$\beta < 0.05\text{–}0.35 \quad \text{for gases with } 0.5 \leq \mathrm{Pr} \leq 0.8 \tag{8.36c}$$

The actual impact of $\bar{e}/D_h \gg 0$ will be quantified more precisely in later numerical analyses. However, it is clear from line (5) of Table 8.7 that the impact of the above variations in β cannot be too significant at Pr $\ll$ 120.

Once the unique in/out SSSF velocity $\bar{V}$ has been computed as indicated above, it is a simple matter to determine the length of travel L and corresponding total single-channel heat transfer surface $A_i = a_i L$, or the dimensionless aspect ratio L/D_h, for a known $(\mathrm{NVH})_f$ from $(\mathrm{NVH})_f \equiv f(L/D_h) \equiv L/(D_h/f) \equiv (L/D_h)/(1/f)$ which can be interpreted either as

$$(\mathrm{NVH})_f = \frac{L}{L_{\mathrm{VH}}} \tag{8.37}$$

with

$$L_{\mathrm{VH}} = \frac{D_h}{f} \equiv \frac{\frac{1}{4}D_h}{\frac{1}{4}f} \equiv \frac{r_h}{f_{rh}}$$

where L_{VH} is the length of travel for a velocity head equal to unity, when $(\mathrm{NVH})_f = 1$ in Eq. (8.37) at $L = L_{\mathrm{VH}}$ or

$$(\mathrm{NVH})_f \equiv \frac{L/D_h}{1/f} \equiv \frac{L/D_h}{(L/D_h)_{\mathrm{VH}}} \tag{8.38}$$

where $(L/D_h)_{\mathrm{VH}}$ is the relative length of travel for a velocity head equal to unity, that is, when $(\mathrm{NVH})_f = 1$ at $L/D_h \equiv (L/D_h)_{\mathrm{VH}}$ with

$$(L/D_h)_{\mathrm{VH}} \equiv 1/f \tag{8.39}$$

Thus, the inverse of the conventional friction factor represents the dimensionless aspect ratio $(L/D)_{\mathrm{VH}}$ for a velocity head of unity and the necessary length of travel L is uniquely defined by

$$L = (\mathrm{NVH})_f L_{\mathrm{VH}} = (\mathrm{NVH})_f (D_h/f) \tag{8.40}$$

with a corresponding dimensionless aspect ratio

$$L/D_h = (\mathrm{NVH})_f (L/D_h)_{\mathrm{VH}} = (\mathrm{NVH})_f (1/f) \tag{8.41}$$

The key parameter $(L/D_h)_{\mathrm{VH}} = 1/f$ can readily be determined by computing the inverse of the friction factor f read in Moody's chart at the known value of Re for given values of $\bar{V}$, μ, ρ, and D_h or directly from the various analytical expressions for f reviewed in Chapter 3.

For graphical solutions based on Moody's chart it is convenient to transform the standard $\log f - \log \mathrm{Re}$ chart into a $\log(1/f) - \log \mathrm{Re}$ chart as indicated in Fig. 8.3 for stable laminar and turbulent regimes.

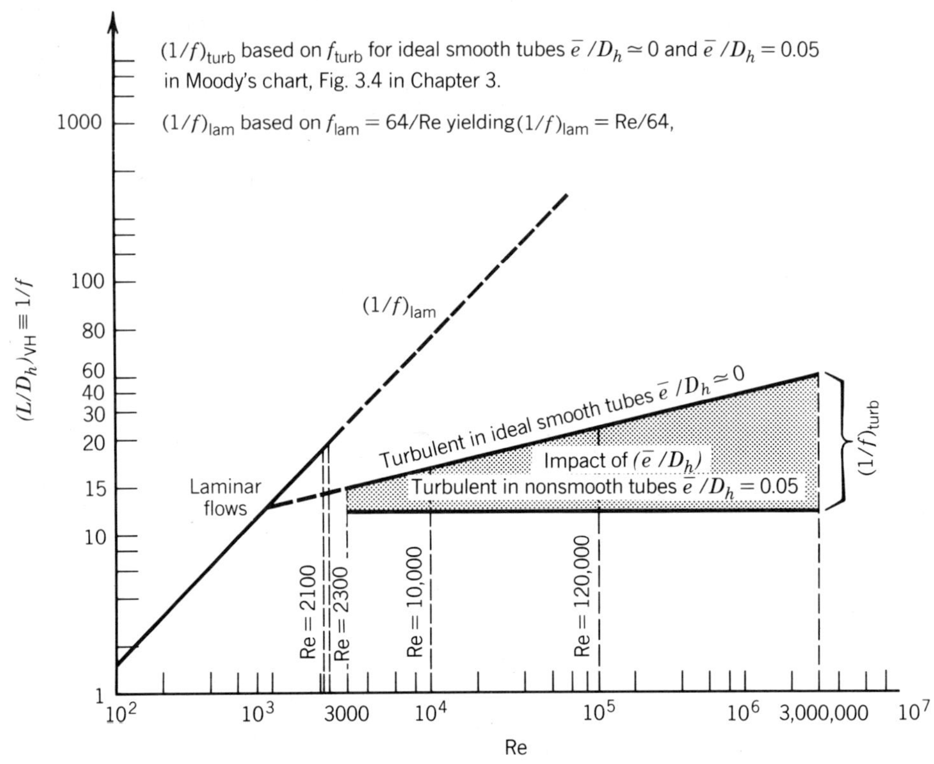

FIGURE 8.3. $(L/D)_{VH} \equiv 1/f$ in terms of Re and $\bar{e}/D$.

For turbulent flows, $(L/D)_{VH}$ can range anywhere between the upper bound defined by the curve $\bar{e}/D_h \simeq 0$ for ideal smooth tubes and the lower bound for $\bar{e}/D_h = 0.05$ shown in Fig. 8.3, depending on actual relative wall roughness. In the context of what was said in Chapter 3 and in Guideline 23 of Chapter 7, it is clearly futile (mainly due to the uncertainties associated with $\bar{e}$) to strive for accuracies substantially better than ± 10–20% in actual heat exchanger design practice.

For numerical analyses it is possible to compute the inverse of the friction factor $1/f = (L/D)_{VH}$ from the generalized Blasius equation (3.44), that is, according to

$$(L/D_h)_{VH} = 1/f = (1/K_n)(\text{Re})^{+n} \tag{8.42}$$

with a suitable pair of constants K_n and n for different ranges of Re numbers, in order to match any natural wall roughness friction factor curve as shown in Chapter 3. However, in normal fluid-to-fluid heat exchangers designed to operate in the relatively narrow range $10{,}000 < \text{Re} < 120{,}000$, it is sufficiently accurate in view of the statistical nature of the wall roughness $\bar{e}$ to use a single pair of constants K_n and n such as the original Blasius constants $K_n = 0.3164$ and $n = 0.25$ per Eq. (3.46) or the McAdams' recommended $K_n = 0.184$ and $n = 0.20$ per Eq. (3.47) for smooth tubes. As shown in Table 3.2, the original Blasius constants are probably slightly more accurate than McAdams' constants at Re < 10,000, whereas McAdams' correlation tends to be more accurate at higher Re > 10,000. However, these differences are insignificant compared to the uncertainties associated with unavoidable wall roughness variations in practical heat exchanger design applications as shown in Chapter 3.

The prime purpose of the following numerical analyses is to determine *adequately accurate* values of the Prandtl exponent β for use in actual forced-convection fluid-to-fluid exchangers, to illustrate more concretely the application of Eqs. (8.20)–(8.36), and to show their practical usefulness in typical heat exchanger design problems by demonstrating, for instance, the significant advantages of small hydraulic tube diameters in typical turbulent flow regimes.

8.1.4 Prandtl Exponent β for Normal Liquids and Gases

To delineate the scope of the following analyses a few key definitions are introduced.

GUIDELINE 25 *Any liquid with a Prandtl number in excess of unity is by definition a* ***normal liquid****. All normal liquids are further divided into two major subgroups arbitrarily defined as follows:* ***low Prandtl*** *normal liquids,* $1 \lesssim \mathrm{Pr} \lesssim 10$; ***high Prandtl*** *normal liquids,* $\mathrm{Pr} \gg 10$.

The above definition of *normal liquids* automatically excludes liquid metals with $\mathrm{Pr} \ll 1$ but includes all other liquids shown in Appendix A.

Water (see Table A.1 in Appendix A), the traditional and most important fluid in thermal power generation cycles, complies substantially with the definition of normal low Pr liquids except in the immediate neighborhood of the freezing (or triple) point at $T_f = T_t = 273.15$ K and the critical thermodynamic point $T_{\mathrm{cr}} = 647.3$ K. The Pr numbers of water are bounded between fairly narrow limits within the domain of existence of the liquid phase at $T_f \simeq T_t < T < T_{\mathrm{cr}}$ corresponding to the following reduced temperatures or pressures: $0.42 < T^+ < 1$, or $2.8 \times 10^{-5} < P^+ < 1$, with a $(\mathrm{Pr})_l = 1$ at $T \simeq 450$ K corresponding to a reduced temperature and pressure of 0.43 and 4.5×10^{-5}, respectively. Similarly, $(\mathrm{Pr})_l \simeq 10$ at $T \simeq 280$ K or $T^+ = 0.70$ and $P^+ = 4.2 \times 10^{-2}$, and it follows that

$$10 \lesssim \mathrm{Pr} \lesssim 1 \tag{8.43a}$$

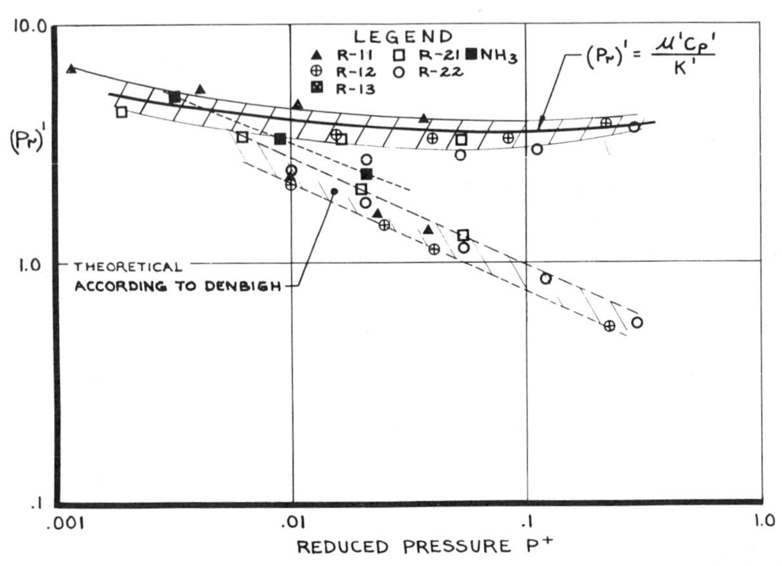

FIGURE 8.4. Liquid Prandtl number in terms of reduced pressure. *Note*: $(\mathrm{Pr})' \equiv (\mathrm{Pr})_l$. [From H. Soumerai, Application of Thermodynamic Similitude, Part II: Heat and Mass Transfer, *ASHRAE J.*, **8**, 38 (July 1966), Fig. 8a. Reproduced with permission.]

at

$$4.5 \times 10^{-5} \lesssim P^+ \lesssim 4.2 \times 10^{-2} \ll 1 \quad \text{for water} \tag{8.43b}$$

It is also clear from Fig. 8.4 that *all* common refrigerants used in reverse power cycle applications also fall well within the definition of normal liquids in the normal refrigeration cycles application range:

$$0.01 \lesssim P^+ \lesssim 0.3 \ll 1 \quad \text{for common refrigerants} \tag{8.43c}$$

By contrast, highly viscous fluids, such as engine oil, clearly fall outside the definition of low Pr liquids with Pr numbers several orders of magnitude higher than Pr = 10 at normal ambient temperatures (even at the highest temperature listed in Table A.2 of Appendix A, Pr = 88 ≫ 10).

In the following numerical analysis we focus first on low-Pr and on high-Pr normal liquids and then on gases.

8.1.4.1 Typical Low* Pr *Normal Liquid: Water. We consider again the archetype constant temperature wall single-channel heat exchanger to find adequately accurate Prandtl exponents β on the basis of Table 8.5, Figs. 8.1–8.3, and the integrated form of the modified Reynolds analogy.

TABLE 8.8. Impact of Pr on $\bar{V}$ and Re for Water

(1) ΔH_f (J / kg)	5				50				Prescribed
(2) $\bar{V}_{virt}$ (m / s)	3.16				10				$(2\Delta H_f)^{1/2}$
(3) AF	0.20 to 0.01				0.20 to 0.01				Prescribed
(4) $\bar{T}_b$ (K)	300 to 400		300 to 400		300 to 400		300 to 400		Prescribed
(5) Pr	5.83	1.34	5.83	1.34	5.83	1.34	5.83	1.34	At $\bar{T}_b$
(6) $(\text{NVH})_f^{1/2}$	3.04	1.96	5.15	3.31	3.04	1.96	5.15	3.31	Eq. (8.29) with $\beta = 0.6$
(7) $\bar{V}$ (m / s)	1.04	1.61	0.61	0.95	3.29	5.10	1.94	3.02	$\bar{V}_{virt} / (\text{NVH})_f^{1/2}$
(8) 10^3 Re at $D = 0.01$ m	12.1	69.6	7.2[a]	41.2	38.4	220.3	22.6	130.5	$\bar{V}D\rho / \mu$
(9) 10^3 Re at $D = 0.04$ m	48.4	278.5	28.6	164.9	153.4	881.2	90.6	521.8	4 × line (8)

[a]Lowest Re = 7200 > 2100 for laminar flows. All other Re > 10,000.

Problem Statement

For representative mean bulk temperatures $300 \le \bar{T}_b \le 400$ K, determine the range of Re in normally effective elementary tubular heat exchangers (i.e., for CF ≥ 80% or AF ≤ 0.2) with circular internal tube diameters ranging from 0.5 to 2 in. (or about 10 to 40 mm) at typical friction head losses $5 \le \Delta H_f \le 50$ J/kg, assuming first ideal smooth tube walls. On the basis of these data, verify that $\beta = 0.6 \pm 0.1$ = constant is adequate in practical design applications for $\bar{e}/D_h \approx 0$ and estimate β conservatively for tubes of any *natural* wall roughness $\bar{e}/D_h \neq 0$.

Solution

The key data summarized in Table 8.8, based on a constant exponent $\beta = 0.6$, confirm that water will typically operate in stable turbulent flow regimes [see Re range on lines (8) and (9)]. It is also clear from Table 8.5 and Figs. 8.1 and 8.2 that the exponent β is substantially independent of the Pr number at a given Re and corresponding n for smooth tubes. However, at the usual friction heads (ΔH_f) available in normal liquid heat exchangers, the Re number will be generally lower at the higher Pr numbers (Pr → 10), the Blasius exponent n correspondingly higher, and β closer to the standard Chilton–Colburn value $\beta \simeq \frac{2}{3}$ than at low Pr numbers (Pr → 1) when Re tends to be higher, n lower, and β correspondingly lower. Obviously, it would be futile to refine recognized correlations for smooth tubes, suggesting exponents β in the range $0.5 < \beta < 0.7$ or $\beta = 0.6 \pm 0.10$ in view of the uncertainties associated with an exact prediction of the friction factor f due to the statistical nature of the relative wall roughness $\bar{e}$. The wide range of variation in $(L/D_h)_{\text{VH}} \equiv 1/f$ due to $\bar{e}/D_h$ shown in Fig. 8.3 demonstrates the overriding importance of the uncertain average relative wall roughness $\bar{e}/D_h$ in real world tubes compared to the relatively minor impact of a ±0.05–0.10 variation in β in the case of normal low Pr liquids.

The family of curves $\beta(n)$ shown in Fig. 8.1 at different values of the parameter Pr are very weak functions of the Prandtl numbers for 1 < Pr < 10 but vary significantly with the Blasius exponent n corresponding to the

Reynolds numbers shown in Table 8.5 for smooth tubes. Although all the empirical $\beta(n)$ curves ultimately converge toward zero at $n \to 0$, as they should for the fully rough wall limits (when the effects of viscosity become insignificant[5]), they are substantially steeper at low Prandtl numbers (Pr $\simeq$ 6) than at the high Prandtl numbers (Pr $\simeq$ 120). This general trend is more apparent in Fig. 8.2, which shows the full variation range of the Blasius exponent for $0 < n < 1$, corresponding to turbulent as well as stable laminar flow regimes with $n = 1$.

To estimate the Prandtl exponent β conservatively at values of the Blasius exponent n approaching the ultimate limit $n \approx 0$, the curve (—•—•—) for $\beta = \beta(n)$ shown in Figs. 8.1 and 8.2 can be used. The main characteristics of this curve can be highlighted as follows:

- The curve tends to *overestimate* β, particularly when $n \to 0$, as indicated in Fig. 8.2 for $\mathrm{Pr} = 6 \pm 4$, and this yields conservative values of α for $\mathrm{Pr} > 1$, since $\alpha \propto 1/(\mathrm{Pr})^{\beta}$.
- It converges as it should (see Sections 7.8–7.10) toward $\beta = 1$ at $n = 1$ for stable laminar flow regimes, as well as toward $\beta = 0$ at $n = 0$ for fully rough wall turbulent flow regimes.

The conservative design values of the exponent β represented by the curve (—•—•—) in Figs. 8.1 and 8.2 can be expressed by the following algorithm:

$$\beta = \left(\frac{1}{2 - n}\right)^{K(\mathrm{Pr})\ln(1/n)} \tag{8.44a}$$

with

$$K(\mathrm{Pr}) = \tfrac{2}{3} = \text{constant} \tag{8.44b}$$

$$\textit{Validity:}\quad 2 < \mathrm{Pr} < 10 \text{ at } 0 < n < 1 \tag{8.44c}$$

It is clear that β does converge toward unity for laminar flows, when $n \to 1$, according to the above equation, and at the other limit, $n \to 0$, corresponding to fully rough turbulent regimes, Eq. (8.44a) yields $\beta = (\frac{1}{2})^{\infty} \to 0$ as it should. By a suitable selection of Prandtl function $K(\mathrm{Pr})$ in Eq. (8.44a), it would be possible, in principle, to match any empirical curve $\beta(n)$.

Comments

1. Since *any* extrapolation involves an element of speculation, we can conclude this analysis with the following extremely cautious general assessment on the basis of the empirical data shown in Figs. 8.1 and 8.2:

$$0 \le \beta \le 0.25 \quad \text{at } n \le 0.05 \text{ for } 1 \le \mathrm{Pr} \le 10 \tag{8.45}$$

By referring to the extrapolated curves $\beta(n)$ shown in Fig. 8.2 at $0 < n < 0.15$, it is clear that the above value $\beta = 0.25$ is probably too high and therefore too conservative for $\mathrm{Pr} > 1$. Although Eq. (8.45) is less explicit than Eq. (8.44), it is certainly less speculative and yet sufficiently accurate to generalize typical two-phase flow heat transfer coefficients on the basis of corresponding states thermodynamic considerations as shown in earlier publications [4.21, 4.22] and highlighted in Section 8.3.

2. It should be emphasized that the terms n and $2 - n$ in Eq. (8.44) have definite physical meanings as shown in Chapter 3. The term n is the Blasius exponent in the generalized Blasius friction factor, Eq. (3.44), and the term $2 - n$ appears in $\Delta H_f \propto \Delta P_f \propto \overline{V}^{2-n} \propto G^{2-n} \propto \dot{m}^{2-n}$. It is thus a simple matter to determine empirically the exponents $2 - n$ and n for a particular channel (of constant cross section) in order to estimate the Prandtl exponent β.

8.1.4.2 *Typical High Pr Normal Liquid: Engine Oil.*

We now consider an extremely viscous fluid—engine oil—with $\mathrm{Pr} \gg 10$, that is by definition a high Pr liquid.

Problem Statement

Repeat the previous analysis with a typical highly viscous fluid such as engine oil (see Table A.2 of Appendix A) instead of water.

Solution

The key data listed in Table 8.9 were obtained exactly in the same manner as in the previous numerical example with water, except that a single value $\mathrm{Pr} = 0.7$ was used to determine the usual range of Re numbers in view of the empirical results shown in Fig. 8.1 and the prior knowledge that highly viscous, high Pr liquids must operate at low Reynolds numbers approaching the turbulent–laminar transition limit.

TABLE 8.9. Impact of Pr on $\overline{V}$ and Re for Engine Oil

(1) ΔH_f (J / kg)	5				50				Prescribed
(2) $\overline{V}_{\mathrm{virt}}$ (m / s)	3.16				10				$(2\Delta H_f)^{1/2}$
(3) AF	0.20 to 0.01				0.20 to 0.01				Prescribed
(4) $\overline{T}_b$ (K)	300 to 400		300 to 400		300 to 400		300 to 400		Prescribed
(5) Pr	6400	152	6400	152	6400	152	6400	152	At $\overline{T}_b$
(6) $(\mathrm{NVH})_f^{1/2}$	38.6	10.4	65.2	17.6	38.6	10.4	65.2	17.6	Eq. (8.29) with $\beta = 0.7$
(7) $\overline{V}$ (m / s)	0.082	0.30	0.048	0.18	0.26	0.96	0.15	0.57	$\overline{V}_{\mathrm{virt}} / (\mathrm{NVH})_f^{1/2}$
(8) Re at $D = 0.01$ m	(1.5)	(28.7)	(0.9)	(16.9)	(4.7)	(907)	(2.7)	(536)	$\overline{V}D\overline{\rho} / \overline{\mu}$
(9) Re at $D = 0.04$ m	(6.0)	(1147)	(3.5)	(678)	(18.8)	3633[a]	(11.2)	2144[a]	4 × line (8)

[a]Note that these are the only Re > 2100 (i.e., marginally above the theoretical laminar turbulent transition Reynolds number). All other Re fall well within the laminar flow regimes, therefore, outside the stated validity limits of Eq. (8.29).

The results summarized in Table 8.9 fully corroborate the key data in Table 7.12, namely, the practical impossibility of having to deal simultaneously with high Prandtl *and* high Reynolds numbers in normal heat exchanger design applications.

Clearly, the standard recommended value of $\beta = 0.7 \simeq \frac{2}{3}$ is quite adequate for high Pr liquid heat exchangers, particularly in view of the overriding impact and the approximate nature of the empirical correction factor $M_{\Delta T}$ which is frequently quite significant in typical chemical process applications with such highly viscous liquids.

8.1.4.3 *Typical Common Gas: Air.* As shown in Table A.3 of Appendix A, the Prandtl numbers of most common gases and saturated vapors fall within the limits $0.7 < \mathrm{Pr} < 1$ or an average $\mathrm{Pr} \simeq 0.85 \pm 0.15$ at normal pressures (i.e., $P^+ \ll 1$). This is further confirmed by the plots of the saturated vapor $(\mathrm{Pr})_g$ numbers in terms of reduced pressure $P^+ = P/P_{\mathrm{cr}}$ shown in Fig. 8.5, which indicates values ranging from 0.8 to 1 or $(\mathrm{Pr})_g = 0.9 \pm 0.1$.

Problem Statement

The problem is the same as stated in Section 8.1.4.1 for water, except that the friction head losses $100 < \Delta H_f < 1000$ are typically over an order of magnitude higher when we consider air at and above normal atmospheric pressure.

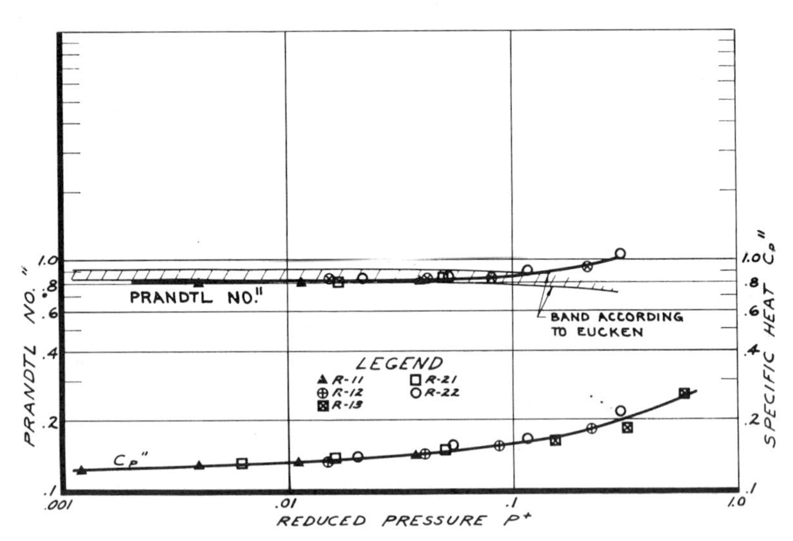

FIGURE 8.5. Vapor Prandtl number at saturation in terms of reduced pressure. *Note:* $(\mathrm{Pr})'' \equiv (\mathrm{Pr})_g$. [From H. Soumerai, Application of Thermodynamic Similitude, Part II: Heat and Mass Transfer, *ASHRAE J.*, **8**, 38 (July 1966), Fig. 7. Reproduced with permission.]

TABLE 8.10. Typical Range of Re for Gases: Air

(1) ΔH_f (J / kg)	100 to 1000								Prescribed
(2) $\bar{V}_{\text{virt}}$ (m / s)	14.14 to 44.72								$(2\,\Delta H_f)^{1/2}$
(3) AF	0.20 to 0.01				0.20 to 0.01				Prescribed
(4) $(\text{NVH})_f^{1/2}$	1.64 to 2.78				1.64 to 2.78				From Eq. (8.29) with Pr = 0.7 = constant and $\beta = 0.5$
(5) $\bar{V}$ (m / s)	8.6		5.1		27.3		16.1		$\bar{V}_{\text{virt}} / (\text{NVH})_f^{1/2}$
(6) $\bar{T}_b$ (K)	300	400	300	400	300	400	300	400	Prescribed
(7) 10^3 Re at $D = 0.01$ m	5.4	3.3	3.2	1.9[a]	17.2	10.3	10.1	6.1	$\bar{V}D\rho / \mu$ at atmospheric pressure[b]
(8) 10^3 Re at $D = 0.04$ m	21.7	13.1	12.8	7.7	68.6	41.3	40.5	24.4	$4 \times$ line (7)

[a]Re = 1900 < 2100 within the laminar flow domain. All other Re > 2100 within the turbulent regimes.

[b]Because μ and Pr are constant for ideal gases, lines (1)–(5) are independent of P. However, since $\rho \propto P$, it follows that $\Delta P_f = \rho\,\Delta H_f \propto \rho \propto P$ as well as Re $= \bar{V}D\rho / \mu \propto \rho \propto P$. Consequently, the Re numbers shown on lines (7) and (8), valid for P = atmospheric pressure, increase in direct proportion to the absolute pressure P > atmospheric pressure!

Solution

Following the same approach as in the case of liquid, we obtain the Reynolds numbers shown on lines (7) and (8) of Table 8.10 at normal atmospheric pressure. As indicated in footnote *b* of Table 8.10, Re is directly proportional to the absolute pressure and it is clear that gases typically operate in stable turbulent regimes at high Re numbers.

In view of the minor impact of the exponent β at Pr $= 0.85 \pm 0.15$ and the fact that by *underestimating* the exponent β one *underestimates* the heat transfer coefficient, the refined analysis carried out for Pr > 1 liquids is not warranted for common gases and vapors. The analysis can be simplified as summarized below without incurring any measurable loss of accuracy in real world heat exchangers with constant cross-section straight channels of any natural wall roughness $\bar{e}/D_h \gg 0$.

For ideal smooth walls at the typical high Re numbers encountered with gases and corresponding *low* Blasius exponent n, Fig. 8.1 and Table 8.5 show that

$$\beta \precsim \tfrac{1}{3} \tag{8.46a}$$

$$\textit{Validity}: \quad \bar{e}/D_h \simeq 0 \quad \text{and} \quad \text{Pr} = 0.85 \pm 0.15 \tag{8.46b}$$

For rougher pipes and channels with a Blasius exponent $0 < n < 0.15$ and a corresponding $2 - n$ term in the range $2 > 2 - n > 1.85$, it is permissible to set

$$\beta \simeq 0 \quad \text{or} \quad (\text{Pr})^{\beta} = 1 \tag{8.47a}$$

$$\textit{Validity}: \quad \text{Pr} = 0.85 \pm 0.15 \text{ for any } \textit{natural} \text{ wall roughness } \bar{e}/D_h \text{ in Moody's chart} \tag{8.47b}$$

It is clear that the impact of an error in β (which can be quite significant for high Pr liquids) becomes hardly measurable at $\mathrm{Pr} \approx 1$ as quantified more precisely in the following sensitivity analysis.

8.1.4.4 Sensitivity Analysis. Because the St number is proportional to $(\mathrm{Pr})^{-\beta}$ according to Eq. (8.12), the error committed by overestimating or underestimating β can be quantified as follows:

$$\frac{\alpha_{\mathrm{act}}}{\alpha_{\mathrm{est}}} \equiv \frac{(\mathrm{St})_{\mathrm{act}}}{(\mathrm{St})_{\mathrm{est}}} = \frac{(\mathrm{Pr})^{-\beta_{\mathrm{act}}}}{(\mathrm{Pr})^{-\beta_{\mathrm{est}}}} = \mathrm{Pr}^{\beta_{\mathrm{est}}-\beta_{\mathrm{act}}} \tag{8.48}$$

where β_{est} = estimated exponent β
β_{act} = actual exponent β

Equation (8.48) shows that, in the case of normal liquids with $\mathrm{Pr} > 1$, overestimating β, that is, when $\beta_{\mathrm{est}} - \beta_{\mathrm{act}} > 0$, yields conservative estimates of α since $\alpha_{\mathrm{act}}/\alpha_{\mathrm{est}} > 1$. As expected, the situation is reversed in the case of gases and vapors with $\mathrm{Pr} < 1$.

With the following definitions and sign conventions

$$\Delta\beta \equiv \beta_{\mathrm{est}} - \beta_{\mathrm{act}} \tag{8.49a}$$

where

$$\Delta\beta > 0 \quad \text{for } \beta_{\mathrm{est}} > \beta_{\mathrm{act}} \tag{8.49b}$$

and

$$\Delta\beta < 0 \quad \text{for } \beta_{\mathrm{est}} < \beta_{\mathrm{act}} \tag{8.49c}$$

Equation (8.48) can be written simply as

$$\frac{\alpha_{\mathrm{act}}}{\alpha_{\mathrm{est}}} = (\mathrm{Pr})^{\Delta\beta} \tag{8.50}$$

Equation (8.50) yields the numerical values shown in Table 8.11 for common gases and saturated vapors, as well as low and high Prandtl normal liquids. To get a better feel of the impact of an error in β on the computed heat transfer coefficient, accuracy bands of ± 5 and 20% have been flagged in Table 8.11.

8.1.4.5 Recommendations for Nonsmooth Tubes. In the context of the above sensitivity analysis, the following adequately accurate recommendations can be made for the Prandtl exponent β by referring to Table 8.5, Figs. 8.1–8.3, and the previous numerical analysis.

TABLE 8.11. Percent Ratio $\alpha_{act}/\alpha_{est}$ per Eq. (8.50) in Terms of $\Delta\beta$ at Different Prandtl Numbers[a]

	Common Gases and Vapors Pr < 1				Normal Liquids Pr > 1						
Pr	1/2	2/3	0.85[b]	1	1/0.85	1.5	2	3	6	12	120[c]
$\Delta\beta + 0.025$	98.3	99.0	99.6	100	100.4	101.0	101.7	102.8	104.6	106.4	112.7
$\Delta\beta - 0.025$	101.7	101.0	100.4	100	99.6	99.0	98.3	97.3	95.6	94.0	88.7
$\Delta\beta + 0.050$	96.6	98.0	99.2	100	100.8	102.0	103.5	105.6	109.3	113.2	127.0
$\Delta\beta - 0.050$	103.5	102.0	100.8	100	99.2	98.0	96.6	94.7	91.4	88.3	78.7
$\Delta\beta + 0.100$	93.3	96.0	98.4	100	101.6	104.1	107.1	111.6	119.6	128.2	161.4
$\Delta\beta - 0.100$	107.1	104.1	101.6	100	98.4	96.0	93.3	89.6	83.6	78.0	62.0
$\Delta\beta + 0.20$	87.1	92.2	96.8	100	103.3	108.4	114.9	124.5	143.1	164.2	260.5
$\Delta\beta - 0.20$	114.9	108.4	103.3	100	96.8	92.2	87.1	80.2	69.9	60.8	38.4

[a]5% accuracy within boundary defined by – – – –; 20% accuracy within boundary defined by ═══.

[b]Average of normal gases and vapors Pr = 0.85 ± 0.15.

[c]High Pr liquids per GL 25.

Normal Gases and Saturated Vapors: Rough Tubes

$$\beta = 0 \quad \text{or} \quad (\text{Pr})^{\beta} = 1 \tag{8.51a}$$

Validity: $0.5 \le \text{Pr} \le 1$ and any relative *natural* wall roughness $\bar{e}/D_h > 0$ in Moody's chart (8.51b)

Estimated Conservative Deviation: 0–5% for $0.5 \le \text{Pr} \le 1$; 0 to $\ll 5\%$ for $0.85 \le \text{Pr} \le 1$ (8.51c)

We recall that the above deviations are well within the percentage deviation of the best friction factor correlations. As the fully rough wall limit is approached, the impact of $\beta \neq 0$ becomes too small for detection in normal laboratory or field tests even at the lowest Pr = 0.5, keeping in mind that the theoretical error committed is on the safe side for gases.

Low Pr Normal Liquids: Rough Tubes

$\beta(n)$ per recommended curve in Fig. 8.3 or Eq. (8.44) with actual Blasius exponent n determined empirically for the constant cross section channel under consideration (8.52a)

Estimated Deviation: Conservative probable/maximum error ±10/20%, becoming vanishingly small as the fully rough wall limit is approached at $n \simeq 0$ (8.52b)

Validity: $1 < \text{Pr} < 7$ (8.52c)

These rather low deviations are plausible for the relatively narrow range of Prandtl number Pr $= 4 \pm 3$ selected. Fortunately, this range of Pr includes most heat transfer media normally used in thermodynamically efficient thermal power generation and heat pumping cycles, as can be seen by referring to Appendix A and Fig. 8.4. This is of course no fortuitous coincidence but the result of a careful selection of heat transfer media for such critical applications as shown (for instance) in the *Handbook of Heat Transfer Media* [8.2] or specialized publications such as ASHRAE's handbooks [8.3].

8.1.5 Multicircuit Forced-Convection Heat Exchangers

So far we have dealt exclusively with single-tube heat exchangers which can be considered the elementary building blocks of large multicircuit single-pass or multipass heat exchangers, as illustrated in Appendix C in the case of water-cooled shell-and-tube condensers. However, multicircuit heat exchangers are by no means limited to the shell-and-tube types of heat exchanger and include all types of design such as multicircuit single-pass or multipass coils and so on (see Appendix B). The necessity of multicircuiting is evident when one recalls that smaller hydraulic diameters are fundamentally more effective (particularly in laminar flow regimes) than larger diameters but have a significantly lower flow capacity for a given velocity $\overline{V}$ and density ρ since the mass flow rate of a single circuit heat exchanger $\dot{m}$ is proportional to $\Omega_i = (\pi/4)D_i^2 \propto D_i^2$ in the case of circular tubes.

In this section the single-circuit analytical methods are extended to the general multicircuit, single-pass and/or multipass arrangements and the economical advantages of a reduction in hydraulic diameter ($D_h = 4r_h = \Omega/a$) are quantified for both laminar and turbulent stable flow regimes. To simplify this analysis we focus on constant wall temperature shell-and-tube heaters or coolers (in the absence of mass transfer) and assume straight circular tubing. However, the proposed methods are basically applicable with some modifications to nonisothermal wall heat exchangers.

For specified design conditions the total flow rate is uniquely defined from Eq. (7.51) according to

$$(\dot{Q}_e)_{\text{tot}} = \dot{m}_{\text{tot}} c_p(\text{LD})(1 - \text{AF}) \tag{8.53}$$

where the subscript tot refers to the total multicircuit heat exchanger capacity and flow rates, whereas the same terms, without subscript, designate the capacity $\dot{Q}_e$ and flow rate $\dot{m}$ for each elementary circuit. The total heat exchanger surface required with different tube hydraulic diameters can be compared as follows.

The total multicircuit heat exchange surface, based on the internal tube surface area, is

$$A_{\text{tot}} = N_c L a_i = N_c L_p N_p a_i \tag{8.54}$$

where a_i = internal surface per unit length of tube
L_p = effective heat exchange tube length per pass
N_c = number of identical circuits in parallel (8.55)
N_p = number of passes
$N_p L_p \equiv L$ = elementary circuit length of travel

From Eq. (8.54) the total heat exchanger surface, A_{tot}, required with a diameter D_h can be compared with $A_{\text{tot, ref}}$ required with an arbitrary reference diameter $(D_h)_{\text{ref}}$ according to

$$\frac{A_{\text{tot}}}{A_{\text{tot, ref}}} = \frac{N_c a_i L}{(N_c)_{\text{ref}}(a_i)_{\text{ref}}(L)_{\text{ref}}} \tag{8.56}$$

where subscript ref refers to all the key data with the reference hydraulic diameter $(D_h)_{\text{ref}}$.

For the same total heat exchanger capacity $(\dot{Q}_e)_{\text{tot, ref}} = (\dot{Q}_e)_{\text{tot}}$ it follows from Eq. (8.53) that $\dot{m}_{\text{tot, ref}} = \dot{m}_{\text{tot}}$ and on the assumption of equal flow distribution per circuit

$$(N_c \Omega_i G)_{\text{ref}} = N_c \Omega_i G \tag{8.57}$$

Therefore,

$$\frac{N_c}{(N_c)_{\text{ref}}} = \frac{(\Omega_i)_{\text{ref}} G_{\text{ref}}}{\Omega_i G} \tag{8.58}$$

Substituting the above expression for $N_c/(N_c)_{\text{ref}}$ in Eq. (8.56) and recalling that $\Omega_i/a_i = r_h = \frac{1}{4}D_h$ yields

$$\frac{A_{\text{tot}}}{A_{\text{tot, ref}}} = \left(\frac{G_{\text{ref}}}{G}\right)\left(\frac{L/D_h}{(L/D_h)_{\text{ref}}}\right) \tag{8.59}$$

By substituting the expression of L/D_h per Eq. (8.21) in Eq. (8.59), we obtain

$$\frac{A_{\text{tot}}}{A_{\text{tot, ref}}} = \left(\frac{G_{\text{ref}}}{G}\right)\left(\frac{2(\text{NTU})(\text{Pr})^{\beta}}{\left[2(\text{NTU})(\text{Pr})^{\beta}\right]_{\text{ref}}}\right)\left(\frac{f_{\text{ref}}}{f}\right) \tag{8.60}$$

which reduces for identical NTU and $(\text{Pr})^{\beta}$ at the same design conditions to

$$\frac{A_{\text{tot}}}{A_{\text{tot, ref}}} = \left(\frac{G_{\text{ref}}}{G}\right)\left(\frac{f_{\text{ref}}}{f}\right) \tag{8.61}$$

With the generalized Blasius expression for f, per Eq. (3.44), the above equation becomes

$$\frac{A_{\text{tot}}}{A_{\text{tot, ref}}} = \left(\frac{G_{\text{ref}}}{G}\right)\frac{(K_n \text{Re}^{-n})_{\text{ref}}}{K_n \text{ Re}^{-n}} \tag{8.62}$$

With $\text{Re} = GD/\mu$ and since $\mu = \mu_{\text{ref}}$, Eq. (8.62) reduces finally to the following most general expression:

$$\frac{A_{\text{tot}}}{A_{\text{tot, ref}}} = \left(\frac{G_{\text{ref}}}{G}\right)^{1-n}\left(\frac{D_h}{(D_h)_{\text{ref}}}\right)^{n}\left(\frac{(K_n)_{\text{ref}}}{K_n}\right) \tag{8.63}$$

which is valid for *stable turbulent as well as laminar flow regimes* with the relevant Blasius constants n and K_n for any constant cross-section channel configuration even when $K_n \neq (K_n)_{\text{ref}}$ in the case of fully developed laminar flows (Table 7.9).

Since our objective is to assess the impact of the size of the characteristic dimension called the hydraulic diameter $D_h = 4r_h = 4\Omega_i/a_i$ rather than the shape of the channel, we compare $A_{\text{tot}}/A_{\text{tot, ref}}$ for fully similar channels when $n = (n)_{\text{ref}}$ and $K_n = (K_n)_{\text{ref}}$. Therefore, the last term on the right-hand side of Eq. (8.63) is equal to unity, and

$$\frac{A_{\text{tot}}}{A_{\text{tot, ref}}} = \left(\frac{G_{\text{ref}}}{G}\right)^{1-n}\left(\frac{D_h}{(D_h)_{\text{ref}}}\right)^{n} \tag{8.64a}$$

Validity: $\bar{e}/D_h \equiv (\bar{e}/D_h)_{\text{ref}} = 0$ or any natural relative wall roughness $\bar{e}/D_h \equiv (\bar{e}/D_h)_{\text{ref}} \gg 0$ for identical[6] cross-section shape (8.64b)
Stable laminar flows with $n = 1$
Stable turbulent flows with $0 < n < 0.30$

For Hagen–Poiseuille laminar flows, with $n = 1$ since $1 - n = 0$ and therefore $[G_{\text{ref}}/G]^0 = 1$, Eq. (8.64) yields

$$\frac{A_{\text{tot}}}{A_{\text{tot, ref}}} = \frac{D_h}{(D_h)_{\text{ref}}} \quad \text{independent of } G \equiv \bar{V}\rho \tag{8.65a}$$

Validity: For *laminar* flows $n = 1$ and identical cross-section shape. (8.65b)

Equation (8.65) clearly shows the significant advantage of small hydraulic diameters in the case of laminar flows, since $A_{\text{tot}} \propto D_h$. It is noteworthy that this limiting case $n = 1$ of the general expression per Eq. (8.64a) is in full agreement with what was shown in Sections 7.8–7.10 for fully developed laminar flows when $\alpha \propto 1/D_h$ and independent of $G = \bar{V}\rho$ or $\bar{V}$ for a fixed density.

For stable turbulent flows the situation is radically different since Eq. (8.64a) yields

$$\frac{A_{\text{tot}}}{A_{\text{tot, ref}}} = \left(\frac{G_{\text{ref}}}{G}\right)^{0.7}\left(\frac{D_h}{(D_h)_{\text{ref}}}\right)^{0.3} \quad \text{to} \quad \left(\frac{G_{\text{ref}}}{G}\right)^{1}\left(\frac{D_h}{(D_h)_{\text{ref}}}\right)^{0} = \frac{G_{\text{ref}}}{G} \tag{8.66a}$$

Validity: Stable *turbulent* flows in channels with Blasius exponent ranging from $n = 0.3$ to 0 (8.66b)

Equation (8.66) shows that at the same mass velocity $G = G_{\text{ref}}$ the hydraulic diameter $D_h \neq (D_h)_{\text{ref}}$ has a relatively minor impact on $A_{\text{tot}}/A_{\text{tot, ref}}$ which vanishes altogether at the fully rough limit when $n \to 0$ and $A_{\text{tot}}/A_{\text{tot, ref}} \to 1$.

For the single-tube heat exchanger with zero inlet and outlet losses in fully developed flows of quasi-incompressible (ρ = constant) fluids, the total head loss from tube inlet to outlet is entirely due to the wall friction ΔH_f; thus,

$$\Delta H_{\text{tot}} = \Delta H_f \tag{8.67}$$

Since for a specified wall friction head ΔH_f, fixed design parameters AF, large temperature difference LD, and fluid properties, the mass velocity G is independent of D_h according to Eq. (8.32), it follows immediately from Eqs. (8.64)–(8.67) that

$$\frac{A_{\text{tot}}}{A_{\text{tot, ref}}} = \left(\frac{D_h}{(D_h)_{\text{ref}}}\right)^{n} \tag{8.68a}$$

Validity: Zero inlet/outlet losses (8.68b)

$$\Delta H_f \equiv \Delta H_{\text{tot}} = \Delta H_{\text{tot, ref}} \equiv (\Delta H_f)_{\text{ref}} \tag{8.68c}$$

since

$$G = G_{\text{ref}} \tag{8.68d}$$

The ratios $A_{\text{tot}}/A_{\text{tot, ref}}$ in terms of $D_h/(D_h)_{\text{ref}}$ computed from Eq. (8.68) are shown in Table 8.12 for Blasius exponents ranging from $n = 1$ for laminar flows to $n = 0.25$ down to $n = 0$ for turbulent flow regimes. The ratios $N_c/(N_c)_{\text{ref}}$, also shown in Table 8.12, are calculated from

$$\frac{N_c}{(N_c)_{\text{ref}}} = \frac{(\Omega_i)_{\text{ref}}}{\Omega_i} = \left(\frac{(D_h)_{\text{ref}}}{D_h}\right)^2 \tag{8.69}$$

obtained from Eq. (8.58) with $G = G_{\text{ref}}$ and $\Omega_i \propto (D_h)^2$.

Table 8.12 clearly shows that the economical advantages of small hydraulic diameters are not very significant in turbulent regimes even in the most favorable case of ideal smooth tubes at typical values of $n = 0.25$ to 0.20. For

TABLE 8.12. Impact of Hydraulic Diameter on Total Heat Exchange Surface $A_{tot} / A_{tot,ref}$ in Percent[a]

$D/(D_h)_{ref}$	1	$\frac{1}{2}$	$\frac{1}{4}$	$\frac{1}{5}$	$\frac{1}{10}$
$N_c/(N_c)_{ref}$	1	4	16	25	100
Turbulent					
$n = 0.25$	100	84	71	67	56
$n = 0.20$	100	87	76	72	63
$n = 0.10$	100	93	87	85	79
$n = 0$	100	100	100	100	100
Laminar					
$n = 1$	100	50.0	25.0	20.0	10.0

[a]Based on Eq. (8.68a), therefore heat exchangers with zero inlet / outlet hydrodynamic losses.

rough wall turbulent flows the advantages of smaller hydraulic diameters diminish further, and in the fully rough limiting case there is no theoretical gain and many practical disadvantages in selecting smaller hydraulic diameters [such as the greater challenge of achieving equal *distribution* in view of $N_c \propto 1/(D_h)^2$ and the increasing difficulty of *mechanically cleaning* smaller tubes, as required in many applications].

By contrast, the advantages of smaller hydraulic diameters are quite significant in laminar flow regimes, keeping in mind, however, that turbulent regimes are generally much more effective than laminar flows as shown in Section 7.11.

The situation changes radically in favor of smaller hydraulic diameters, even in the worst limiting case of fully rough turbulent flows, when all inlet and outlet hydraulic losses are considered in real world multicircuit heat exchangers, as shown in the next sections.

8.1.5.1 Individual Hydraulic Losses in Multicircuit Heat Exchangers. For incompressible fluid flows (which includes elastic gases and vapors at $\Delta P/P \ll 1$), the total head loss from heat exchanger inlet to outlet is the sum of the *wall friction* and *individual* losses; that is,

$$\Delta H_{tot} = \Delta P_{tot}/\rho = \Delta H_f + \Sigma(\Delta H)_{indiv} \tag{8.70}$$

where ΔH_f = tube wall (skin) friction loss = $\Delta P_f/\rho$
$\Sigma(\Delta H)_{indiv}$ = sum of individual friction losses = $(\Sigma K)_{indiv}\overline{V}^2/2$
$(\Sigma K)_{indiv}$ = sum of individual loss coefficients K_{indiv} with reference to $\overline{V}^2/2$ where $\overline{V}$ is the average heat exchange tube velocity

The total head loss per Eq. (8.70) can be expressed most generally with $\Delta H_f = (\text{NVH})_f \overline{V}^2/2$:

$$(\Delta H)_{tot} = (\Delta H)_f[1 + (\Sigma K)_{rel}] \tag{8.71a}$$

where the term $(\Sigma K)_{\text{rel}}$ is defined as follows:

$$(\Sigma K)_{\text{rel}} \equiv \frac{(\Sigma K)_{\text{indiv}}}{(\text{NVH})_f} = \frac{(\Sigma\,\Delta H)_{\text{indiv}}}{(\Delta H)_f}$$

$$= \text{sum of individual losses relative to } (\Delta H)_f \qquad (8.71\text{b})$$

The term $(\Sigma K)_{\text{rel}}$ thus quantifies the sum of all the individual hydraulic losses in terms of ΔH_f. When $(\Sigma K)_{\text{rel}}$ is known, $(\Sigma\,\Delta H)_{\text{indiv}}$ can be determined from

$$(\Sigma\,\Delta H)_{\text{indiv}} = (\Sigma K)_{\text{rel}}(\Delta H)_f = (\Sigma K)_{\text{rel}}(\text{NVH})_f \overline{V}^2/2 \qquad (8.72\text{a})$$

or expressed nondimensionally in terms of the velocity head $\overline{V}^2/2$,

$$\frac{(\Sigma\,\Delta H)_{\text{indiv}}}{\overline{V}^2/2} = (\Sigma K)_{\text{rel}}(\text{NVH})_f \qquad (8.72\text{b})$$

In forced-convection heat exchanger analysis it is essential to recognize that the wall friction component ΔH_f of the total dissipation $(\Delta H)_{\text{tot}}$ plays a constructive role (because it also enhances the rate of heat transfer since $\alpha \propto f$ according to the modified Reynolds analogy) in contrast to the other irreversible flow losses $\Sigma(\Delta H)_{\text{indiv}}$ which are pure losses from the heat exchanger designer's point of view.[7] To make a sharp distinction between these two radically different components of ΔH_{tot}, the following definition is introduced

$$\eta_{\text{wf}} \equiv \frac{\Delta H_f}{(\Delta H)_{\text{tot}}} = \textbf{\textit{wall friction effectiveness}} \qquad (8.73)$$

Since the system engineer (who, in principle, is not concerned about the internal details of individual system components) normally specifies only the total head, $(\Delta H)_{\text{tot}} = (\Delta P)_{\text{tot}}/\rho$, that may be dissipated in the heat exchanger, the useful part $(\Delta H)_f$ of the total head available $(\Delta H)_{\text{tot}}$ can be determined from

$$(\Delta H)_f = \eta_{\text{wf}}(\Delta H)_{\text{tot}} \qquad (8.74)$$

whenever η_{wf} is known. The available head $(\Delta H)_{\text{tot}}$ is clearly most effectively utilized by the heat exchanger designer when $(\Sigma K)_{\text{rel}} \to 0$. This is the case for the archetype single-tube heat exchanger with zero inlet and outlet flow losses discussed so far, when $(\Sigma K)_{\text{rel}} = 0$ and $\eta_{\text{wf}} = 1$, and therefore $(\Delta H)_{\text{tot}} = (\Delta H)_f$. Thus, η_{wf} measures how effectively the heat exchanger designer has been able to utilize the total head dissipation allotted by the thermal system engineer responsible for the total system thermal performance (see also Guideline 8 in Chapter 4).

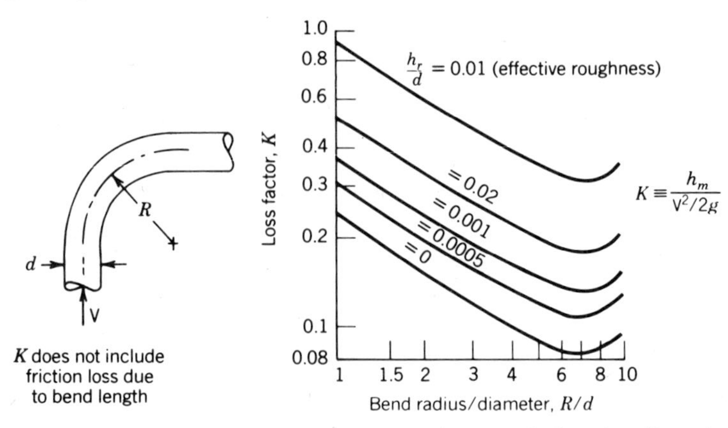

FIGURE 8.6. Loss factor K for a 90° constant radius bend, including the effect of surface roughness. [From F. W. Schmidt et al., *Introduction to Thermal Sciences*, Sec. 1.2, p. 2, Fig. 8.5, Wiley, New York, 1984. Reproduced with permission.]

8.1.5.2 Advantages of Small Hydraulic Diameters in Multicircuit Heat Exchangers. Most of the available empirical data published for the individual flow loss coefficients K_{indiv} are valid for turbulent flows only.[8] Some of the currently accepted values of K_{indiv}, which are particularly relevant in heat exchanger optimization studies, are reproduced here as Figs. 8.6 and 8.7.

Referring to the typical single-pass and multipass shell-and-tube heat exchangers depicted in Fig. 8.8 and Fig. 11.3 of Appendix B, we can compute the ratio $A_{\text{tot}}/A_{\text{tot, ref}}$ with full consideration of the impact of individual losses as outlined below. The individual losses from heat exchanger inlet pipe to inlet chamber, as well as outlet chamber to heat exchanger pipe outlet, are essentially independent of the arbitrary size of the heat exchanger tube diameter, so that it is sufficient to compare two heat exchangers with two different tube

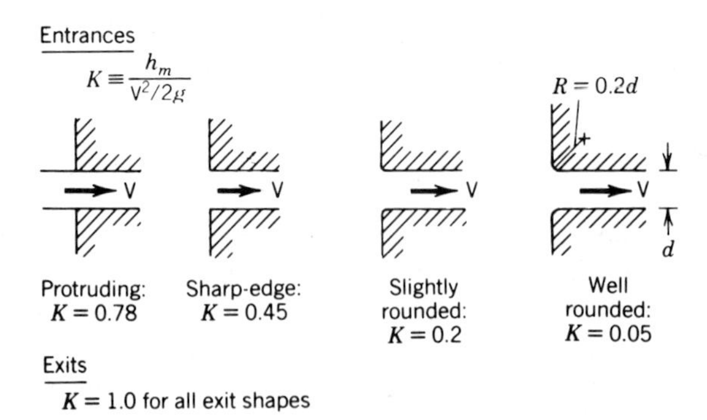

FIGURE 8.7. Loss factor K for pipe entrances and exits. [From F. W. Schmidt et al., *Introduction to Thermal Sciences*, Sec. 1.2, p. 2, Fig. 8.6, Wiley, New York, 1984. Reproduced with permission.]

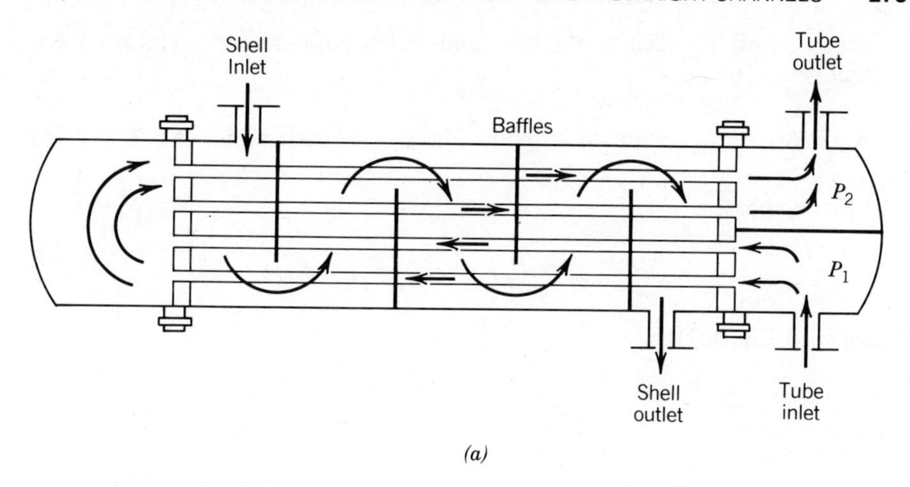

(a)

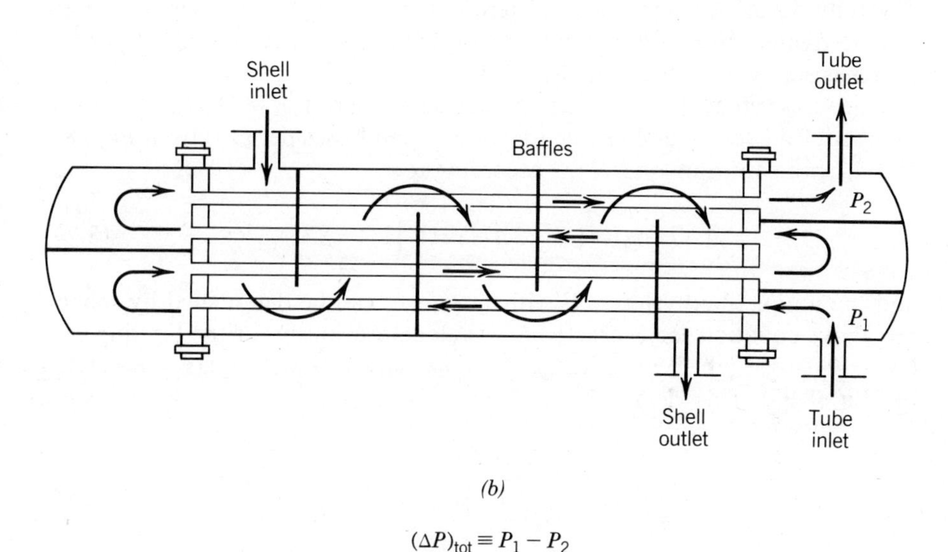

(b)

$$(\Delta P)_{\text{tot}} \equiv P_1 - P_2$$
$$(\Delta H)_{\text{tot}} = (\Delta P)_{\text{tot}}/\rho$$

FIGURE 8.8. Definition of $(\Delta H)_{\text{tot}}$ in shell-and-tube heat exchangers optimization per Numerical Analysis 8.3. (*a*) Shell-and-tube heat exchanger with one shell pass and two tube passes. (*b*) Shell-and-tube heat exchanger with one shell pass and four tube passes.

diameters [D_h and $(D_h)_{\text{ref}}$] on the basis of the total dissipation defined according to

$$(\Delta H)_{\text{tot}} \equiv (\Delta H)_{\text{tot, ref}} = \text{total head loss from inlet to outlet chamber} \quad (8.75)$$

as highlighted in Fig. 8.8.

Let K_{in} and K_{ex} denote the inlet and exit head loss coefficients per pass, respectively. Then

$$K_p = K_{in} + K_{ex} = \text{sum of individual head loss coefficients per pass} \quad (8.76)$$

and for N_p passes

$$(\Sigma K)_{indiv} \equiv N_p K_p = N_p(K_{in} + K_{ex}) \quad (8.77)$$

Thus, with Eq. (8.71b),

$$(\Sigma K)_{rel} \equiv \frac{N_p(K_{in} + K_{ex})}{(\text{NVH})_f} \quad (8.78)$$

The fully developed zero inlet and zero outlet loss single-tube heat exchanger can be treated as a special limiting case of Eq. (8.78) when the number of passes is set at $N_p = 0$ and $(\Sigma K)_{rel} = 0$.

Once the actual $(\Sigma K)_{rel}$ has been calculated from Eq. (8.78) for a selected number of passes N_p and known individual head losses per pass from Figs. 8.6 and 8.7, we can express $(\Delta H)_{tot}$ according to

$$(\Delta H)_{tot} = (\overline{V}^2/2)(\text{NVH})_f[1 + (\Sigma K)_{rel}] \quad (8.79)$$

and the unique fluid velocity $\overline{V}$ with N_p passes can be determined by solving the above equation for $\overline{V}$ exactly as was done previously for the limiting case ($N_p = 0$) single-tube heat exchanger with the known $(\text{NVH})_f = 2(\text{Pr})^\beta(\text{NTU})_{des}$ per Eq. (8.23). This yields

$$\overline{V} = \frac{[2(\Delta H)_{tot}]^{1/2}}{[2(\text{Pr})^\beta(\text{NTU})_{des}]^{1/2}} \frac{1}{[1 + (\Sigma K)_{rel}]^{1/2}} \quad (8.80a)$$

or, with the definition $[2(\Delta H)_{tot}]^{1/2} = \overline{V}_{virt}$

$$\overline{V} = \frac{\overline{V}_{virt}}{[2(\text{Pr})^\beta(\text{NTU})_{des}]^{1/2}} \frac{1}{[1 + (\Sigma K)_{rel}]^{1/2}} \quad (8.80b)$$

For $N_p = 0$ and $(\Sigma K)_{rel} = 0$, the above equation reduces, (since $\text{NTU}_{des} = \ln(1/(\text{AF})_{des})$), to the single-pass velocity per Eq. (8.32). Therefore, the velocity $\overline{V}$ for the multicircuit arrangement per Eq. (8.80b) may be written

$$\overline{V} = \overline{V}_{zero} \frac{1}{[1 + (\Sigma K)_{rel}]^{1/2}} \quad (8.81a)$$

or, in dimensionless terms,

$$\frac{\overline{V}}{\overline{V}_{\text{zero}}} = \frac{1}{[1 + (\Sigma K)_{\text{rel}}]^{1/2}} \tag{8.81b}$$

where

$$\overline{V}_{\text{zero}} \equiv \frac{\overline{V}_{\text{virt}}}{\left[2(\text{Pr})^{\beta}(\text{NTU})_{\text{des}}\right]^{1/2}} \equiv \textit{zero-pass ideal velocity} \tag{8.82a}$$

and

$$\overline{V}_{\text{virt}} \equiv [2(\Delta H)_{\text{tot}}]^{1/2} \tag{8.82b}$$

Recalling that the heat exchange capacity *per circuit* $\dot{Q}_e \propto \overline{V} \propto G$ for the specified design conditions, we can expand Eq. (8.81b) as follows:

$$\frac{\overline{V}}{\overline{V}_{\text{zero}}} \equiv \frac{G}{G_{\text{zero}}} \equiv \frac{\dot{Q}_e}{(\dot{Q}_e)_{\text{zero}}} = \frac{1}{[1 + (\Sigma K)_{\text{rel}}]^{1/2}} \tag{8.82c}$$

where the subscript zero refers to the ideal zero-pass limiting case when $(\Sigma K)_{\text{rel}} = 0$.

With Eq. (8.81c) it is now possible to quantify the reduction in velocity and corresponding capacity per circuit at a prescribed $(\Delta H)_{\text{tot}}$ in terms of $(\Sigma K)_{\text{rel}}$ as shown in Table 8.13. It is clear from Table 8.13 that the individual hydraulic losses can have a massive impact on the size of the heat exchanger at $(\Sigma K)_{\text{rel}} \gg 1$, that is, when $\eta_{\text{wf}} \to 0$.

Equation (8.81c) can also be used to assess the effect of the hydraulic diameter on A_{tot} according to Eq. (8.64a) in the most general case when the relative individual head loss terms $(\Sigma K)_{\text{rel}}$ are different, that is, $(\Sigma K)_{\text{rel}} \neq (\Sigma K)_{\text{rel, ref}}$. Since under specified design conditions $\overline{V}_{\text{zero}}$ and G_{zero} are always identical for D_h and $(D_h)_{\text{ref}}$, the ratio G_{ref}/G obtained from Eq. (8.81c) is:

$$\frac{G_{\text{ref}}}{G} \equiv \left(\frac{1 + (\Sigma K)_{\text{rel}}}{1 + (\Sigma K)_{\text{rel, ref}}}\right)^{1/2} \tag{8.83}$$

TABLE 8.13. Impact of $(\Sigma K)_{\text{rel}}$ at Prescribed $(\Delta H)_{\text{tot}}$ per Eq. (8.81c)

$(\Sigma K)_{\text{rel}}$	0	0.10	0.50	1.00	5	10	100	∞
$1 + (\Sigma K)_{\text{rel}}$	1	1.1	1.5	2.0	6	11	101	∞
$\frac{\dot{Q}_e}{(\dot{Q}_e)_{\text{zero}}} = \frac{1}{[1 + (\Sigma K)_{\text{rel}}]^{1/2}}$	1	0.95	0.82	0.71	0.41	0.30	0.10	0
$\eta_{\text{wf}} \equiv \frac{(\Delta H)_f}{(\Delta H)_{\text{tot}}} \equiv \frac{1}{1 + (\Sigma K)_{\text{rel}}}$	1	0.91	0.67	0.5	0.17	0.091	0.01	0

Introducing the above expression for G_{ref}/G in Eq. (8.64a) yields the most general solution:

$$\frac{A_{\text{tot}}}{A_{\text{tot, rel}}} = \left(\frac{D_h}{(D_h)_{\text{ref}}}\right)^n \left(\frac{1 + (\Sigma K)_{\text{rel}}}{1 + (\Sigma K)_{\text{rel, ref}}}\right)^{(1-n)/2} \tag{8.84a}$$

Validity: Multicircuit heat exchangers under same assumptions as Eqs. (8.64a)–(8.64b) when $(\Sigma K)_{\text{rel}} \neq (\Sigma K)_{\text{rel, ref}}$ (8.84b)

As expected, Eq. (8.84a) reduces to Eq. (8.68a) for $(\Sigma K)_{\text{rel}} \equiv (\Sigma K)_{\text{rel, ref}}$ when the same number of passes can be used with both hydraulic diameters. However, significantly lower numbers of passes are generally possible with the smaller hydraulic diameters whenever physical limitations impose some upper bounds on the overall shell-and-tube heat exchanger length (or the width of the heat exchanger coil in the case of multicircuit coils) and this can have a major impact on $A_{\text{tot}}/A_{\text{tot, ref}}$ as shown in the next numerical analysis.

NUMERICAL ANALYSIS 8.3

Impact of hydraulic diameter in shell-and-tube gas heaters.

General background

It is desired to heat 2200 kg/h = 0.611 kg/s of compressed air (typical gas) from $T_{1,C} = 295$ K over a rather narrow temperature range $T_{2,C} - T_{1,C} = 8$ K up to an exit temperature $T_{2,C} = 303$ K with low pressure steam maintaining the inner tube wall temperature at 305 K, that is, at a LD $= T_w - T_{1,C} = 10$ K. It is agreed to use standard circular copper tubing in nominal outside diameters 0.5–2 in. with inside tube diameters ranging from $D = 0.43$ in. $= 1.092 \times 10^{-2}$ m to $D_{\text{ref}} = 1.87$ in. $= 4.75 \times 10^{-2}$ m depending on the required tube wall thickness. This range of internal diameter corresponds to an extreme ratio $D/D_{\text{ref}} = 0.43/1.87 = 0.23$. For the large heat exchange capacity under consideration, it is necessary to circuit several tubes in parallel even in the case of the larger reference tube diameter D_{ref}. This logically leads to the shell-and-tube type of heat exchanger design with the low-pressure condensing steam on the shell side and the compressed gas (air in this case) flowing inside the tubes.

The main objective of this case study is to assess the impact of the size of the hydraulic diameter ($D_h = D$ for circular tubes) on the total internal heat transfer surface, and the resulting overall configuration of the shell-and-tube heat exchanger, that is, the required number of tubes circuited in parallel (N_c), the number of passes (N_p), the straight tube length per pass (L_p), and the total internal surface required ($N_c N_p L_p \pi D$). Two design cases are considered in the following problem statements.

Problem Statement

Design Case (a). The total head available from tube sheet inlet to outlet chamber, $(\Delta H)_{\text{tot}} = 200$ J/kg, must be identical in all cases. The specified inlet chamber air pressure is $P_1 = 2 \times$ (normal atmospheric pressure) $= 2 \times 1.013 \times 10^5 \simeq 2 \times 10^5$ N/m^2. There are no other geometric constraints in this design case.

Design Case (b). Same as case (a) plus one important additional geometric constraint: the tube length per pass, L_p, may *not* exceed 1.5 m.

Solution

Since there are no restrictions on overall heat exchanger lengths for design case (a), it is evident that the single-pass arrangement will be most economical, since the available $(\Delta H)_{\text{tot}}$ is more effectively used in a single-pass than in a multipass arrangement (i.e., η_{wf} is closer to unity). The key data are summarized in Table 8.14, columns (a)–(c), for the smallest and largest internal diameters under consideration. The calculations are based on the following general considerations and the use of relevant properties for air shown in Appendix A.

TABLE 8.14. Advantages of Small Hydraulic Diameters

Key parameter	Units	(a)	(b)	(c) Column (a) / Column (b)	(d)	(e) Column (a) / Column (d)	Remarks on Key Data in Columns (a), (b), (d)
(1) $10^2 D$	m	**1.092**	**4.75**	**0.23**	**4.75**	**0.23**	Independent parameter
(2) N_p		**1**	**1**	**1**	**5**	**0.20**	Number of passes
(3) $1 + (\Sigma K)_{\text{rel}}$		2.5	2.5	1	8.5	0.294	$1 + N_p(1.5)$
(4) $\overline{V}/\overline{V}_{\text{virt}}$		0.369	0.369	1	0.20	0.542	$(\text{NVH})_f^{-1/2}[1 + (\Sigma K)_{\text{rel}}]^{-1/2}$
(5) $\overline{V}$	m / s	7.39	7.39	1	4.0	0.542	$20(\overline{V}/\overline{V}_{\text{virt}})^d$
(6) 10^3 Re		10.15	44.13	4.35	23.92	2.36	$\overline{V}D\bar{\rho}/\bar{\mu}$
(7) $1/f$		34.39	46.15	0.745	40.8	0.843	$(1/0.184)\text{Re}^{0.2}$
(8) L/D		100.8	135.2	0.745	119.5	0.843	$(1/f)(\text{NVH})_f$
(9) L	m	1.101	6.423	0.171	5.676	0.194	$(L/D) \times (D)$
(10) L_p	m	**1.101**[a]	**6.423**[b]	**0.171**	**1.135**[c]	**1.03**	$(L)/(N_p)$
(11) $10^3 \dot{m}$	kg / s	1.61	30.4	0.053	16.45	0.098	$(\overline{V}\rho(\pi/4)D^2)$
(12) N_c		380.0	20.1 (~ 20)	18.90	37.11 (~ 37)	10.2	$0.611/\dot{m}^e$
(13) $N_c N_p$		380.0	20.1 (~ 20)	18.91	185.6 (~ 186)	2.05	Total number tubes
(14) A_{tot}	m^2	**14.35**	**19.27**	**0.745**	**31.44**	**0.456**	$N_c N_p L_p \pi D$
(15) $\overline{\Delta\theta}$	s	0.15	0.87	0.171	1.42	0.105	$N_p L_p/\overline{V}$

[a] $L_p < 1.5$ m, within allowable limit of design case (b).
[b] $L_p > 1.5$ m, beyond allowable limit of design case (b).
[c] $L_p < 1.5$ m, within allowable limit of design case (b).
[d] $\overline{V}_{\text{virt}} = 20$ m / s.
[e] $\dot{m}_{\text{tot}} = 0.611$ kg / s.

1. At an absolute pressure $P_1 = 2 \times 10^5$ N/m^2 and $(\Delta H)_{\text{tot}} = 200$, the relative pressure drop $(\Delta P)_{\text{tot}}/\overline{P}_1 = \rho(\Delta H)_{\text{tot}}/P_1 \simeq 0.002 \ll 1$ is negligibly small. Thus, the average density $\bar{\rho}$ may be computed at $\overline{T}_b$ at an average pressure $\overline{P} = P_1 - \frac{1}{2}(\Delta P)_{\text{tot}} \simeq P_1 \simeq P_2$ as if the gas was fully incompressible without any significant error.

2. The key numerical data that are independent of the hydraulic diameters selected are conveniently computed first. This includes lines (3)–(5) of Table 8.14 (note that $\overline{V}_{\text{virt}} = [2(\Delta H)_{\text{tot}}]^{1/2} = (400)^{1/2} = 20$ m/s). The approach factor $\text{AF} = (T_w - T_{2,C})/(T_w - T_{1,C}) = 2/10 = 0.2$; therefore, $CF \equiv \varepsilon = 1 - \text{AF} = 1 - 0.2 = 0.8$. We are thus dealing with a *normally effective* heat exchanger according to Guideline 24.

3. It is, of course, necessary first to verify that the Reynolds numbers fall within the normal validity limits $10{,}000 \leq \text{Re} \leq 120{,}000$ of all the equations used in Table 8.14. At an approach factor AF = 0.2, the mean logarithmic temperature difference $\text{LMTD} \simeq \frac{1}{2}\text{LD}$ according to Eq. (7.54a) as shown in Table 7.2. Thus, for LD = 10 K, $\text{LMTD} = (\frac{1}{2})10 = 5$ K and $\overline{T}_b = T_w - \text{LMTD} = 305 - 5 = 300$ K. At $\overline{T}_b = 300$ K and $\overline{P} = 2 \times$ atmospheric pressure, the density $\bar{\rho} = 2 \times \rho_{\text{atm}} = 2 \times (1.1614) = 2.323$, where ρ_{atm} is the value shown in Table A.3 at $\overline{T}_b = 300$ K (note that the same value could be computed from the ideal gas laws). At the average bulk temperature $\overline{T}_b = 300$ K the other two relevant physical properties (which are substantially independent of absolute pressure, are also obtained from Table A.3, namely, $\bar{\mu} = 184.6 \times 10^{-7}$ N · s/m^2 and $\overline{\text{Pr}} = 0.707$. The Reynolds numbers can thus be computed from $\text{Re} = \overline{V}D\bar{\rho}/\bar{\mu}$. The numerical values shown on line (6) indicate that the validity limits $10{,}000 \leq \text{Re} \leq 120{,}000$ are satisfied.

4. To compute the required $(\text{NVH})_f = 2(\text{Pr})^{\beta}\ln(1/\text{AF})$, a safe constant value $\beta = 0.25$ was used (see Fig. 8.1) on lines (4) and (8) to calculate $\overline{V}/\overline{V}_{\text{virt}}$ and the required aspect ratio L/D.

5. The calculations are based on equal flow per circuit, that is, $\dot{m} = \dot{m}_{\text{tot}}/N_c$ and $N_c = (\dot{Q}_e)_{\text{tot}}/\dot{Q}_e = \dot{m}_{\text{tot}}/\dot{m}$. This tacitly assumes a perfect flow distribution system (see Guideline 8).

6. McAdams' recommended friction factor correlation for smooth tubes $f = 0.184(\text{Re})^{-0.2}$, Eq. (3.40a), was used to compute $1/f$ as shown on line (7). The basic advantages of small hydraulic diameters are thus theoretically slightly understated when compared with the numerical results that would be obtained on the basis of the original Blasius equation with an exponent $n = 0.25 > 0.2$. However, the differences are insignificant in the light of what was said previously, particularly in view of the uncertainties associated with unavoidable variations in wall roughness, wall thickness tolerances, and so on in real world heat exchangers (see also Guideline 23 in Chapter 7).

7. The individual friction dissipation per pass $K_p = K_{in} + K_{ex} = 1.5$ used on line (3) to compute $1 + (\Sigma K)_{rel} = 1 + N_p(1.5)$ is obtained from the empirical data shown in Fig. 8.7, indicating the following ranges of values:

$$K_{ex} = 1 \tag{8.85a}$$

$$K_{in} = 0.05\text{–}0.78 \tag{8.85b}$$

By selecting an average representative value of $K_{in} = 0.5$, we obtain

$$K_p = K_{in} + K_{ex} = 1.5 \tag{8.85c}$$

8. The smaller hydraulic diameter satisfies the maximum length limitation requirement $L = L_p = 1.101 < 1.5$ m for design case (b) in a single-pass arrangement as shown on line (10), column (a), of Table 8.14. By contrast, the length of travel $L = L_p = 6.42$ shown on line (10), column (b), for the larger hydraulic diameter is roughly five times too high and requires a multipass arrangement with a minimum $N_p = 5$ to meet the design specification for design case (b), as shown in column (d) of Table 8.14.

9. The results obtained with the two extreme tube diameters selected for this study are compared line by line in columns (c) and (e) of Table 8.14 mainly for didactic purposes. The prime objective of this numerical analysis could be achieved much more simply and succinctly on the basis of Eq. (8.84a), as shown below.

For the design case (a), Eq. (8.84a) yields, with $N_p = 1$ in both cases,

$$\frac{A_{tot}}{A_{tot,\,ref}} = \left(\frac{1.092}{4.75}\right)^{0.2} = 0.745$$

which is in agreement with the numerical values shown on line (14), column (c), of Table 8.14.

For design case (b), with $N_p = 5$ and $1 + (\Sigma K)_{rel,\,ref} = 8.5$ for the larger reference tube diameter against $N_p = 1$ and $1 + (\Sigma K)_{rel} = 2.5$ for the smallest diameter, we obtain directly from Eq. (8.84a)

$$\frac{A_{tot}}{A_{tot,\,ref}} = (0.745)\left(\frac{2.5}{8.5}\right)^{0.4} = 0.456$$

the same numerical value (within hand calculator accuracy) as shown on line (14), column (d).

Comments

1. *Design Case (a).* When the longer single-pass shell-and-tube heat exchanger length is acceptable with the largest tube diameter, it is clear from line (14), column (c), of Table 8.14 that the theoretical reduction in A_{tot} is not very striking for turbulent flows even in the more favorable case of smooth clean tubes. As a matter of fact, when all costs and other practical factors are considered (e.g., lower relative labor costs with fewer tubes, better distribution, and impact of fouling factor), the largest or an intermediate size diameter might prove more economical.

The above analysis is based on the assumption of ideal smooth clean tubes. For fully rough circular tubes of equal relative wall roughness $[\bar{e}/D = (\bar{e}/D)_{ref}]$, the same total surface would be required regardless of channel hydraulic diameter since $f = f_{ref}$ = constant and the term $(L/V)_{VH} = 1/f$ shown on line (7) would be identical for all diameters. This is confirmed by Eq. (8.84a) which shows that $A_{tot}/A_{tot,\,rel} \to 1$ for $n \to 0$ with $(\Sigma K)_{ref} = (\Sigma K)_{rel,\,ref}$. In this case the larger diameter with fewer tubes (N_c) in parallel would generally prove more economical, particularly when maldistribution problems are considered.

2. *Design Case (b).* When space limitations prohibit the larger diameter solution, the advantages of smaller hydraulic diameters can become quite significant as indicated by the ratio $A_{tot}/A_{tot,\,ref} = 0.456$ shown on line (14), column (e). It is evident from Eq. (8.81c) and the key data shown in Table 8.14 that the superiority of smaller hydraulic diameters is due in large part to the significant reduction of individual friction losses or a more effective utilization of the head available $(\Delta H)_{tot}$; that is, $(\eta_{wf})_{D_h} > (\eta_{wf})_{D_h,\,ref}$ when $D_{h,\,ref} \gg D_h$.

3. *Major Impact of Individual Friction Losses.* The major impact of individual friction losses is systematically ignored in many analytical solutions proposed in the technical literature which focus exclusively on local friction factors and heat transfer coefficient comparisons. This may lead to totally erroneous conclusions in real world compact tubular heat exchangers [8.4–8.6 and references cited therein] when all practical factors are considered, including the significant impact of individual friction losses in multipass, multicircuit heat exchangers of any type (shell-and-tube, coils, single-phase fluids, as well as heat exchangers with phase changes).

4. *Residence Time.* As a matter of general interest, the "in tube" residence time $N_p L_p/\bar{V} = \overline{\Delta\theta}$ is shown on line (15) of Table 8.14 and compared under columns (c) and (d).

5. *Balanced Heat Exchangers.* The key data summarized in Table 8.14 are *exactly* applicable only for the archetype case under consideration when $T_{1,\,H} = T_{2,\,H} = T_w$ = constant (see Section 7.6). For well-balanced heat exchangers, that is, when $\alpha_o a_o \approx \alpha_i a_i$, the overall thermal resistance would be

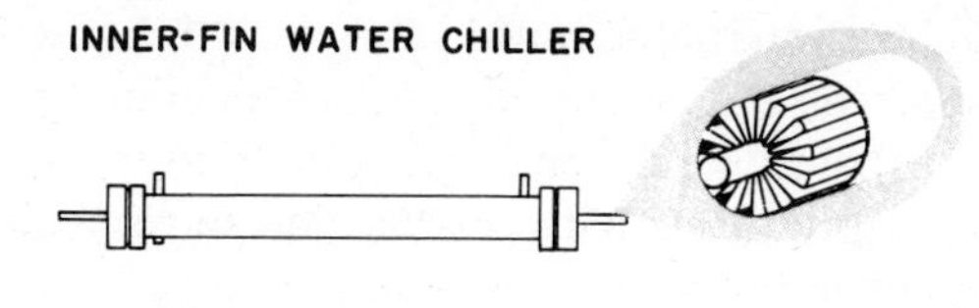

BARE TUBE WATER CHILLER

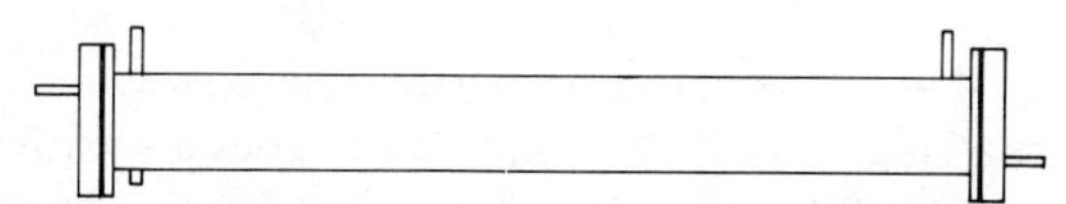

FIGURE 8.9. Relative size comparison between inner-fin and bare tube air conditioning water chillers. [From C. Boling et al., Heat Transfer Evaporating Freon with Inner-Fin Tubing, *Refrig. Eng.*, (Feb. 1954). Reproduced with permission from ASHRAE.]

doubled since $1/U_i a_i \approx 2/\alpha_i a_i$. Thus, the required $(\text{NVH})_f$ and resulting length of travel (or residence time) would be twice as high as the values obtained in this numerical analysis. Consequently, well-balanced heat exchangers would require more passes under otherwise prescribed conditions than the values N_p shown on line (2) of Table 8.14 and this would further enhance the advantages of smaller channel hydraulic diameters. In addition, small hydraulic diameters also yield higher external heat transfer coefficients (α_o) on the shell-side (outside) of the tubes as shown in Refs. 1.2, 1.3, and 8.3. It is thus clear that the above analysis constitutes a *worst* case analysis, underestimating the inherent advantages of small hydraulic diameters for turbulent flow regimes inside channels of any natural wall roughness.

6. *Forced-Convection Heat Exchangers With Phase Changes*. The same basic analytical approach for forced-convection evaporators (with heat addition) or condensers (with heat removal) yields a similar trend in favor of small hydraulic diameters. In fact, the additional penalty due to the saturation temperature drop associated with ΔP_{tp} in two-phase flow (see Chapter 4 and Guideline 9) further enhances the inherent advantages of small hydraulic diameters in compact forced-convection evaporators and/or condensers. This is the main reason for the spectacular size and cost reductions achieved by Boling et al. [8.4] in the 1950s, as highlighted in Fig. 8.9, with the type of proprietary internally finned tubes shown in Fig. 4.6. These internally finned tubes typically have hydraulic diameters ranging down to values as low as $1\text{–}2 \times 10^{-3}$ m, that is, one order of magnitude smaller than standard bare circular tubes. The impressive size reduction shown in Fig. 8.9, which compares a *single-pass* compact water inner fin chiller with a typical competitive (in the 1950s) *multipass* bare tube chiller design, is due in great part to the much *more effective utilization of the total available friction head* in the single-pass arrangement made possible with the significantly smaller hydraulic

diameters of these internally finned tubes. In this context it should be stressed again that some of the enhanced tubes evaluation criteria published in the literature [8.5], based solely on local wall friction factors and heat transfer coefficients at a given Reynolds number, fail to recognize the significant gains attainable in single-phase *and* two-phase compact heat exchangers due to a more effective utilization of the total available friction head and thus may grossly underestimate the inherent advantages of small hydraulic diameter enhanced surfaces.

7. *Fluid Cooling Mode*. The above numerical analysis was arbitrarily based on $T_w > T_{1,C}$ when a single-phase "coolant" undergoes a positive temperature change as it receives heat from an external constant hotter temperature source $T_H = T_w =$ constant (i.e., fluid heating mode when $\dot{Q}_e > 0$). However, the same methodology is applicable in the fluid cooling mode when $T_C = T_w =$ constant $< T_{1,H}$ and the hotter fluid, acting as the heat source, discards heat ($\dot{Q}_e < 0$) to a lower-temperature sink, as highlighted in the following numerical analysis.

NUMERICAL ANALYSIS 8.4

Symmetry between cooling and heating modes for single-phase fluids.

Background

When the nonisothermal correction factor $M_{\Delta T}$ defined by Eq. (8.3) does not differ measurably from unity, it is clear, since $\alpha_{\text{cooling mode}} = \alpha_{\text{heating mode}}$, that the total heat transfer rate $|\dot{Q}_e|$ is independent of the direction of heat flow. The main purpose of this numerical analysis is to illustrate more concretely this quasi-mirror-image behavior in the simple case of single-phase fluids, keeping in mind that this mirror-image concept is particularly useful in more complex forced-convection two-phase heat exchanger applications with phase changes.

The constant wall temperature sink condition, $T_w < T_{1,H}$, can be realized in practice when a suitable refrigerant (such as ammonia, with $\alpha_{\text{evaporation}}$ of the same order of magnitude as boiling water; see Table 7.1) evaporates at a constant temperature outside the tubes. Alternatively, a constant wall temperature $T_w < T_{1,H}$ can also be achieved on the "hot" side of pure single-phase parallel flow heat exchangers as in the case of the mirror-image heat exchanger system [8.6] described in Chapter 9.

Problem Statement

Recheck the calculations carried out in the previous fluid heating mode numerical analysis $(T_w)_{\text{FHM}} > T_{1,C}$ under the same general conditions for the fluid cooling mode case $(T_w)_{\text{FCM}} < T_{1,H}$, except that the key temperatures $T_{1,C}$, $(T_w)_{\text{FHM}}$, $T_{1,H}$, and $(T_w)_{\text{FCM}}$ are "switched" in such a manner that T_{1C} now becomes $(T_w)_{\text{FCM}}$ and $(T_w)_{\text{FHM}} = T_{1,H}$, and compare the results.

Solution

Since $\overline{T}_b = T_w + \text{LMTD} = T_w + \frac{1}{2}\text{LD}$ it is evident that the average bulk (or thermodynamic) temperature $(\overline{T}_b)_{\text{cooling mode}} = (\overline{T}_b)_{\text{heating mode}}$ under the stated design conditions. Therefore, the heat transfer coefficient $(\overline{\alpha}_i)_{\text{cooling mode}} = \overline{\alpha}_{\text{heating mode}}$ for $M_{\Delta T} = 1$. As shown in Section 8.1.1 and highlighted in Table 8.3, the condition $M_{\Delta T} = 1$ is fully satisfied (well within the accuracy of normal laboratory test codes) for quasi-ideal gases even at $T_w - \overline{T}_b$ values far in excess of the present design specification $\text{LD} = T_{1,H} - T_w = 10$ K. Furthermore, the coupled fluid flow phenomena are substantially independent of the heat flow direction, particularly in the case of stable turbulent flow regimes and ideal gases as shown in Section 3.11. Consequently, all the calculation steps and results summarized in Table 8.14 remain unchanged in absolute values for the *fluid cooling mode*. It is thus only necessary to acknowledge that the sign of the fluid's temperature change, as well as $\dot{Q}_e$, is reversed. In other words, $T_{2,C} - T_{1,C} = T_{1,H} - T_{2,H}$ and $\dot{m}_C c_p(T_{2,C} - T_{1,C}) = -\dot{m}_H c_p(T_{2,H} - T_{1,H})$, whereas the head losses $(\Delta H)_{\text{tot}} \equiv (\Delta P)_{\text{tot}}/\rho$ are identical at the same $\dot{V}$ in absolute value and sign, since $(P_1 - P_2)_{\text{FCM}} \equiv (P_1 - P_2)_{\text{FHM}} > 0$.

Comments

1. It is clear in the context of what was said in Section 8.1.1 regarding the nonisothermal correction factor $M_{\Delta T}$ (see Table 8.3) that all the conclusions derived above for common gases and vapors remain valid for liquids at sufficiently low $|\overline{T}_w - \overline{T}_b|/\overline{T}_b$, since at the limit $|\overline{T}_w - \overline{T}_b|/\overline{T}_b \to 0$ the symmetry between the cooling and heating modes becomes totally exact regardless of types of single-phase fluid.

2. The same basic symmetry between the fluid cooling and heating modes makes it possible to extend the above conclusions to forced-convection two-phase flow heat exchangers (with bulk evaporation/condensation) designed to operate efficiently in low thermal lift thermodynamic systems when nucleate boiling regimes are suppressed, as further discussed in Section 8.2.

8.2 CONVECTIVE HEAT TRANSFER WITH PHASE CHANGES

For the reasons outlined in Section 4.1 we focus on *low thermal lift* power generation and heat pumping systems with special emphasis on annular-spray two-phase flow regimes typically encountered in stable forced-convection

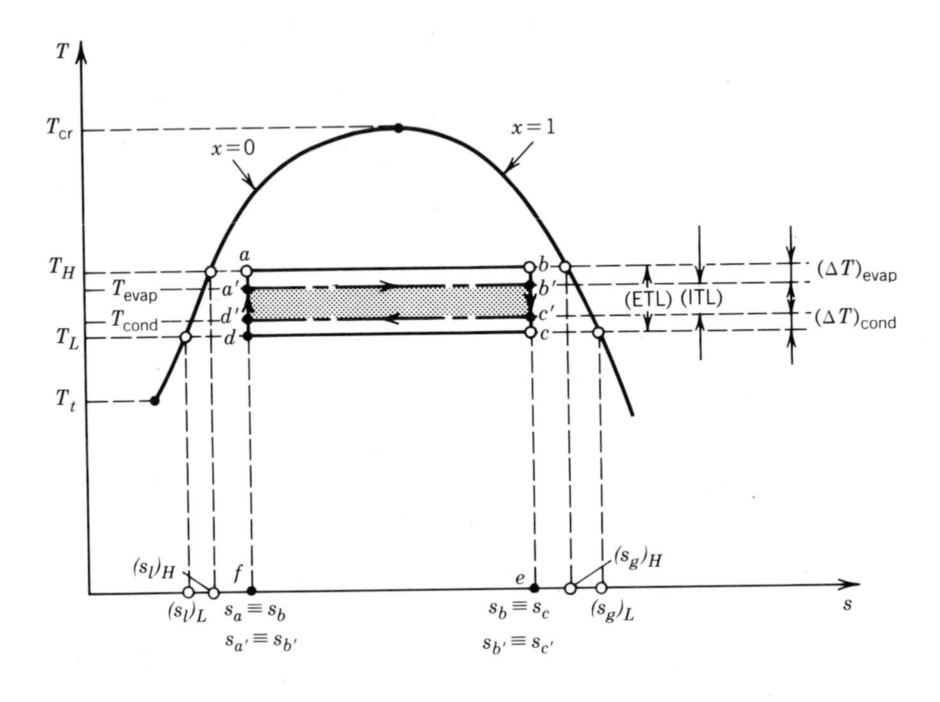

(a)

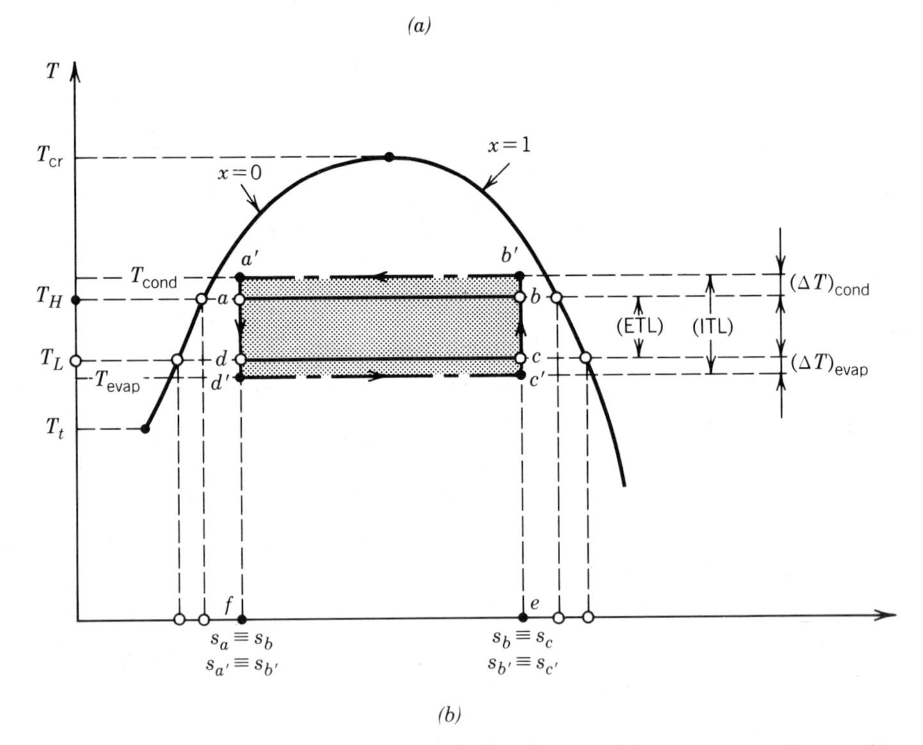

(b)

FIGURE 8.10. Low thermal lift vapor cycles in *T*-*s* diagram: (*a*) power generation mode and (*b*) heat pumping mode.

evaporators and condensers, which include the turbulent–turbulent (tt) Martinelli–Nelson dissipative flow regimes [3.4, 3.32, 4.67, 8.3] reviewed in Chapter 5. These two-phase flow regimes play the same central role in forced-convection evaporators and condensers just as turbulent regimes do in single-phase flow heat exchangers.

In the next section, the somewhat vague notion of low thermal lift (LTL) cycles is defined quantitatively in order to identify the unique characteristics of low thermal lift two-phase flow heat exchangers that make it possible to simplify their analysis on rational grounds without sacrificing accuracy.

8.2.1 Low Thermal Lift Heat Exchangers

To delineate and quantify approximately the severe design constraints imposed on the main heat exchangers in low thermal lift power generation or reverse power (heat pump and/or refrigeration) cycles, we refer to the ideal reversible Carnot cycle depicted in the skeleton temperature–entropy charts shown in Fig. 8.10.

Although a thermodynamic analysis based on ideal Carnot cycles is admittedly rudimentary, it will suffice to answer the key question addressed in this section: How much of the available external temperature lift (ETL) may be reasonably dissipated in the two main heat exchangers of vapor power generation (or heat pumping) cycles, namely, the high-temperature evaporator and the low-temperature condenser of vapor power generation systems or the low-temperature evaporator and the high-temperature condenser of reverse power cycles. We consider first the power generation mode.

8.2.1.1 Power Generation Mode. For a given

$$\text{External thermal lift (ETL)} \equiv T_H - T_L \qquad (8.86)$$

and since heat can flow only from a warmer to a colder region, in accordance with the second law of thermodynamics:

$$\text{Internal thermal lift (ITL)} \equiv T_{\text{evap}} - T_{\text{cond}} < T_H - T_L = \text{ETL} \qquad (8.87a)$$

where

T_H, T_L are the constant *external* heat source and heat sink temperatures, respectively (8.87b)

whereas

$T_{\text{evap}}, T_{\text{cond}}$ are the working fluid's *internal* evaporating and condensing temperatures, respectively (8.87c)

Therefore, $T_{evap} < T_H$ and $T_L < T_{cond}$, or $T_H - T_{evap} \equiv (\Delta T)_{evap} > 0$ and $T_{cond} - T_L = (\Delta T)_{cond} > 0$. Consequently, the internal thermal lift can be expressed (see Fig. 8.10a) according to

$$\begin{aligned} \text{ITL} &= T_H - T_L - \left[(\Delta T)_{evap} + (\Delta T)_{cond}\right] \\ &= (T_H - T_L)\left[1 - \frac{(\Delta T)_{evap} + (\Delta T)_{cond}}{T_H - T_L}\right] \end{aligned} \tag{8.88}$$

and with Eq. (8.86)

$$\text{ITL} = (\text{ETL})[1 - (\Delta t)_{rel}] \tag{8.89a}$$

or, in dimensionless terms as a fraction of the external thermal lift,

$$\frac{\text{ITL}}{\text{ETL}} = 1 - (\Delta t)_{rel} \tag{8.89b}$$

where by definition

$$(\Delta t)_{rel} \equiv \frac{(\Delta T)_{cond} + (\Delta T)_{evap}}{T_H - T_L} \equiv \frac{(\Delta T)_{cond} + (\Delta T)_{evap}}{\text{ETL}} \tag{8.89c}$$

The dimensionless term $(\Delta t)_{rel}$ thus represents the fraction of the available external thermal lift that is deliberately dissipated in the evaporator and condenser to reduce the required heat transfer surfaces to economically acceptable levels, and the limiting case $(\Delta t)_{rel} = 0$ corresponds to the classical ideal Carnot cycle with infinitely large heat exchanger surfaces and zero friction dissipation.

To quantify the impact of $(\Delta t)_{rel}$ on the thermal efficiency we consider first the ideal Carnot cycle with $(\Delta t)_{rel} = 0$ operating fully in the saturated liquid and vapor (two-phase) region or *wet domain*. In principle, any pure substance or azeotropic mixture that changes phase at constant temperature can be considered as a working fluid provided the following necessary conditions are satisfied (see Fig. 8.10a).

$$T_t < T_L \le T_{cond} < T_{evap} < T_H < T_{cr} \tag{8.90}$$

Therefore,

$$T_{evap} - T_{cond} \le T_H - T_L < T_{cr} - T_t \tag{8.91a}$$

as well as

$$s_a \ge (s_l)_H \quad \text{and} \quad s_d \ge (s_l)_L \tag{8.91b}$$

whereas

$$s_b \leq (s_g)_H \quad \text{and} \quad s_c \leq (s_g)_L \tag{8.91c}$$

so that the vapor quality x at any point of the vapor Carnot cycle remains within the

$$\text{wet domain} \quad 0 \leq x \leq 1 \tag{8.91d}$$

where T_t and T_{cr} are the triple and critical points, respectively, of the working fluid under consideration.

Since we are only concerned here with Carnot cycles operating fully in the wet domain, the conditions expressed by Eq. (8.90) can be stated most generally in terms of reduced temperatures after dividing all terms by T_{cr} (see Chapter 6) according to

$$T_t^+ < T_L^+ \leq T_{\text{cond}}^+ < T_{\text{evap}}^+ \leq T_H^+ < 1 \tag{8.92a}$$

or, alteratively, in terms of reduced pressures

$$P_t^+ < P_L^+ \leq P_{\text{cond}}^+ < P_{\text{evap}}^+ \leq P_H^+ < 1 \tag{8.92b}$$

Similarly, Eq. (8.91a) yields

$$T_{\text{evap}}^+ - T_{\text{cond}}^+ \leq T_H^+ - T_L^+ < 1 - 1/T_t^+ \tag{8.93}$$

According to Carnot's theorem, the ideal Carnot cycle thermal efficiency $\eta_C = (T_H - T_L)/T_H$ is applicable for any working fluid satisfying all necessary conditions expressed by Eqs. (8.90)–(8.93). Since η_C is by definition the ratio of the ideal Carnot cycle work output per unit of thermal energy input at the high temperature level T_H, and recalling that the areas in a T-s diagram for any fully reversible cycles are a measure of energy, we see by referring to Fig. 8.10a that

$$\eta_C \equiv \frac{T_H - T_L}{T_H} \equiv \frac{\text{rectangular area } abcda}{\text{rectangular area } abefa} \equiv \frac{\text{work output}}{\text{thermal energy input}} \tag{8.94}$$

for all ideal Carnot cycles operating between $T_H = T_{\text{evap}}$ and $T_L = T_{\text{cond}}$ with $(\Delta t)_{\text{rel}} = 0$ within the domain defined by the saturated liquid ($x = 0$) and vapor ($x = 1$) lines in the T-s diagram.

Since the usual average yearly sink temperature $T_{\text{sk}} \equiv T_L$ on planet earth is substantially constant at about 300 K, it is convenient for our present purpose to modify Eq. (8.94) to assess η_C in terms of $(T_H - T_L)/T_L$ as outlined below.

$$\eta_C = \frac{T_H - T_L}{T_H}\left(\frac{T_L}{T_L}\right) \equiv \frac{T_H - T_L}{T_L}\left(\frac{T_L}{T_H}\right) \tag{8.95}$$

and with $T_H = T_L + (T_H - T_L)$

$$\eta_C = \frac{T_H - T_L}{T_L}\left(\frac{T_L}{T_L + (T_H - T_L)}\right) \equiv \frac{T_H - T_L}{T_L}\left(\frac{1}{1 + (T_H - T_L)/T_L}\right) \tag{8.96}$$

or

$$\eta_C = (\mathrm{ETL})_{\mathrm{rel}}\left(\frac{1}{1 + (\mathrm{ETL})_{\mathrm{rel}}}\right) \tag{8.97a}$$

where by definition

$$(\mathrm{ETL})_{\mathrm{rel}} \equiv \frac{T_H - T_L}{T_L} \equiv \textit{relative external temperature lift} \tag{8.97b}$$

When the relative external temperature lift approaches zero, it is evident that $\eta_C \to (T_H - T_L)/T_L \equiv (\mathrm{ETL})_{\mathrm{rel}}$ since the terms in the large parentheses on the right-hand side of Eq. (8.96) is of the form $1/(1 + a)$, which converges rapidly toward unity as $a \to 0$; see Table 8.15. This means that it is permissible to ignore the insignificant difference in absolute temperature levels between T_L and T_H and to calculate η_C according to

$$\eta_C \simeq \frac{T_H - T_L}{T_L} \tag{8.98a}$$

with an accuracy of better than

$$100\left(\frac{T_H - T_L}{T_L}\right)\% \quad \text{at} \quad \left(\frac{T_H - T_L}{T_L}\right) < 0.20 \tag{8.98b}$$

The ratio obtained by dividing the exact Carnot cycle efficiencies $\eta_C = (T_H - T_L)/T_H$ by the approximate value $\eta_C \simeq (T_H - T_L)/T_L = (\mathrm{ETL})_{\mathrm{rel}}$ per Eq. (8.98a)

$$\frac{\eta_C}{(\mathrm{ETL})_{\mathrm{rel}}} = \frac{T_L}{T_H} = \frac{1}{1 + (T_H - T_L)/T_L} = \frac{1}{1 + (\mathrm{ETL})_{\mathrm{rel}}} \tag{8.99}$$

TABLE 8.15. Low Thermal Lift Ideal Carnot Power Cycle Efficiency η_C in Terms of Relative External Thermal Lift

(1)	$(\mathrm{ETL})_{\mathrm{rel}} \equiv (T_H - T_L)/T_L$ =	0.20	0.15	0.10	0.05	0.02	0.01	0.001
(2)	Exact η_C (%) =	16.66	13.04	9.09	4.76	1.96	0.99	0.0999
(3)	$100(\mathrm{ETL})_{\mathrm{rel}} \simeq \eta_C$ (%) =	20.00	15.00	10.00	5.00	2.00	1.00	0.100
(4)	Line (2) / line (3) =	0.833	0.870	0.909	0.952	0.98	0.99	0.999

quantifies the error committed when η_C is computed according to Eq. (8.98a). The numerical values on line (4) of Table 8.15 clearly show that this error becomes negligibly small for an order of magnitude analysis of the impact of $(\Delta t)_{rel}$ particularly at extremely low $(ETL)_{rel}$ of the order of 0.01–0.05, typical of low thermal lift cycles encountered in some power generation cycles, particularly those being considered for ocean thermal energy conversion (OTEC) systems with maximum temperature lifts of the order of less than $T_H - T_L = 10$ K or $(ETL)_{rel} \simeq 10/300 = 0.03$!

To differentiate high thermal lift (HTL) power generation (or heat pumping) systems from LTL thermodynamic cycles, the following arbitrary definition is introduced.

GUIDELINE 26 *Thermal cycles designed to operate at relative external lifts* $(ETL)_{rel} = (T_H - T_L)/T_L < 0.2$ *are by definition* ***low thermal lift*** *cycles and* $\eta_C \simeq (T_H - T_L)/T_L = (ETL)_{rel}$ *with a maximum error of less than* 17%.

For operating and economic reasons, it is frequently preferable to select the working fluids of LTL thermal cycles in such a manner that they never operate below atmospheric pressure. This means that $T_b < T_L$, where T_b is the boiling temperature of the working fluid at normal atmospheric pressure.

The maximum conceivable relative external thermal lift $(ETL)_{rel, max}$ for any working fluid when $T_L > T_b$ can be estimated as outlined below. It is clear that

$$(ETL)_{rel} < (ETL)_{rel, max} = (T_{cr} - T_b)/T_b \tag{8.100}$$

since $(ETL)_{max} = T_{cr} - T_b$ within the domain of existence of the liquid phase and the ratio $ETL/T_L = (ETL)_{rel}$ must decrease with increasing $T_L > T_b$.

With $T_b^+ \equiv T_b/T_{cr} \equiv$ *reduced normal boiling temperature*, Eq. (8.100) can be "generalized" as follows:

$$(ETL)_{rel} < (ETL)_{rel, max} \equiv \frac{1}{T_b^+} - 1 \tag{8.101a}$$

$$\textit{Validity}: \quad T_L > T_b \quad \text{and} \quad T_H < T_{cr} \tag{8.101b}$$

Because the reduced normal boiling temperature T_b^+ is roughly constant for all common working fluids, including polar substances such as water, $(ETL)_{rel, max}$ is also nearly constant according to Eq. (8.101a). This is verified in Table 8.16 for 16 different chemical species on the basis of the thermophysical data in Table A.6 of Appendix A.

The interesting conclusion from this rough analysis is that $(ETL)_{rel, max} \simeq 0.61 =$ constant within an accuracy of better than $\pm 20\%$, as shown in Table 8.16. However the *absolute* value of the maximum external thermal lift $(ETL)_{max}$ is roughly proportional to the absolute normal boiling temperature

TABLE 8.16. Maximum Relative External Thermal Lift of Common Fluids

Fluid Designation[a]	Boiling Point (°C)	Critical Temperature (°C)	$(ETL)_{rel, max}$ per Eq. (8.101a)
702*n*	−252.8	−259.2	0.63
720	−246.1	−228.7	0.64
728	−198.8	−146.9	0.70
732	−182.9	−118.4	0.71
14	−127.9	−45.7	0.57
13	−81.4	−28.8	0.57
502	−45.4	82.2	0.56
22	−40.8	96.0	0.59
115	−39.1	79.9	0.51[b]
717	−33.3	133.0	0.69
12	−29.8	112.0	0.58
114	3.8	145.7	0.51[b]
21	8.9	178.5	0.60
11	23.8	198.0	0.59
113	47.6	214.1	0.52
718	100.0	374.2	0.73[c]
		Arithmetic average:	0.61

[a]See Appendix A, Table A.6, for name and chemical formula.

[b]Lowest value of $(ETL)_{rel, max}$.

[c]Highest value of $(ETL)_{rel, max}$ for R718, that is, water.

T_b since, according to the definition of $(ETL)_{rel, max}$ for $T_b < T_L$,

$$\frac{(ETL)_{max}}{T_b} \equiv (ETL)_{rel, max} \simeq 0.61 \simeq \text{constant} \tag{8.102}$$

therefore,

$$(ETL)_{max} \simeq 0.61 T_b \propto T_b \tag{8.103}$$

It is thus evident from Eq. (8.103) that the $(ETL)_{max}$ of a low boiling substance such as normal hydrogen (702*n*) will be significantly lower than that of water (R718). As can be seen from Table 8.16 $(ETL)_{max} = T_{cr} - T_b = t_{cr} - t_b = 259.2 - 252.8 = 6.4°C/K$ for R702*n*, that is normal hydrogen, whereas $(ETL)_{max} = 374.2 - 100 = 274.2°C/K$ for water (R718). This indicates again that water is an exceptional and therefore atypical working fluid[9] [highest boiling temperature t_b and largest $(ETL)_{rel, max} = 0.73$ among all the substances listed in Table 8.16].

Referring to Fig. 8.10*a*, we now address the constant temperature power generation cycle depicted by the rectangle $(a'b'c'd'a')$, which is an ideal Carnot cycle operating without any internal irreversible losses. However, the fluid evaporates at a constant temperature $T_{evap} < T_H$ and condenses at a constant temperature $T_{cond} > T_L$. It should be noted that this *internally* *reversible* Carnot cycle has been so selected that $s_{a'} = s_{b'} = s_a = s_b$ and $s_{c'} = s_{d'} = s_c = s_d$, as shown on the skeleton T-s diagram. Clearly, this internal Carnot cycle (ICC) converges to the classical external Carnot cycle (ECC) when $T_H - T_{evap} \equiv (\Delta T)_{evap} \to 0$ and $T_{cond} - T_L = (\Delta T)_{cond} \to 0$, that is, when $(\Delta t)_{rel} \to 0$. The classical external Carnot cycle can thus be treated as the limiting ideal case of the more realistic reversible internal Carnot cycle operating with finite temperature differences in the two heat exchangers when $(\Delta t)_{rel} \to 0$.

It is now a simple matter to estimate the impact of $(\Delta t)_{rel}$, in LTL Carnot cycles [i.e., when Eq. (8.98) is applicable] as follows.

Since the areas in the T-s diagram are a measure of energy, we may write, by referring to Fig. 8.10*a*,

$$\frac{W_{ICC}}{W_{ECC}} = \frac{\text{rectangular area } a'b'c'd'a'}{\text{rectangular area } abcda} = \frac{(T_H - T_L) - \left[(\Delta T)_{evap} + (\Delta T)_{cond}\right]}{(T_H - T_L)} \tag{8.104}$$

which reduces with Eqs. (8.89b) and (8.89c) to

$$\frac{W_{ICC}}{W_{ECC}} \equiv \frac{\text{ITL}}{\text{ETL}} = 1 - (\Delta t)_{rel} \tag{8.105}$$

where W_{ICC} = maximum work output per unit mass of fluid for the *internal* Carnot cycle $(a'b'c'd'a')$ when $(\Delta t)_{rel} > 0$
W_{ECC} = maximum work output per unit mass of fluid for the classical *external* Carnot cycle $(abcda)$ when $(\Delta t)_{rel} = 0$

Solving Eq. (8.105) for W_{ICC} yields

$$W_{ICC} = W_{ECC}\left[1 - (\Delta t)_{rel}\right] = W_{ECC} - W_{ECC}(\Delta t)_{rel} \tag{8.106a}$$

or

$$W_{ECC} - W_{ICC} = W_{ECC}(\Delta t)_{rel} \equiv \text{AWL} \tag{8.106b}$$

where AWL = *available work loss* due to $(\Delta t)_{rel}$ and

$$\frac{W_{ECC} - W_{ICC}}{W_{ECC}} = (\Delta t)_{rel} \equiv \text{RAWL} \tag{8.107a}$$

where RAWL = *relative available work loss* for a given external thermal lift ETL = $T_H - T_L$, due solely to the finite temperature differences in the condenser and the evaporator expressed in nondimensional terms by $(\Delta t)_{rel}$.

Validity: LTL power generation cycles when $(\text{ETL})_{rel} \leq 0.20$ (8.107b)

According to Eq. (8.107a), $(\Delta t)_{\mathrm{rel}}$ represents the fraction of the maximum potential work available when $(\Delta t)_{\mathrm{rel}} = 0$ which has become unavailable because $(\Delta t)_{\mathrm{rel}} > 0$. Since $(\Delta t)_{\mathrm{rel}}$ is inversely proportional to the external thermal lift $(T_H - T_L)$ according to the defining Eq. (8.89c), it is clear that what matters is not the absolute values of $(\Delta T)_{\mathrm{evap}}$ and $(\Delta T)_{\mathrm{cond}}$ but rather their relative values $(\Delta T)_{\mathrm{evap}}/(T_H - T_L)$ and $(\Delta T)_{\mathrm{cond}}/(T_H - T_L)$. This means that for a given RAWL lower temperature driving forces $(\Delta T)_{\mathrm{evap}}$ and $(\Delta T)_{\mathrm{cond}}$ must be selected for the design of low thermal lift heat exchangers than in the case of high thermal lift cycles. In fact, the situation is even worse than would appear from Eq. (8.107a) alone, since the ideal Carnot cycle efficiency itself is roughly proportional to the external thermal lift and therefore much lower in LTL than in HTL cycles. Thus, it would be desirable to offset the low thermal efficiency associated with LTL cycles by selecting even lower $(\Delta t)_{\mathrm{rel}}$ than in HTL applications and this would further compound the heat exchanger design problem. The situation is less critical in the case of heat pumping systems since extremely low lift conditions coincide with extremely low power consumption for a given heat rejection rate, as highlighted in the next section.

8.2.1.2 Heat Pumping Mode. The ideal Carnot reverse power cycle (*adcba*) depicted in Fig. 8.10*b* operates as indicated, in the reverse (counterclockwise) direction at the same absolute temperature levels T_H and T_L with the same fluid and through the same equilibrium states, a, b, c, d, a, as the power generation cycle shown in Fig. 8.10*a*.

Since by definition the ideal Carnot cycle operates without any irreversible losses and is physically fully reversible, everything that was said for the power mode remains valid in absolute value for the heating pumping mode; however, all the signs must be reversed. The ideal (maximum) work output $|W|_{\mathrm{ECC}}$ now becomes the ideal (minimum) work input $|W|_{\mathrm{ECC}}$ represented by the area *adcba* which must be expended to absorb the thermal energy $T_L\,\Delta s$ = area *dcefd* from the "low side" heat source at $T_{\mathrm{evap}} = T_L$ as the working fluid partially evaporates from d to c and to reject the total heat $T_H|\Delta s| \equiv (T_H - T_L)|\Delta s| + T_L\,\Delta s = |\vec{Q}|$ represented by the area *afeba* to the "high side" heat sink at $T_H \equiv T_{\mathrm{cond}}$ as the fluid partially condenses from point b to a in the skeleton T-s diagram.

This ideal reverse power cycle thus yields the same ratio $|\Delta s|(T_H - T_L)/(|\Delta s|T_H) \equiv (T_H - T_L)/T_H$ as in the power generation mode: that is,

$$\eta_C \equiv \frac{T_H - T_L}{T_H} \equiv \frac{\text{minimum work input per unit mass}}{\text{heat rejected per unit mass at } T_H} \equiv \frac{\overset{\swarrow}{W}_{\mathrm{ECC}}}{|\vec{Q}|} \qquad (8.108)$$

However, it is customary to rate heat pumps in terms of the so-called *coefficient of performance* (COP) defined below:

$$\text{COP} \equiv \frac{\text{thermal energy transferred at } T_H}{\text{work input}} \equiv \frac{\left|\overset{\nearrow}{Q}\right|}{\overset{\swarrow}{W}_{\text{ECC}}} \qquad (8.109)$$

which reduces in the case of any ideal Carnot cycle with Eq. (8.108) under SSSF conditions to

$$(\text{COP})_C \equiv \frac{T_H}{T_H - T_L} \equiv \frac{1}{\eta_C} \equiv \frac{\left|\overset{\nearrow}{Q}\right|}{\left|\overset{\swarrow}{W}_{\text{ECC}}\right|} \equiv \frac{\left|\overset{\nearrow}{\dot{Q}}\right|}{\left|\overset{\swarrow}{\dot{W}}_{\text{ECC}}\right|} \qquad (8.110)$$

The minimum work (or power) that must be expended to reject a specified amount of heat $|\overset{\nearrow}{Q}|$ (or $|\overset{\nearrow}{\dot{Q}}|$) to the heat sink at $T_H = T_{\text{cond}}$ is obtained by solving Eq. (8.110) for $|W_{\text{ECC}}|$ (or $|\dot{W}_{\text{ECC}}|$), that is,

$$\left|\overset{\swarrow}{W}_{\text{ECC}}\right| = \left|\overset{\nearrow}{Q}\right| \left(\frac{T_H - T_L}{T_H}\right) \equiv \left|\overset{\nearrow}{Q}\right| \eta_C \qquad (8.111\text{a})$$

or, under SSSF conditions,

$$\left|\overset{\swarrow}{\dot{W}}_{\text{ECC}}\right| = \left|\overset{\nearrow}{\dot{Q}}\right| \left(\frac{T_H - T_L}{T_H}\right) \equiv \left|\overset{\nearrow}{\dot{Q}}\right| \eta_C \qquad (8.111\text{b})$$

As in the case of ideal Carnot power generation cycles, it is convenient to express η_C and $(\text{COP})_C$ in terms of $(\text{ETL})_{\text{rel}} = (T_H - T_L)/T_L$. Equation (8.108) yields, with Eq. (8.97a),

$$\eta_C \equiv \frac{T_H - T_L}{T_L} \frac{1}{1 + (T_H - T_L)/T_L} \equiv (\text{ETL})_{\text{rel}} \left(\frac{1}{1 + (\text{ETL})_{\text{rel}}}\right) \qquad (8.112)$$

and the COP of the ideal Carnot heat pump cycle becomes, with Eq. (8.110),

$$(\text{COP})_C = \frac{1}{\eta_C} = \frac{1}{(\text{ETL})_{\text{rel}}} [1 + (\text{ETL})_{\text{rel}}] \qquad (8.113)$$

These COP values are shown in Table 8.17, which focuses mainly on low thermal lift cycles [i.e., $(\text{ETL})_{\text{rel}} < 0.2$ according to Guideline 26] when it is permissible to write:

$$(\text{COP})_C \simeq \frac{1}{(\text{ETL})_{\text{rel}}} \qquad (8.114\text{a})$$

Validity: LTL systems when $(\text{ETL})_{\text{rel}} < 0.2$ (8.114b)

TABLE 8.17. Coefficients of Performance $(\mathrm{COP})_C$ of Low Thermal Lift Ideal Carnot Heat Pump Cycles

$(\mathrm{ETL})_{rel} = \dfrac{T_H - T_L}{T_L}$	= 0.20	0.15	0.100	0.075	0.05	0.04	0.03	0.02	0.01	0.001
$(\mathrm{COP})_C = \dfrac{100}{\eta_C} \simeq \dfrac{100}{(\mathrm{ETL})_{rel}}$	= 5.0	6.7	10.0	13.3	20.0	25.0	33.3	50	100	1000

The fundamental difference between the power generation mode and heat pumping mode of operation highlighted by the key data summarized in Tables 8.15 and 8.17 can be stated succinctly as follows. The incentive is to maximize $(\mathrm{ETL})_{rel}$ and thus maximize the useful power output ($\overset{\nearrow}{\dot{W}}$) in the power generation mode, whereas the prime incentive in the heat pumping mode is to minimize $(\mathrm{ETL})_{rel}$ and thereby minimize the required power consumption $|\overset{\swarrow}{\dot{W}}|$ for a given heat rejection rate.

As in the power generation mode case, it is possible to estimate the impact of the finite temperature differentials in the *low side evaporator* ($T_{evap} < T_L$) and the *high side condenser* ($T_{cond} > T_H$) as shown in Fig. 8.10*b*. Since heat can only flow from a warmer to a colder region, it is clear, by referring to Fig. 8.10*b*, that the internal thermal lift (ITL) = $T_{cond} - T_{evap}$ must exceed the external thermal lift (ETL) = $T_H - T_L$ in the heat pumping mode; that is,

$$\mathrm{ITL} \equiv T_{cond} - T_{evap} \geq T_H - T_L = \mathrm{ETL} \tag{8.115a}$$

or

$$\mathrm{ITL}/\mathrm{ETL} > 1 \tag{8.115b}$$

in the heat pumping mode of operation.

Following the same reasoning as in the power generation mode, the impact of $(\Delta t)_{rel} \geq 0$ in LTL heat pumping systems when $\eta_C \simeq (\mathrm{ETL})_{rel}$ can be assessed as follows:

$$\overset{\swarrow}{\dot{W}}_{ICC} = [1 + (\Delta t)_{rel}]\left|\overset{\swarrow}{\dot{W}}\right|_{ECC} = \left|\overset{\swarrow}{\dot{W}}\right|_{ECC} + \left|\overset{\swarrow}{\dot{W}}\right|_{ECC}(\Delta t)_{rel} \tag{8.116}$$

Therefore,

$$\frac{\left|\overset{\swarrow}{\dot{W}}\right|_{ICC} - \left|\overset{\swarrow}{\dot{W}}\right|_{ECC}}{\overset{\swarrow}{\dot{W}}_{ECC}} \equiv (\Delta t)_{rel} \equiv \mathrm{REW} \tag{8.117a}$$

where REW = *relative excess work* due to $(\Delta t)_{rel}$.

Validity: LTL systems when $(\mathrm{ETL})_{rel} = (T_H - T_L)/T_L < 0.2$ (8.117b)

In other words, the relative excess work (REW) consumption, due exclusively to the sum of the finite temperature differences in the evaporator and condenser, is directly proportional to $(\Delta t)_{rel}$ in LTL heat pumping systems.

8.2.1.3 Numerical Order of Magnitude Analyses. The severe constraints imposed by Eqs. (8.107a) and (8.117a) in the power generation and heat pumping modes of operation are assessed in the following analyses to establish reasonable orders of magnitude for $(\Delta T)_{cond} + (\Delta T)_{evap}$. We focus particularly on $(\Delta T)_{evap}$ since this has a most significant impact on the types of *boiling regime* that can physically occur in thermodynamically efficient LTL forced-convection evaporators.

NUMERICAL ANALYSIS 8.5

Low thermal lift vapor power generation cycles.

Background

The prime objective is to establish reasonable levels for ΔT_{evap} and ΔT_{cond} in typical low thermal lift power generation cycles.

Problem Statement

Assuming that the maximum loss of power output due exclusively to the finite temperature differentials in the two main heat exchangers may not exceed 20%, determine the allowable levels of $(\Delta t)_{rel}$, $(\Delta T)_{cond}$, and $(\Delta T)_{evap}$ for typical (i.e., earthbound) power plant applications with heat sink temperatures $T_{sk} \equiv T_L = 300$ K for the following cases:

(a) *Case 1.* A (geothermal) heat source is available at a moderate constant temperature $T_H = 350$ K.

(b) *Case 2.* $T_H \leq 310$ K as in the case of an ocean thermal energy conversion (OTEC) system.

Solution

(a) *Case 1.* $T_H - T_L = 50$ and $(\text{ETL})_{rel} \equiv (T_H - T_L)/T_L = (350 - 300)/300 = \leq 0.17 < 0.20$. Therefore, we are dealing with a typical LTL system and Eq. (8.107a) is applicable. Thus, RAWL $= (\Delta t)_{rel} = (\Delta T_{cond} + \Delta T_{evap})/(T_H - T_L) \simeq 0.17$ or, solving for $(\Delta T)_{cond} + (\Delta T)_{evap}$ with $T_H - T_L = 50°$C,

$$\Delta T_{cond} + \Delta T_{evap} < (0.17)50 = 8.5°\text{C} \tag{8.118a}$$

If the same differential is allocated to each heat exchanger,

$$(\Delta T)_{\text{cond}} = (\Delta T)_{\text{evap}} = \tfrac{1}{2}(8.5) = 4.25°\text{C} \tag{8.118b}$$

If one of the two main two-phase heat exchangers is favored at the expense of the other, then the following extreme values are possible: either

$$(\Delta T)_{\text{cond}} \leq 8.5 \quad \text{and} \quad (\Delta T)_{\text{evap}} \simeq 0°\text{C} \tag{8.118c}$$

or

$$(\Delta T)_{\text{evap}} \leq 8.5 \quad \text{and} \quad (\Delta T)_{\text{cond}} \simeq 0°\text{C} \tag{8.118d}$$

(b) *Case 2*. ETL $= T_H - T_L = 10°\text{C}$ and the relative external lift $(\text{ETL})_{\text{rel}} = 10/300 = 0.033 \ll 0.20$. This is an extremely low thermal lift system and Eq. (8.107a) is valid with an accuracy of about 3% (see Table 8.15). Repeating the same calculating steps as under Case 1 with ETL = 10 instead of 50 [i.e., at $(\text{ETL})_{\text{Case 2}}/(\text{ETL})_{\text{Case 1}} = \frac{1}{5}$) immediately leads to the following answers:

$$(\Delta T)_{\text{cond}} + (\Delta T)_{\text{evap}} \leq 1.7°\text{C} \tag{8.119a}$$

Therefore,

$$(\Delta T)_{\text{cond}} = (\Delta T)_{\text{evap}} \leq 0.85°\text{C} \tag{8.119b}$$

or

$$(\Delta T)_{\text{cond}} \leq 1.7 \quad \text{when } (\Delta T)_{\text{evap}} \simeq 0°\text{C} \tag{8.119c}$$

and

$$(\Delta T)_{\text{evap}} \leq 1.7 \quad \text{when } (\Delta T)_{\text{cond}} \simeq 0°\text{C} \tag{8.119d}$$

Comments

1. For properly balanced fluid-to-fluid heat exchangers, the average hot fluid to wall temperature differential is equal to the average wall to cold fluid differential. Therefore,

$$(\overline{\Delta T_w})_{\text{evap}} \equiv \overline{T}_w - \overline{T}_{\text{sat}} = \tfrac{1}{2}(\Delta T)_{\text{evap}} \tag{8.120a}$$

Validity: Typical *balanced* fluid-to-fluid heat exchangers (8.120b)

This corresponds to typical $(\overline{\Delta T_w})_{\text{evap}} \leq \frac{1}{2}(4.25) \simeq 2.12$ in Case 1, and $(\overline{\Delta T_w})_{\text{evap}} < \frac{1}{2}(0.85) \simeq 0.42$ in Case 2 per Eqs. (8.118a) and (8.119b), respectively. Such low $(\overline{\Delta T_w})_{\text{evap}}$ are generally *below* the minimum *threshold* required to initiate *nucleate boiling* regimes, as highlighted in Eq. (7.27d) and further discussed in Section 8.2.2. This rough order of magnitude analysis indicates that nucleate boiling regimes are extremely unlikely in forced-convection evaporators in such LTL power generation cycles.

2. In typical high thermal lift (HTL) power cycles, such as fossil-fired steam power plants, $T_H \gg T_{\text{evap}}$ and $(\Delta T_w)_{\text{evap}}$ is generally selected on the basis of safe wall material operating temperatures rather than thermodynamic considerations. In other words, the maximum potentially available temperature differential between combustion products and evaporating steam on the high-pressure side of the cycle is deliberately dissipated for safety reasons only [the actual $(\Delta T)_{\text{evap}}$ is in fact so high that the radiation rather than the convective mode becomes dominant]. This explains why most published steam power plant computer-aided heat exchanger thermodynamic (second law) *optimization* programs focus on the low side condenser and the feed heaters but *rarely* on the *high side evaporator* of steam power cycles!

NUMERICAL ANALYSIS 8.6

Low thermal lift vapor heat pump cycles.

Problem Statement

The problem is the same as in the previous numerical analysis except that 20% now represents the maximum allowable excess power consumption in the heat pumping mode of operation.

(a) *Case 1*. A constant high-temperature sink temperature $T_{\text{sk}} \equiv T_H = 293$ K $\simeq$ typical room temperature ($\simeq 20°$C) with an external lower temperature source $T_{\text{so}} = 273$ K ($\simeq 0°$C).

(b) *Case 2*. Milder external source temperature

$$T_{\text{so}} = T_L = 283 \text{ K } (\simeq +10°\text{C}),$$

otherwise the same as Case 1.

Solution

(a) *Case 1*. $T_H - T_L = 20$ and $(\text{ETL})_{\text{rel}} = 20/293 = 0.066 < 0.20$. Therefore, Eq. (8.117a) for LTL systems is applicable with an excellent degree of accuracy. From Eq. (8.117a) it follows that

$$\text{REW} = (\Delta t)_{\text{rel}} \equiv \frac{(\Delta T)_{\text{cond}} + (\Delta T)_{\text{evap}}}{20} = 0.066 \qquad (8.121)$$

Therefore,

$$(\Delta T)_{\text{cond}} + (\Delta T)_{\text{evap}} \leq 1.33°\text{C} \qquad (8.122\text{a})$$

Following the same reasoning as in Numerical Analysis 8.5 yields

$$(\Delta T)_{\text{cond}} = (\Delta T)_{\text{evap}} < 0.66°\text{C} \tag{8.122b}$$

and

$$(\Delta T)_{\text{cond}} \leq 1.33 \quad \text{with } (\Delta T)_{\text{evap}} \simeq 0°\text{C} \tag{8.122c}$$

or

$$(\Delta T)_{\text{evap}} < 1.33 \quad \text{with } (\Delta T)_{\text{cond}} \simeq 0°\text{C} \tag{8.122d}$$

(b) *Case 2*. Since $T_H - T_L = 10 = \frac{1}{2}(T_H - T_L)_{\text{Case 1}}$, it is only necessary to reduce all the temperature differentials estimated in Case 1 by one-half; therefore,

$$(\Delta T)_{\text{cond}} + (\Delta T)_{\text{evap}} \leq 0.66°\text{C} \tag{8.123a}$$

Consequently,

$$(\Delta T)_{\text{cond}} = (\Delta T)_{\text{evap}} < 0.33°\text{C} \tag{8.123b}$$

and

$$(\Delta T)_{\text{cond}} \leq 0.66 \quad \text{with } (\Delta T)_{\text{evap}} \simeq 0°\text{C} \tag{8.123c}$$

or

$$(\Delta T)_{\text{evap}} \leq 0.66 \quad \text{with } (\Delta T)_{\text{cond}} \simeq 0°\text{C} \tag{8.123d}$$

Comments

1. It is clear that the preliminary conclusions derived under comment 1 of Numerical Analysis 8.5 for LTL power generation cycles (i.e., the unlikelihood of nucleate boiling regimes) is applicable to typical LTL heat pumping cycles. This conclusion is further reinforced by the next comment.

2. In typical mechanical heat pump/refrigeration systems, the constant saturation temperature at the evaporator outlet $T_{\text{sat},2}$ corresponding to P_2 (the low-pressure side of the heat pumping system) is maintained at the desired level below $T_{\text{so}} \equiv T_L$ by a compressor as shown in Chapter 4. When $(\Delta T)_{\text{evap}}$ is increased to reduce the size and capital cost of the evaporator, it is necessary to reduce $T_{\text{sat},2}$ for a given $T_{\text{so}} \equiv T_L$. Therefore, a larger compressor displacement is required since a reduction in $T_{\text{sat},2}$ is associated with a reduction in the saturation pressure P_2 and density of the vapor at the compressor intake. This means that a higher $(\Delta T)_{\text{evap}}$ is always associated with an increase in compressor size and capital costs in addition to the negative impact of increased power consumption. Conversely, a lower-cost smaller compressor is required in addition to a desirable reduction in power consumption when $(\Delta T)_{\text{evap}}$ is reduced.

It should be noted that in low side pressure (condensers) optimization studies of steam power generation cycles an increase in $(\Delta T)_{\text{cond}}$ coincides with an increased steam condensing pressure and a decrease in steam turbine size and capital cost. There is therefore more incentive to increase the low side

$(\Delta T)_{\text{cond}}$ in steam power plants since the condenser *and* turbine capital costs, weight, and size *decrease* with increasing $(\Delta T)_{\text{cond}}$ and conversely.

By contrast, the decreasing capital costs of the low-side evaporator in heat pumping cycles are (partially) offset by increased compressor capital costs when $(\Delta T)_{\text{evap}}$ increases. This may explain why heat pump/refrigeration system manufacturers could agree on standard rating conditions based on rather low $(\Delta T)_{\text{evap}}$ long before the first oil crisis of the 1970s.

8.2.2 Two-Phase Heat Transfer With Heat Addition: Evaporation

To provide an authoritative broad brush overview of this important and complex segment of the science (and the art) of heat transfer, we rely heavily on an excellent survey paper, with 83 different references, entitled "Heat Transfer with Boiling" [4.28] by a world renowned authority on the subject, Warren Rohsenow, head of the MIT Heat Transfer Laboratory. Several excerpts are quoted, with only a few minor modifications (shown in square brackets) to comply with the terminology and reference numbers used in this book.

In his introductory section, Rohsenow delineates the various types of boiling regime:

> The process of evaporation, as considered here, results in the conversion of a liquid into a vapor. When this conversion occurs with a liquid, forming vapor bubbles, it is called boiling. Of primary interest is the evaporation associated with the transfer of heat from a solid surface to a fluid.
>
> If the liquid is greatly subcooled, for example, below the boiling-point temperature, the processes of evaporation by boiling may occur locally at the heating surface and may be accompanied by subsequent condensation in the colder bulk of the fluid, resulting in no net evaporation.
>
> In ordinary boiling of a pot of water at its saturation temperature, evaporation occurs at the free surface without the formation of bubbles when the heating-surface temperature T_w is only a few degrees above saturation temperature T_{sat}. Then as $T_w - T_{\text{sat}}$ is increased, vapor bubbles form and agitate the liquid in the vicinity of the heating surface. This type of boiling is called nucleate boiling. These bubbles rise and break through the free surface. Eventually, as $T_w - T_{\text{sat}}$ is increased, the amount of heating surface covered with vapor bubbles is increased until the entire surface becomes vapor blanketed. This results in a process called [vapor] film boiling. The heat transfer rate associated with vigorous nucleate boiling is very high because of the agitation by the bubbles of the fluid near the heating surface; the heat transfer rate associated with film boiling is much lower because of the insulating effect of the vapor film. An example of [vapor] film boiling is the Leidenfrost phenomenon of water droplets dancing on a very hot surface. The droplets do not evaporate rapidly, because an insulating vapor film forms between the hot surface and the droplets.

We recall that the first step in single-phase forced-convection heat transfer analysis is to determine whether stable operation is possible in the laminar or

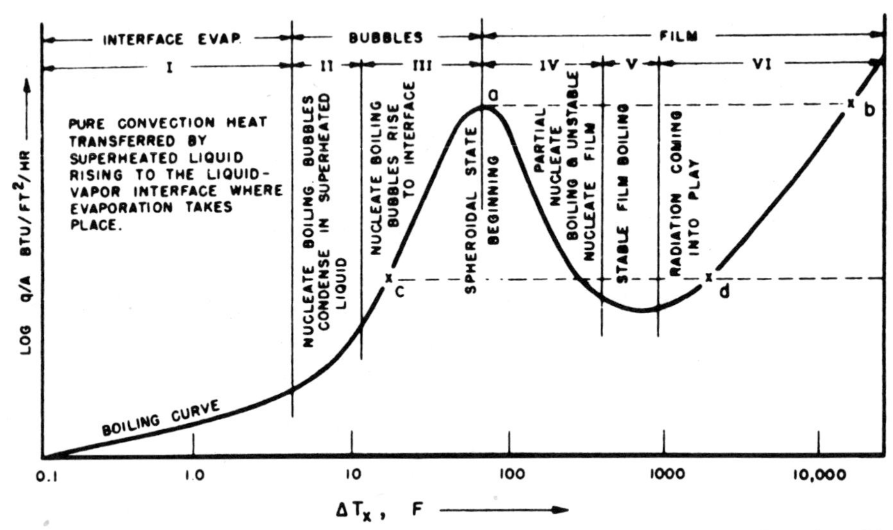

FIGURE 8.11. Characteristic boiling curve. [From W. M. Rohsenow, Heat Transfer with Boiling, *ASHRAE Trans.*, **72**, Pt. 1, 7 (1966), Fig. 1. Reproduced with permission.]

turbulent type of flow regime, in order to use the correct heat transfer coefficient equation. We have seen in Chapter 7 that the turbulent flow regimes are generally much more effective than the laminar flows and that the heat exchanger designer will endeavor to circuit forced-convection single-phase heat exchangers in such a manner that the controlling fluid will flow in the more favorable turbulent regimes. Indeed, specific heat transfer media [8.2] suitable for different applications are usually so selected that they can operate in the more favorable turbulent flow regimes at acceptable levels of friction dissipation $\Delta P_f/\rho = \Delta H_f$. Similarly, the first step in heat transfer analysis with bulk evaporation is the identification of the type of flow regime one can expect or should strive for (if there is a choice) in real world evaporators. Indeed, the first analytical section in Rohsenow [4.28] addresses "Regimes in Boiling" on the basis of the characteristic boiling curve reproduced here as Fig. 8.11.

Because nucleate boiling regimes are far more effective than any other regime, this is the preferred choice for high heat flux applications, such as nuclear reactors, where it is economical to pack as much cooling capacity per unit volume as possible and strive for extremely high operating heat fluxes q'' of the order of 10^6 W/m^2 without any attempt at "minimizing" the wall-to-fluid temperature differentials on thermodynamic grounds, as long as the nuclear channel wall can be kept below a specified maximum safe upper limit. This explains the emphasis on *critical heat fluxes* in nuclear reactor oriented boiling research.

Physical channel damage due to excessive wall temperature is obviously not conceivable in low thermal lift fluid-to-fluid heat interchangers since T_w will, in any case, remain below the finite moderately warmer maximum, or inlet temperature $T_{1,H}$ of the fluid acting as the heat source. Consequently, we confine this brief overview of Ref. 4.28 to those aspects that are particularly relevant in LTL forced-convection evaporation with a main objective that can be formulated as follows. Granted that nucleate boiling regimes are far more effective than any other modes of evaporation at extremely high heat fluxes, the relevant question in our present context is: Can one expect nucleate boiling to play a significant role in *low-pressure* forced-convection LTL evaporators at the typical small temperature differentials $(\Delta T)_{\text{evap}}$ estimated in the previous numerical analyses?

In an attempt to answer this central question, we focus first on pool boiling.

8.2.2.1 Pool Boiling. Referring to Fig. 8.11, we can see that nucleate boiling seems to occur at values of $T_w - T_{\text{sat}}$ of the order of somewhere between 10 and 100°F (or 5.5–55°C). Furthermore, in evaluating different mathematical models to predict the minimum "superheat required to initiate nuclear boiling," Rohsenow states that "some experiments have shown that, at a heated surface in water at atmospheric pressure, boiling begins at around 30°F [16.7°C] above saturation." Thus, it would seem that the typical low values of ΔT_w estimated in the previous numerical analyses for LTL system evaporators are about one order of magnitude too small to initiate nucleate boiling. As a matter of interest, Fig. 8.12 [8.7] shows typical ΔT_w of the order of 10–100 K for nucleate boiling at representative nuclear reactor heat fluxes $q'' \approx 10^6$, that is, q'' and ΔT_w several orders of magnitude higher than those

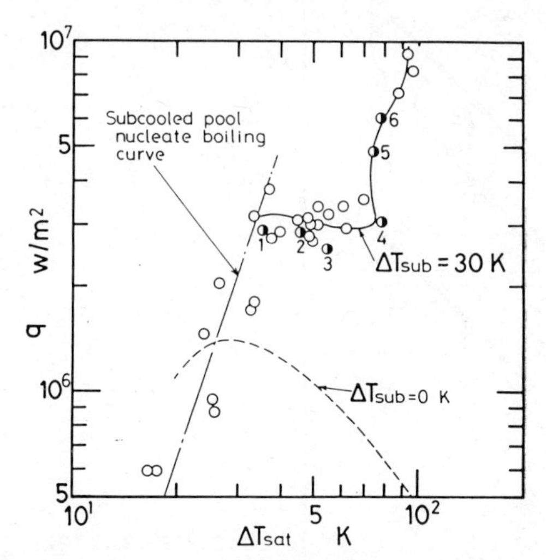

FIGURE 8.12. Boiling curve and the measured boiling regime. [From S. Inada et al., Liquid–Solid Contact State in Subcooled Pool Transition Boiling System, *Trans. ASME J. Heat Transfer*, **108**, 219–221 (Feb. 1986), Fig. 3. Reproduced with permission.]

encountered in LTL forced-convection evaporators typically operating slightly above the normal boiling point.

8.2.2.2 Forced-Convection Boiling. It is important to keep in mind that the characteristic boiling curves shown in Fig. 8.11 are based, according to Rohsenow, on "experiments in pool boiling with an electrically heated horizontal wire submerged in a tank at saturation temperature." This is in accordance with the test method proposed by Nukiyma [4.3] who first showed this now well-known characteristic boiling regime curve in the 1930s (as highlighted in Section 4.2). In other words, Fig. 8.11 depicts *natural-convection* boiling regimes in the absence of significant forced bulk liquid and gas motion typical of forced-convection tubular evaporators. The following additional excerpts from Rohsenow [4.28 pp. 8, 9] are still particularly relevant.

> The process of boiling inside a tube with a positive flow of fluid through the tube is more complicated than the pool-boiling processes.
>
> If the fluid is heated by passing an electric current through the tube wall, the I^2R energy must be transferred away from the tube; hence, the tube-wall temperature

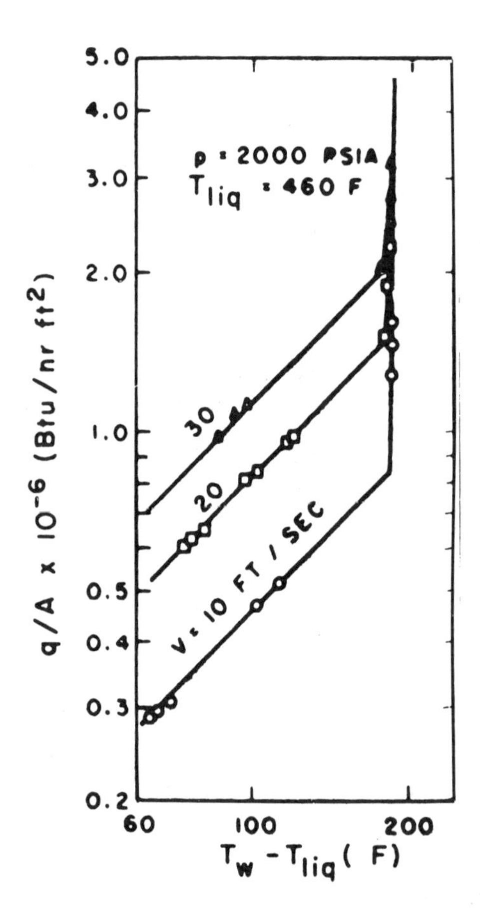

FIGURE 8.13. Effect of velocity in forced-convection boiling. [From W. M. Rohsenow, Heat Transfer with Boiling, *ASHRAE Trans.*, **72**, 8 (1966), Fig. 5. Reproduced with permission.]

will increase until this amount of heat is transferred.... [This is what differentiates electric or nuclear applications from normal fluid-to-fluid two-phase heat exchangers. In the latter case $(\Delta T)_{\text{evap}}$ and, therefore, the smaller wall to evaporating fluid temperature differentials $\Delta T_w = T_w - T_{\text{sat}}$ are the independent variables whereas the heat flux q'' is the dependent variable!]

If the liquid entering at the bottom of a heated [vertical] tube is well below the saturation temperature, boiling may take place at the surface and the bubbles travel out into the main stream where they condense. This process has been called local boiling or surface boiling. No net vapor leaves the tube, but the rate of transfer is greatly increased because of the stirring action of the bubbles. Curves showing the effect of fluid velocity at a particular value of liquid sub-cooling are shown in [Fig. 8.13]. The lines with a slope of unity are in the nonboiling region and show a velocity effect of $[q'' \equiv]\ (q/A) \sim V^{0.8}$, as expected in nonboiling. When Δt is increased enough to cause boiling to begin, the curve bends sharply upward. [Figure 8.13] shows that, in the boiling region, the curves tend to merge on this type of graph....

A summary presentation of all of these effects is shown in [Fig. 8.14]. The lower solid line BCDE represents pool boiling of liquid at its saturation temperature. The lower dotted line shows the effect of subcooling the liquid; the curve in the natural-convection region is raised, and the peak heat flux is raised.

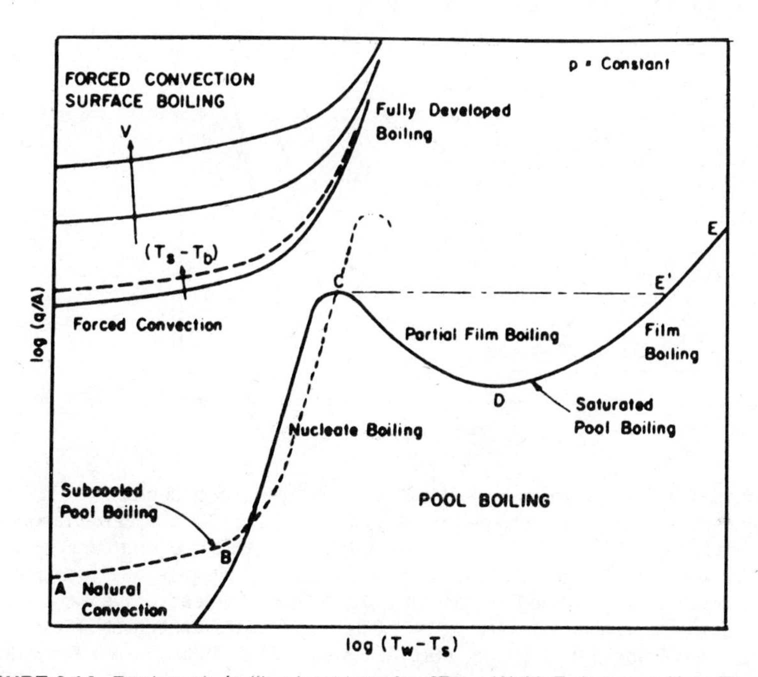

FIGURE 8.14. Regimes in boiling heat transfer. [From W. M. Rohsenow, Heat Transfer with Boiling, *ASHRAE Trans*, **72**, 9 (1966), Fig. 7. Reproduced with permission.]

The curves at the upper left represent the effect of forced convection. At the far left, where T_w approaches T_s, these curves asymptotically approach single-phase forced-convection heat transfer rates for sub-cooled liquids $[q/A \equiv q'' \equiv \alpha(T_s - T_b)]$, where $[\alpha]$ is the forced convection, non-boiling heat transfer coefficient. At the right, as $[q/A \equiv q'']$ increases, the curves on a log–log plot appear to merge into a single curve with a slope of approximately 3 for most commercial surfaces.

From a cursory look at Figs. 8.13 and 8.14 it is clear in the context of the above assessment that increasing fluid bulk velocities tend to suppress nucleate boiling and significantly higher levels of wall-to-fluid superheat are required to initiate nucleation. Consequently, *nucleate boiling is even less likely in* LTL *forced-convection* evaporators than in LTL *natural-convection* or so-called[10] flooded evaporators [8.8].

8.2.2.3 Effect of Pressure on Nucleate Boiling. The significant impact of pressure can be assessed from empirical results such as those shown in Fig. 8.15 or theoretically on the basis of Eq. (5) in Ref. 4.28.

$$T_v - T_{\text{sat}} \simeq 2\left(\frac{R_v T_{\text{sat}}^2}{h_{lg} P_l}\right)\left(\frac{\sigma}{r}\right) \tag{8.124}$$

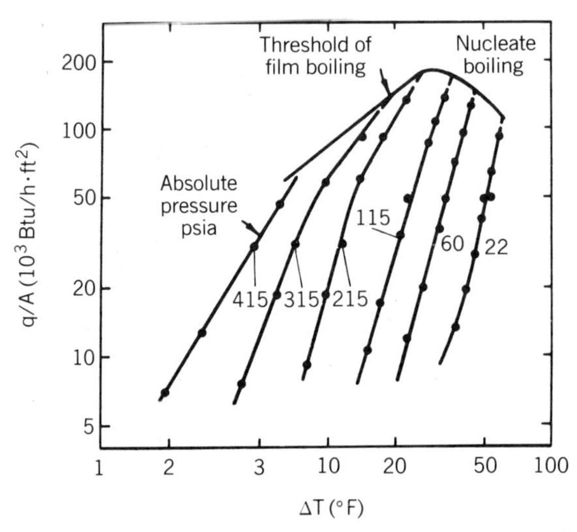

FIGURE 8.15. Effect of pressure on pool boiling curve. With a critical pressure $P_{cr} = 3226$ psia for water, the absolute pressures of 415, 315, 215, 115, 60, and 22 correspond to reduced pressures $P^+ = P/P_{cr}$ of 0.129, 0.098, 0.067, 0.036, 0.019, and 0.0068, respectively. By comparison, the reduced pressure at the normal boiling point P_b^+ of common halogenated refrigerants falls in the range of 0.020 to 0.032, whereas for the two dissimilar polar substances ammonia and water $P_b^+ \simeq 0.0089$ and 0.0046, respectively. [From W. M. Rohsenow, Heat Transfer with Boiling, *ASHRAE Trans.*, **72**, 14 (1966), Fig. 16. Reproduced with permission.]

where T_v is the absolute temperature of the superheated vapor bubble, R_v is the gas constant of the particular chemical species, σ is the surface tension of the liquid–vapor interface, and r is the radius of an assumed spherical gas bubble. In Rohsenow's words, Eq. (8.124) "represents an approximate expression for the superheat required for equilibrium of a bubble of radius r. Nuclei of radius greater than r from [Eq. (8.124)] should become bubbles and grow; those of smaller radius should collapse."

To determine the impact of pressure on nucleation, it is convenient to modify Eq. (8.124) by substituting $\mathscr{R}/M$ for R_v according to the ideal gas laws which are applicable at $P^+ = P/P_{cr} \ll 1$:

$$T_v - T_{sat} \simeq \frac{2}{Mh_{lg}/T_{sat}}\left(\frac{\mathscr{R}T_{sat}}{P_l}\right)\frac{\sigma}{r} \tag{8.125a}$$

which reduces to

$$T_v - T_{sat} \simeq \frac{2}{\text{Trouton entropy}}\left(\frac{\mathscr{R}T_{sat}}{P_l}\right)\left(\frac{\sigma}{r}\right) \tag{8.125b}$$

with Mh_{lg}/T_{sat} = Trouton entropy.

Since $\mathscr{R}T/P_l \simeq Mv_g = Mv_{cr}(v_g/v_{cr}) = Mv_{cr}v_g^+$, the above equation can be expressed most generally as follows:

$$T_v - T_{sat} \simeq \frac{2}{\text{Trouton entropy}} Mv_{cr}v_g^+\left(\frac{\sigma}{r}\right) \tag{8.126}$$

Focusing on conditions typical of LTL heat pump/refrigeration systems operating at "low side" reduced pressures $P^+ \ll 1$, we see from Fig. 6.4 that the Trouton entropy has approximately the same numerical value for all common fluids and remains substantially constant. Therefore, Eq. (8.126) can be written

$$T_v - T_{sat} \simeq \text{constant}(Mv_{cr})\left(\frac{\sigma}{r}\right)v_g^+ \tag{8.127a}$$

which indicates that, for a fixed bubble radius r,

$$T_v - T_{sat} \propto Mv_{cr} \tag{8.127b}$$

$$T_v - T_{sat} \propto \sigma \tag{8.127c}$$

$$T_v - T_{sat} \propto v_g^+ \tag{8.127d}$$

Equation (8.127) further suggests that

$$T_v - T_{sat} \to 0 \quad \text{as } P^+ \to 1 \tag{8.128}$$

since $\sigma \to 0$ as $P^+ \to 1$. Furthermore, much higher wall-to-fluid temperature differentials are necessary to initiate nucleate boiling regimes at low reduced pressure $P^+ \ll 1$ typical of low side reverse power evaporators than at the higher reduced pressures $P^+ \lessapprox 1$ typical of high side (nuclear and/or fossil) steam power generation cycles, since $v_g^+ = 1/\rho_g^+$ is several orders of magnitude larger at $P^+ \ll 1$ than at $P^+ \lessapprox 1$, as shown in Fig. 6.3. This somewhat theoretical conclusion is further reinforced by referring to the empirical nucleate boiling curves shown in Fig. 8.15 (see corresponding reduced pressures $P^+ = P/P_{cr}$ shown in Figure caption).

These semianalytical conclusions can be summarized in the form of a general guideline as follows:

> ***GUIDELINE 27*** *Nucleate boiling regimes are most improbable in* LTL *forced-convection evaporators designed to operate in stable annular-spray flow regimes at low reduced pressures* $P^+ \ll 1$ *at the typical low wall-to-fluid temperature differentials estimated in numerical analyses* 8.5 *and* 8.6.

Guideline 27 is supported by empirical data obtained under actual conditions typical of reverse power cycles, namely, with bulk evaporation in the so-called vapor quality region as discussed below.

8.2.2.4 Forced-Convection Evaporation in the Quality Region. We deliberately ignore the section in Ref. 4.28 entitled "Forced Convection—Boiling Correlations Subcooled Boiling" (which is mostly of interest in specialized HTL power generation applications) to focus on conditions more typical of LTL forced-convection evaporators with bulk evaporation at average flowing qualities of the order of $x_m = \frac{1}{2}(x_1 + x_2) \approx 0.5$. Rohsenow calls this two-phase domain the *quality region* (to differentiate it from the subcooled boiling region where the equilibrium state of the fluid is in the subcooled domain, that is, at a *bulk* enthalpy $h < h_l$ or an apparent quality $x < 0$) and adds the following significant comments [4.28, pp. 20–21] which are still relevant [8.9]:

> Many correlations of heat transfer data have been proposed for this higher quality region.... Common to most of them is the concept of modifying the single-phase heat transfer prediction equation by an additive term to account for the increased velocity in the quality region....
>
> Clearly the question of which correlation should be used in this high quality region is not resolved at the present time.

This general conclusion is based on a brief assessment of representative empirical results and/or correlations proposed by Bo Pierre [5.9], and other researchers.

In the following analysis we focus on the empirical two-phase heat transfer coefficient shown in Fig. 8.16, because these data are especially emphasized in

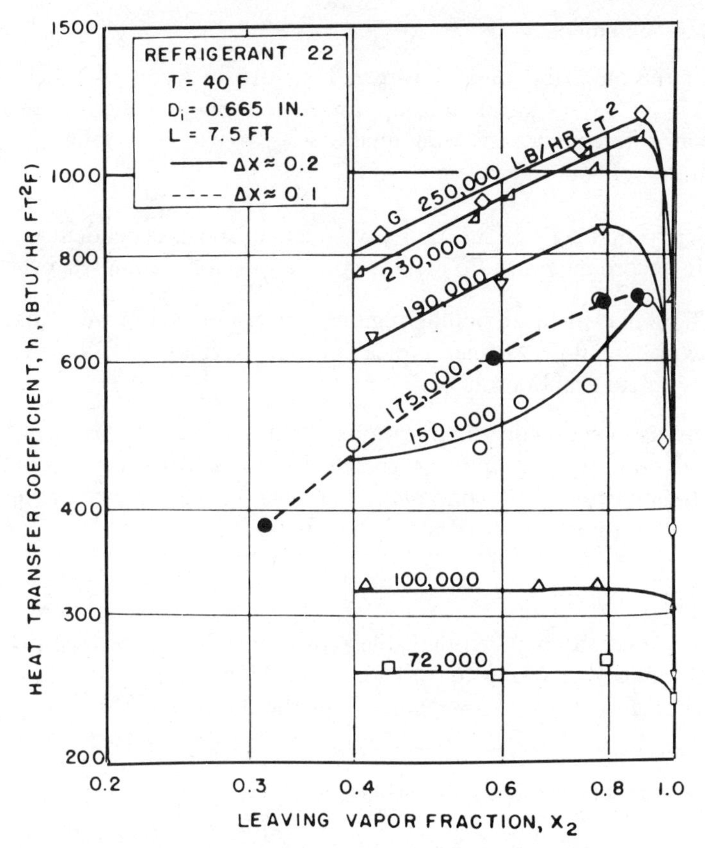

FIGURE 8.16. Heat transfer coefficient versus vapor fraction for partial evaporation. [From S. W. Anderson et al., Evaporation of Refrigerant 22 in a Horizontal 3/4-in. OD Tube, *ASHRAE Trans.*, **72**, Pt. 1, 34 (1966), Fig. 4. Reproduced with permission.]

Rohsenow's assessment [4.28] as well as in the most recent (1985) edition of the *ASHRAE Handbook, Fundamentals* [8.8].

NUMERICAL ANALYSIS 8.7

Alternative interpretation of Fig. 8.16.

Background

The prime intent of this analysis is to verify the most far-reaching (and generally ignored) conclusion derived by Bo Pierre [5.9] for forced-convection LTL direct-expansion evaporators with incomplete evaporation and confirm Guideline 27.

Problem statement

Recast the empirical data shown in Fig. 8.16 in terms of a heat flux $\bar{q}''$ versus $\overline{\Delta T}_w = \bar{T}_w - \bar{T}_{sat}$ relationship for a prescribed quality change $\Delta x = x_2 - x_1$ at different mean flowing qualities $x_m = \frac{1}{2}(x_1 + x_2)$. On the basis of these data:

(a) Verify that $\overline{\Delta T}_w$ is substantially constant and independent of the heat flux $\bar{q}''$ in accordance with Bo Pierre's correlations for incomplete evaporation.

(b) Show that nucleate boiling regimes are suppressed in wet wall annular-spray two-phase flow regimes typical of direct-expansion LTL evaporators with incomplete evaporation.

(c) Discuss the practical implications of the sharp drop shown in Fig. 8.16 for the two-phase heat transfer coefficient α_{tp} at exit vapor qualities x_2 approaching unity and a fortiori at $x_2 > 1$ when the effluent vapor equilibrium state is well within the superheated vapor region.

Solution

The key data summarized in Table 8.18 are based on the empirical results shown graphically in Fig. 8.16 (in U.S. technical units) and the following fixed parameters: $L = 7.5$ ft, $D_i = 0.665/12$ ft, therefore, $L/D_i = 7.5 \times 12/0.655$

TABLE 8.18. Alternative Interpretation of Fig 8.16

	(a)	(b)	(c)	(d)	(e)	(f)
(1) $10^2 G$ lb / h · ft^2	720	1000	1500	1900	2300	2500
(2) $\bar{q}''$ Btu / hr · ft^2	**2307**	**3204**	**4806**	**6088**	**7369**	**8010**
For constant average quality $x_m = 0.5$ at constant $x_2 = 0.6$, $\Delta x = 0.2$, and $x_1 = 0.4$						
(3) $\bar{\alpha}_{tp}$ Btu / h · ft^2 · °F	255	320	510	750	940	980
(4) $\bar{T}_w - \bar{T}_{sat}$ (°F)	**9.05**	**10.01**[a]	**9.42**	**8.12**	**7.84**[b]	**8.17**
For constant average quality $x_m = 0.3$ at constant $x_2 = 0.4$, $\Delta x = 0.2$, and $x_1 = 0.2$						
(5) $\bar{\alpha}_{tp}$ Btu / h · ft^2 · °F	255	320	475	620	760	800
(6) $\bar{T}_w - \bar{T}_{sat}$ (°F)	**9.05**[b]	**10.01**	**10.11**[a]	**9.82**	**9.70**[b]	**10.01**
(7) $\dfrac{\bar{q}''}{\bar{q}'' \text{ of column (a)}}$	1	1.38	2.08	2.64	3.19	3.47
(8) Theoretical[c] $\dfrac{\bar{T}_w - \bar{T}_{sat}}{(\bar{T} - \bar{T}_{sat}) \text{ at } \bar{q}'' \text{ of column (a)}}$	1	1.11	1.28	1.38	1.47	1.51
(9) Theoretical[d] ($\bar{T}_w - \bar{T}_{sat}$) (°F)	**9.05**	**10.07**	**11.6**	**12.50**	**13.28**	**13.65**

[a]Maximum $\bar{T}_w - \bar{T}_{sat}$ on lines (4) and (6).

[b]Minimum $\bar{T}_w - \bar{T}_{sat}$ on lines (4) and (6).

[c]Per Eq. (8.136a).

[d]Calculated from line (8) with empirical $\bar{T}_w - \bar{T}_{sat} = 9.05$ shown under column (a), lines (4) and (6), per Eq. (8.136b).

= 135.4. From the U.S. technical unit version of Table A.7 in Appendix A for refrigerant R-22 we obtain $h_{lg} = 86.72$ Btu/lb at the constant saturation temperature of $T_{sat} = 40°F$ shown in Fig. 8.16, which is typical of high back pressure air conditioning applications. The corresponding saturation pressure $P_{sat} = 83.21$ psia, or a reduced pressure $P^+ \equiv P_{sat}/P_{cr} = 83.21/721.91 = 0.115$. Note that the reduced evaporating pressure would be significantly lower in lower-temperature pure refrigeration as opposed to air conditioning applications.

The total elementary single-tube heat exchanger cooling capacity $\dot{Q}_e$ can be expressed according to

$$\dot{Q}_e = \dot{m} h_{lg} \, \Delta x = G \Omega_i h_{lg} \, \Delta x \tag{8.129}$$

on the basis of the energy conservation law. Dividing both sides of Eq. (8.129) by $A_i = a_i L$ yields the average heat flux $\bar{q}''$:

$$\bar{q}'' \equiv \frac{\dot{Q}_e}{a_i L} = \frac{G \Omega_i h_{lg} \, \Delta x}{a_i L} \tag{8.130}$$

The above equation reduces, with $\Omega_i / a_i = r_{h,i} = \frac{1}{4} D_i$ for circular tubes, to

$$\bar{q}'' = G \left(\frac{h_{lg} \, \Delta x}{4L/D_i} \right) \propto G \tag{8.131}$$

since all the terms within the parentheses are constant for a fixed T_{sat}, Δx, and L/D_i. The numerical values shown on line (2) of Table 8.18 were calculated from Eq. (8.131) with the stated constant test parameters which yield for the constant term in parentheses

$$\left(\frac{h_{lg} \, \Delta x}{4L/D_i} \right) = \frac{(86.72)(0.2)}{4(135.4)} = 3.202 \times 10^{-2}$$

or

$$\bar{q}'' = 3.202 G \times 10^{-2} \text{ Btu/h} \cdot \text{ft}^2 \tag{8.132}$$

The average wall to evaporating fluid temperature driving force $\overline{\Delta T}_w \equiv \bar{T}_w - \bar{T}_{sat}$ is obtained from the defining equation for the heat transfer coefficient

$$\bar{q}'' = \bar{\alpha}_{tp} \left(\bar{T}_w - \bar{T}_{sat} \right) \tag{8.133}$$

which yields the actual temperature differential

$$\bar{T}_w - \bar{T}_{sat} = \frac{\bar{q}''}{\bar{\alpha}_{tp}} \tag{8.134}$$

for known empirical values of $\bar{q}''$ and $\bar{\alpha}_{tp}$ or, after introducing the expression for $\bar{q}''$ per Eq. (8.131) in Eq. (8.134),

$$\bar{T}_w - \bar{T}_{sat} = \frac{G}{\bar{\alpha}_{tp}}\left(\frac{h_{lg}\,\Delta x}{4L/D_i}\right) \tag{8.135a}$$

For the case under consideration, the above equation reduces, with Eq. (8.132), to

$$\bar{T}_w - \bar{T}_{sat} = 3.202\frac{G}{\bar{\alpha}_{tp}}10^{-2} \tag{8.135b}$$

The numerical values for $\bar{T}_w - \bar{T}_{sat}$ shown on line (4) for $x_m = 0.5$ and $\Delta x = 0.2$ were obtained from Eq. (8.135b) with the empirical values of $\bar{\alpha}_{tp}$ shown on line (3), which were read directly (with an estimated accuracy of about $\pm 10\%$) from the $\Delta x = 0.2$ curves shown on Fig. 8.16 at different G for $x_2 = 0.6$; therefore, $x_1 = 0.6 - \Delta x = 0.4$ and $x_m = \frac{1}{2}(x_1 + x_2) = 0.5$.

The numerical values shown on line (6) for a lower average flowing quality $x_m = 0.3$ at the same $\Delta x = 0.2$ were computed in a similar manner with the empirical values of $\bar{\alpha}_{tp}$ shown on line (5) read from Fig. 8.16 at different G for $x_2 = 0.4$; therefore, $x_1 = 0.2$ and $x_m = 0.3$.

(a) ΔT_w = *Constant*, *Independent of* q''. Referring to line (4) in Table 8.18, we see that $\overline{\Delta T}_w = \bar{T}_w - \bar{T}_{sat}$ is practically constant and independent of G or q'' in spite of the fact that $\bar{q}''$ varies by a factor of 3.47 over the full empirical range analyzed [as shown on line (7)]. For $x_m = 0.5$ = constant and $\Delta x = 0.20$ = constant, the arithmetic mean of columns (a)–(f) is $(\bar{T}_w - \bar{T}_{sat})_{am} = 8.8°\text{F}$ (or 4.9°C). Line (4) shows a maximum variation from this arithmetic mean of $+14.1\%/-10.5\%$, which is roughly in line with the accuracy of the empirical data. Note that Anderson et al. [4.29 p. 32] state that "of the 43 tests, 24 had a heat balance agreement of less than 5%, an additional 18 had a value between 5 and 10%, and one test had a value of 11%." To this inherent heat balance inaccuracy must be added additional uncertainties associated with the calculation of the overall heat transfer coefficient U and breaking down U into its components (see Section 7.6).

The situation is quite similar at the lower average flowing quality $x_m = 0.3$ = constant for the same $\Delta x = 0.2$ = constant as indicated by the empirical value of $\bar{T}_w - \bar{T}_{sat}$ shown on line (6). The arithmetic mean $(\bar{T}_w - \bar{T}_{sat})_{am} = 9.8°\text{F}$ (or 5.4°C) with a maximum/minimum variation of only $+3.3\%/-1.0\%$, that is, well within the accuracy limits of the original test data.

These results are fully in line with the most significant and astonishing conclusion derived by Bo Pierre for *incomplete* evaporation [5.9], namely, that for a given refrigerant, fixed x_m, Δx, T_{sat}, and channel geometry, $\overline{\Delta T}_w$ remains constant and independent of q''.

(b) *Suppression of Nucleate Boiling Regimes.* As shown previously, nucleate boiling regimes are characterized by a heat flux versus $T_w - T_{sat}$ relationship of the general form $q'' \propto (T_w - T_{sat})^a$ for a given geometry, wall material, fluid, and T_{sat} with an exponent a typically of the order of 3 for most commercial surfaces [4.28]. This means that $q'' \propto (\Delta T_w)^3$ or conversely $(\Delta T_w) \propto (q'')^{1/3}$. If nucleate boiling regimes were controlling, it is clear that the theoretical ratio shown on line (8) in Table 8.18 should vary with the ratio $\bar{q}''/(\bar{q}'')_{\text{column (a)}}$ shown on line (7) according to

$$\frac{\bar{T}_w - \bar{T}_{sat}}{(\bar{T}_w - \bar{T}_{sat}) \text{ at } \bar{q}'' \text{ of column (a)}} = \left(\frac{\bar{q}''}{\bar{q}'' \text{ of column (a)}} \right)^{1/3} \tag{8.136a}$$

and, one would therefore expect $T_w - T_{sat}$ to vary according to

$$\bar{T}_w - \bar{T}_{sat} = 9.05 \left(\frac{\bar{q}''}{\bar{q}'' \text{ of column (a)}} \right)^{1/3} \tag{8.136b}$$

with the empirical value $\bar{T}_w - \bar{T}_{sat} = 9.05°\text{F}$ shown under column (a), lines (4) and (6). The numerical values obtained from Eq. (8.136b) are shown on line (9).

Comparing the constant and even slightly *decreasing* empirical actual values of $\bar{T}_w - \bar{T}_{sat}$ shown on lines (4) and (6) with the increasing theoretical $\bar{T}_w - \bar{T}_{sat}$ values typical of nucleate boiling shown on line (9) clearly infers that nucleate boiling regimes have no measurable impact within the test range analyzed in Table 8.18. Recalling that the data reported by Anderson et al. [4.29] and in Fig. 8.16 focus on air conditioning applications, that is, on the highest reduced pressure encountered in reverse power cycles, it is clear in the context of Section 8.2.2.3 that nucleate boiling regimes are even less likely to be a factor in lower-temperature refrigeration applications (i.e., lower reduced pressures P^+ for a given refrigerant). This further reinforces the semianalytical conclusions summarized in Guideline 27.

(c) *Partially Dry Wall Regimes.* As shown in Fig. 8.16 the heat transfer coefficient drops sharply when $x_2 > 0.9$ at the higher mass flow rate G and corresponding higher heat fluxes $\bar{q}'' \propto G$ per Eq. (8.131). This massive deterioration is due to a liquid deficiency at the wall, resulting in *partially dry wall* flow regimes. When x_2 approaches a critical limit $x_{\text{dry wall}} \lessapprox 1$, a major portion of the evaporator tube operates in the fully dry wall regimes at local α_{tp} values converging rapidly toward the much lower single-phase (dry wall) gas heat transfer coefficients. This provides a more fundamental rationale for the significant advantages of the compact chiller system designs described in Section 4.3.3. Clearly, it is far more effective to break down the forced-convection evaporating system in two sections. The first one operates in the fully *wet wall* (annular + spray) regimes with incomplete evaporation at outlet, that is,

$x_2 < x_{\text{dry wall}}$. In the second section of the system, the postevaporator sketched in Fig. 4.5, the slightly wet vapor is fully "dried" by raising its enthalpy to a level corresponding to the desired bulk vapor superheat typically of the order of 10°C (~ 18°F), using the available "hotter" refrigerant coming from the high-pressure condenser as the heat source. In this manner the external cooling load $\dot{Q}_e$ can be handled in the far more effective fully wet wall evaporating regimes at a wall-to-fluid temperature difference $(\overline{T}_w - \overline{T}_{\text{sat}})$ as low as a fraction of a degree, if one wishes, but certainly significantly lower than the minimum temperature driving force necessary to fully dry and superheat the outlet vapor within the evaporator by some 10°C. The compact postevaporator heat interchanger can be designed conservatively by treating the slightly wet vapor ($x_2 \lessapprox 0.97$ in the compact internally finned chiller system described in Chapter 4) as if it were dry. The resulting combined package consisting of an evaporator and postevaporator heat interchanger will be significantly more compact and competitive than the conventional single evaporator bundle operating with fully superheated exit vapor at $x_2 > 1$. The massive size reductions highlighted in Fig. 8.9 are due in large part to the fact that *fully wet* wall regimes are possible at exit vapor qualities as high as 0.95–0.98 with these internally finned small hydraulic tube diameters [4.4, 4.6]. There is indeed no surer way to waste heat transfer surface in compact LTL evaporators than to impose excessively high *dry wall* to fluid temperature differentials by superheating the refrigerant within the evaporator proper!

8.2.3 Low Thermal Lift Forced-Convection Evaporators

For forced-convection evaporation inside tubes in LTL systems under conditions typical of single-stage refrigeration and/or heat pump applications, the authors of the 1985 edition of *ASHRAE Handbook, Fundamentals* state that Bo Pierre's equation "is recommended for broadest application to refrigerant evaporation in tubes. It fits a wide range of R-12 and R-22 data." Consequently, we focus on the empirical correlations proposed some 40 years ago by the Swedish researcher Bo Pierre, with special emphasis on incomplete evaporation at tube outlet since this leads to more compact and/or more economical evaporator designs.

8.2.3.1 Bo Pierre's General Correlation. Bo Pierre's proposed correlation has the following general form for complete *and* incomplete evaporation

$$(\text{Nu})_l = c\left[(\text{Re})_l^2 K_f\right]^m \propto (\text{Re})_l^{2m}(K_f)^m \tag{8.137a}$$

where

$$(\text{Nu})_l \equiv \frac{\bar{\alpha}_{\text{tp}} D}{k_l} = \text{liquid-based Nusselt number} \tag{8.137b}$$

Note that $(\mathrm{Nu})_l$ is based on the conductivity k_l of the liquid phase. The liquid Reynolds number $(\mathrm{Re})_l$ and Bo Pierre boiling number K_f in the above correlations are defined by Eqs. (5.1b) and (5.1c).

Bo Pierre could correlate his extensive heat transfer test data with the following constant exponents:

$$m = 0.4 = \text{constant for complete evaporation,}$$
$$x_2 > 1\ (5\text{–}10°\mathrm{C}\ \text{superheat}) \tag{8.137c}$$
$$m = 0.5 = \text{constant for incomplete evaporation}$$
$$x_2 \le 0.9 \tag{8.137d}$$

and two different constants c in Eq. (8.137a) for complete and incomplete evaporation.

8.2.3.2 Discussion of Bo Pierre's Empirical Correlation for Incomplete Evaporation.

For a fixed tube geometry when D_i, L/D_i, $\bar{e}/D_i$, $A_i = a_i L =$ constant, and $\Omega_i =$ constant, a prescribed refrigerant, evaporator temperature, and *given* quality change $\Delta x = x_2 - x_1 =$ constant, all the terms in Eq. (8.137a) are fixed except the flow rate of the "cold" fluid $\dot{m}_C \equiv G\Omega_i$. Furthermore, $\dot{Q}_e = \dot{m}_C h_{lg} \Delta x \propto \dot{m}_C$ since h_{lg} is also constant. It follows that the average heat flux $\bar{q}_i'' \equiv \dot{Q}_e / a_i L \propto \dot{m}_C$, where $\bar{q}_i''$ is based on the internal (refrigerant side) total heat exchanger surface. Therefore,

$$\bar{q}_i'' \propto \dot{m}_C \tag{8.138}$$

In the specific case of incomplete evaporation at $x_2 \le 0.9$, with an exponent $m = 0.5$ according to Eq. (8.137d), it is evident from Eq. (8.137a), since the liquid conductivity k_l and viscosity μ_l are also fixed at a prescribed evaporating temperature, that the following relationship must also be satisfied:

$$\bar{\alpha}_{\mathrm{tp}} \propto (\mathrm{Re})_l^1 \propto \dot{m}_C \tag{8.139}$$

This indicates that the two-phase heat transfer coefficient with incomplete evaporation is directly proportional to the flow rate under these prescribed conditions. By definition,

$$\bar{\alpha}_{\mathrm{tp}} \equiv \frac{\bar{q}_i''}{\overline{\Delta T_i}} \equiv \frac{\bar{q}_i''}{\overline{\Delta T_w}} \tag{8.140a}$$

where Bo Pierre's term for the internal temperature difference $\overline{\Delta T_i}$ is identical to $\overline{\Delta T_w} = \bar{T}_w - \bar{T}_{\mathrm{sat}}$ used in this text; therefore,

$$\overline{\Delta T_i} \equiv \left(\overline{\Delta T}\right)_w \equiv \bar{T}_w - \bar{T}_{\mathrm{sat}} \tag{8.140b}$$

It is evident that Eqs. (8.138)–(8.140) can be satisfied simultaneously only if $\overline{\Delta T}_i$ = constant and is independent of $\dot{m}$ and consequently $\bar{q}_i''$ under the prescribed conditions. This rather startling conclusion is based essentially on the fact that the empirical exponent m in Eq. (8.137a) is equal to unity for incomplete evaporation within the range of conditions investigated by Bo Pierre (and confirmed by subsequent researchers [4.4, 4.6, 4.52] as summarized below:

$$\overline{\Delta T}_i \equiv \left(\overline{\Delta T}\right)_w \equiv \bar{T}_w - \bar{T}_{\text{sat}} = \text{constant and independent of } \dot{m}_C \text{ or } \bar{q}_i'' \quad (8.141a)$$

Validity: Prescribed refrigerant, evaporating temperature, inlet and outlet vapor quality (x_1 and x_2), and fixed tube geometry (D_i, L/D_i, $\bar{e}/D_i$) (8.141b)

within the following empirical limits (Bo Pierre's test range in [5.9]).

$x_1 \geq 0.08$ and $x_2 \leq 0.90$; $930 \lesssim q_i'' \lesssim 34900$ W/m^2 (or $295 \lesssim \bar{q}_i'' < 11{,}000$ Btu/h $\cdot$ ft^2 in U.S. technical units)

Refrigerants R-12 and R-22, evaporating temperatures of 0 to -20°C corresponding to reduced pressures $P^+ = (0.0366)$ to (0.0749) for R-12 and $P^+ = 0.0492$ to 0.100 for R-22 and traces of oil (volumetric fraction measured in condensed liquid of less than 0.5% but greater than 0%). (8.141c)

Geometry: Circular smooth copper tubes of 12 and 18 mm internal diameters, $L = 4.08$–9.5 m, and typical aspect ratios $117 \lesssim L/D \lesssim 796$

For the specific case of refrigerant R-12 with incomplete evaporation, Bo Pierre derived a relationship of the following type:

$$\overline{\Delta T}_i = \bar{T}_w - \bar{T}_{\text{sat}} = \text{constant}\,(D_i)\left(\frac{h_{lg}\,\Delta x}{L}\right)^{1/2} \qquad (8.142)$$

where the constant is a function of fluid properties only (and dimensional constants in the European technical system of units used by Bo Pierre [5.9]). He concluded (italics added) that

> it will be seen that with incomplete evaporation, we get the possibly surprising result that for a given condition $\overline{\Delta T}_i$ *depends only on the dimension of the evaporator tube*... and, using [Eq. (8.142)] we can determine the value of $\overline{\Delta T}_i$ prevailing on the refrigerant side of a given evaporator.

8.2.3.3 The Bo Pierre Paradox. Although many research workers have since substantially confirmed Bo Pierre's correlations for the average heat transfer coefficient, little attention has been given to his significant finding that $\overline{\Delta T}_i \equiv \overline{\Delta T}_w = \bar{T}_w - \bar{T}_{\text{sat}}$ = constant and independent of $\dot{m}_C$ at specified T_{sat}, Δx for a given refrigerant and tube geometry (D_i, $\bar{e}/D_i$, L/D_i) and

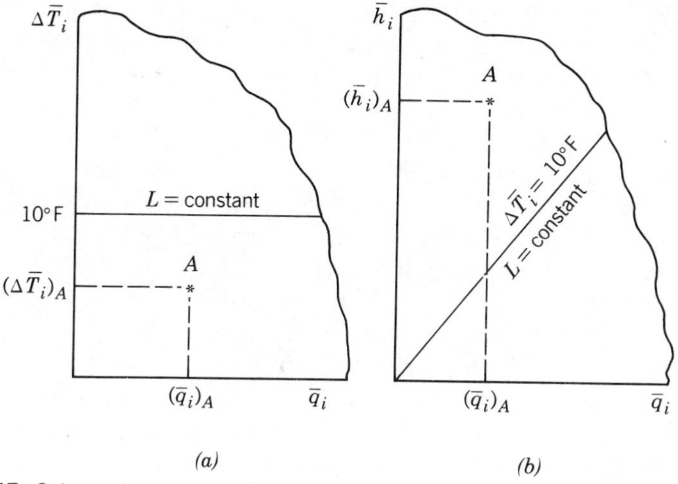

FIGURE 8.17. Schematic representation of Bo Pierre's Eq. (8.142) for a given refrigerant. (*a*) $\overline{\Delta T}_i \equiv \overline{T}_w - \overline{T}_{\text{sat}}$ versus $\overline{q}_i \equiv \overline{q}_i''$. (*b*) $\overline{h}_i \equiv \overline{\alpha}_{\text{tp}}$ versus $\overline{q}_i \equiv \overline{q}_i''$.

therefore independent of $\overline{q}_i'' = \dot{m}_C \Delta x\, h_{lg}/a_i L$. We may accept this empirical conclusion as an observed fact on the basis of the extensive work of Bo Pierre and others (including the empirical data of Anderson et al. [4.29] discussed in the previous section). However, this conclusion logically leads to the following corollary: The discrete values of $\overline{\Delta T}_i$ = constant under the specified conditions must also be dependent in some fashion on external conditions [external coupling $(\dot{m}c_p)_H$, α_H, a_H, and the average hotter water temperature $\overline{T}_H$ that prevailed during the single-tube tests]. Although Bo Pierre's equations correlate the results of his and other researchers' tests under typical normal refrigeration design conditions, they cannot be generally valid, as is evident from the following considerations.

The existence of other actual $\overline{\Delta T}_i$ values smaller than those computed from Eq. (8.142) for any given fixed geometry T_{sat}, x_1, Δx, and refrigerant can be explained readily in the following manner. Let us select a length L short enough to obtain from Eq. (8.142) a relatively large $\overline{\Delta T}_i$ = 10°F as shown in Fig. 8.17*a*. We may now choose to conduct our tests at a water inlet temperature $T_{1,H}$ sufficiently low so that $\Delta T_{\max} = T_{1,H} - T_{2,C} < 10$°F but at high values of $\dot{m}_H$ and α_H (the latter depends on the external geometry and $\dot{m}_H$). Under these conditions a finite quantity of heat ($\overline{q}_i'' \neq 0$) can be transferred from the external hot fluid to the refrigerant evaporating inside the tube. Since the average wall-to-refrigerant temperature must be smaller than $\Delta T_{\max}$, we have established the existence of a lower value of $\overline{\Delta T}_i$, identified by point A in Fig. 8.17*a*,

$$\overline{\Delta T}_{i,A} \leq (\Delta T)_{\max} < 10°\text{F} \quad \text{obtained from Eq. (8.142) at } (\overline{q}_i'')_A \neq 0 \qquad (8.143)$$

For a fixed saturation temperature inside a given tube, we can conceive, by varying $T_{1,H}$, $(\dot{m}c_p)_H$, and α_H, of an infinite number of $\overline{\Delta T_i}$. The $\overline{\Delta T_i}$ = constant obtained from Eq. (8.142) is therefore a discrete value within a yet not fully defined domain of conceivable ΔT_i, which may exist under in/out SSSF conditions.

Since by definition $\bar{\alpha}_{tp} \equiv \bar{q}_i''/\overline{\Delta T_i}$, it is clear that point A at $(q_i'')_A$ in Fig. 8.17*a* must yield a higher average heat transfer coefficient $(\bar{\alpha}_{tp})_A$ as highlighted in Fig. 8.17*b* than the value computed according to Bo Pierre's Eq. (8.137a) or (8.142) which would predict a $\overline{\Delta T_i} = 10°F$ = constant > $(\overline{\Delta T_i})_A$. It is clear that the existence of a wide domain of $\overline{\Delta T_i}$ at any fixed value of $\bar{q}_i''$ will as a corollary require the existence of a wide domain of $\bar{\alpha}_{tp}$. Thus, different numerical values of $\bar{\alpha}_{tp}$ may be obtained at a prescribed $\bar{q}_i''$ and the practical usefulness of such an ambiguous heat transfer coefficient becomes questionable. Before verifying empirically the above theoretical conclusions, it seems appropriate to discuss the practical usefulness of conventional correlations for $\bar{\alpha}_{tp}$ in typical low thermal lift evaporators.

8.2.3.4 Discussion of Conventional Correlations for $\bar{\alpha}_{tp}$. It may seem surprising that, in spite of the apparent fundamental limitations of the $\bar{\alpha}_{tp}$ correlations proposed by Bo Pierre [5.9] and others [5.5, 5.6], these empirical relations have produced useful results and contributed to significant advances in design practice.

A probable explanation is that most direct-expansion evaporators are still designed for operation with superheated vapor at the evaporator outlet and Bo Pierre's correlation for complete evaporation [Eq. (8.137c)] is used. The refrigerant is usually metered by a thermal expansion valve set to maintain a relatively large superheat of 5–10°C generated within the evaporator and thus reducing the "wet portion" of the internal surface of the tube. Under these conditions a relatively large part of the internal surface "dries up," and the two-phase heat transfer coefficient quickly converges to the normally lower values of a single-phase vapor heat transfer coefficient (as discussed previously), and the average heat transfer coefficient $\bar{\alpha}_{tp}$ can thus yield reasonably good results in conjunction with the conventional logarithmic mean temperature difference. Furthermore, to minimize potential errors caused by major extrapolations beyond the test range, pragmatic heat transfer engineers/researchers conducted their tests in a range close to the intended practical applications. Thus, within certain limits of similitude, the results obtained on single-tube tests could be applied successfully to full size forced-convection evaporators even in the case of compact chillers with incomplete evaporation operating in the fully wet wall regimes. However, deviations from calculated values under these conditions can become more significant; this was shown by Boling et al. [5.5] who ran their tests with wet vapor outlet (this is achieved by positioning the thermal bulb of the expansion valve on the downstream side of the heat exchanger as shown in Fig. 4.5) and compared the results obtained on a complete chiller with those determined from single-tube tests. The results [5.5, Fig. 6] indicated a good agreement in "normal applications," but the deviations became much more significant when operating at lower chilled water flow rates (lower α_H) and greater water temperature cooling range ($T_{1,H} - T_{2,H}$), and consequently a greater $\Delta T_{\max} \equiv T_{1,H} - T_{2,C}$.

The limitations of conventional methods, particularly in the case of compact thermodynamically efficient chillers operating with incomplete evaporation (in fully wet wall regimes), can best be explained by reviewing their application in actual refrigeration design practice and pointing out the major discrepancies [4.4–4.6] between the conventional analytical methods and the results obtained in numerous tests (see Soumerai [4.6], "Application of Conventional Method of Calculation in Refrigeration Design Practice—Historical Background").

Discrepancy 1: Parallel Versus Counterflow Arrangement. Contrary to what would be expected on the basis of the log mean temperature differences, exactly the *same total capacity* $\dot{Q}_e$ is obtained consistently in parallel flow and counterflow arrangements under otherwise identical conditions.

Discrepancy 2: Effect of External Constraints (Heat Source) on $\bar{\alpha}_{tp}$. The average heat transfer coefficient $\bar{\alpha}_{tp}$ is *not* a unique function of refrigerant flow rate $\dot{m}_C$ for a prescribed tube geometry ($D_{h,i}$, $L/D_{h,i}$, and $\bar{e}/D_{h,i}$) under otherwise specified operating conditions on the refrigerant side, $P_{2,C}$, and corresponding saturation temperature, x_2, and x_1, as would be expected on the basis of the correlations proposed by *all* research workers in the 1960s and illustrated in the typical design chart in Ref. 4.6 (p. 132). In other words, in two-phase flow regimes typical of refrigeration, the average heat transfer coefficient $\bar{\alpha}_{tp}$ with incomplete evaporation cannot be correlated like a conventional single-phase forced-convection heat transfer coefficient for a given tube geometry in terms of mass velocity or flow rate alone. On the contrary, the average length integrated refrigerant heat transfer coefficient $\bar{\alpha}_{tp}$ is strongly coupled to external parameters on the heat source side of the heat exchanger such as:

> *Direction of Flow*: $\bar{\alpha}_{tp}$ in parallel flow is less than $\bar{\alpha}_{tp}$ in counterflow since the corresponding LMTD in parallel flow is greater than LMTD in counterflow and $\dot{Q}_e$ in parallel flow equals $\dot{Q}_e$ in counterflow
>
> *Maximum Temperature Difference*: $(\Delta T)_{\max} \equiv \text{LD} = T_{1,H} - T_{2,C}$. For a fixed direction of flow, the average heat transfer coefficient $\bar{\alpha}_{tp}$ increases with a decrease of LD and, conversely, decreases with increasing LD. The corresponding mean wall-to-refrigerant temperature difference $\overline{\Delta T_i}$ increases with increasing LD and conversely.

This led to our concept of a wide domain of $\bar{\alpha}_{\mathrm{tp}}$ and corresponding domain of $\overline{\Delta T}_i \equiv \overline{\Delta T}_w = \bar{T}_w - \bar{T}_{\mathrm{sat}}$ already mentioned in the discussion of Bo Pierre's correlations for incomplete evaporation in fully wet wall flow regimes.

Discrepancy 3: Effect of Total Circuit Length L on $\bar{\alpha}_{\mathrm{tp}}$. Contrary to the then (1960s) widely accepted theory, the average heat transfer coefficient, under otherwise fixed conditions, *is strongly dependent on total circuit length L.*

The assumption that, in typical LTL forced-convection evaporators, the average heat transfer coefficient $\bar{\alpha}_{\mathrm{tp}}$ would behave like a fully developed forced-convection single-phase heat transfer coefficient proved erroneous. Under otherwise specified conditions it was found that $\bar{\alpha}_{\mathrm{tp}}$ decreases with increasing circuit length L and conversely. Qualitatively, these findings agree with Bo Pierre's Eq. (8.142). Bo Pierre was apparently the first research worker to recognize the significant effect of length, or L/D, on the average length integrated heat transfer coefficient $\bar{\alpha}_{\mathrm{tp}}$ under typical refrigeration conditions. All other research workers cited in [4.6] failed to consider the circuit length L as an independent variable in their studies of full length evaporator tubes (measurements of average $\bar{\alpha}_{\mathrm{tp}}$). The possible impact of L is totally ignored in all so-called local studies. In fact, fairly recent [8.10] average heat transfer coefficient design charts for use in refrigeration practice still fail to recognize, among other things, the strong effect of length on $\bar{\alpha}_{\mathrm{tp}}$ under typical design conditions. It is therefore not surprising that the results derived from such design charts may contradict predictions based on Bo Pierre's correlations. These apparent contradictions are understandable when one recognizes that the correlations proposed in [8.10] for the average heat transfer coefficients were not supported by actual tests of full length evaporator tubes operating under normal conditions; these are *theoretical average* values based on *local* heat transfer measurements with constant (electrical) heat fluxes which were mathematically integrated on the assumption of constant heat flux along the full length of the evaporator tube from inlet quality $x_1 = 0$ to exit quality $x_2 = 1$. The assumption of a constant heat flux independent of length is very atypical of LTL fluid-to-fluid heat exchanger applications when one is dealing with constant or monotonously decreasing heat source temperatures and this can lead to significant errors, as shown in Appendix A.3 of Ref. 4.6. This is just one of the several limitations of so-called local studies carried out with abnormally short tube lengths or L/D_h ratios and constant electrical heat fluxes (see Section 3.5.3 and very recent (1985) confirmation by Steiner [8.11] of the significant impact of conditions prior to tube entrance).

8.2.3.5 Empirical Verification of the Bo Pierre Paradox. To verify empirically the concept of a wide domain of $\bar{\alpha}_{\mathrm{tp}}$ and the corresponding domain of $\overline{\Delta T}_i \equiv \overline{\Delta T}_w = \bar{T}_w - \bar{T}_{\mathrm{sat}} = \text{constant}$ (discussed in Section 8.2.3.3), we refer to test series No. 3 and No. 4 in Tables 1 of Refs. 4.4 and 4.6, which yield the

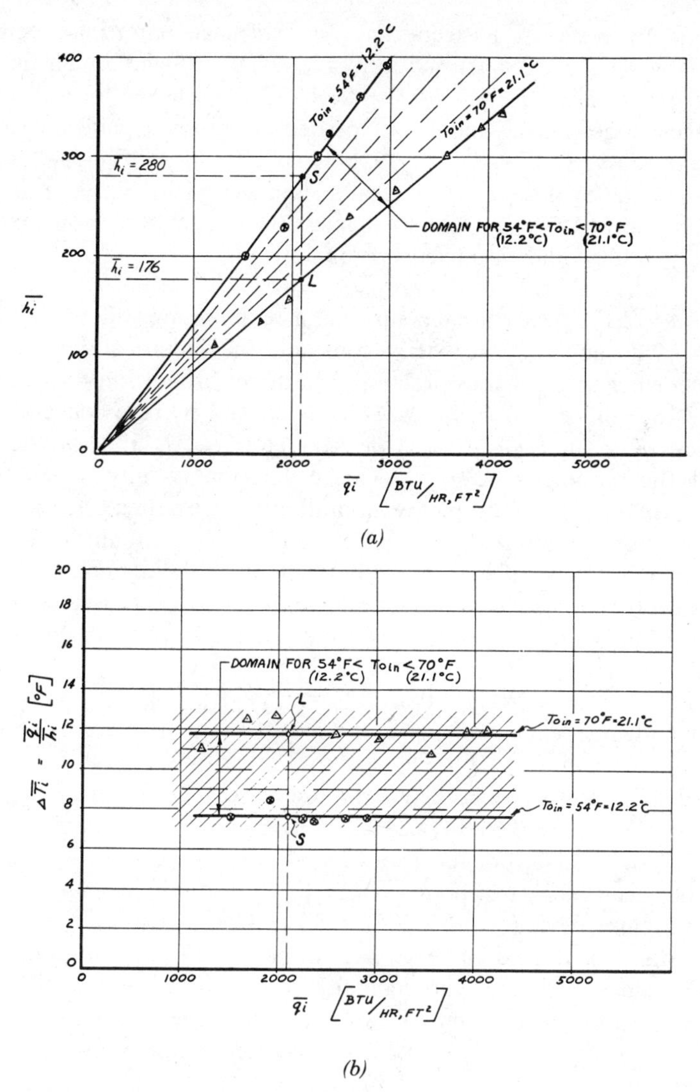

FIGURE 8.18. Effect of large temperature difference LD at fixed Δx, x_2, and exit saturation temperature $T_{2,\text{sat}}$. (*a*) $\bar{h}_i$ versus $\bar{q}_i$. (*b*) $\overline{\Delta T}_i$ versus $\bar{q}_i$. *Note*: According to the nomenclature used in this book, $\bar{q}_i'' \equiv q_i$, $\bar{h}_i \equiv \bar{\alpha}_{\text{tp}}$, $T_{0,\text{in}} \equiv T_{1,H}$, LD $\equiv (\Delta T)_{\max} = T_{1,H} - T_{2,C}$, with $T_{2,C} \equiv T_{2,\text{sat}}$ = constant = 35°F in the above test series.

empirical heat transfer coefficient $\bar{\alpha}_{\text{tp}}$ shown graphically in terms of the average heat flux $\bar{q}_i''$ in Fig. 8.18*a* and the corresponding $\overline{\Delta T}_i \equiv \overline{\Delta T}_w = \bar{T}_w - \bar{T}_{\text{sat}} = q_i''/\bar{\alpha}_{\text{tp}}$ in Fig. 8.18*b*. It is important to note that these two test series were conducted with the same elementary single-tube evaporator (5 ft tube in counterflow arrangement) under exactly the same operating conditions (i.e., $T_{2,C}$ = 35°F = 1.66°C = saturation temperature at channel outlet, identical

x_2, x_1 and $\Delta x = x_2 - x_1$), except that a *normal* heat source inlet temperature $T_{1,H} = 54°F = 12.22°C$ was used in test series No. 3, whereas an *abnormally high* inlet temperature $T_{1,H} = 70°F = 21.11°C$ was selected for test series No. 4. Thus, test series Nos. 3 and 4 were carried out at two significantly different $(\Delta T)_{max} = LD = T_{1,H} - T_{2,C}$ of 19°F (10.6°C) and 35°F (19.4°C), respectively. We analyze the empirical results summarized in Fig. 8.18 for a prescribed constant $(\Delta T)_{max} = LD$, then assess the impact of a significant change in $(\Delta T)_{max}$ under otherwise fixed text conditions.

Prescribed $(\Delta T)_{max}$. At a fixed maximum temperature differential $(\Delta T)_{max} = LD = 19°F$ and $T_{2,C} = 35°F =$ constant, the empirical two-phase heat transfer coefficient $\bar{\alpha}_{tp}$ varies practically in direct proportion to $\bar{q}_i''$ as shown in Fig. 8.18*a*. Since $a_i L = A_i$, $\Delta x = x_2 - x_1$, and h_{lg} are constant in these test series, $\bar{q}_i'' = \dot{m}_C \Delta x h_{lg}/a_i L \propto \dot{m}_C$; therefore, $\bar{\alpha}_{tp} \propto \bar{q}_i'' \propto \dot{m}_C$ for a fixed value of the parameter $(\Delta T)_{max}$. This conclusion is fully in line with Bo Pierre's correlation for incomplete evaporation at tube outlet. Equation (8.137d) is thus verified for internally finned channels with hydraulic diameters one order of magnitude smaller than the bare circular tube diameters used by Bo Pierre and other researchers, noting that similar results are obtained with the internally finned tube of smaller hydraulic diameter $D_h = 4r_h = 1.1$ mm as reported in [4.4, 4.6].

It follows from $\bar{\alpha}_{tp} \propto \bar{q}_i''$ that $\overline{\Delta T}_i = \bar{T}_w - \bar{T}_{sat} = \bar{q}_i''/\bar{\alpha}_{tp}$ must also be constant and independent of $\bar{q}_i''$ at a fixed $(\Delta T)_{max}$ in accordance with Eq. (8.141) and as confirmed by Fig. 8.18*b*. The arithmetic average of all the measured $\overline{\Delta T}_i$ shown in Fig. 8.18*b* for test series No. 3 at $(\Delta T)_{max} = 19°F$ (10.6°C) is $(\overline{\Delta T}_i)_{am} = 7.64°F$ (4.24°C) with a maximum/minimum variation of only $+9.8/-3.8\%$. For test series No. 4 at $(\Delta T)_{max} = 35°F$ (19.4°C), the arithmetic mean of all test points is $(\overline{\Delta T}_i)_{am} = 11.77°F$ (6.5°C) with a maximum/minimum deviation of $+7.8/-8.7\%$. Consequently, $\bar{T}_w - \bar{T}_{sat}$ is indeed constant for a fixed value of the parameter $(\Delta T)_{max}$ within an accuracy of better than $\pm 10\%$. This can generally be considered quite satisfactory for empirical correlations of complex two-phase flow phenomena, particularly at the low $\bar{T}_w - \bar{T}_{sat}$ typical of LTL forced-convection two-phase heat exchangers.

These findings can readily be explained at a fixed value of the parameter $(\Delta T)_{max}$ for the typical *fully rough wet wall* turbulent two-phase flow regimes when the effective friction factor $\bar{f}_{tp}$ is constant and independent of the Reynolds number, or $\dot{m}_C$, as highlighted in Section 5.2.2. This limiting case, $\bar{T}_w - \bar{T}_{sat} =$ constant and independent of $\bar{q}_i''$ and $\dot{m}_C$, is also fully consistent with the results of the variation analysis of single-phase fluids flowing in fully rough wall tubes summarized in Table 7.7. Indeed, the fundamental coupling between $\bar{\alpha}_{tp}$ and $\bar{f}_{tp}$ could be used to postulate that $\bar{f}_{tp}$ must be substantially constant and independent of $\dot{m}_C$ from the empirical fact that $\overline{\Delta T}_i = \bar{T}_w - \bar{T}_{sat}$ is substantially constant and independent of $\bar{q}_i''$ and $\dot{m}_C$ under the stated operating conditions.

Effect of External Coupling: $(\Delta T)_{\max}$ Variations. Both $\bar{\alpha}_{tp}$ and the corresponding unique value of $\overline{\Delta T_i} = \bar{T}_w - \bar{T}_{sat}$ are *clearly strongly dependent* on the parameter $(\Delta T)_{\max}$ = LD, that is, the difference between the maximum external heat source temperature $T_{1,H}$ and the evaporating fluid's exit saturation temperature $T_{2,C}$ according to Fig. 8.18. Contrary to what would be expected in nucleate boiling regimes, $\bar{\alpha}_{tp}$ actually *decreases* with *increasing* $\overline{\Delta T_i} = \bar{T}_w - \bar{T}_{sat}$ and conversely. This further confirms that nucleate boiling regimes are suppressed in typical LTL direct-expansion evaporators according to Guideline 27. It is further evident from Fig. 8.18*a* that $\bar{\alpha}_{tp}$ is *not* uniquely defined at a prescribed average heat flux $\bar{q}_i''$. The effective heat transfer coefficient $\bar{\alpha}_{tp}$ is therefore coupled to the external system in some fashion [4.4]. Note that this coupling and associated concept of a wide domain of possible $\bar{\alpha}_{tp}$ and corresponding $\overline{\Delta T_i} = \bar{T}_w - \bar{T}_{sat}$ were verified in numerous unpublished (proprietary) elementary single-tube and complete multicircuit chillers [4.4, 4.6].

The corollary to Eq. (8.137d), the so-called Bo Pierre paradox, discussed in Section 8.2.3.3, is thus empirically verified and it is clear that the constant values $(\bar{T}_w - \bar{T}_{sat})$ obtained from such correlations as Bo Pierre's Eq. (8.142) are only *discrete solutions within a much wider domain* of possible $\overline{\Delta T_i} = \bar{T}_w - \bar{T}_{sat}$ under in/out SSSF conditions.

8.2.3.6 Thermodynamic Interpretation of the Bo Pierre Paradox. The prime objective of the following (nonequilibrium) thermodynamic considerations is to provide a plausible simple explanation for the empirically validated Bo Pierre paradox.

We consider again the classical constant temperature wall boundary $\bar{T}_w = T_w$ = constant along the full tube length from $l = 0$ to $l = L$ and analyze the heat transfer problem in exactly the same manner as in Sections 7.8–7.10 for single-phase fluids flowing inside straight constant cross-section channels. This constant wall interface temperature boundary condition in two-phase flow applications is actually satisfied, for instance, in cascade condensers/evaporators when the evaporating fluid flowing inside the tubes is heated by another fluid, condensing on the external (bare or finned) surface[11] as described schematically in Fig. 7.4.

The two-phase fluid flowing inside the heat exchanger channel is treated like a homogeneous mixture (see Guideline 22, Chapter 7) at prescribed equilibrium states $(ES)_1$ and $(ES)_2$. We neglect the pressure drop (since this has no impact on the basic conclusions derived from this analysis) and assume that $\Delta P_{tp} = 0$; therefore, $P_1 = P_2 = P$ = constant with a corresponding saturation temperature $T_{1,sat} = T_{2,sat} = T_{sat}$ = constant. Thus, the inlet equilibrium state $(ES)_1$ is fully defined at the prescribed inlet quality x_1 (or enthalpy h_1) and saturated temperature T_{sat} corresponding to P. This is shown schematically in Fig. 8.19*a* for $x_1 = 0.2$ and $1 - x_1 = y_1$ = inlet flowing liquid quality = 0.8. Similarly, the evaporator outlet equilibrium state $(ES)_2$ is fully defined at any prescribed quality $x_2 = x_1 + \Delta x < 1$ for incomplete evaporation on the isotherm T_{sat} corresponding to the isobar $P = P_1 = P_2$ as indicated

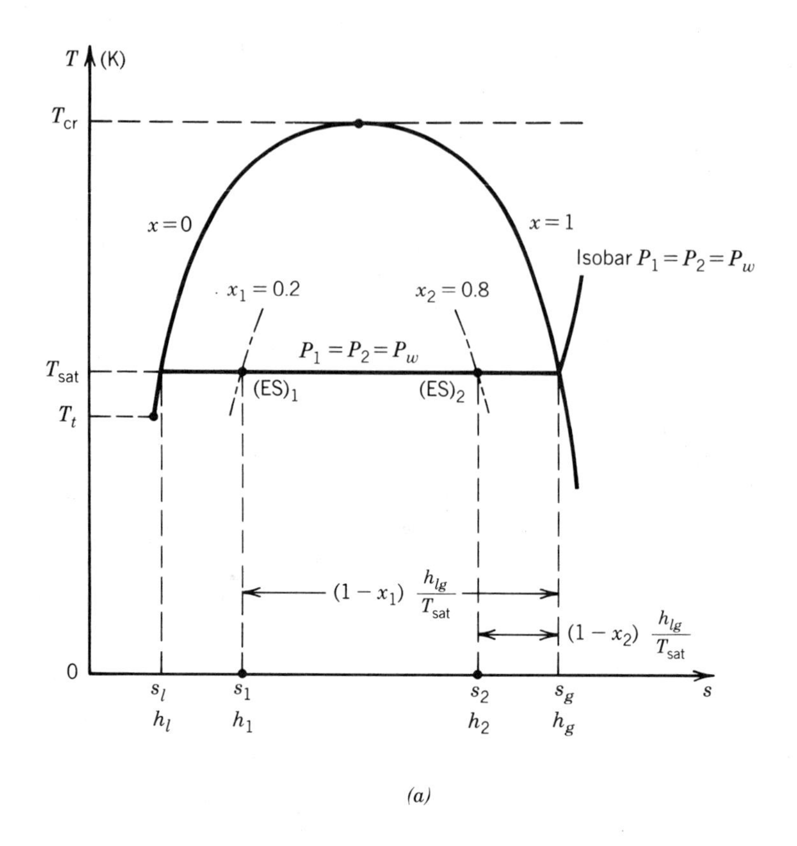

(a)

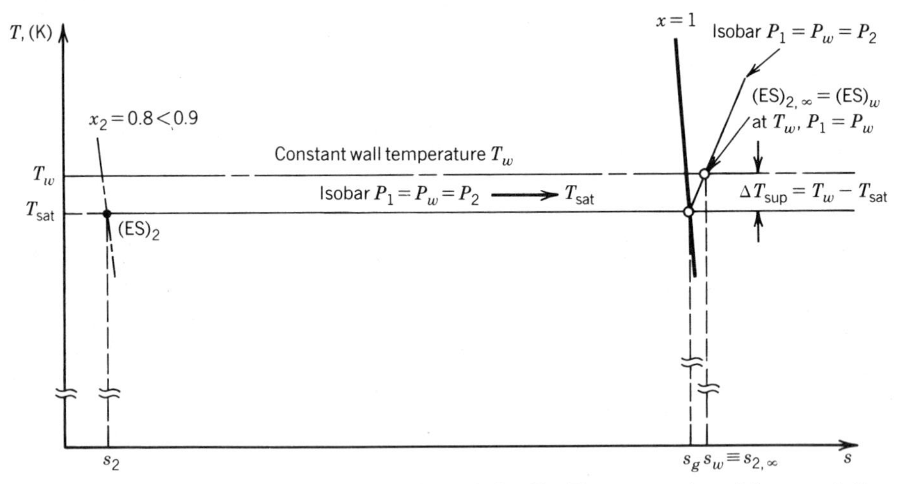

FIGURE 8.19. Thermodynamic interpretation of the Bo Pierre paradox; (*a*) normal *T-s* diagram and (*b*) exploded *T-s* diagram for tube outlet region.

for $\Delta x = 0.6$, $x_2 = 0.8$, and $1 - x_2 = y_2$ = exit flow liquid quality = 0.2, that is, well within the validity limits of Bo Pierre's correlations with incomplete evaporation per Eq. (8.137d) and Eq. (8.141c).

For a very long length of travel L [or extremely long average residence time $(\overline{\Delta\theta})$], the two-phase mixture must ultimately reach the higher wall temperature $T_w > T_{sat}$ on the $P = P_1 = P_2 = P_w$ isobar. In other words, the end equilibrium state $(ES)_w = (ES)_{2,\infty}$ shown in Fig. 8.19*b* must be superheated by an amount ΔT_{sup}

$$\Delta T_{sup} = T_w - T_{sat} \tag{8.144}$$

In thermodynamically efficient low thermal lift heat exchangers, $\Delta T_{sup}/T_w \approx 0$ and it is necessary to explode the *T-s* diagram to identify the end equilibrium state $(ES)_w$ on the isobar $P = P_2 = P_1 = P_w$ = constant, as shown in Fig. 8.19*b*. Referring to the equilibrium states $(ES)_1$, $(ES)_2$, and $(ES)_w$ identified in Figs. 8.19*a* and 8.19*b*, the enthalpies h_w, h_2, and h_1 on the isobar $P = P_1 = P_2 = P_w$ can be expressed according to

$$h_w = h_l + h_{lg} + (c_p)_g(\Delta T)_{sup} \tag{8.145a}$$

$$h_2 = h_l + x_2 h_{lg} \tag{8.145b}$$

$$h_1 = h_l + x_1 h_{lg} \tag{8.145c}$$

The maximum potential enthalpy change at a fixed wall temperature T_w for extremely large L and $(\overline{\Delta\theta})$ can be computed with Eqs. (8.145a) and (8.145c) according to

$$(\Delta h)_{max} = h_w - h_1 = h_{lg}\left[(1 - x_1) + \frac{(c_p)_g(\Delta T)_{sup}}{h_{lg}}\right] \tag{8.146}$$

The corresponding maximum rate of heat exchange $(\dot{Q}_e)_{max}$ at a prescribed $\dot{m}_C$ and constant T_w becomes

$$(\dot{Q}_e)_{max} = \dot{m}_C h_{lg}\left[(1 - x_1) + \frac{(c_p)_g(\Delta T)_{sup}}{h_{lg}}\right] \tag{8.147a}$$

or

$$(\dot{Q}_e)_{max} = \dot{m}_C h_{lg}(1 - x_1)\left[1 + \frac{(c_p)_g(\Delta T)_{sup}}{(1 - x_1)h_{lg}}\right] \tag{8.147b}$$

For thermodynamically efficient direct-expansion LTL evaporators under typical operating conditions—$x_1 = 0.1$–$0.3 \simeq 0.2$ therefore $1 - x_1 \simeq 0.8$, $(\Delta T)_{sup}$ of the order of 1°C at "low side" evaporator reduced pressures $P^+ = P/P_{cr} \ll 1$—it is easy to see (by referring to thermophysical charts and/or tables such as A.7 in Appendix A for R-22) that the dimensionless term $(c_p)_g(\Delta T)_{\text{sup}}/h_{lg}$ is negligibly small compared to $1 - x_2$. It follows that Eq. (8.147b) reduces to

$$(\dot{Q}_e)_{\max} \simeq \dot{m}_C h_{lg}(1 - x_1) \tag{8.148a}$$

valid for typical direct-expansion evaporators with incomplete evaporation within operating limits stated in Eq. (8.141c) when

$$\frac{(c_p)_g(\Delta T)_{\text{sup}}}{(1 - x_1)h_{lg}} < \frac{(c_p)_g(\Delta T)_{\text{sp}}}{(1 - x_2)h_{lg}} \approx 0 \tag{8.148b}$$

At the same total mass flow rate $\dot{m}_C$ and a suitably shorter length of travel L, or shorter residence time $(\overline{\Delta\theta})$, the two-phase mixture can be heated from the same inlet vapor quality x_1 at h_1 to the specified exit vapor quality $x_2 < 1$ at $h_2 < h_g$ corresponding to an enthalpy change

$$\Delta h = h_{lg}(x_2 - x_1) \tag{8.149}$$

and an actual rate of heat exchange

$$\dot{Q}_e = \dot{m}_C h_{lg}(x_2 - x_1) \tag{8.150}$$

Following the same reasoning as in the case of single-phase heat exchangers, we define with Eqs. (8.148)–(8.150) a two-phase contact factor $(\text{CF})_{\text{tp}}$,

$$(\text{CF})_{\text{tp}} \equiv \frac{\dot{Q}_e}{(\dot{Q}_e)_{\max}} \simeq \frac{x_2 - x_1}{1 - x_1} \tag{8.151}$$

By substituting $(1 - x_1) - (1 - x_2)$ for $x_2 - x_1$ in the above equation, the following equivalent expression is obtained for the two-phase contact factor

$$(\text{CF})_{\text{tp}} \simeq \frac{(1 - x_1) - (1 - x_2)}{1 - x_1} \equiv 1 - \frac{1 - x_2}{1 - x_1} \equiv 1 - \frac{y_2}{y_1} \tag{8.152a}$$

and for the corresponding two-phase approach factor

$$(\text{AF})_{\text{tp}} \equiv 1 - (\text{CF})_{\text{tp}} \simeq \frac{y_2}{y_1} \tag{8.152b}$$

where, by definition,

$$\begin{aligned} y_2 &\equiv 1 - x_2 = \text{flowing liquid quality at channel outlet} \\ y_1 &\equiv 1 - x_1 = \text{flowing liquid quality at channel inlet} \end{aligned} \tag{8.152c}$$

Recalling that in the two-phase equilibrium region ($0 < x < 1$) the total entropy of the liquid and vapor mixture on the isotherm T_{sat} corresponding to the isobar $P = P_1 = P_2 = P_w$ can be expressed according to

$$s = s_l + x s_{lg} \tag{8.153}$$

and an enthalpy change Δh according to

$$\Delta h = (\Delta s) T_{sat} \tag{8.154}$$

it follows, since

$$\frac{\Delta h}{(\Delta h)_{max}} \equiv \frac{(\Delta s) T_{sat}}{(\Delta s)_{max} T_{sat}} \equiv \frac{\Delta s}{(\Delta s)_{max}} \tag{8.155}$$

that Eqs. (8.152a) and (8.152b) can be expressed more generally according to

$$(\mathrm{CF})_{tp} \equiv \frac{h_2 - h_1}{h_w - h_1} \equiv \frac{s_2 - s_1}{s_w - s_1} \simeq 1 - \frac{y_2}{y_1} \tag{8.156a}$$

and

$$(\mathrm{AF})_{tp} \equiv \frac{h_w - h_2}{h_w - h_1} \equiv \frac{s_w - s_2}{s_w - s_1} \simeq \frac{y_2}{y_1} \tag{8.156b}$$

with the validity limits of Eq. (8.148b).

As discussed more fully in Refs. 6.5, 6.7, and 6.8 and Chapter 9, it is convenient in two-phase flow heat exchanger applications to use alternate enthalpy-based or more generally valid entropy-based heat transfer driving forces in lieu of the conventional temperature driving forces first proposed by Newton in 1701 (i.e., long before the fundamental laws of classical thermodynamics had been firmly established).

For *single-phase* fluids an alternate enthalpy-based driving force and corresponding empirical heat transfer coefficient α_h can be derived from the conventional convective heat rate equation $q'' = \alpha_t(T_w - T)$, defining the usual single-phase heat transfer coefficient α_t, by multiplying and dividing the above equation by the specific heat at constant pressure, c_p, as follows:

$$q'' = \frac{\alpha_t}{c_p}\left[c_p(T_w - T)\right] \equiv \alpha_h(h_w - h) \equiv \alpha_h(\Delta h)_w \tag{8.157a}$$

where, by definition,

$$(\Delta h)_w \equiv h_w - h \equiv \text{enthalpy driving force} \tag{8.157b}$$

with

$$\frac{\alpha_t}{c_p} = \alpha_h \equiv \text{enthalpy-based heat transfer coefficient} \tag{8.157c}$$

and

$$\alpha_t \equiv \text{temperature-based heat transfer coefficient} \tag{8.157d}$$

The conventional relationships, Eqs. (7.47)–(7.49) valid for single-phase fluids at a constant wall interface temperature T_w reviewed in Chapter 7, can thus be expressed with Eqs. (8.157a) and (8.157b) according to

$$\frac{1}{\text{AF}} \equiv \frac{h_w - h_1}{h_w - h_2} = e^{\text{NTU}} \tag{8.158a}$$

or, in logarithmic form,

$$\ln\left(\frac{1}{\text{AF}}\right) = \ln\left(\frac{h_w - h_1}{h_w - h_2}\right) = \text{NTU} \tag{8.158b}$$

From the definition of the term NTU per Eq. (7.49) with $\dot{m} = G\Omega_i$, $\Omega_i/a_i = \frac{1}{4}D_{h,i}$, and $\bar{\alpha}_t/c_p = \bar{\alpha}_h$,

$$\text{NTU} = \frac{\bar{\alpha}_t}{c_p G}\left(\frac{L}{\Omega_i/a_i}\right) \equiv \frac{\bar{\alpha}_h}{G}\left(\frac{4L}{D_{h,i}}\right) \tag{8.158c}$$

By analogy we can write Eq. (8.156b) for single-component two-phase fluids with heat addition (i.e., bulk evaporation *without* nucleate boiling as is typical for low thermal lift forced-convection evaporators per Guideline 27) as follows

$$\frac{y_1}{y_2} \simeq \frac{1}{(\text{AF})_{\text{tp}}} \equiv \frac{h_w - h_1}{h_w - h_2} \equiv \frac{s_w - s_1}{s_w - s_2} = e^{(\text{NTU})_{\text{tp}}} \tag{8.159a}$$

or

$$\ln\left(\frac{y_1}{y_2}\right) \simeq \ln\left(\frac{1}{(\text{AF})_{\text{tp}}}\right) \equiv \ln\left(\frac{h_w - h_1}{h_w - h_2}\right) \equiv \ln\left(\frac{s_w - s_1}{s_w - s_2}\right) = (\text{NTU})_{\text{tp}} \tag{8.159b}$$

with

$$(\text{NTU})_{\text{tp}} = \frac{(\bar{\alpha}_h)_{\text{tp}}}{G}\left(\frac{4L}{D_{h,i}}\right) \tag{8.159c}$$

It is now possible to provide a plausible explanation for the Bo Pierre paradox in the light of Eqs. (8.159a)–(8.159c) by referring to Fig. 8.19 as shown below.

1. *Impact of $\dot{m}$ Variation at Fixed*: *Geometry*, $(\Delta T)_{sup} \equiv T_w - T_{sat}$, T_{sat}, x_1, *and* x_2. In this case the only independent variable is the total mixture mass flow rate $\dot{m}$ or the mass velocity $G = \dot{m}/\Omega_i$ for a fixed geometry. All the terms appearing in Eqs. (8.159a)–(8.159c) are fixed except for G and the coupled average two-phase heat transfer coefficient $(\bar{\alpha}_h)_{tp}$. Since $(\text{NTU})_{tp}$ must be constant for incomplete evaporation in accordance with the Bo Pierre experimental data, it is clear from Eq. (8.159c) that $(\bar{\alpha}_h)_{tp}$ must vary in direct proportion with G. This conclusion is in full agreement with Bo Pierre's Eqs. (8.137d) and (8.142) for incomplete evaporation and consistent with the fully rough wet wall flow model [4.21, 4.67] associated with a constant two-phase flow friction factor $\bar{f}_{tp}$ as discussed previously.

2. *Impact of* $(\Delta T)_{sup} = T_w - T_{sat}$ *Variation at Fixed*: *Geometry*, $\dot{m}$, T_{sat}, x_1, *and* x_2. Since for fixed values of x_1 and x_2 the corresponding values of y_1 and y_2 are also fixed, it is clear from Eq. (8.159a) and (8.159b) that the approach factor $(\text{AF})_{tp}$ remains substantially constant and the contact factor $(\text{CF})_{tp} = 1 - (\text{AF})_{tp})$ is *not* measurably affected by changes in the wall superheat $(T_w - T_{sat})$ within the stated validity limits of Eq. (8.148). The same total rate of heat exchange $\dot{Q}_e = (\text{CF})_{tp}(\dot{Q}_e)_{max}$, and consequently the same average heat flux $\dot{Q}_e/a_iL = \bar{q}_i''$, can thus occur at different levels of wall superheat $(\Delta T)_{sup} = T_w - T_{sat}$ as shown empirically in Fig. 8.18b. This provides a rational explanation for the empirically validated Bo Pierre paradox and at the same time confirms the usefulness, particularly in two-phase flow applications, of the alternate thermodynamic heat transfer driving forces discussed in Chapter 9.

It should be stressed again that the above thermodynamic considerations[12] are based on (fictive) equilibrium states at T_{sat} corresponding to measured pressures and *not* on empirical heat transfer coefficients based on measured nonequilibrium two-phase fluid temperatures ($T_{meas} \neq T_{sat}$). This is fully consistent with the conventional analytical methods used in single-phase forced-convection heat exchangers (see Section 7.6 and Guideline 22 in Chapter 7) and it is the only rational approach in actual evaporator design practice [4.4, 4.6, 6.7].

8.3 GENERALIZATION OF HEAT TRANSFER DATA ON THE BASIS OF REDUCED PRESSURE

In view of the coupling between heat and momentum transfer in single- and two-phase, natural- and forced-convection heat exchangers, it is clear that the methods of generalizing fluid flow data described in Chapter 6 can readily be extended to generalize theoretical and/or empirical heat transfer data [6.13].

The extended corresponding states principle (ECSP) method of generalization could be applied, for instance, to generalize Bo Pierre's forced-convection two-phase correlations, based on empirical data obtained with R-12 and R-22, to other refrigerants. Such a generalization, though extremely useful to the refrigeration specialist, would be of limited interest in other two-phase flow applications in view of the relatively narrow validity limits of Bo Pierre's correlations. Consequently, we focus here on more general and widely accepted two-phase flow local (or average) heat transfer coefficient correlations α_{tp} (or $\bar{\alpha}_{tp}$) in order to take full advantage of the main feature of all classical thermodynamic tools, namely, their universal[13] validity, since basic equilibrium thermodynamics laws are completely independent of physical configurations, flow regimes, and chemical species (see Guideline 14, Chapter 7).

8.3.1 Generalization of Two-Phase Flow Heat Transfer Coefficients in Fluid Heating Mode: Evaporation

The majority of the more sophisticated correlations (slip-flow and void fraction models) recommended for α_{tp} for forced convection with heat addition inside horizontal or vertical tubes and channels, such as those listed in Ref.

TABLE 8.19. Most General Expression for Multiplier $M_h = h_{tp} / h_L$

Investigators	$M_h = h_{tp} / h_L$
1. Dengler and Addoms	$3.5(1 / \bar{X}_{tt})^{0.5}$
2. Schrock and Grossman	$7400[B + 1.5 \times 10^{-4}(1 / \bar{X}_{tt}^{2/3})]$ where $B \equiv$ boiling number $\equiv \dfrac{Q / A_i}{G h_{fg}} = \dfrac{\Delta x}{4L / D_H} = \dfrac{\Delta x}{L / r_H}$
3. Sani and Wright	Same as 2, except constants and B changed to $B_M \equiv B\left(\dfrac{v''}{v'}\right) = \dfrac{\Delta x}{L / r_H}\left(\dfrac{v''}{v'}\right)$
Most general formulation for above equations	$M_h = fct\left(\bar{X}_{tt}; \dfrac{\Delta x}{L / r_H}; \dfrac{v''}{v'}\right)^a$
	$M_h \equiv \left.\begin{matrix} G \\ p^+ \\ \Delta x \\ D_H \\ L / D_H \end{matrix}\right\} \equiv$ Similar fluids

[a] M_h identical at given P^+, G, D_H, $\Delta x / (L / D_H)$ since v'' / v' and $\bar{X}_{tt}$ are also identical at given P^+.

Source: Reproduced with permission from "Application of Thermodynamic Similitude, Part II: Heat and Mass Transfer," *ASHRAE Journal*, 1966. Note that according to the nomenclature adopted in this book $\alpha_{tp} / \alpha_L \equiv h_{tp} / h_L$ and the subscripts l and g are equivalent to the superscripts $'$ and $''$, respectively.

4.53, are based on an empirical multiplier $M_h = \alpha_{tp}/\alpha_L$ [4.21, Part II, Table V] highlighted in Table 8.19. Thus, the unknown two-phase heat transfer coefficient α_{tp} is simply computed from $\alpha_{tp} = M_h \alpha_L$, where α_L is the conventional single-phase heat transfer coefficient calculated as if the total[14] mass flow rate $\dot{m}$ were to flow as a single-phase liquid (equivalent to a liquid quality $y = 1 - x = 1$ with a vapor quality $x = 0$).

It is remarkable that, in spite of the general *lack* of agreement among the many different empirical mathematical expressions for the empirical correction factor M_h shown in Table 8.19, these correlations nevertheless do have one essential feature in common: They confirm the basic conclusion derived from the application of the ECSP, that is, the value of the multiplier M_h is identical for all similar fluids at the same G, reduced pressure P^+, flowing qualities, and configuration, or

$$M_h = M_{h,\text{ref}} \quad \text{at } identical\ P^+, G, x_1, \Delta x = x_2 - x_1, x_m, D_H, L/D_H, \text{ and } \bar{e}/D_H \tag{8.160}$$

since the Martinelli–Nelson parameters are identical, or $X_{tt} = X_{tt,\text{ref}}$ according to Eq. (6.23) and v_g/v_l is a universal function of P^+.

NUMERICAL ANALYSIS 8.8

Quantitative assessment of the accuracy of the identity $M_h = M_{h,\text{ref}}$ per Eq. (8.160).

Background

This is essentially an extension of the two-phase fluid flow generalization methods described in Chapter 6 and it is assumed that the reader is familiar with this methodology and the key results derived in Numerical Analyses 6.1–6.4.

Problem Statement

(a) Verify over the full design saturation temperature range of single-stage refrigeration/heat pump (low-pressure side) evaporators that $X_{tt,\text{R-502}} = X_{tt,\text{R-22}}$ per Eq. (8.160) and quantify the actual percentage deviations on the basis of the latest thermophysical property data [4.24].

(b) Assess the impact of the above deviations on the multiplier M_h per Eq. (8.160) and the accuracy of α_{tp} computed from the conventional single-phase heat transfer coefficient α_L according to the recommended equation $\alpha_{tp} = M_h \alpha_L$ on the basis of $M_{h,\text{R-502}} = M_{h,\text{R-22}}$.

Solution

(a) The actual percentage deviations: $100(X_{tt,\text{R-502}} - X_{tt,\text{R-22}})/X_{tt,\text{R-22}}$ are shown on lines (i) and (j) in Table 6.1 for a Blasius friction factor exponent $n = 0.2$ as well as for the fully rough wall limiting case $n = 0$. It is evident that these deviations are insignificant compared to typical deviations in f_{tp} (or $\bar{f}_{\text{tp}}$) of the order of ± 30–40% according to Eq. (5.54b) and confirmed in a recent (1986) research paper [8.11] (indicating that *at least* 50–60% of all actual ΔP_{tp} test points fall outside a $\pm 30\%$ band around the values predicted by the *best* known correlations for ΔP_{tp}) or even compared to the *normal statistical variations* due exclusively to test inaccuracies/manufacturing tolerances which are typically of the order of no less than $\pm 10\%$ in normal heat exchangers.

(b) Since the empirical deviations of the best semianalytical two-phase flow heat transfer correlations proposed to date are at best equal to those of the coupled two-phase friction factor f_{tp} (or $\bar{f}_{\text{tp}}$), that is, generally in *excess* of ± 30–40%, it is clear that the accuracy of the identity expressed by Eq. (8.160) is totally adequate in actual evaporator heat exchanger design practice for similar halogenated refrigerants. As shown in Chapter 6, Eq. (6.23), and consequently Eq. (8.160), is also applicable even in the case of *extremely dissimilar polar* substances such as water and ammonia in the limiting case of *fully rough* flow regimes since (1) the Blasius exponent $n \to 0$, and the impact of the viscosity vanishes, and (2) the numerical values of v_g^+, as well as v_g/v_l, of these dissimilar fluids also fall on the universal curves shown in Fig. 6.3.

Comments

This is essentially a brief update based on additional more recent correlations not considered in earlier (1960s) papers for forced-convection, as well as nucleate boiling, regimes.

1. The analytical model proposed by Chen for forced-convection evaporation in vertical tubes [8.12] and mentioned in Rohsenow's survey paper on boiling [4.28], which was not included in the original version of Table 8.19, will be briefly discussed here since it is one of the correlations still recommended in the 1985 edition of the *ASHRAE Handbook, Fundamentals*. It is easy to demonstrate that Chen's correlation can also be generalized according to the ECSP method as follows.

Chen's model assumes that the effective α_{tp} is the sum of two components—a convective and a nucleate boiling heat transfer coefficient. His convective two-phase heat transfer coefficient is based on the type of correlations shown in Table 8.19, which can be generalized on the basis of reduced pressure as shown above. The second component,—the nucleate boiling heat transfer coefficient—can also be generalized according to the ECSP method as already demonstrated in the 1960s by Borishansky et al. [4.64, 4.66] and recently (1982/84) confirmed by Cooper [4.56–4.58]. It is thus evident that the resulting sum of these two components, that is, α_{tp}, can also be generalized according to the ECSP method in terms of reduced pressure in spite of the mathematical complexity of Chen's semianalytical correlations.

2. It is noteworthy that Cooper [4.56–4.58] arrives at the same basic conclusions for *nucleate boiling* as those derived in the 1960s by Soumerai [4.21, 4.22] for *forced-convection evaporation* as indicated by the following excerpts from Cooper's abstract [4.56] (note that comments and/or modifications are shown in square brackets):

> Many correlations for pool boiling heat transfer give similar answers and trends, although they look very different ... They can also be formulated... using just $[P^+]$ and $(-\log_{10}[P^+]$ [therefore, some universal function of reduced pressure $P^+ = P/P_{cr}$ in accordance with [[4.21, 4.22]] ... with accuracy again a few % over the narrow range generally used, e.g., from 1 atm to 0.9 of critical pressure [in reverse power cycles the low side evaporator must operate at $P^+ \ll 0.9$ in contrast to steam power cycles to ensure an acceptable reverse power cycle coefficient of performance or COP]. These re-formulations imply no further assumptions concerning the nature of boiling [no physical model is required].... The reduced-property formulation of correlations enable us to see why similar numerical results come from very different-looking multiproperty formulations.

8.3.2 Generalization of Two-Phase Heat Transfer Coefficients in Fluid Cooling Mode: Condensation

It is evident, in view of the fundamental coupling, regardless of the direction of heat flow, between heat and momentum transfer in two-phase natural as well as forced convection, that the same generalization methods highlighted for the fluid heating mode (evaporation) can also be applied usefully for the two-phase flow with heat removal, that is, condensation. For example, the Nusselt correlations for condensation on falling condensate films (in laminar or turbulent flow regimes) can readily be generalized according to the methodology described in Section 6.3 and in earlier publications [4.21, 4.22]. This conclusion is almost self evident for low thermal lift applications in view of the inherent symmetry between the condensing and evaporating modes of heat transfer in the absence of nucleate boiling regimes (see Guideline 27 in Section 8.2.2.3, and Fig. 4.10 in Section 4.7).

8.3.3 Conclusions

The main advantages of the ECSP generalization methods summarized in Chapter 6 for the fluid flow aspects of two-phase forced-convection heat exchangers are applicable (and even more significant) to the coupled heat transfer aspects since these are inherently less accurately predictable than the fluid flow phenomena. We reemphasize below only two among the many practical and fundamental advantages that are particularly relevant in the context of the questions addressed in Chapter 9, namely, the universal validity and the second law implications of the ECSP methodology.

8.3.3.1 Universal Validity. The ECSP generalization method, like all classical thermodynamic tools, is totally independent of the detailed nature of the phenomena and the physical configuration of the system. The universal validity of the ECSP generalization method is self-evident in the case of ideal van der Waal fluids or thermodynamically "fully similar" fluids obeying exactly the ECSP laws outlined in Chapter 6. Fortunately, deviations from the ECSP become insignificant in the more favorable turbulent stable flow regimes typically encountered in heat transfer engineering, and this is what makes the ECSP thermodynamic generalization method so effective in complex two-phase flow heat exchanger applications. Thus, the generalization methods based on reduced pressures can be applied beyond the few illustrative examples considered here to include any heat exchanger geometry, with evaporation or condensation in natural or forced convection inside or outside channels, as well as for the limiting cases of two-phase flow when $x > 1$ or $x < 0$, that is, single-phase flow heat exchangers. Indeed, there are preliminary indications[15] that the ECSP generalization method can also be applied to multicomponent evaporators and condensers [8.13].

To the applied heat transfer researchers whose prime objective is to predict, with an adequate level of accuracy, the illusive heat transfer coefficients (see Guideline 12 in Chapter 7), the ECSP generalization method should also prove a valuable tool, as stated in the concluding remarks of Ref. 4.21, Part II, and quoted here:

> The theory of thermodynamic similitude [ECSP] should also provide a useful tool to those engaged in basic heat, mass transfer, and fluid flow research, since it makes it possible to minimize the number of similar fluids that need to be investigated in order to establish generally valid explicit correlations over a wide range of operating conditions.

8.3.3.2 Second Law Implications. The remarkable phenomenological laws—that the molar or Trouton[16] entropy Mh_{lg}/T and the ratio $v_g/v_l = \rho_l/\rho_g$ are for all practical purposes universal functions of reduced pressure (see Figs. 6.3 and 6.4)—provide a more fundamental rationale for the empirically validated ECSP generalization methods when interpreted in the light of the alternate enthalpy-based and/or entropy-based heat transfer driving forces highlighted in this chapter. The main practical advantage of an entropy-based heat transfer driving force in two-phase flow applications is due to the fact that the molar entropy is a universal function of reduced pressure for most common heat transfer media. Conversely, the mere fact that two-phase heat flow generalizations can be explained more rationally on the basis of entropy-based heat transfer driving forces provides additional support for the fundamental validity and, more importantly, the *usefulness* of alternate *second law driving forces*, the main topic addressed in the concluding chapter.

NOTES

1. It is clear that an extremely effective power plant water-cooled condenser with an effectiveness $\varepsilon \cong 100\%$ (or a perfect approach factor AF $\cong 0$), operating at a maximum temperature differential $T_H - T_{1,C} = (\Delta T)_{max}$ of say 30°C, will obviously be thermodynamically far less efficient, that is, cause more irreversible losses, than an *ineffective* condenser with an effectiveness ε of only 50% designed to operate at the same cold water inlet temperature $T_{1,C}$ at $(\Delta T)_{max} = T_H - T_{1,C} = 10$°C. Therefore, a high exchanger effectiveness does not necessarily mean high thermodynamic efficiency. However, if the two condensers were designed for operation at the same $(\Delta T)_{max}$, the $\varepsilon \cong 100\%$ solution would certainly be thermodynamically more efficient than the $\varepsilon = 50\%$ alternative designed to operate at the same water side pumping power. Similarly, a steam power plant cycle incorporating seven or eight feedwater heaters each with an effectiveness of $\varepsilon = 70\%$ can be thermodynamically far more efficient than with a single perfectly effective $\varepsilon \cong 100\%$ feedwater heater yielding the same feedwater temperature. It is therefore evident that the heat exchanger effectiveness ε should not be confused with thermodynamic efficiency. It is worth mentioning that the "older" method of rating heat exchangers in terms of a contact factor CF (or approach factor AF = 1 − CF) is less misleading in this respect.
2. The mere fact that Petukhov [3.42] is the author of a major chapter on this topic in *Advances in Heat Transfer* confirms this assessment.
3. This excludes exotic liquid metals with $\mathrm{Pr} \ll 1$ (see Table A.4 in Appendix A).
4. Petukhov's most accurate data makes it possible to quantify more precisely the (semi)analytical conclusions presented by Soumerai in earlier (1960s) papers [4.21, 4.22].
5. Keep in mind that $\mathrm{Pr} = c_p\mu/k \propto \mu$ because c_p and k are roughly constant within the usual application range of "normal liquids" in power generation and reverse power cycles (see Appendix A).
6. In reality $\bar{e}/D_h > (\bar{e}/D_h)_{ref}$ when $D_h < D_{h,ref}$ as shown in Chapter 3. Thus, this analysis underestimates the advantages of smaller hydraulic diameters on two counts. First, $f > f_{ref}$ for $\bar{e}/D_h > (\bar{e}/D_h)_{ref}$. Second, the Blasius exponent n

would tend to be lower for the smaller D_h; therefore, $\beta < \beta_{ref}$ according to Fig. 8.1, thus reducing the negative impact of the $(Pr)^{\beta} > 1$ in the case of liquids. For $Pr = 1$ fluids, such as gases and vapors (steam) at moderate pressures, the improvement is only due to $f > f_{ref}$ since $(Pr)^{\beta} \simeq 1$.

7. The thermal system project engineer (as opposed to the heat exchanger designer) makes no such distinction. From the project engineer's point of view all friction losses are equally bad.
8. This illustrates once more that turbulent flow regimes are generally more easy to deal with in real world heat exchanger design applications than the orderly laminar flow regimes (see Guideline 10 in Chapter 6).
9. This discussion is confined to single fluid cycles, excluding cascade refrigeration and heat pump cycles [4.7] or binary vapor power cycle plants using liquid metals such as mercury [8.14].
10. It is well known that boiling on the outside of bare, or externally finned, tube bundles in so-called flooded evaporators is enhanced by the superimposed convective flow. See, for instance, a recent paper by Hahne and Muller [8.9].
11. It is only necessary to substitute on the *tube side* an evaporating halogenated refrigerant for water in Fig. 1 of Appendix C, and, for instance, ammonia condensing on the *shell side* at a temperature slightly higher (low thermal lift) than the evaporating fluid flowing inside the tube to achieve a constant wall temperature on the cold (evaporating) fluid's side.
12. Keep in mind that nucleate boiling regimes are suppressed in these typical low thermal lift, low-pressure side ($P^+ \ll 1$) forced-convection evaporators. These conclusions are not applicable in nucleate boiling regimes typical of high heat flux, high-pressure side ($P^+ \lesssim 1$) steam power plant applications. This again highlights the importance of making a sharp distinction between thermodynamically efficient fluid-to-fluid two-phase LTL heat exchanger applications and atypical nuclear reactor core power generation applications.
13. It is tacitly assumed that the continuum hypothesis is satisfied, as is typically the case in earthbound macroscopic thermal components, except in rather exotic high-vacuum heat exchanger (and aerospace) applications.
14. Some recommended correlations are based on the liquid flowing fraction $(1 - x)\dot{m}$ rather than $\dot{m}$. However, this has no impact on the results of the generalization methods discussed here.
15. This was confirmed by Schmitt [8.13] in the discussion period following the oral presentation of this paper at the yearly (1986) meeting of GVC, Gesellschaft Verfahrenstechnik und Chemieingenieurwesen (chemical engineering group) of VDI.
16. The rule of thumb indicating that the molar entropy at normal atmospheric pressure is roughly constant is known in the German and English literature as *Trouton's rule*. However, this relationship was first reported by Raoul Pierre Pictet who published his findings in 1876 [8.15], that is, 8 years before Trouton, as pointed out by Rudolf Plank [8.16].

CHAPTER 9

NONEQUILIBRIUM THERMODYNAMIC CONSIDERATIONS

We focus in the first part on linear heat flux fluid-to-fluid heat exchangers (Guidelines 19 and 20 in Chapter 7) operating under conditions *near equilibrium* [9.1]. Second law thermodynamic considerations are then extended beyond the current state of the art to include applications *far from equilibrium*, that is, to the nonlinear domain of modern *nonequilibrium thermodynamics* [9.2]. The chapter includes a broad assessment of the practical usefulness of thermodynamic tools in complex heat transfer engineering applications.

9.1 LOW THERMAL LIFT MIRROR-IMAGE HEAT EXCHANGERS

The useful concept of a mirror-image heat exchanger system (MIHES) is defined on the basis of the old contact factor-NTU heat exchanger calculation method as the special case of a pure parallel flow arrangement, when the hot and cold fluids are identical and flow at the same mass velocity inside identical channels. The key characteristics of the MIHES are first analyzed according to conventional methods, then in the light of classical equilibrium thermodynamic considerations with the help of a normalized temperature–entropy diagram, focusing again on linear heat flux, normally effective heat exchangers meeting the criteria defined by Guidelines 19 and 24 in Chapters 7 and 8, respectively.

The main conclusions and practical applications of this thermodynamic analysis can be summarized as follows:

- The results corroborate the fundamental validity of the novel dimensionless semiempirical correlations and the thermodynamic mixing model initially proposed by Soumerai [6.8] and highlighted in this Chapter.[1]

- These technical thermodynamic considerations provide a simple bridge with the linear branch of nonequilibrium thermodynamic theory and extends its practical validity to include nonlinear turbulent flow regimes as anticipated by Guideline 20c in Chapter 7, thus demonstrating the relevancy of the new (twentieth century) nonequilibrium branch of thermodynamics in heat, mass, and fluid flow engineering applications.

9.1.1 Elementary Parallel Flow Heat Exchangers: Forced-Convection Single-Phase Tube Flow Applications

As shown in Appendix B, the key dimensionless analytical relations for the archetype elementary parallel flow, double pipe heat exchanger effectiveness are

$$\varepsilon = \frac{1 - \exp[-\mathrm{NTU}(1 + C)]}{1 + C} \tag{9.1}$$

$$C \equiv \frac{(\dot{m}c_p)_{\min}}{(\dot{m}c_p)_{\max}} \tag{9.2}$$

$$\mathrm{NTU} \equiv \frac{UA}{(\dot{m}c_p)_{\min}} \tag{9.3}$$

$$\varepsilon = \frac{\dot{Q}_e}{(\dot{Q}_e)_{\max}} = \frac{(\dot{m}c_p)_{\min}|T_2 - T_1|}{(\dot{m}c_p)_{\min}(T_{1,H} - T_{1,C})} = \frac{|T_2 - T_1|}{T_{1,H} - T_{1,C}} \tag{9.4}$$

The effectiveness thus represents the absolute temperature change normalized by convention in terms of the maximum total temperature differential betwen the two fluids:

$$(\Delta T)_{\max} = T_{1,H} - T_{1,C} \tag{9.5}$$

for the fluid undergoing the larger temperature change, that is, the one having the minimum value of $\dot{m}c_p$.

It is evident from Eq. (9.2) that the numerical value of the dimensionless ratio C has a lower bound $C = 0$, when $(\dot{m}c_p)_{\min}$ is negligibly small compared to $(\dot{m}c_p)_{\max}$, and an upper bound $C = 1$, when the fluids capacity rates are identical on the cold and hot sides of the heat exchanger. Therefore, $0 \leq C \leq 1$ and the term $1 + C$ appearing twice in Eq. (9.1) must be confined between the following numerical values: $1 \leq 1 + C \leq 2$. Thus, the maximum effectiveness ε_∞, when the exchange surface becomes infinitelly large, can only vary between the following limits:

$$1 \leq \varepsilon_\infty \equiv \frac{|T_{2,\infty} - T_1|}{T_{1,H} - T_{1,C}} = \frac{1}{1 + C} \leq \frac{1}{2} \tag{9.6}$$

The maximum effectiveness ε_∞ of a pure parallel flow heat exchanger is therefore solely a function of the ratio C and will always be less than unity, even with an infinitely large heat exchanger surface, except in the limiting case $C = 0$ as illustrated below.

$C = 0$	0.25	0.5	0.75	1
$\varepsilon_\infty = 1.000$	0.800	0.666	0.571	0.500

For parallel flow heat exchangers it is convenient and physically more meaningful to reintroduce on the basis of Eqs. (9.1) and (9.6) the older heat exchanger performance factor, called the contact factor (CF), defined as follows:

$$\mathrm{CF} \equiv \frac{\varepsilon}{\varepsilon_\infty} = 1 - \exp[-(1 + C)(\mathrm{NTU})] \tag{9.7}$$

From Eqs. (9.4)–(9.7) we can also write

$$\mathrm{CF} \equiv \frac{(\dot{m}c_p)_{\min}|T_2 - T_1|}{(\dot{m}c_p)_{\min}|T_{2,\infty} - T_1|} \equiv \frac{\dot{Q}_e}{(\dot{Q}_e)_{\max,\infty}} \equiv \frac{|T_2 - T_1|}{|T_{2,\infty} - T_1|} \tag{9.8}$$

The main disadvantage of the conventional effectiveness is that Eq. (9.4) defines ε in terms of a *non*attainable maximum capacity in the case of parallel flow heat exchangers. By contrast, the CF defined by Eqs. (9.7) and (9.8) is physically more meaningful for parallel flow heat exchangers because it represents the absolute temperature change normalized in terms of a physically *attainable* maximum temperature change

$$|\Delta T_{\max,\infty}| = |T_{2,\infty} - T_1| \tag{9.9}$$

and the CF $\to 1$ when NTU $\to \infty$ according to Eq. (9.7) as a true *heat exchanger* performance factor should; that is,

$$0 \le \mathrm{CF} \le 1 \tag{9.10}$$

For the two limiting cases of special interest in this thermodynamic analysis, $C = 1$ and $C = 0$, Eq. (9.7) yields

$$(\mathrm{CF})_{C=1} = 1 - \exp[-2(\mathrm{NTU})] \tag{9.11}$$

$$(\mathrm{CF})_{C=0} = 1 - \exp(-\mathrm{NTU}) \tag{9.12a}$$

Noting that the same expression as Eq. (9.12a) is obtained from Eq. (9.1) for the limiting case $C = 0$, we may write

$$\varepsilon_{C=0} = (\mathrm{CF})_{C=0} = 1 - \exp(-\mathrm{NTU}) \tag{9.12b}$$

It is noteworthy that CF $\equiv \varepsilon$ in pure counterflow heat exchangers according to Eqs. (9.4) and (9.8) since $|\Delta T_{\max,\infty}| \equiv \Delta T_{\max} \equiv T_{1,H} - T_{1,C}$ in this case.

As the contact factor concept is physically more meaningful than the effectiveness at $C \neq 0$ for parallel flow heat exchangers, we shall focus exclusively on the CF concept (rather than ε) from now on. It is then useful to reintroduce the old concept of a bypass factor (BF) defined as follows:

$$\mathrm{BF} \equiv 1 - \mathrm{CF} \tag{9.12c}$$

which is, of course, another name[2] for the approach factor defined in Chapter 7, since AF $\equiv$ 1 $-$ CF $\equiv$ BF.

According to the above definition the BF $\to$ 0, as CF $\to$ 1, when the outlet temperatures reach the end equilibrium limit $T_{2,C} = T_{2,H} = T_\infty$ and the bypass factor BF (or the approach factor AF) thus represents the *terminal temperature difference (TTD) normalized in terms of the true maximum potential temperature change* $|\Delta T_{\max,\infty}| \equiv |T_{2,\infty} - T_1|$ attainable in pure parallel flow heat exchangers.

9.1.2 Theoretical Validity Limits of First Law Analytical Methods

Since the ε and CF versus NTU as well as the logarithmic mean temperature difference (LMTD) analytical methods are based on the same fundamental assumptions and first law thermodynamic considerations (see Section 7.6), we shall simply refer to them as the *conventional* and/or *first law* methods for the balance of this chapter.

Like all analytical expressions of physical laws [Guideline 2(a)], the above equations are based on a number of simplifying assumptions (highlighted in Section 7.6 and Appendix B). These assumptions are listed below with a few additional comments that are particularly relevant in the context of this chapter.

A.1 Absence of chemical reactions and mass transfer. Although evident to the practicing heat transfer engineer, it is necessary to state this fundamental assumption in order to situate this archetype elementary heat transfer design problem in the context of the vastly broader and more complex field of nonequilibrium thermodynamics.[3]

A.2 Steady-state steady-flow conditions *throughout the heat exchanger* from inlet state (1) at axial position $x = 0$ to outlet state (2) at $x = L$. It should be noted that this is more restrictive than the alternative in/out SSSF definition introduced in Chapter 7 which does not preclude a priori local fluctuations within the heat exchanger at any position x within $0 \leq x \leq L$ as is typical in many practical single- and two-phase forced-convection heat exchangers.

A.3 Incompressible fluids: ρ_H and ρ_C are constant. This condition is readily met with single-phase fluids except in the neighborhood of the thermodynamic critical point. Common compressible gases and vapors can also be treated as incompressible fluids provided

$$\left(\frac{\Delta P}{P}\right)_C \quad \text{and} \quad \left(\frac{\Delta P}{P}\right)_H \ll 1 \tag{9.13}$$

and

$$\frac{(T_2 - T_1)_C}{T_C} \quad \text{and} \quad \frac{(T_1 - T_2)_H}{T_H} \ll 1 \tag{9.14}$$

These conditions are usually met in thermodynamically efficient heat exchangers (or suitable average values ρ_C and ρ_H used if necessary).

A.4 Straight uninterrupted channels with constant Ω, a, and $r_h = \Omega/a$; that is,

$$\Omega_H, \Omega_C, a_H, a_C, D_H, \text{ and } D_C \text{ are constant for } all \text{ values } 0 \leq x \leq L \tag{9.15}$$

From assumptions A.2, A.3, and Eq. (9.15) it follows, with $G = \dot{m}/\Omega$ and $\overline{V} = G/\rho$, that

$$\overline{V}_C \text{ and } \overline{V}_H \text{ are constant for } all \text{ values of } 0 \leq x \leq L \tag{9.16}$$

A.5 Constant specific heat fluids: $(c_p)_C$ and $(c_p)_H$. This requirement can always be met within a narrow absolute temperature range typical of low thermal lift heat exchangers except in the immediate neighborhood of the thermodynamic critical point (which is, however, an inherently unstable and/or thermodynamically inefficient operating region).

A.6 Perfectly insulated system: $\dot{Q}_{\text{loss}} = 0$.

A.7 No axial heat flow: $\dot{Q}_{\text{axial}} = 0$.

A.8 Negligible total friction dissipation or

$$\frac{\dot{Q}_{f,\text{total}}}{\dot{Q}_e} = \frac{(\dot{m}\,\Delta H_f)_C + (\dot{m}\,\Delta H_f)_H}{\dot{Q}_e} = 0 \tag{9.17}$$

neglecting individual losses (see Section 8.1.5).

The energy conservation law yields, with the axiomatic assumptions $\dot{Q}_{\text{loss}} = \dot{Q}_{\text{axial}} = \dot{Q}_{f,\text{tot}} = 0$, the *conventional heat balance* for incompressible ($\rho =$ constant) fluids

$$(\dot{Q}_e)_C \equiv (\dot{m}c_p)_C (T_2 - T_1)_C = (\dot{m}c_p)_H (T_1 - T_2)_H \equiv (\dot{Q}_e)_H \tag{9.18}$$

Equation (9.18) is used to compute $\dot{Q}_e$ on either the cold or hot side of the heat exchanger in design practice. It is also put to practical use in most test codes to verify the test accuracy by measuring both $(\dot{Q}_e)_C$ and $(\dot{Q}_e)_H$ and specifying that $(\dot{Q}_e)_H = (\dot{Q}_e)_C$ within an acceptable deviation of a few percent (typically 1–5%). This heat balance is also employed in reliable laboratory (or field) tests to verify the accuracy of the measured heat rate and estimate the resulting accuracy (at best $\pm 5\%$, usually worse) of the empirical U, α_C, and/or α_H determined from such tests. Because these empirical coefficients are then used in design practice under operating conditions "not too far" from the laboratory test conditions in conjunction with the same Eq. (9.18), the method is self-consistent and any conceivable effects of $\dot{Q}_{f,\text{total}}/\dot{Q}_e \neq 0$, will usually remain undetected. This ambiguity is generally harmless in normal heat exchanger applications, but Eqs. (9.17) and (9.18) are no longer applicable under extreme conditions such as:

- Extremely low $\dot{Q}_e$ typically encountered in some refrigeration applications as highlighted in Refs. 6.7, 6.8, 7.83, and 9.1–9.5 and discussed further in the thermodynamic section of the present chapter.
- High (near sound) gas velocities [7.87] when the total temperature rise of the fluid caused by *friction dissipation*: $\Delta T_f = \Delta H_f/c_p = \Delta P_f/\rho c_p \gg 0$ is no longer negligible [6.8, 9.6].

A.9 Linear relationships between heat fluxes and driving forces. This linearity is expressed by the constancy of each term in the expression

$$\Sigma R_{\text{individual}} \equiv R = R_H + R_C + R_M + R_{F,C} + R_{F,H}$$
$$= \text{sum of individual thermal resistances connected in series} \quad (9.19)$$

at any value $0 \leq x \leq L$, regardless of the intensity of the driving forces $(T_H - T_C)$, $(T_H - T_{w,H})$, $(T_{w,C} - T_C)$, and so on. This axiomatic assumption, though not generally correct, is valid with an adequate level of accuracy for linear heat flux heat exchangers designed to operate at $(T_w - \overline{T})/\overline{T} \ll 1$.

To simplify this discussion (and this has no impact on the main conclusions derived in the present analysis), we shall assume surgically clean channels and negligible metal resistance so that Eq. (9.19) and the previous assumptions reduce to:

$$\alpha_C \text{ and } \alpha_H \text{ are constant at any axial position } x, \; 0 \leq x \leq L, \text{ and independent of } \Delta T_{\max} \quad (9.20a)$$

and

$$Ua = \frac{1}{1/(\alpha a)_C + 1/(\alpha a)_H} \quad (9.20b)$$

when

$$R_{F,H} = R_{F,C} = 0 \quad \text{and} \quad \frac{R_M}{R_H + R_C} = 0 \quad (9.20c)$$

9.1.3 Practical Validity Limits

The millions of fluid-to-fluid heat exchangers successfully produced to date, which were designed on the basis of the first law analytical methods, confirm that the above assumptions are satisfied with an adequate level of accuracy in most *normal* fluid-to-fluid heat exchangers and with an excellent accuracy in the case of low thermal lift cycle applications. It is nevertheless essential for the heat transfer engineer to recognize situations when some of the above-mentioned fundamental assumptions are so flagrantly violated that the conventional methods are no longer applicable within an acceptable level of accuracy. The following remarks are intended to provide a proper perspective on the accuracy of conventional analytical methods in a few relevant forced-convection applications. For the sake of brevity, the various fundamental assumptions are simply identified by the paragraph numbers A.1–A.9 in the previous section.

Two-Phase Channel Flow at Extremely High Heat Fluxes. These conditions are typical of high-pressure side steam power generation evaporators when the fundamental assumption of linearity (A.9) is grossly violated since for nucleate boiling the heat fluxes are proportional to $(\Delta T)^a_{\text{sup}}$ with exponents a as high as 3–5. This is the central theme of a series of challenging monographs by Adiutori entitled *The New Heat Transfer* [7.56]. Although Adiutori correctly diagnoses an abnormal behavior in the case of nuclear applications in the (subcooled) nucleate boiling regions, it is important to keep a proper sense of perspective since such nuclear heat transfer channels represent an atypical[4] and insignificant fraction (of the order of 1 ppm) of the total population of forced-convection fluid-to-fluid heat exchangers successfully produced to date and designed on the basis of generally accepted first law analytical methods.

Two-Phase Channel Flow in Low Thermal Lift Systems. These are the conditions typically encountered in refrigeration evaporators (or condensers) with heat addition (or removal) and in any thermodynamically efficient two-phase forced-convection heat exchangers such as condensers and feed-water heaters in steam power plants and the main heat exchangers of ocean thermal energy conversion systems, in view of their extremely low thermal lift.

The most grossly violated assumptions in this case are those expressed by Eq. (9.16) and, as a corollary, Eq. (9.20), since α_{tp} increases significantly with the velocity V_g from channel inlet to outlet as the vapor content increases from x_1 to $x_2 = x_1 + \Delta x$. This is further complicated by the fact that the saturation pressure loss ΔP_{tp} is associated with a corresponding saturation temperature drop ΔT_{sat}. Thus, the thermal and hydrodynamic phenomena are directly coupled and the actual driving force $(T_w - T_{\text{sat}})$ is strongly affected by ΔP_{tp}. In other words, the hydrodynamic and thermodynamic problems cannot be uncoupled as in single-phase heat exchangers.[5] Therefore, the conventional methods are no longer valid on fundamental grounds and their application can lead to totally erroneous conclusions. In principle, the above problems could be resolved by resorting to modern computerized finite difference techniques. However, another still poorly understood fundamental physical problem—the occurrence of nonequilibrium conditions [3.32, 4.4]—may frustrate such attempts.[6]

Because of these highly complex physical phenomena, it was natural to seek alternative thermodynamic solutions involving entropy dissipation and the extremely useful corresponding states principle discussed previously. In effect, the thermodynamic methods outlined here for *single-phase* flow are a logical *outgrowth* of those originally devised to solve more complex *two-phase* flow problems in compact internally finned refrigeration evaporators and condensers. Thus, it is not altogether surprising that the most important practical conclusions presented in Refs. 6.5 and 6.8 include the possibility of extending the proposed methods to two-phase flow applications! To the writer this does not represent a long shot extrapolation into the unknown, but rather a logical "retour aux sources" [3.32, 4.4].

Common Gases at $|T_2 - T_1|/\bar{T}$ and $|\bar{T}_w - \bar{T}|/\bar{T} \ll 1$ in Compact Heat Exchangers With $\Omega \neq$ Constant. Most of the basic assumptions are satisfied with only one notable exception, A.4, since the *free flow cross section is discontinuous and far from constant* in many compact heat exchangers so that the gas velocity varies significantly from tube row to tube row. The empirical α and f presented graphically in terms of $(\mathrm{Re})_{\max}$, which is based on $\bar{V}_{\max}$, are therefore *core-averaged* empirical values rather than truly local coefficients and valid only for a *specific type* of core geometry. This explains the popularity of *Compact Heat Exchangers* [9.7], since this book provides reliable empirical data for 88 different core geometries. However, in spite of the use of the familiar looking Chilton–Colburn type of correlation ("...in order to maintain a uniform methodology..."), the empirical curves $(\alpha/c_pG)(\mathrm{Pr})^{2/3}$ and f in terms of a $(\mathrm{Re})_{\max}$ based on average hydraulic diameter[7] but a maximum velocity $\bar{V}_{\max}$ at the minimum flow cross section $\Omega_{\min}$ *cannot* possess the *universal validity* of the original Chilton–Colburn correlation for constant cross section (Ω *and* $4r_h \equiv D_h$ = constant) in long ($L/D \gg 1$) straight channels according to Eq. (7.10a) for stable fully developed turbulent flow[8] in normally effective low thermal lift heat exchangers, namely,

$$\overline{\mathrm{St}}(\mathrm{Pr})^{-\beta} \equiv \frac{\bar{\alpha}}{c_p\rho\bar{V}}(\mathrm{Pr})^{-\beta} = \tfrac{1}{2}f_{r_h} = \tfrac{1}{8}f \tag{9.21}$$

The impact of flow discontinuities and corresponding velocity changes due to the significant flow cross-section variations in compact heat exchangers is particularly significant on f_{r_h} which includes (for good pragmatic reasons) in the words of Kays and London [9.7] "the effect of viscous shear (skin friction) and pressure force (form drag)."

In contrast to these typical compact heat exchangers, the empirical friction factor f_{r_h} in straight constant cross section (Ω) and long channels ($L/D_h \gg 1$) is *entirely due to skin friction* and the Reynolds number used in conjunction with Eq. (9.21) is based on the physically meaningful constant $D_h = 4r_h$ and $G = \bar{V}\rho$ (i.e., $\bar{V}$ = constant) and essentially fully developed turbulent flow regimes when $M_{L/D}$ per Eq. (8.5) is equal to unity. This explains why bare circular, noncircular, as well as internally finned straight channels of constant cross section Ω and $L/D > 60$ can be correlated with adequate accuracy according to the Chilton–Colburn type of Reynolds analogy, Eq. (9.21), for practical application in typical shell-and-tube heat exchangers as highlighted previously in Chapter 7 and discussed more thoroughly in Ref. 3.49 (and references cited therein).

Ideal Reference Case: Turbulent Flow of Gases in Long ($L/D_h > 60$ or $L/r_h > 240$) Straight Channels of Constant Cross Section at $|T_2 - T_1|/\bar{T}$ and $|T_w - \bar{T}|/\bar{T} \ll 1$. Assumptions A.1–A.9 are most faithfully respected as one would expect since this case duplicates the double pipe model originally used to derive the conventional NTU analytical method. The agreement between conventional theory and actual performance is particularly good for normally effective (ε = CF > 80%) low thermal lift heat exchangers, especially in the case of turbulent flow of common gases since the $(\mathrm{Pr})^{-\beta}$ term in the *universal Eq. (9.21)* remains substantially *independent of P and T* and the key variable on the right-hand side of Eq. (9.21), the friction factor f, is also quite insensitive to temperature variations in turbulent regimes [6.5, 6.7] as shown in previous chapters.

We shall focus on this ideal reference case in the balance of this section for two pragmatic reasons: First, it is the best known and most accurately documented case, and second, ideal gases are the most convenient guinea pigs for thermodynamic analyses. Consequently, the following relations, valid for perfect gases, are assumed:

$$Z = 1 \quad \text{in} \quad \frac{1}{\rho} = \frac{Z(\mathscr{R}/M)T}{P} \tag{9.22a}$$

$$c_p \quad \text{and} \quad c_v \text{ are constant} \tag{9.22b}$$

$$T_2 \text{ on } \textit{isotherm } T_1 \text{ for an adiabatic irreversible throttling process when } \Delta P_f = P_1 - P_2 > 0 \tag{9.22c}$$

It should be stressed, however, that the basic conclusions derived from these perfect gas analyses can be extended to real gases and liquids, as well as two-phase fluids, on the basis of general thermodynamic concepts such as the application of the extended corresponding states principle along the lines described in Chapters 6 and 8.

9.1.4 Limiting Cases: $C = 0$ and $C = 1$

We consider the two limiting cases of the CF-NTU method when $C = (\dot{m}c_p)_{\min}/(\dot{m}c_p)_{\max} = 0$ and $C = 1$.

9.1.4.1 First Limiting Case $C = 0$ and $Ua = (\alpha a)_C$. In this particular limiting case, the cold fluid is heated by a constant temperature source T_H *and* it is implicitly assumed from $Ua = (\alpha a)_C$ that all other thermal resistances are negligibly small compared to $R_C = 1/(\alpha a)_C$ according to Eq. (9.20b).

Problem Statement

(a) Find the NTU required for design values of the CF ranging from 0.5 to 0.999 and corresponding BF = 1 − CF = 0.5 to 0.001.

(b) Find a general dimensionless expression for the required heat exchanger length, L_{rel}, at any fixed value of $(\alpha a)_C$ and $(\dot{m}c_p)_C$.

(c) Find the temperature change $T_{2,C} - T_{1,C}$ as well as $T_H - T_w$ as a function of L_{rel} for $\Delta T_{\max} = 10$ K.

Solution

The results summarized in Table 9.1 can be highlighted as follows:

(a) The required BF and NTU computed from Eqs. (9.12a)–(9.12c) are shown in Table 9.1 on lines (2) and (3) for any prescribed value of the CF.

TABLE 9.1. Key Data for Conventional and Thermodynamic Numerical Analyses of MIHES

(a)	(b)	(c)	(d)	(e)	(f)	(g)	(h)	(i)
(1) $\mathrm{CF} \equiv \lvert\Delta t_n\rvert \equiv \lvert\Delta s_e\rvert_n$	0	0.50	0.63	0.70	0.80	0.90	0.99	0.999
(2) $\mathrm{BF} \equiv 1 - \lvert\Delta t_n\rvert \equiv 1 - \lvert\Delta s_e\rvert_n$	**1**	**0.50**	**0.37**	**0.30**	**0.20**	**0.10**	**0.01**	**0.001**
(3) $L_{rel} \equiv (\mathrm{NTU})_{C=0} \equiv (\mathrm{NTU})_{\text{one-side}, C=1}$	**0**	**0.69**	**1**	**1.20**	**1.61**	**2.30**	**4.61**	**6.91**
(4) L_{rel} at BF = 0.1, 0.01, 0.001	—	—	—	—	—	ln(10)	2 ln(10)	3 ln(10)
(5) $T_{2,C} - T_{1,C}$ (K)	0	+5.00	+6.32	+7.00	+8.00	+9.00	+9.90	+9.99
(6) $T_{2,H} - T_{1,H}$ (K)	0	−5.00	−6.32	−7.00	−8.00	−9.00	−9.90	−9.99
(7) $(T_{2,C} - T_{1,C}) + (T_{2,H} - T_{1,H})$ (K)	0	0	0	0	0	0	0	0
(8)[a] $(\mathrm{NVH})_f/(\mathrm{Pr})^{\beta}$, one-side	**0**	**1.39**	**2**	**2.41**	**3.22**	**4.61**	**9.21**	**13.82**
(9)[b] $(\mathrm{NVH})_f$, one-side	0	1.51	1.66	2.00	2.67	3.82	7.64	11.47
(10)[c] ΔH_f, one-side (J / kg)	0	151	166	200	267	382	764	1147
(11)[d] $(\Delta s_f)_n$, one-side	**0**	**+0.114**	**+0.165**	**+0.199**	**+0.266**	**+0.380**	**+0.761**	**+1.141**
(12)[d] $(s_{2,C} - s_o) = -\lvert\mathrm{BF}\rvert + (\Delta s_f)_n$	−1	−0.386	−0.203	−0.101	+0.066	+0.280	+0.751	+1.140
(13)[d] $(s_{2,C} - s_o) = +\lvert\mathrm{BF}\rvert + (\Delta s_f)_n$	+1	+0.614	+0.533	+0.499	+0.466	+0.480	+0.771	+1.142
(14)[d] $\dot{S}/\dot{m}$ = lines (12) + (13) ≡ 2 × line (11)	0	+0.288	+0.33	+0.398	+0.532	+0.76	+1.522	+2.282

[a] $(\mathrm{Pr}) \geq 0.7$
[b] $(\mathrm{Pr})^{\beta} = 0.83$
[c] $(\mathrm{Pr})^{\beta} = 0.83$ and $\frac{1}{2}\overline{V}^2 = 100$
[d] $(\mathrm{Pr})^{\beta} = 0.83$, $\frac{1}{2}\overline{V}^2 = 100$, and $T_w - T_{1,C} = 1$ K

(b) The heat exchanger length required at any NTU, shown on line (3), is calculated from Eq. (9.3) with $Ua = (\alpha a)_C$ according to

$$L = (\mathrm{NTU})\frac{(\dot{m}c_p)_C}{(\alpha a)_C} \tag{9.23}$$

Since L is directly proportional to NTU, it is possible[9] to express the required length in terms of an arbitrary reference length L_{ref} for constant values of $(\alpha a)_C$ and $(\dot{m}c_p)_C$ as

$$\frac{L}{L_{\mathrm{ref}}} = \frac{L/D}{(L/D)_{\mathrm{ref}}} = \frac{A}{A_{\mathrm{ref}}} = \frac{\mathrm{NTU}}{(\mathrm{NTU})_{\mathrm{ref}}} \tag{9.24}$$

As suggested by Le Goff [7.86] it is convenient to use the heat exchanger length required for NTU = 1 as a reference when AF = BF = $1/e$ or, with Eqs. (9.23) and (9.24),

$$L_{\mathrm{rel}} \equiv \frac{L}{L_{1/e}} = \frac{\mathrm{NTU}}{1} = \mathrm{NTU} \tag{9.25}$$

According to Eq. (9.23) when NTU = 1

$$L_{1/e} = \frac{(\dot{m}c_p)_C}{(\alpha a)_C} \tag{9.26}$$

In other words, line (3) in Table 9.1 shows the required dimensionless heat exchanger lengths, L_{rel}, normalized in terms of the reference heat exchanger length per Eq. (9.26).

It is noteworthy that the ultimate equilibrium temperature $T_{2,C} = T_\infty$ is achieved with an error of *less than 1%* at $L_{\mathrm{rel}} > 4.605$ and BF < 0.01 per columns (h) and (i) of Table 9.1. To reduce the BF by a factor of 10, an additional $L_{\mathrm{rel}} = \ln(10) = 2.303$ is required as illustrated by line (4). In normal heat exchanger design practice, an accuracy of 1% is totally adequate because most performance test codes and/or heat transfer coefficient correlations are based on allowable test deviations of several percent. We shall therefore assume, in line with chemical engineering design practice [7.86], that the end equilibrium temperature T_∞ has been reached for all practical purposes at $L_{\mathrm{rel}} > 4.6$ or, to use a round number, $L_{\mathrm{ref}} > 5$.

(c) From the definition of the CF it is evident that $T_{2,C} - T_{1,C} = 10$ CF, as shown on line (5) in terms of NTU = L_{rel}. From $(\alpha a)_C/(\alpha a)_H = 0$, it follows immediately that $T_H - T_w = 0$ and $T_w = T_H$ = constant and independent of L_{rel} since T_H = constant. This limiting case is the classical example of an isothermal wall boundary condition mentioned in most heat transfer textbooks and discussed previously in Chapters 7 and 8.

Comments

Consider the opposite limiting case when the thermal resistance on the constant temperature side of the heat exchanger is controlling, that is, $(\alpha a)_H/(\alpha a)_C = 0$. Following the same reasoning as under (c), $T_w - T_C = 0$; therefore, $T_w = T_C$ varies with L_{rel} exactly as shown on line (5). This limiting case is approached in many process and refrigeration heat exchanger applications. Indeed, most water-cooled shell-and-tube halogenated refrigerant condensers use some form of extended surface on the condensing side of the tubes to reduce the overall thermal resistance as shown in Appendix C. It is therefore evident that the condition $C = 0$ is not sufficient to ensure an isothermal wall boundary. This fact needs to be kept in mind because the basic assumption A.7 (i.e., $\dot{Q}_{axial} = 0$) may be grossly violated when axial wall temperature variations are significant.

9.1.4.2 Second Limiting Case: $C = 1$ and $R_H = R_C$. Assume, as in the previous analysis, zero metal and fouling thermal resistances.

Problem Statement

(a) Find the end equilibrium temperature T_∞.

(b) Find the wall temperature T_w in terms of L_{rel}.

(c) Find the absolute temperature change on the *cold fluid side* $(T_{2,C} - T_{1,C})$ for $\Delta T_{max,tot} \equiv T_{1,H} - T_{1,C} = 20$ K as well as the normalized change $(\Delta t_C)_n = (T_{2,C} - T_{1,C})/(T_\infty - T_{1,C})$ as a function of L_{rel}.

(d) Same as (c) for the *hot fluid side*.

(e) Plot the results in terms of L_{rel}.

Solution

(a) By reasons of symmetry [and per Eq. (9.6)], T_∞ must be at the midpoint between $T_{1,H}$ and $T_{1,C}$; that is,

$$T_\infty = \tfrac{1}{2}(T_{1,H} + T_{1,C}) \tag{9.27}$$

(b) For this fully balanced heat exchanger, with $(\alpha a)_C = (\alpha a)_H$, per Eq. (9.20b), the local wall temperature at any point along the axial length is at the midpoint between T_H and T_C. Since $(T_2 - T_1)_C = -(T_2 - T_1)_H$ for $C = 1$, it follows that $T_w = T_\infty$ from inlet to outlet. Therefore, with Eq. (9.27),

$$T_w = T_\infty = \tfrac{1}{2}(T_{1,H} + T_{1,C}) = \text{constant} \tag{9.28}$$

(c) With $(\alpha a)_C = (\alpha a)_H \equiv (\alpha a)_{one\text{-}side}$ and Eq. (9.20b), it follows that

$$Ua = \tfrac{1}{2}(\alpha a)_{one\text{-}side} \tag{9.29}$$

and since in this case the exponent in Eq. (9.7) is

$$2(\text{NTU})_{\text{tot}} = 2(\tfrac{1}{2})(\text{NTU})_{\text{one-side}} = (\text{NTU})_{\text{one-side}} \tag{9.30}$$

we obtain with Eq. (9.11)

$$\text{CF} = 1 - \exp[-(\text{NTU})_{\text{one-side}}] \tag{9.31}$$

In other words, the CF is identical (as could be expected) to the CF obtained in the first limiting case for each side of the $C = 1$ balanced heat exchanger. Consequently, line (3) of Table 9.1 remains valid for the second limiting case with $(\text{NTU})_{\text{one-side}}$ as shown. For $T_{1,H} - T_{1,C} = 20$ K and $|\Delta T_{\max,\infty}| = T_w - T_{1,C} = T_{1,H} - T_w = 10$ K, exactly the same positive temperature change $(T_{2,C} - T_{1,C})$ is thus obtained as in the first limiting case $\Delta T_{\max,\infty} = 10$ K according to line (5) of Table 9.1.

(d) In accordance with the first law of thermodynamics and assumption A.8 (i.e., $\dot{Q}_f = 0$), the term $T_{2,H} - T_{1,H}$ is negative and equal in absolute value to $T_{2,C} - T_{1,C}$ as highlighted by lines (6) and (7) of Table 9.1.

(e) The above results are plotted in terms of L_{rel} with the origin of the temperature scale arbitrarily selected at $T_w = T_\infty$ as shown in Fig. 9.1, using

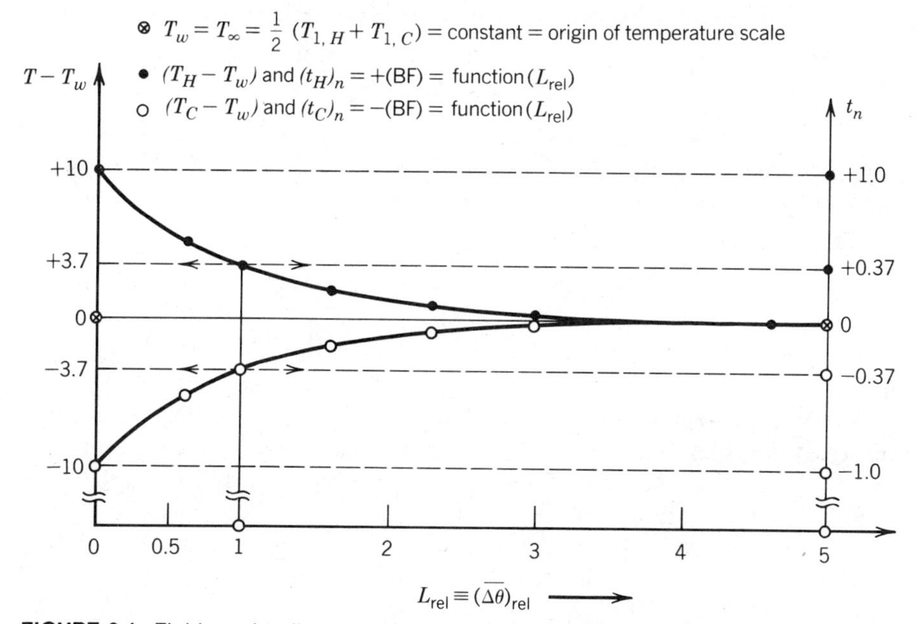

FIGURE 9.1. Fluids and wall temperatures in terms of relative length of travel L_{rel}. [From H. Soumerai, "Thermodynamic Considerations on a 'Mirror Image' Concept—Forced Convection Single-Phase Tube Flow Applications," ASME Winter Annual Meeting, Miami, Florida, Nov. 1985. Reproduced with permission.]

the conventional temperature scale on the left-hand side and a *normalized* (*dimensionless*) *scale* Δt_n on the right-hand side, defined according to

$$|\Delta t_n| \equiv \frac{|T_{2,C} - T_{1,C}|}{|\Delta T_{\max,\infty}|} \equiv \frac{|T_{1,H} - T_{2,H}|}{|\Delta T_{\max,\infty}|} \equiv \mathrm{CF} \tag{9.32}$$

which is identical to the contact factor CF in view of the definition of the CF. It can readily be seen from Fig. 9.1 that BF $\equiv 1 -$ CF represents, in absolute value, the normalized terminal temperature difference since, with Eq. (9.32),

$$1 - \mathrm{CF} = \mathrm{BF} = 1 - |\Delta t_n| \equiv |t|_n \tag{9.33}$$

Comments

1. In contrast to the first limiting case ($C = 0$), the second limiting case ($C = 1$ and $R_H = R_C$) *always represents the classical isothermal wall* temperature *boundary* condition exactly for *any finite* value of αa in stable turbulent or laminar flow regimes. This second limiting case clearly meets all criteria derived previously for the isothermal heat exchanger surface boundary condition[10] with the additional unique feature that the constant wall temperature T_w is exactly at the midpoint between T_H and T_C because of the full symmetry of all key parameters on the hot and cold sides of the separating wall.

2. The performance of a limiting Case No. 1 heat exchanger operating at $(\Delta T_{\max,\infty})_{\text{No. 1}}$ is exactly duplicated by *one-side* of a Case No. 2 heat exchanger operating at

$$|\Delta T_{\max,\infty}|_{\text{No. 2, one-side}} = |\Delta T_{\max,\infty}|_{\text{No. 1}} \tag{9.34}$$

3. Recalling that the wall surrounding the fluids is at a uniform temperature $T_w = T_\infty$, it is clear by referring to Fig. 9.1 and Eqs. (9.32)–(9.34) that the BF provides a direct normalized *measurement* of nonequilibrium conditions prevailing *inside* the heat exchanger. As the hot (or cold) fluid penetrates the heat exchanger at $L_{\text{rel}} = 0$ [or a relative residence time $\overline{\Delta\theta}_{\text{rel}} = 0$ per Eq. (7.78)] and a maximum bypass factor $(\mathrm{BF})_{\max} = 1$, it is at its maximum state of nonequilibrium, that is, $+1$ (or -1) normalized degrees in relation to the end equilibrium temperature $T_\infty = T_w$; at $L_{\text{rel}} \equiv \overline{\Delta\theta}_{\text{rel}} = 1$ and BF $= 0.37$ this nonequilibrium condition has already decreased to a level of $+0.37$ (or -0.37) normalized degrees; at $L_{\text{rel}} \equiv \overline{\Delta\theta}_{\text{rel}} > 5$ the end equilibrium state $T_{2,C} = T_{2,H} = T_w = T_\infty$ has been reached for all practical purposes with an accuracy of better than 1% (or 0.1% at $L_{\text{rel}} > 6.91$).

As we move into the channels and at any position $0 < L_{\text{rel}} < 5$ we are entering "will-nilly" the field of twentieth century irreversible nonequilibrium thermodynamics. It is therefore important to remember, as shown in Chapter

7, that the laws of classical equilibrium thermodynamics, based on the axiomatic assumption of equilibrium states, are not rigorously applicable *within the heat exchanger* because of the *nonequilibrium* conditions highlighted above. It is nevertheless permissible to make useful overall statements on the equilibrium states preceding and following the heat exchanger on the basis of classical thermodynamic laws such as the well-known nineteenth century Gouy–Stodola theorem, discussed later.

9.1.5 Mirror-Image Heat Exchanger Systems (MIHES)

We define the MIHES first analytically, then describe one of several conceivable concrete physical models satisfying the analytical definition. This physical model makes it possible to *visualize* some subtle points of technical equilibrium thermodynamics and very abstract concepts of nonequilibrium thermodynamics addressed in this chapter and provides a simple test setup to obtain empirical heat transfer coefficients under truly isothermal wall conditions.

9.1.5.1 Analytical Definition. The MIHES can be defined analytically as a pure parallel flow heat exchanger satisfying the following requirements:

- The hot and cold fluids are identical, that is, $\rho_H = \rho_C \equiv \rho$, $(c_p)_H = (c_p)_C \equiv c_p$, and $(\mathrm{Pr})_H = (\mathrm{Pr})_C \equiv \mathrm{Pr}$.
- Identical channels $\Omega_C = \Omega_H \equiv \Omega$ and $a_C = a_H \equiv a$.
- Identical flow rates: $\dot{m}_H = \dot{m}_C \equiv \dot{m}$; therefore, $G_H = G_C \equiv G$ since $\Omega_C = \Omega_H$.
- $\alpha_C = \alpha_H \equiv \alpha$ at any axial position x and, as a corollary, the length average $\bar{\alpha}_C = \bar{\alpha}_H \equiv \bar{\alpha}$. Note that this most general definition of the MIHES allows axial variations of α and f due to inlet effects [6.5, 6.7]. These identities are permissible in view of the full symmetry between the hot and cold sides and the empirical fact that the conventional forced-convention heat transfer coefficients are independent of the direction of heat flow in low thermal lift cycles, when the correction factor $M_{\Delta T}$ [defined by Eq. (8.3a)] is equal to unity, as shown in Chapter 8.

It is evident that the general MIHES thus defined satisfy all the requirements of the second limiting case $C = 1$ and $(\alpha a)_C = (\alpha a)_H$ and all the characteristics described in the previous analysis, including Fig. 9.1, are therefore also valid for any MIHES.

9.1.5.2 Physical Models. Figure 9.2 depicts one physical MIHES model (among others) complying with the above analytical definition. Essentially, the model consists of two identical parallel channels embedded in, or extruded from, a good thermal (metal) conductor and perfectly insulated so that

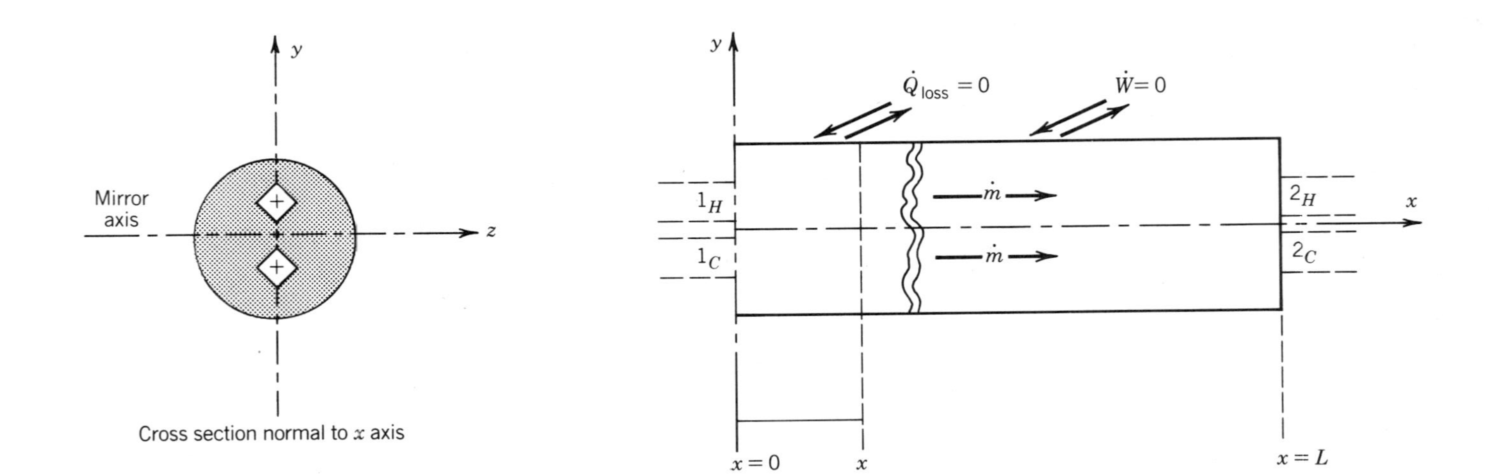

FIGURE 9.2. Physical model of a mirror-image heat exchanger system. [From H. Soumerai, "Thermodynamic Considerations on a 'Mirror Image' Concept—Forced Convection Single-Phase Tube Flow Applications," ASME Winter Annual Meeting, Miami, Florida, Nov. 1985. Reproduced with permission.]

$\dot{Q}_{\text{loss}} = 0$. We consider first the most general model when the metal wall thermal resistance $R_M \neq 0$, then the simple MIHES when $R_M = 0$.

General MIHES: $R_M \neq 0$. With the temperature and spatial coordinates defined in Figs. 9.1 and 9.2, respectively, the following key characteristics are easily identified:

(a) The vertical plane through the x, z axes represents the *vertical mirror plane* (the choice of z as the vertical coordinate is only significant in two-phase flow applications as addressed in Ref. 7.83). By reason of symmetry the mirror plane must be at a uniform temperature $T_\infty = \frac{1}{2}(T_{1,H} + T_{1,C}) = \text{constant}$ at any position in this plane. However, $T_w \neq T_\infty$ at $L_{\text{rel}} \ll 5$ for this general case since the metal thermal resistance $R_M \neq 0$, even though $T_{2,H} = T_{2,C} = T_\infty = T_w$ at $L_{\text{rel}} > 5$.

(b) Referring to the channel cross sections shown for an arbitrary axial position x on the left side of Fig. 9.2, we see that the "picture" on the hot side is the exact *mirror image* of the "picture" on the cold side and the three-dimensional temperature field on the hot side must also be exactly mirrored (because of full symmetry) on the cold side. Since this statement applies at any axial position, it follows that the three-dimensional temperature field on one side is *exactly mirrored* (i.e., same temperature but *opposite signs*) on the other side of the mirror plane, thus the name mirror-image heat exchanger system.

Simple MIHES: $R_M = 0$. The temperature differences between T_w and T_∞ vanish when Eq. (9.20c) is satisfied (with $R_M = 0$ and zero fouling resistance) and the *solid core* is at a *uniform temperature* $T_w = T_\infty = \textit{constant}$ from inlet to outlet. The solid thermal conductor thus constitutes at the same time the constant temperature *source* ($T_w > T_C$) for the cold fluid and the constant temperature *sink* ($T_w < T_H$) for the hot fluid. Although the requirements expressed by Eq. (9.20c) could be simply postulated for a thermodynamic analysis, it is interesting to note that this condition is particularly well satisfied in the case of common gases with relatively low α values and normal metallic wall materials in tubular heat exchangers when the thermal resistance on the gas side is controlling. Focusing again on common gases, it is immediately apparent that the requirement $\rho_H = \rho_C = \rho$ with Eq. (9.22) and the definition of the MIHES means that the total pressure drop due to wall friction is identical on the hot and cold sides; that is,

$$(\Delta P_f)_C = (\Delta P_f)_H > 0 \tag{9.35}$$

In other words, the *energy dissipation* terms $\Delta H_f = \Delta P_f/\rho$ are also *identical* on both sides of the MIHES in *absolute value and in sign, both being positive.*

9.1.5.3 Conventional Analysis of a Simple MIHES. The main purpose of this analysis is to provide some useful numerical benchmarks for the otherwise excessively abstract thermodynamic analyses discussed in the next sections.

Problem Statement

Find the following on the basis of the methods derived in Chapter 8.

(a) The dimensionless friction head dissipation $(\text{NVH})_f = \Delta H_f/(\bar{V}^2/2)$ required on either side of the MIHES at any design CF for single-phase fluids flowing in straight ideal smooth circular tubes in stable turbulent regimes at $10{,}000 \le \text{Re} \le 120{,}000$.

(b) Apply the results obtained under (a) to air with $\text{Pr} = 0.69$ and $c_p = 1.005$ kJ/kg · K at 300 K.

Solution

(a) The required number of velocity heads on one side can be most generally expressed in terms of the required $(\text{NTU})_{\text{one-side}}$ for a given Pr according to Eq. (8.23):

$$\frac{(\text{NVH})_f}{(\text{Pr})^\beta} = 2(\text{NTU})_{\text{one-side}} \tag{9.36}$$

The values thus obtained are listed on line (8) of Table 9.1.

(b) For ideal smooth circular tubes and $0.5 < \text{Pr} < 1$, the generally[11] accepted Pr exponent $\beta = 0.5$; therefore, $(\text{Pr})^\beta = (0.69)^{0.5} = 0.83$ and the values shown on line (9) of Table 9.1 are obtained.

Comments

1. As already emphasized in the discussion of the integrated modified Reynolds analogy in Section 8.1.3, it is most remarkable that the required $(\text{NVH})_f$ is fully defined for any given fluid (i.e., $\text{Pr} = \text{constant}$) at any design $\text{CF} = 1 - \text{BF}$ and therefore completely independent of the spatial dimension $D_h = 4r_h$ in the case of constant cross-section straight channels. This is the essential feature that makes it possible to treat the problem most generally according to classical thermodynamic laws which are also independent of spatial dimensions [see Guideline 14(a)].

2. It is noteworthy that the numerical values shown on line (8) also represent the actual $(\text{NVH})_f$ required for all fluids with a $(\text{Pr})^\beta = 1$ and they are therefore valid with an adequate accuracy (slightly conservative theoretical error) for all common gases as shown in Section 8.1.

9.2 EQUILIBRIUM THERMODYNAMIC RELATIONSHIPS

We address first high thermal lift (HTL) applications typical of power generation cycles making extensive use (mainly for didactic reasons) of temperature–entropy diagrams.[12]

Since these equilibrium thermodynamic relationships are discussed at length in all introductory engineering thermodynamic textbooks (such as those cited in Chapter 7), we omit specific references and very rigorous explanations in this brief overview.

9.2.1 High Thermal Lift Systems (HTLS)

Because the specific entropy change $(s_2 - s_1)$ from an initial equilibrium state at $s_1(P_1, T_1)$ to a second equilibrium state at $s_2(P_2, T_2)$ is independent of the *reversible thermodynamic path* selected, it is convenient for our present purpose to compute $s_2 - s_1$ in two steps: first, the change from T_1 to T_2 on the isobar P_1 = constant denoted by $(\Delta s)_P = s(P_1, T_2) - s(P_1, T_1)$; second, the change on the isotherm T_2 = constant from P_1 to P_2 denoted by $(\Delta s)_T = s(P_2, T_2) - s(P_1, T_2)$.

For ideal gases the change along the isobar can be derived from the Clausius definition of the differential entropy change $(ds)_P$:

$$(ds)_P \equiv \left(\frac{dh}{T}\right)_P = c_p \frac{dT}{T} = c_p d \ln T \tag{9.37}$$

Since c_p = constant for ideal gases, the entropy change $(\Delta s)_P$ from T_1 to T_2 at $P = P_1$ = constant becomes

$$(\Delta s)_P = \int_{T_1}^{T_2} (ds)_P = c_p \int_{T_1}^{T_2} d \ln T = c_p \ln\left(\frac{T_2}{T_1}\right) \tag{9.38}$$

Substituting $T_1 + (T_2 - T_1)$ for T_2 in the above equation yields, with $T_2/T_1 = 1 + (T_2 - T_1)/T_1$, the following equivalent analytical expressions for $(\Delta s)_P$:

$$(\Delta s)_P = c_p \ln\left(\frac{T_2}{T_1}\right) \equiv c_p \ln\left(1 + \frac{T_2 - T_1}{T_1}\right) \tag{9.39a}$$

Validity: Particular perfect gas

or in dimensionless terms,

$$\frac{(\Delta s)_P}{c_p} = \ln\left(\frac{T_2}{T_1}\right) = \ln\left\{1 + \frac{T_2 - T_1}{T_1}\right\} \tag{9.39b}$$

Validity: All ideal gases

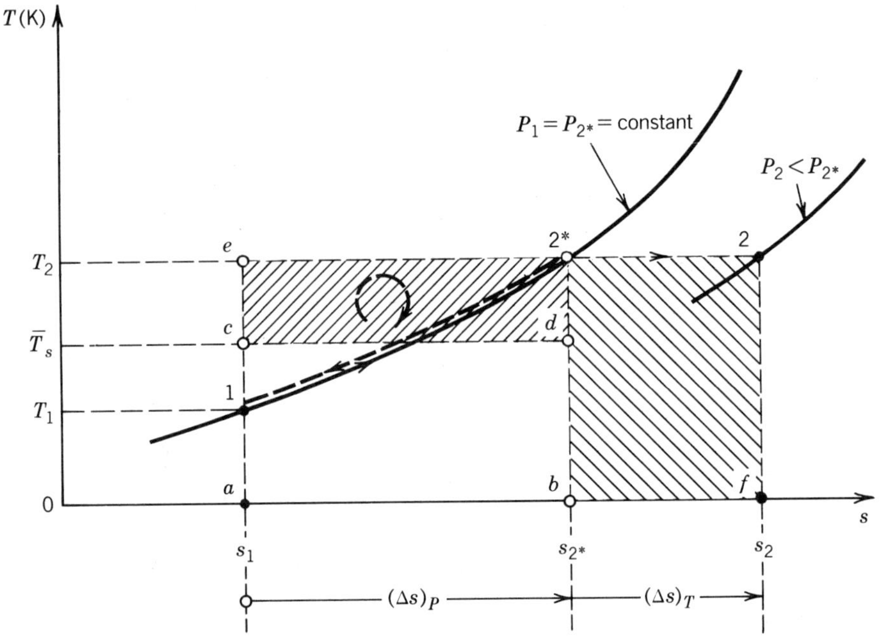

FIGURE 9.3. Skeleton *T-s* diagram for high thermal lift systems (HTLS).

It is thus possible with Eqs. (9.39a) and (9.39b) to calculate the entropy change from any equilibrium temperature T_1 to a specified equilibrium temperature T_2 for all ideal gases on any isobar, that is, whenever $P_1 = P_2$ or $|\Delta P| = P_2 - P_1 = 0$ as illustrated by point 2* in the skeleton *T-s* diagram shown in Fig. 9.3.

When $P_2 \neq P_1$ and, more specifically for the cases of practical interest here, when $P_2 \leq P_1$, or $P_1 - P_2 \geq 0$, the additional positive entropy change $(\Delta s)_T$ during an isothermal expansion from point 2* at $s_{2*}(P_1, T_2)$ to the equilibrium point 2 at $s_2(P_2, T_2)$ can be expressed, with R = gas constant of the particular gas under consideration, according to the well-known relationships[13]

$$(\Delta s)_T = -R \ln\left(\frac{P_2}{P_1}\right) = +R \ln\left(\frac{P_1}{P_2}\right) \tag{9.40}$$

since $\ln(P_1/P_2) \equiv -\ln(P_2/P_1)$.

Equation (9.40) can be expanded with $P_1 = P_2 + (P_1 - P_2)$ and $\Delta P = P_1 - P_2 > 0$ as follows:

$$(\Delta s)_T = R \ln\left(\frac{P_1}{P_2}\right) \equiv R \ln\left(1 + \frac{P_1 - P_2}{P_2}\right) \tag{9.41a}$$

or, more generally,

$$\frac{(\Delta s)_T}{c_p} = \frac{R}{c_p} \ln\left(1 + \frac{P_1 - P_2}{P_2}\right) \tag{9.41b}$$

Since for ideal gases c_p and c_v are constant, the adiabatic exponents $c_p/c_v \equiv \kappa = \text{constant}$ and $R = c_p - c_v = c_p(1 - 1/\kappa) = c_p(\kappa - 1)/\kappa$. Equation (9.41b) can thus be reduced to the following most general expression:

$$\frac{(\Delta s)_T}{c_p} = \left(\frac{\kappa - 1}{\kappa}\right) \ln\left(1 + \frac{P_1 - P_2}{P_2}\right) \tag{9.42}$$

bearing in mind that the adiabatic exponent κ of many quasi ideal gases are practically identical at moderate pressures.[14]

With Eqs. (9.39)–(9.42) the total entropy change from $s_1(P_1, T_1)$ to $s_2(P_2, T_2)$ becomes

$$s_2 - s_1 = (\Delta s)_P + (\Delta s)_T = c_p \ln\left(1 + \frac{T_2 - T_1}{T_2}\right) + R \ln\left(1 + \frac{P_1 - P_2}{P_2}\right) \tag{9.43a}$$

or, with Eq. (9.42),

$$\frac{s_2 - s_1}{c_p} = \ln\left(1 + \frac{T_2 - T_1}{T_1}\right) + \left(\frac{\kappa - 1}{\kappa}\right) \ln\left(1 + \frac{P_1 - P_2}{P_2}\right) \tag{9.44a}$$

which can be generalized further by multiplying and dividing the left-hand side of Eq. (9.44a) by M to express the entropy change per mole, $M(s_2 - s_1)$, according to

$$\frac{M(s_2 - s_1)}{\mathscr{C}_P} = \ln\left(1 + \frac{T_2 - T_1}{T_1}\right) + \left(\frac{\kappa - 1}{\kappa}\right) \ln\left(1 + \frac{P_1 - P_2}{P_2}\right) \tag{9.44b}$$

where

$$Mc_p \equiv \mathscr{C}_P \equiv \text{molar specific heat at constant pressure} \tag{9.44c}$$

Validity: All ideal gases (9.44d)

We now take advantage of the most useful feature of the temperature–entropy diagrams to define the physical meanings of a few relevant areas in the skeleton T-s chart shown in Fig. 9.3 and to derive graphically some useful analytical relationships.

The trapezoidal area ($a12^*ba$) in Fig. 9.3 is identical to the enthalpy change ($h_{2*} - h_1$) which is equal to $c_p(T_2 - T_1)$. The identities,

$$\begin{aligned}&\text{Trapezoidal area } (a12^*ba) \equiv c_p(T_2 - T_1)\\&= (s_{2*} - s_1)\overline{T}_s \equiv \text{rectangular area } (acdba)\end{aligned} \tag{9.45}$$

define the constant equivalent thermodynamic temperature $\overline{T}_s$. An analytical expression for $\overline{T}_s$ is obtained by solving Eq. (9.45) for $\overline{T}_s$ with $s_{2*} - s_1 \equiv (\Delta s)_P$ per Eq. (9.39); that is,

$$\overline{T}_s \equiv \frac{T_2 - T_1}{\ln(T_2/T_1)} \equiv \frac{T_2 - T_1}{\ln[1 + (T_2 - T_1)/T_1]} \tag{9.46a}$$

Conversely,

$$T_2 - T_1 \equiv \overline{T}_s \ln\left(\frac{T_2}{T_1}\right) \equiv \overline{T}_s \ln\left(1 + \frac{T_2 - T_1}{T_1}\right) \tag{9.46b}$$

The semitriangular area ($1e2^*1$) in Fig. 9.3 represents the maximum specific (i.e., per kilogram of gas) reversible work output $(\mathrm{RW})_P$ that could be produced while raising reversibly the temperature of one unit mass of gas from T_1 to T_2 along the isobar $P_1 = P_{2*} =$ constant, when an external heat source at a constant temperature $T_2 \neq T_1$ is available (in the fluid heating case depicted in Fig. 9.3, $T_2 > T_1$). This can be explained by considering the reversible *closed* cycle formed by the three thermodynamic processes, defined by the dotted lines $1e$, $e2^*$, and 2^*1. First step: 1 kg of gas is compressed isentropically [i.e., $s_1 =$ constant $= s(P_1, T_1)$] from point 1 to point e at the heat source temperature T_2 and corresponding pressure $P_e > P_1$. Second step: isothermal expansion from P_e at $T_2 =$ constant to point 2^* on the isobar $P_1 =$ constant at T_2. Third step: closing the cycle by cooling the closed cycle gas from point 2^* to the initial point 1 at T_1, P_1 by reversible heat interchange (i.e., ideal counterflow heat exchanger with zero temperature differential and zero friction) with an equal amount of the same type of gas flowing in an open cycle and heated from T_1 to T_2 along the same isobar P_1. The positive specific work output $(\mathrm{RW})_P$ in the closed cycle, per unit mass of gas heated in the open cycle from points 1 to 2^*, thus represents the maximum specific work attainable with an external heat source at a constant temperature $T_2 > T_1$ when $P_2 = P_1$. The area ($1e2^*1$) in Fig. 9.3 is equal to the rectangular area ($ae2^*ba$) minus the trapezoidal area ($a12^*ba$) which is also identical to $c_p(T_2 - T_1)$ according to Eq. (9.45). Thus

$$(\mathrm{RW})_P \equiv \text{area}(1e2^*1) = T_2(s_{2*} - s_1) - c_p(T_2 - T_1) \tag{9.47}$$

Substituting the expression $(\Delta s)_P$ per Eq. (9.39a) for the term $s_{2*} - s_1$ in the above equation yields

$$(\mathrm{RW})_P = c_p T_2 \ln\left(\frac{T_2}{T_1}\right) - c_p(T_2 - T_1) \tag{9.48a}$$

Alternatively, with Eq. (9.46b),

$$(\text{RW})_P = c_p(T_2 - \overline{T}_s)\ln\left(\frac{T_2}{T_1}\right) \tag{9.48b}$$

The more general expressions are

$$\frac{(\text{RW})_P}{c_p} = T_2\ln\left(\frac{T_2}{T_1}\right) - (T_2 - T_1) \tag{9.49a}$$

and

$$\frac{(\text{RW})_P}{c_p} = (T_2 - \overline{T}_s)\ln\left(\frac{T_2}{T_1}\right) \tag{9.49b}$$

Keeping in mind the definition of $\overline{T}_s$ and Eqs. (9.48) and (9.49), we see, by referring to Fig. 9.3, that the rectangular area $(ce2^*dc)$ = triangular area $(1e2^*1) = (\text{RW})_P$. The designation *equivalent thermodynamic temperature* for $\overline{T}_s$ is thus justified on two counts: first, because the rectangular area $(acdba) \equiv \overline{T}_s(\Delta s)_P$ = trapezoidal area $(a12^*ba) \equiv c_p(T_2 - T_1)$—this might be called a first law identity; and second, because the rectangular area $(ce2^*dc) = (T_2 - \overline{T}_s)(\Delta s)_P \equiv$ semitriangular area $(1e2^*1) = (\text{RW})_P$—that is, a second law identity ensuring that the maximum reversible work available under the specified conditions remains unchanged when the sloping temperature curve along the isobar is replaced by the constant temperature $\overline{T}_s$.

The above analytical and graphical expressions for the maximum available work $(\text{RW})_P$ are applicable for an exit state 2* at $P_2 = P_1$. When a lower pressure sink $P_2 < P_1$ is available for the open system in which the gas is heated from T_1 to T_2, an additional amount of positive work can be obtained by letting the gas expand reversibly at the constant external source temperature T_2 along the isotherm T_2 from point 2* at P_1 to point 2 at $P_2 < P_1$. This additional positive work per unit mass of fluid flowing in the open system is called $(\text{RW})_T$ (the subscript T identifying a T = constant, i.e., isothermal process) and is represented by the rectangular area $(b2^*2fb)$ in Fig. 9.3. Thus, the following relationships are applicable:

$$\begin{aligned}(\text{RW})_T &\equiv \text{rectangular area } (b2^*2fb) \\ &\equiv T_2(\Delta s)_T = T_2\left[R\ln\left(\frac{P_1}{P_2}\right)\right]\end{aligned} \tag{9.50}$$

Consequently, the total work available $(\text{RW})_{\text{tot}}$ for a reversible thermodynamic process from an initial equilibrium state $(\text{ES})_1$ at T_1, P_1 to a final equilibrium state $(\text{ES})_2$ at T_2 = temperature of the constant external source (or sink in the fluid cooling mode) and a prescribed exit pressure $P_2 < P_1$ will be

$$(\mathrm{RW})_{\mathrm{tot}} = (\mathrm{RW})_P + (\mathrm{RW})_T \tag{9.51}$$

with $(\mathrm{RW})_P$ and $(\mathrm{RW})_T$ expressed analytically, or graphically in a T-s diagram, according to the previously mentioned relationships. The total available (maximum) reversible work output is thus equivalent to the sum of the following two rectangular areas in the T-s diagram:

$$(\mathrm{RW})_{\mathrm{tot}} = (ce2^*dc) + (b2^*2fb) \tag{9.52}$$

which are shown cross-hatched in Fig. 9.3. The first rectangular area $= (\mathrm{RW})_P$ is entirely due to the maximum temperature differential $T_2 - T_1$ available between the oncoming lower (higher) temperature fluid in the fluid heating (cooling) mode when the constant source (sink) temperature T_2 is greater (smaller) than T_1. The second rectangular area $= (\mathrm{RW})_T$ is entirely due to the lower available exit pressure $P_2 < P_1$. For the limiting case $\Delta P = P_1 - P_2 \to 0$, this second contribution vanishes since point 2 converges to 2* as $P_1/P_2 \to 1$ and $\ln(1) \to 0$.

Although we have focused so far exclusively on the fluid heating mode, when $T_2 > T_1 \equiv T_{1,C}$, it is evident that the previous graphical and analytical relationships remain valid for fully reversible processes, exactly as stated, when the ideal gas is cooled (reversibly) from an initial equilibrium state at $T_1 = T_{1,H} > T_2$ down to T_2 at a constant pressure $P_1 = P_2$. It is only necessary to observe the recognized engineering thermodynamic sign conventions to make a distinction between the fluid cooling and heating modes as highlighted in Fig. 9.4 for an initial equilibrium gas temperature $T_{1,H} > T_2$ on the same isobar $P_{1,H} = P_{1,C}$ and the same end equilibrium temperature T_2, that is, the same end equilibrium state 2* as in the fluid heating mode described in Fig. 9.3. To identify more readily the most significant differences between the heating and cooling modes at high absolute temperature ratios, identical reversible temperature changes, $T_{1,H} - T_2 = -(T_{1,C} - T_2) = T_2 - T_{1,C}$, have been chosen in Fig. 9.4 for both heat exchange modes.

Referring to Fig. 9.4 and noting that the points in the cooling modes designated by $1'$, a', b', c', d', and e' (i.e., with a prime) correspond to the points 1, a, b, c, d, and e in the fluid heating mode (i.e., without a prime) and keeping in mind that point 2* is identical in both modes, we can draw the following major conclusions by following exactly the same reasoning as for the fluid heating mode.

Conclusion 1. The enthalpy change $h_{2*} - h_{1,C} = c_p(T_2 - T_{1,C}) = -(h_{2*} - h_{1,H}) = h_{1,H} - h_{2*} = c_p(T_{1,H} - T_2)$ for $T_2 - T_{1,C} = T_{1,H} - T_2$ since $c_p =$ constant for all ideal gases. Consequently, the following rectangular areas are identical in absolute value for the cooling and heating modes:

$$\mathrm{area}(a'c'd'b'a') = \mathrm{area}(acdba) \tag{9.53}$$

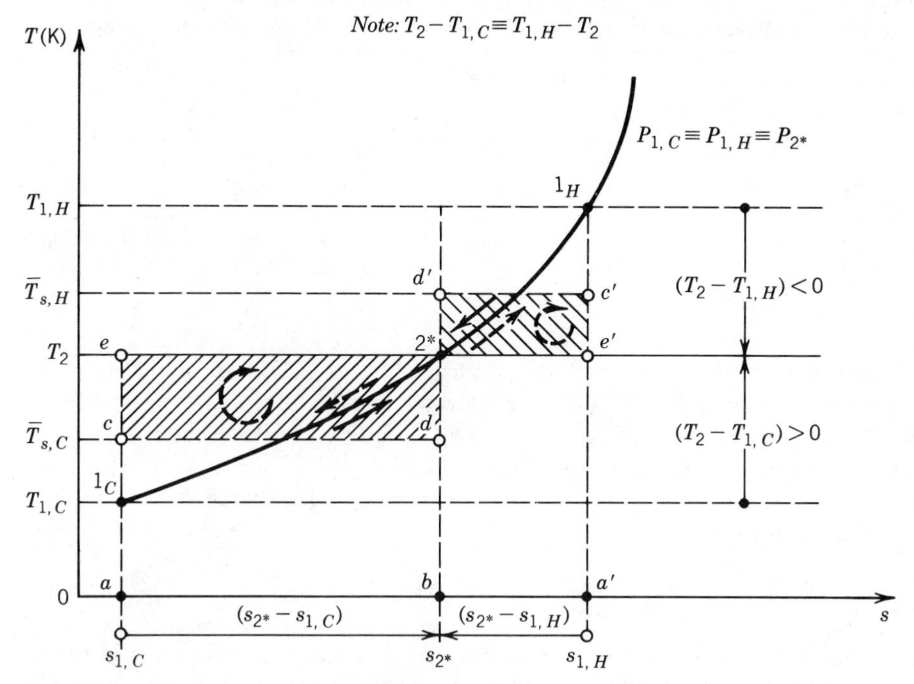

FIGURE 9.4. Reversible work potential $(RW)_P$ in a *T-s* diagram for HTLS: fluid heating and cooling modes.

This essentially confirms what was said previously, namely, that first law relationships are independent of the direction of heat flow. In other words, the enthalpy changes remain identical in absolute value, although of opposite signs, for the same positive or negative temperature change along an isobar. By mere convention the signs of the enthalpy changes are defined as positive in the fluid heating and negative in the fluid cooling mode.

Conclusion 2. Because the entropy change $(\Delta s)_P$ for an ideal gas along an isobar (and along an isochoric or v = constant line) is a logarithmic function of the absolute temperature ratio (T_2/T_1), it is evident that the isobar $P_{1,H} = P_{1,C} = P_{2*}$ is represented by a logarithmic curve in a *T-s* diagram. Consequently, the entropy change $(\Delta s)_P$ in the fluid cooling mode for $T_{1,H} - T_2 \equiv T_2 - T_{1,C}$ in the fluid heating mode for the same equilibrium temperature T_2 in both heat exchange modes must be generally different in sign and absolute value or, using the subscripts FHM and FCM to identify the fluid heating and cooling modes, respectively, we can write

$$(\Delta s)_{P,\,\mathrm{FHM}} \neq -(\Delta s)_{P,\,\mathrm{FCM}} \tag{9.54a}$$

Validity: $T_2 - T_{1,C} \equiv T_{1,H} - T_2$ and identical T_2; same isobar $P_{1,C} = P_{1,H} = P_{2*}$; ideal gases (9.54b)

TABLE 9.2. Nonlinear Variation of $(\Delta s)_P$ in Terms of $T_2 - T_1$ at T_2 / T_1 Far From Unity

	Fluid Cooling Mode $T_2 < T_1$				Fluid Heating Mode $T_2 > T_1$		
T_2 / T_1	$1/\infty$	$1/10$	$1/e$	1	e	10	∞
$(\Delta s)_P / c_p = \ln(T_2 / T_1)$	$-\infty$	**−2.30**	**−1.0**	**0**	**+1.0**	**+2.30**	$+\infty$
T_1 / T_2	∞	10	e	1	$1/e$	$1/10$	$1/\infty$
$(T_2 - T_1) / T_2 \equiv 1 - (T_1 / T_2)$	$-\infty$	**−9.00**	**−1.72**	**0**	**+0.63**	**+0.9**	**+1.0**
T_2 (K)	300	300	300	300	300	300	300
$T_2 - T_1$ (K)	$-\infty$	**−2700**	**−515.5**	**0**	**+189.6**	**+270**	**+300**
T_1 (K)	$+\infty$	3000	815.5	300	110.4	30	0

Furthermore, it is clear from the original Clausius definition of the entropy differential $ds = dQ_{\text{rev}}/T$ that the entropy change $|\Delta s|_{P,\text{FHM}}$ in the fluid heating mode will be most generally greater in absolute value than $|\Delta s|_{P,\text{FCM}}$ in the fluid cooling mode as shown in Fig. 9.4, since $\overline{T}_{s,C} < \overline{T}_{s,H}$ and $|Q_{\text{rev}}|_{\text{FHM}} = c_P|T_2 - T_{1,C}| = c_p|T_2 - T_{1,H}| = |Q_{\text{rev}}|_{\text{FCM}}$. Therefore, the following inequality holds:

$$|\Delta s|_{P,\text{FHM}} > |\Delta s|_{P,\text{FCM}} \tag{9.54c}$$

This last conclusion is quantified in Table 9.2 for extremely large positive temperature ratios $(T_2/T_{1,C})$ in the fluid heating mode and extremely small ratios $(T_2/T_{1,H})$ in the fluid cooling mode, that is, when T_2/T_1 departs significantly from unity.

Conclusion 3. The maximum reversible work potential in the fluid cooling mode $(\text{RW})_{P,\text{FCM}}$, represented by the triangular area $(1_H e' 2^* 1_H)$, is also positive (work output is defined by arbitrary convention as positive in technical engineering thermodynamics) in the clockwise direction, as illustrated in Fig. 9.4; therefore,

$$(\text{RW})_{P,\text{FCM}} \quad \textit{and} \quad (\text{RW})_{P,\text{FHM}} > 0 \tag{9.55a}$$

However, in view of Eq. (9.54c) and as highlighted in Fig. 9.4, it is clear that

$$\text{area}(ce2^*dc) = (\text{RW})_{P,\text{FHM}} > (\text{RW})_{P,\text{FCM}} = \text{area}(c'e'2^*d'c') \tag{9.55b}$$

$$\textit{Validity}: \quad \text{Same as Eq. (9.54b)} \tag{9.55c}$$

The most fundamental aspects of the above analysis are summarized in the following guideline.

GUIDELINE 28 For ***high thermal lift*** *thermodynamic systems, when the absolute temperature ratios and the corresponding absolute pressure ratios depart significantly from unity: (a) the* ***symmetry*** *between the cooling and heating mode, which is typical of first law relationships,* ***is broken for all***

second law relationships *involving the state property entropy; (b) in contrast to typical first and second law relationships, the maximum specific reversible work* $(\mathrm{RW})_P$ *per Eqs.* (9.48) *and* (9.49) *remains positive regardless of the direction of heat flow, but* $(\mathrm{RW})_{P,\,\mathrm{FHM}}$ *in the fluid heating mode always exceeds* $(\mathrm{RW})_{P,\,\mathrm{FCM}}$ *in the fluid cooling mode according to Eq.* (9.55b).

9.2.2 Low Thermal Lift Systems (LTLS)

All the graphical and analytical relationships, as well as Guideline 28 derived in the previous section, are valid, exactly as stated, for high thermal lift systems typical of most conventional thermal power generation applications, in which the absolute temperature and pressure ratios are significantly different from unity. We focus now on low thermal lift systems in which the absolute temperature and corresponding pressure ratios approach unity, that is, under conditions typically encountered in thermodynamically efficient heat exchangers in LTL systems.

We recall that the series expansion of the logarithmic expression $\ln(1+a)$ yields, for any real number $-1 < a < 1$,

$$\ln(1+a) = a - \tfrac{1}{2}a^2 + \tfrac{1}{3}a^3 - \tfrac{1}{4}a^4 + \cdots \tag{9.56}$$

or, neglecting higher-order terms,

$$\ln(1+a) \simeq a \tag{9.57}$$

Thus, the logarithmic expressions for the entropy changes $(\Delta s)_P$ and $(\Delta s)_T$ per Eqs. (9.39) and (9.41), respectively, can be reduced with Eq. (9.57) to the following simpler *linear relationships*:

$$\frac{|\Delta s|_P}{c_p} \simeq \frac{|T_2 - T_1|}{\overline{T}_s} \simeq \frac{|T_2 - T_1|}{T_1} \simeq \frac{|T_2 - T_1|}{T_2} \tag{9.58a}$$

when

$$\frac{|T_2 - T_1|}{T_1} \simeq \frac{|T_2 - T_1|}{T_2} \ll 1 \tag{9.58b}$$

so that the difference between the absolute temperatures T_1 and T_2 appearing as the denominator on the right-hand side of Eqs. (9.58a) and (9.58b) becomes insignificant and it is permissible to choose the end temperature T_2 to calculate $(\Delta s)_P$ according to

$$\frac{|\Delta s|_P}{c_p} = \frac{T_2 - T_1}{T_2} = 1 - \frac{T_1}{T_2} \tag{9.59a}$$

TABLE 9.3. Accuracy of Linearized Expression for $(\Delta s)_P / c_p$ at $0.9 < T_2 / T_1 < 1.1$

(1) $\dfrac{T_2}{T_1}$	(2) $\dfrac{(\Delta s)_P}{c_p} = \ln\left(\dfrac{T_2}{T_1}\right)$	(3) $\dfrac{(\Delta s)_P}{c_p} \simeq \dfrac{T_2 - T_1}{T_2} = 1 - \dfrac{T_1}{T_2}$	(4) Deviation[a] (%)
0.900	−0.10536	−0.1111	+5.5
0.950	−0.05129	−0.0526	+2.6
0.990	−0.01005	−0.01010	+0.5
0.995	−0.005013	−0.005025	+0.25
0.999	−0.0010005	−0.001001	+0.05
1.000	0	0	0
1.001	+0.0009995	+0.000999	−0.05
1.005	+0.0049875	+0.004975	−0.25
1.010	+0.00995	+0.00901	−0.5
1.050	+0.04879	+0.04762	−2.4
1.100	+0.09531	+0.0909	−4.6

[a]Deviation = 100 × [column (3) − column (2)] / column (2) (in %), or, with sufficient accuracy within above T_2 / T_1 range,

$$\text{Deviation} \simeq \tfrac{1}{2}|1 - T_1 / T_2| \times 100 \text{ (in \%)}.$$

Equation (9.59a) is most generally valid for all ideal gases as well as real gases and vapors satisfying substantially Eqs. (9.22a) and (9.22b) within the application range of interest (this excludes applications in the vicinity of the critical thermodynamic point where c_p becomes highly variable).

Validity: Linear domain when $|T_2 - T_1|/T_2 \ll 1$ (9.59b)

Accuracy: Maximum deviation $= \frac{1}{2}(1 - T_1/T_2)100$ in % as shown more precisely in Table 9.3 (9.59c)

The percent deviations indicated in column (4) of Table 9.3 were obtained by comparing the numerical results calculated with the simpler linear relationship, Eq. (9.59a), with those computed from the accurate logarithmic Eq. (9.39b).

Similarly, the following linear expressions are obtained from Eqs. (9.41a) and (9.41b) for $(\Delta s)_T$ and $(\Delta s)_T/c_p$ by neglecting higher-order terms:

$$(\Delta s)_T = R\left(\frac{P_1 - P_2}{P_2}\right) \tag{9.60a}$$

$$\frac{(\Delta s)_T}{c_p} = \frac{R}{c_p}\frac{(P_1 - P_2)}{P_2} \tag{9.60b}$$

when

$$\frac{P_1 - P_2}{P_2} = \frac{|\Delta P|}{P_2} \ll 1 \tag{9.60c}$$

In forced-convection heat exchanger design applications, the positive pressure drop $P_1 - P_2 = \Delta P_f > 0$ is due entirely to fluid friction and can be conveniently expressed in terms of the friction head $\Delta H_f \equiv \Delta P_f/\rho$. For ideal gases which are treated as incompressible fluids ($\bar{\rho} = \rho_1 = \rho_2$) in conventional heat exchanger analytical methods, the density term ρ_2 can be expressed according to the perfect gas law Eq. (9.22a) as $\rho_2 = 1/v_2 = P_2/RT_2$. Thus, $\Delta P_f = \rho_2 \Delta H_f = P_2 \Delta H_f/RT_2$. Introducing this expression for $P_1 - P_2 = \Delta P_f$ in Eqs. (9.60a) and (9.60b) finally yields the following simple linear relationships:

$$(\Delta s)_T \equiv \Delta s_f = \frac{\Delta H_f}{T_2} \tag{9.61a}$$

or

$$\frac{(\Delta s)_T}{c_p} \equiv \frac{\Delta s_f}{c_p} = \frac{\Delta H_f}{c_p T_2} \equiv \frac{\Delta T_f}{T_2} \tag{9.61b}$$

where ΔT_f is the fictive (for ideal gases) friction temperature rise defined by Eq. (7.98b).

The ratio $(\Delta s)_f/|\Delta s|_P$ obtained from Eqs. (9.61b) and (9.59a),

$$\frac{\Delta s_f}{|\Delta s|_P} = \frac{\Delta H_f}{c_p(T_2 - T_1)} \equiv \frac{\Delta T_f}{T_2 - T_1} \tag{9.62}$$

is typically small compared to unity in normally efficient heat exchanger applications. Consequently, if the linear approximation for $(\Delta s)_P$ is sufficiently accurate in practical heat exchanger applications, the linear approximation for $(\Delta s)_T = (\Delta s)_f$ per Eq. (9.61) is sufficiently accurate for design purposes and it is only necessary to check the accuracy of $(\Delta s)_P$ as highlighted in the following discussion.

Discussion. Since normally efficient heat exchangers must operate at $|T_2 - T_1|/T_2 \ll 1$ as shown previously, it is clear from Table 9.3 that the maximum error committed by calculating $(\Delta s)_P$ on the basis of the linearized Eq. (9.59a) is not very significant. In low thermal lift systems, such as low side steam power plant condensers, or low side refrigeration evaporators, the heat exchangers operate in the neighborhood of normal ambient temperatures of about 300 K at maximum differentials $(T_H - T_C)_{\max}$ of the order of 10 K. This yields, in typical well-balanced designs, a maximum temperature differential on either side of the separating wall of the order of $1/2(10) = 5$ K; therefore, $(T_2 - T_1)/T_2 \simeq 5/300 = 0.017$ corresponding to a maximum error due to the linearization of $\simeq 0.8\% < 1\%$. A maximum deviation of less than 1% is well within the allowable industrial performance code test tolerances (typically 1–5%) and negligibly small compared to deviations of the order of $\pm$10–20%

for the best single-phase heat transfer coefficient calculations. It is thus permissible to compute $(\Delta s)_P$ and $(\Delta s)_T \equiv \Delta s_f$ on the basis of the simpler linear relationships in typical LTL heat exchanger applications, *without* incurring a *measurable* loss of accuracy and Guideline 28 may be amended for low thermal lift heat exchanger systems as follows:

GUIDELINE 29 *For* ***low thermal lift forced-convection*** *heat exchangers designed to operate at absolute temperature ratios approaching unity, that is,* ***not far from equilibrium****, within the linear domain of nonequilibrium thermodynamics (see Guideline* 20, *Chapter* 7) *and within the validity limits of the simpler linear expressions for* $(\Delta s)_P$ *and* $(\Delta s)_T$: (*a*) *the* ***symmetry*** *between the cooling and heating modes is* ***maintained*** *for both first law and second law relationships; and* (*b*) *within normal industrial test code accuracies, the maximum reversible work* $(\mathrm{RW})_P$ *in the fluid cooling and heating modes are identical, that is,*

$$(\mathrm{RW})_{P,\mathrm{FHM}} = (\mathrm{RW})_{P,\mathrm{FCM}} \tag{9.63}$$

It is thus permissible to replace the $>$ *sign in Eq.* (9.55b) *by the* $=$ *sign*!

We make extensive use of the simpler *linearized solutions* per Eqs. (9.59)–(9.63) and the above guideline in the following thermodynamic analysis of low thermal lift forced-convection MIHES.

9.3 THERMODYNAMIC ANALYSIS OF MIHES

We consider first the idealized limiting case of zero friction dissipation, then the situation encountered in real world heat exchangers when $\Delta H_f = \Delta P_f/\rho > 0$ and the corresponding internal entropy dissipation due to friction $\Delta s_f = \Delta H_f/T$ [according to Eq. (9.61)] has a finite nonnegative value.

9.3.1 Idealized Heat Exchangers: ΔH_f and $\Delta s_f = 0$

Referring to the conventional analysis of the MIHES, particularly Table 9.1 and Figs. 9.1 and 9.2, and making use of the linear relationships described in the previous sections and Guideline 29, we can express the entropy change Δs_e due *exclusively* to the external heat interchange according to

$$\Delta s_e = (s_2 - s_1)_e = c_p(T_2 - T_1)/T_w \tag{9.63a}$$

$$\textit{Validity}: \quad \text{Linear domain when } |T_w - T_1|/T_w \ll 1 \tag{9.63b}$$

From now on we abandon the classical (equilibrium) thermodynamic designation $(\Delta s)_P$ per Eq. (9.38) with subscript P and use instead the subscript e (standing for external heat exchange) to differentiate this postive or negative component of the total entropy change Δs_{tot} from the radically different component $\Delta s_f > 0$ due *exclusively to internal friction* dissipation.[15] With this new terminology, which is more practical in normal fluid-to-fluid heat exchanger applications, Eq. (9.43a) becomes

$$s_2 - s_1 \equiv \Delta s_{\text{tot}} = \Delta s_e + \Delta s_f \tag{9.43b}$$

and it is only necessary to substitute the term Δs_e for $(\Delta s)_P$ in all the previous second law relationships, noting that this substitution does not affect in any way the validity of these graphical or analytical relationships.

For the sake of completeness, it should also be noted that we have taken advantage of Eq. (9.58) in the above analytical expression for Δs_e and selected for the absolute temperature appearing in the denominator on the right-hand side of Eq. (9.63) the wall temperature $T_w = (T_{2,C})_{\max} = (T_{2,H})_{\min} = T_\infty$ for reasons that will become clearer later.

Within the linear domain, it is permissible to normalize Δs_e to obtain universal $\Delta s_n - \Delta t_n$ relationships, applicable to any constant specific heat single-phase fluids (and by extension to two-phase flow situations [4.50, 9.4]) as follows.

With $(T_{2,C})_{\max} = (T_{2,H})_{\min} = T_\infty = T_w$, Eq. (9.63a) yields on the cold side of the MIHES

$$(\Delta s_e)_{\max} = (s_w - s_{1,C})_e = c_p(T_w - T_{1,C})/T_w \tag{9.64}$$

Dividing Eq. (9.63a) by (9.64) and in view of the definition of the contact factor CF, we obtain the following dimensionless relationships on both sides of the MIHES:

$$|\Delta s_e|_n \equiv \frac{(s_2 - s_1)_e}{(s_w - s_1)_e} \equiv \frac{T_2 - T_1}{T_w - T_1} \equiv \text{CF} \tag{9.65}$$

Since by definition 1 − CF = BF,

$$\text{CF} \equiv |\Delta s_e|_n \equiv |\Delta t|_n \tag{9.66a}$$

$$1 - \text{CF} \equiv \text{BF} \equiv 1 - |\Delta s_e|_n \equiv 1 - |\Delta t|_n \equiv |t|_n \tag{9.66b}$$

$$\textit{Validity}: \quad \text{Linear domain} \tag{9.66c}$$

where $|t|_n = 1 - |\Delta t|_n$ represents the dimensionless terminal temperature difference (TTD) on each side of the MIHES, normalized in terms of $T_w - T_{1,C} = T_{1,H} - T_w$. Because the origin $s_o = 0$ of the entropy scale is completely arbitrary in the absence of chemical reactions, it is permissible to select for the origin of the normalized entropy and temperature scales the end equilibrium state at $(T_{2,C})_{\max} = (T_{2,H})_{\min} = T_\infty = T_w$ and $P = P_1$. We thus obtain the normalized s_n versus t_n diagram bounded between $-1 \le s_n \le +1$ and $-1 \le$

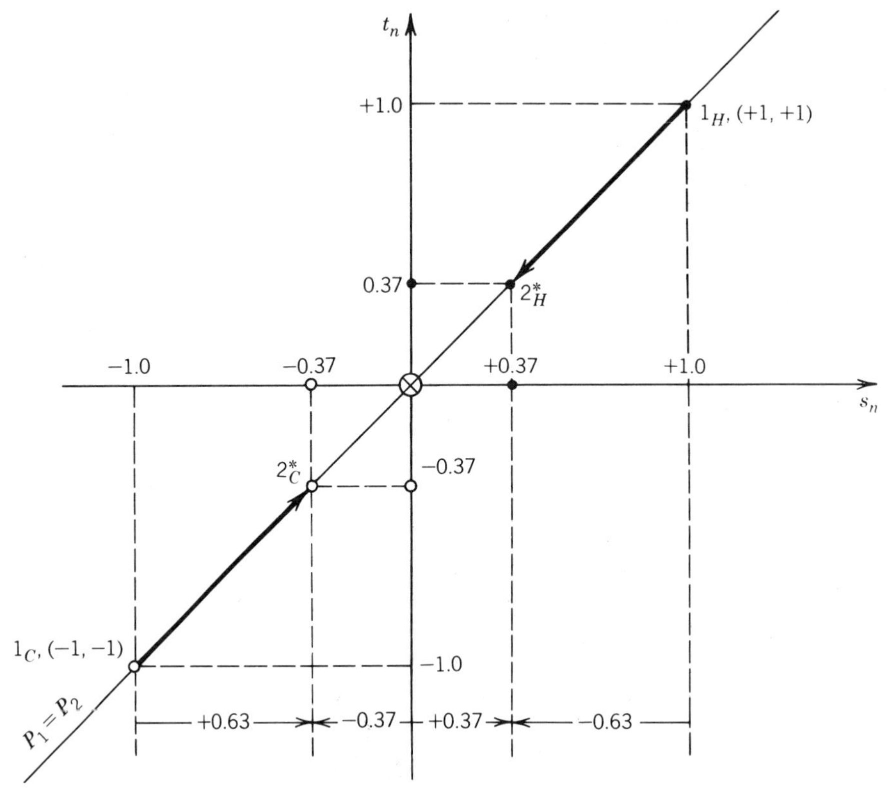

FIGURE 9.5. Idealized MIHES, $\Delta s_f = 0$, in normalized s_n versus t_n diagram. (Outlet states 2^*_C and 2^*_H for $L_{rel} = 1$ at CF = 0.63 and BF = 0.37.) [From H. Soumerai, "Thermodynamic Considerations on a 'Mirror Image' Concept—Forced Convection Single-Phase Tube Flow Applications," ASME Winter Annual Meeting, Miami, Florida, Nov. 1985.]

$t_n \leq +1$ shown in Fig.9.5. Consequently, the cold fluid and hot fluid inlet coordinates in Fig. 9.5 must be at $(-1, -1)$ and at $(+1, +1)$, respectively. For the limiting case of zero dissipation, that is, $\Delta s_f = 0$, any conceivable exit state 2^*_C on the cold fluid side of the MIHES must be located on the isobar $P = P_1$, that is on the *straight mixing* line connecting point 1_C and the origin O. Similarly, the straight mixing line connecting the origin O to point 1_H represents the locus of all possible end states 2^*_H on the hot fluid side of the MIHES. The straight lines $\overline{O1}_C$ and $\overline{O1}_H$ are called mixing lines for reasons that will become apparent when we later discuss the integrated thermodynamic mixing model of the classical forced-convection single-tube heat exchanger.

As an illustration, the end state 2^*_C at $L_{rel} = \overline{\Delta\theta}_{rel} = 1$ and corresponding BF = 0.37 per column (d) of Table 9.1 is shown at coordinates $(-0.37, -0.37)$ in the *negative quadrant* of Fig. 9.5. By reason of symmetry, point 2^*_H is at the coordinates $(+0.37, +0.37)$ in the *positive quadrant* of the s_n versus t_n chart.

The symmetrical arrows, or *vectors*, are intentionally shown in bold lines in Fig. 9.5 to emphasize the important fact that, in the language of irreversible thermodynamics, the *origin* O, at $T_w = T_{C,\infty} = T_{H,\infty}$, represents the *attractor state* on the isobar $P = P_1 = P_2$ for the idealized case $\Delta P_f/\rho = \Delta H_f = 0$ and $\Delta s_f = 0$.

Keeping in mind the ideal gases relationships per Eq. (9.22), we see that, since $\Delta u = c_v \, \Delta T$ and $\Delta h = c_p \, \Delta T$, Eqs. (9.65) and (9.66) can be expanded by following the same reasoning as above according to

$$\mathrm{CF} = (\Delta t)_n = (\Delta u)_n = (\Delta h)_n = (\Delta s_e)_n \tag{9.67a}$$

$$\mathrm{BF} = 1 - (\Delta t)_n = 1 - (\Delta u)_n = 1 - (\Delta h)_n = 1 - (\Delta s_e)_n \tag{9.67b}$$

or

$$\mathrm{BF} \equiv t_n \equiv u_n \equiv h_n \equiv (s_e)_n \equiv \mathrm{NTDF} \tag{9.67c}$$

where

$$t_n \equiv \frac{T_w - T_2}{T_w - T_1} \qquad u_n \equiv \frac{u_w - u_2}{u_w - u_1} \qquad h_n \equiv \frac{h_w - h_2}{h_w - h_1} \qquad (s_e)_n \equiv \frac{(s_w - s_2)_e}{(s_w - s_1)_e} \tag{9.67d}$$

$$\textit{Validity}: \quad \text{Linear domain} \tag{9.67e}$$

We use the following definition for the term NTDF

$$\mathrm{NTDF} \equiv \frac{\mathrm{TDF}}{\mathrm{LDF}} = \text{normalized terminal driving force} \tag{9.68a}$$

where

$$\mathrm{DF} \equiv \text{any linear driving force} \tag{9.68b}$$

$$\mathrm{TDF} \equiv \text{terminal driving force} \tag{9.68c}$$

$$\mathrm{LDF} \equiv \text{largest driving force} \tag{9.68d}$$

Equations (9.67b)–(9.67e) can now be stated more succinctly with the above definition of NTDF as

$$\mathrm{BF} \equiv \mathrm{NTDF} \tag{9.69a}$$

Validity: Linear domain with any linear heat transfer driving force (9.69b)

Thus, the term BF provides a direct normalized measure of the nonequilibrium conditions within the MIHES in relation to the end equilibrium state when BF = 0 at $T_\infty = T_w$ and corresponding u_w, h_w, and $(s_e)_w$ on the isobar $P = P_1$, at any $L_{rel} = \overline{\Delta\theta}_{rel} < 5$.

Thus the ordinate t_n in Fig. 9.5 also represents the normalized ordinate u_n or h_n and this fact has no impact on what was said regarding the attractor state represented by the end equilibrium state at $L_{rel} = \overline{\Delta\theta}_{rel} > 5$. It is permissible to conclude this analysis with the following statements:

- Any one of the following alternative *heat transfer driving forces* (DF) could be used in lieu of the conventional $|\Delta T|_w$ originally postulated by Newton in 1701, namely, $|\Delta u|_w = u_w - u$, $|\Delta h|_w = h_w - h$, and, as proposed by Soumerai [6.5, 6.7, 6.8], $|\Delta s_e|_w = (s_w - s)_e$ with appropriate changes to the definition of the corresponding empirical heat transfer coefficient as discussed later. As a matter of fact, an enthalpy driving force has been successfully used[16] in applications involving simultaneous heat and mass transfer, such as humidification or dehumidification of vapor–gas mixtures in surface heat exchangers and direct contact heat exchangers [9.8, 9.9].
- The significant and *unique* advantages of the simplest *second law driving force*, namely, $(\Delta s_e)_w$, in the linear domain, will become more evident in the next section when we depart from the extreme idealization represented by the assumption $\Delta H_f = 0$ and recognize that ΔH_f has a finite positive value in real world heat exchangers. In the language of modern thermodynamicists and physicists [2.17, 3.6, 7.17, 7.18, 7.63–7.66], we move conceptually from reversible to irreversible (macro) physics in the next section, when the extreme idealization $\Delta H_f = 0$ is discarded.
- It should be reemphasized (see Chapter 7) in this thermodynamic context that all *conventional* forced-convection analytical methods axiomatically based on the usual assumption (A.8) that $\dot{Q}_{f,\text{tot}}/\dot{Q}_e = 0$ (including the most sophisticated mathematical solutions of the Navier–Stokes equations derived from the assumption of *zero dissipation* [2.12]) are still fundamentally embedded in classical *reversible physics*.

9.3.2 Real Heat Exchangers: ΔH_f and $\Delta s_f > 0$

The irreversible friction head loss penalty that must be paid to increase the rate of heat transfer in forced convection fluid-to-fluid surface heat exchangers can be expressed [see Eq. (9.61)] as $\Delta H_f = T_w \, \Delta s_f$ and the *internal* irreversible entropy change due *exclusively* to *friction dissipation* is thus uniquely defined for a known value of ΔH_f or ΔP_f by

$$\Delta s_f = \frac{\Delta H_f}{T_w} = \frac{\Delta P_f/\rho}{T_w} \geq 0 \tag{9.70}$$

where the sign $>$ indicates that the *nonnegative quantity* Δs_f can only be positive[17] or vanishingly small but never negative in real systems.

Although ΔH_f and therefore Δs_f remain substantially independent of $|\Delta s_e|_{max}$ for single-phase fluids within the linear domain, it is also useful, as illustrated more precisely in the next graphical/numerical analysis, to normalize Δs_f by dividing Eq. (9.70) by Eq. (9.64). This yields the following dimensionless expressions, valid for either the cold or hot side of a simple MIHES:

$$(\Delta s_f)_n = \frac{\Delta s_f}{|\Delta s_e|_{max}} = \frac{\Delta H_f}{c_p|T_w - T_1|} = \frac{\dot{Q}_{f,\,\text{one side}}}{|\dot{Q}_e|_{max}} = \frac{\Delta T_f}{|T_w - T_1|} \tag{9.71a}$$

with

$$\Delta T_f \equiv \frac{\Delta H_f}{c_p} \tag{9.71b}$$

Note again that ΔT_f represents a *virtual* friction temperature increase in the case of ideal gases in view of Eq. (9.22), whereas it does represent a physical temperature increase in the case of ideal liquids as highlighted in Refs. 6.5 and 6.7.

9.3.3 Entropy Production $\dot{S}$ for a Simple MIHES

Applying the well-known Gouy–Stodola equation for steady flow (see, for instance, Bejan [9.10]) to the *fully insulated* MIHES depicted in Fig. 9.2 yields the following algebraic equation:

$$\dot{S} = \dot{m}_C\left[(\Delta s_e)_C + (\Delta s_f)_C\right] + \dot{m}_H\left[(\Delta s_e)_H + (\Delta s_f)_H\right] \tag{9.72}$$

Considering the unique characteristics (full symmetry) of the MIHES, respecting the sign conventions, and substituting the expressions derived previously for Δs_e and Δs_f in Eq. (9.72), one obtains

$$\dot{S} = \dot{m}(\text{CF})(c_p)\frac{T_w - T_{1,C}}{T_w}\left\{\left(1 - \frac{|\Delta s_e|_{max,\,H}}{(\Delta s_e)_{max,\,C}}\right) + \left(\frac{2\,\Delta H_f}{(\text{CF})c_p(T_w - T_{1,C})}\right)\right\} \tag{9.73}$$

where $\dot{m} \equiv \dot{m}_C = \dot{m}_H$. For a prescribed value of $T_w - T_{1,C} = T_{1,H} - T_w = \frac{1}{2}(T_{1,H} - T_{1,C})$, the ratio $|\Delta s_e|_{max,\,H}/(\Delta s_e)_{max,\,C}$ can be measurably smaller than unity in the *nonlinear* domain. Therefore, the first positive term in the large parentheses of Eq. (9.73), namely, $1 - |\Delta s_e|_{max,\,H}/(\Delta s_e)_{max,\,C} \equiv \delta$, has a finite value $\gg 0$ and contributes to the total entropy production $\dot{S}$ along with the second term, $2\,\Delta H_f/(\mathrm{CF})c_p(T_w - T_{C,1}) = 2\,\Delta s_f/\Delta s_{e,C} = \dot{Q}_{f,\mathrm{tot}}/\dot{Q}_e$, where $\dot{Q}_{f,\mathrm{tot}} \equiv \dot{Q}_{f,C} + \dot{Q}_{f,H} = 2\dot{m}\,\Delta H_f$. In the *linear* domain, however, when $0.90 \le T_w/T_1 \le 1.10$, the first term $\delta \to 0$ and its contribution to $\dot{S}$ becomes vanishingly small compared to the second contribution due to the total internal friction dissipation on *both* sides of the MIHES at the typical *finite* ΔH_f values encountered in real world heat exchangers. Consequently, Eq. (9.73) reduces to

$$\dot{S} = \dot{m}(2)\,\Delta s_f = \frac{\dot{m}(2)\,\Delta H_f}{T_w} \tag{9.74}$$

and according to the Gouy–Stodola theorem, the total power dissipated within the MIHES is

$$T_w(\dot{S}) = 2\dot{m}\,\Delta H_f = \dot{Q}_{f,\mathrm{tot}} \tag{9.75}$$

as could be expected within the *linear domain*.

It is important to note that $T_w = \frac{1}{2}(T_{1,H} + T_{1,C})$ used in Eq. (9.75) is the *only physically meaningful absolute temperature* for the MIHES since any other temperatures that may exist in a complete thermal system have no bearing whatsoever on the physical phenomena occurring *within* the insulated system. To express this another way, *relative to the fluids* entering the simple MIHES, there *exists no other equilibrium* temperature besides the constant surrounding wall temperature $T_w = T_\infty = \frac{1}{2}(T_{1,H} + T_{1,C})$ *created by the prior interaction of the two fluids* themselves under steady-state conditions (note the deterministic implications since this is the result of prior events).

9.3.4 Thermodynamic Mixing Model

Forced convection in turbulent tube flow being essentially a macroscopic mixing process,[18] it is logical to consider a model based on global, or integrated, mixing laws [6.8, 9.3] of technical thermodynamics in a manner similar to the methods shown in Bosnjakovic [9.11] and Plank [9.12] for injector analyses. Because the excellent graphical analytical methods proposed by Bosnjakovic [9.11] are rarely mentioned in standard American (mechanical engineering) thermodynamic texts, they will be highlighted here, to the minimum extent necessary for our present purpose.

We begin with high thermal lift applications then focus on the simpler linear solutions applicable in low thermal lift linear heat flux fluid-to-fluid heat exchanger systems, such as the MIHES just described.

9.3.4.1 High Thermal Lift Systems. The prime objective of the following thermodynamic analyses is to determine the equilibrium state $(\mathrm{ES})_2$ of an effluent stream $\dot{m}_{\mathrm{tot}} \equiv \dot{m}_2$ resulting from the mixing of two streams $\dot{m}_1$ and $\dot{m}_w$ of the same fluid at different equilibrium states $(\mathrm{ES})_1$ and $(\mathrm{ES})_w$ as shown

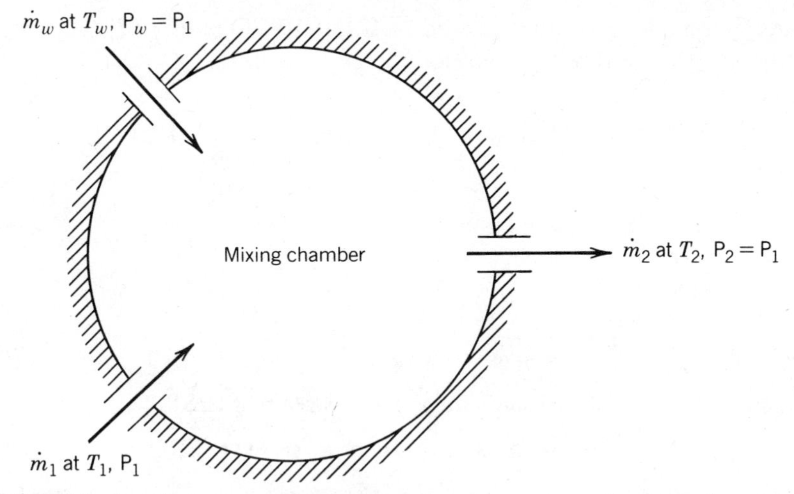

FIGURE 9.6. Constant pressure thermodynamic mixing: schematic physical model.

schematically in Fig. 9.6. Most generally, the equilibrium state is uniquely defined by the numerical values of two independent state properties such as T and P which are more readily measurable in normal engineering applications. Consequently, the equilibrium states $(\text{ES})_1$ and $(\text{ES})_w$ are uniquely defined by (T_1, P_1) and (T_w, P_w), respectively, and what we are seeking is the effluent equilibrium state $(\text{ES})_2$ defined by T_2 and P_2. The problem is treated in two steps: first, the *idealized* limiting case of zero pressure drop in the mixing chamber, that is a constant pressure mixing process when $P_1 = P_w = P_2$; and second, the real case $P_1 - P_2 = \Delta P_f > 0$.

Constant Pressure Case: $P_1 - P_2 \equiv \Delta P_f = 0$. From the mass and energy conservation laws

$$\dot{m}_1 + \dot{m}_w = \dot{m}_{\text{tot}} \equiv \dot{m}_2 \tag{9.76}$$

and

$$\dot{m}_1 h_1 + \dot{m}_w h_w = \dot{m}_{\text{tot}} h_2 \equiv \dot{m}_2 h_2 \tag{9.77}$$

on the assumption that the following conditions are satisfied during the mixing process:

- In/out SSSF conditions are possible.
- The kinetic energy terms are negligible.
- There is no *external* heat exchange ($\dot{Q}_e = 0$).

In mixing process analyses it is convenient to express all thermodynamic relationships on the basis of a unit mass of effluent stream $\dot{m}_{\text{tot}} \equiv \dot{m}_2$. This is readily done by dividing all sides of Eqs. (9.76) and (9.77) by $\dot{m}_2$ and introducing the following definitions:

$$\text{Mass flow fraction of stream } \dot{m}_1 \equiv \frac{\dot{m}_1}{\dot{m}_2} \equiv \text{BF} \tag{9.78a}$$

$$\text{Mass flow fraction of stream } \dot{m}_w \equiv \frac{\dot{m}_w}{\dot{m}_2} \equiv \text{CF} \tag{9.78b}$$

where

$$\dot{m}_1 = \text{mass flow rate of stream at } (\text{ES})_1 \tag{9.78c}$$

$$\dot{m}_w = \text{mass flow rate of stream at } (\text{ES})_w \tag{9.78d}$$

$$\dot{m}_2 = \text{mass flow rate of mixture at } (\text{ES})_2 \tag{9.78e}$$

Substituting $(\text{BF})\dot{m}_2$ and $(\text{CF})\dot{m}_2$ obtained from the defining Eqs. (9.78a) and (9.78b) for $\dot{m}_1$ and $\dot{m}_w$, respectively, in Eq. (9.76) yields

$$(\text{BF})\dot{m}_2 + (\text{CF})\dot{m}_2 = \dot{m}_2 \tag{9.79}$$

which reduces to the following equivalent expressions of the mass conservation law:

$$\text{BF} + \text{CF} = 1 \tag{9.80a}$$

or

$$\text{CF} = 1 - \text{BF} \tag{9.80b}$$

The numerical values of the mass fraction BF can vary, according to the defining Eq. (9.78a), between unity and zero; that is,

$$+1 \leq \text{BF} \leq 0 \tag{9.80c}$$

with a corresponding range for the mass fraction $\text{CF} = 1 - \text{BF}$ of

$$0 \leq \text{CF} \leq +1 \tag{9.80d}$$

In a similar fashion, the energy balance Eq. (9.77) can be expressed by

$$(\text{BF})\dot{m}_2 h_1 + (\text{CF})\dot{m}_2 h_w = \dot{m}_2 h_2 \tag{9.81}$$

which reduces, after eliminating $\dot{m}_2$ in the above equation, to

$$(\text{BF})h_1 + (\text{CF})h_w = h_2 \tag{9.82}$$

Finally, by substituting $1 - \text{BF}$ for CF in the above equation, in accordance with Eq. (9.80b), and solving for BF we obtain

$$\mathrm{BF} = \frac{h_w - h_2}{h_w - h_1} \tag{9.83a}$$

Since $\Delta h = c_p \, \Delta T$, $\Delta u = c_v \, \Delta T$, and c_p and c_v are constant in the case of ideal gases, it is evident that Eq. (9.83a) can also be expressed according to

$$\mathrm{BF} = \frac{u_w - u_2}{u_w - u_1} \tag{9.83b}$$

or

$$\mathrm{BF} = \frac{T_w - T_2}{T_w - T_1} \tag{9.83c}$$

In line with the terminology introduced in Bosnjakovic [9.11] and Plank [9.12], we define a (generally fictitious) reversible mixing process in which the entropy would be conserved, in other words, a *reversible mixing process* with a corresponding effluent equilibrium state $(\mathrm{ES})_{2_r}$ at point 2_r in the T-s diagram shown in Fig. 9.7 for a rather high ratio $T_w/T_1 = 2$, noting that the subscript r identifies the effluent state of the reversible mixing process.

Following exactly the same reasoning as previously for the energy conservation law, it is easy to see that the following alternative expression is applicable for BF in terms of s_w, s_1, and $s_{2,r}$

$$\mathrm{BF} = \frac{s_w - s_{2,r}}{s_w - s_1} \tag{9.84a}$$

where

$$s_{2,r} = \text{specific entropy of } \textit{reversible mixing} \text{ effluent} \tag{9.84b}$$

It is now possible to determine graphically the reversible and real effluent equilibrium states $(\mathrm{ES})_{2,r}$ and $(\mathrm{ES})_{2*}$ for the constant pressure (isobaric) case in a temperature–entropy diagram as illustrated in Fig. 9.7. (Although Bosnjakovic [9.11] and Plank [9.12] both use the Mollier—h versus s—diagram, it is evident in view of Eqs. (9.83a)–(9.83c) and (9.84a) that the same graphical solutions are also valid in u versus s or T versus s diagrams.)

The reversible mixing state 2_r is obtained by drawing a *straight line* between the equilibrium states 1 and w at T_1 and T_w on the isobar $P_1 = P_w$ in the T-s diagram as shown. The straight line $\overline{1w}$ is by definition the *linear mixing line*. It is further evident from the defininig equations (9.78a) and (9.78b) for BF and CF = 1 − BF, as well as the conservation law expressions per Eqs. (9.83) and (9.84), that point 2_r must be located on the linear mixing

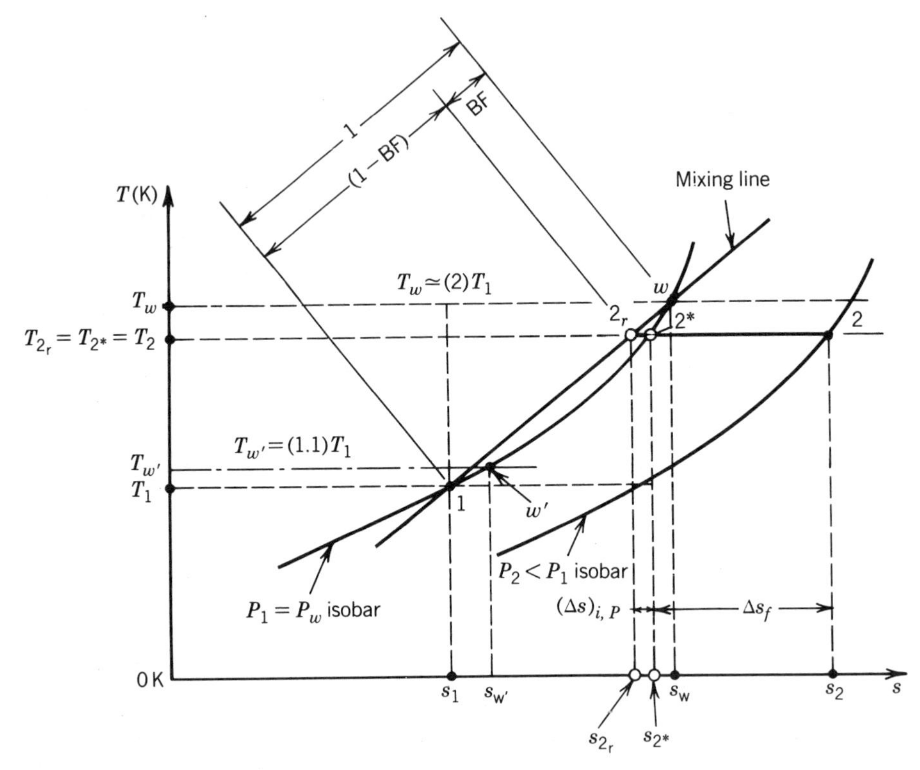

FIGURE 9.7. Thermodynamic mixing in a *T-s* diagram for high thermal lift systems. The mixing line $\overline{1w}$ is shown for $T_w / T_1 \simeq 2$. As $T_w \to T_1$ and $T_w / T_1 \simeq 1$ the mixing line coincides with the logarithmic isobar $P_1 = P_w =$ constant as highlighted by line $\overline{1w'}$ for $T_{w'} = (1.1)T_1$.

line and must divide the distance $\overline{1w}$ in such a manner that

$$\frac{\text{distance } \overline{2_r w}}{\text{distance } \overline{1w}} \equiv \text{BF} \equiv \frac{T_w - T_{2_r}}{T_w - T_1} \equiv \frac{s_w - s_{2_r}}{s_{w} - s_1} \tag{9.85}$$

$$\frac{\text{distance } \overline{12_r}}{\text{distance } \overline{1w}} \equiv \text{CF} \equiv \frac{T_{2_r} - T_1}{T_w - T_1} \equiv \frac{s_{2_r} - s_1}{s_w - s_1} \tag{9.86}$$

The real effluent state $(\text{ES})_{2*}$, point 2* in Fig. 9.7, at a specific entropy s_{2*} can readily be determined graphically in the *T-s* diagam at the intersection of the isotherm $T_{2_r} = T_{2*}$ with the isobar $P_1 = P_w = P_2 =$ constant. The entropy increase $\Delta s_{\text{irrev}} = s_{2*} - s_{2_r} > 0$ thus obtained is due entirely to the inherent irreversibility associated with mixing two streams of the same fluid at different initial equilibrium temperatures $T_1 \neq T_w$ on an isobar ($P_1 = P_w = P_2 =$ constant) in the absence of any additional internal dissipation due to fluid friction when $\Delta P_f = 0$; that is,

$$\Delta s_{\text{irrev}, P=\text{constant}} \equiv \Delta s_{i,P} = s_{2*} - s_{2_r} \tag{9.87}$$

General Case: $P_1 - P_2 = \Delta P_f > 0$. In this most general case, the equilibrium state $(\text{ES})_2$ of the effluent mixture, point 2 in Fig. 9.7, can be read directly in the *T-s* diagram at the intersection of the exit pressure isobar $P_2 = P_1 - \Delta P_f < P_1$ with the isotherm $T_{2_r} \equiv T_{2*} \equiv T_2 = \text{constant}$ (because $h = \text{constant}$ in an adiabatic throttling process and accordingly $T = \text{constant}$ in the case of ideal gases [9.11, 9.12]) as shown. The effluent equilibrium state 2 in Fig. 9.7 is thus shifted further to the right by an amount $s_2 - s_{2*} > 0$ due exclusively to internal friction dissipation; that is,

$$(\Delta s)_{\text{irrev}, T=\text{constant}} \equiv \Delta s_f = s_2 - s_{2*} > 0 \tag{9.88}$$

Therefore, the total irreversible entropy change $(\Delta s)_{\text{irrev, tot}}$ is the sum of two independent positive components $s_{2*} - s_{2_r} \equiv \Delta s_{i,P}$ and $s_2 - s_{2*} \equiv \Delta s_f$ per Eqs. (9.87) and (9.88), respectively; that is,

$$(\Delta s)_{\text{irrev, tot}} = (s_{2*} - s_{2_r}) + (s_2 - s_{2*}) = (\Delta s)_{i,P} + \Delta s_f \tag{9.89}$$

Equation (9.89) can be expressed conveniently in terms of $(\Delta s)_{i,P}$ when this component of $(\Delta s)_{\text{irrev, tot}}$ is dominant, that is, in the *nonlinear* domain when T_w/T_1 differs significantly from unity and the internal friction dissipation component Δs_f becomes less significant, according to

$$(\Delta s)_{\text{irrev, tot}} = (\Delta s)_{i,P}\left(1 + \frac{\Delta s_f}{(\Delta s)_{i,P}}\right) \tag{9.90a}$$

which converges toward $(\Delta s)_{i,P}$ as $\Delta s_f/(\Delta s)_{i,P} \to 0$; that is,

$$(\Delta s)_{\text{irrev, tot}} \to (\Delta s)_{i,P} \quad \text{when} \quad \frac{\Delta s_f}{(\Delta s)_{i,P}} \to 0 \tag{9.90b}$$

Conversely, when the fluid flow friction dissipation component Δs_f is dominant, that is, when $T_w/T_1 \approx 1$, and Δs_f has a significant finite value, Eq. (9.89) can be written

$$(\Delta s)_{\text{irrev, tot}} = \Delta s_f\left(1 + \frac{(\Delta s)_{i,P}}{\Delta s_f}\right) \tag{9.91a}$$

which converges toward Δs_f, when $(\Delta s)_{i,P}/\Delta s_f \to 0$; that is,

$$(\Delta s)_{\text{irrev, tot}} \to \Delta s_f \quad \text{when} \quad \frac{(\Delta s)_{i,P}}{\Delta s_f} \to 0 \tag{9.91b}$$

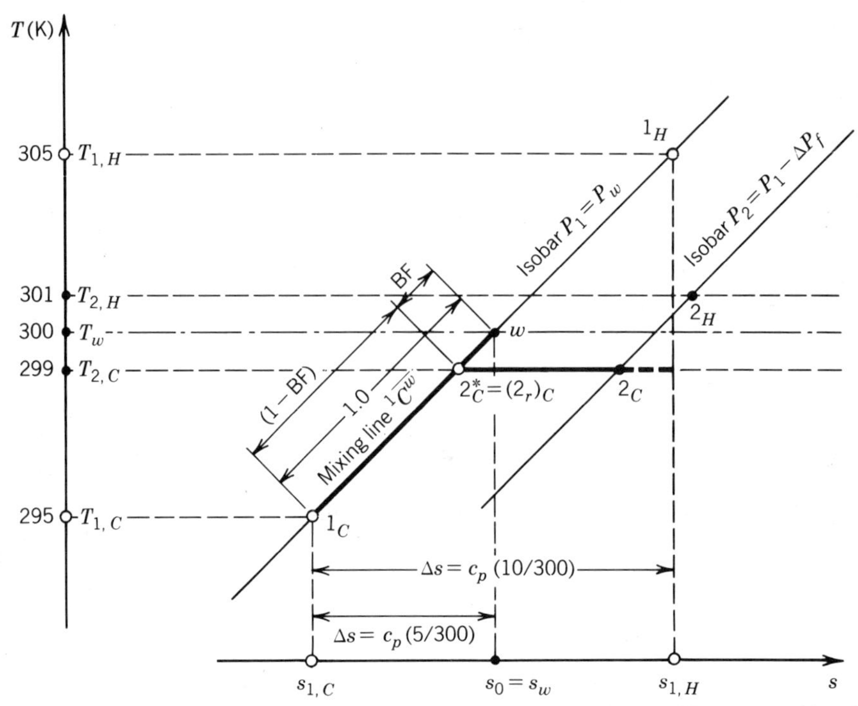

FIGURE 9.8. Thermodynamic mixing in a *T-s* diagram for low thermal lift systems. (Graphical example drawn for BF = 0.2, CF = 1 − BF = 0.8 and $\Delta T_{max} / T_w = 10 / 300 < 0.10$, i.e., within validity limits of the simplified linear expressions for Δs_p and Δs_f.)

9.3.4.2 Low Thermal Lift Systems. Within the validity limits $0.9 < T_w/T_1 < 1.10$ of the linear entropy relationships, when $T_w/T_1 \approx 1$, the isobar P_1 = constant becomes linear as shown previously, and it is clear by referring to Fig. 9.7 that the linear mixing line $\overline{1w}$ merges with the linearized isobar.[19] Thus, points 2_r and 2^* coincide as shown in Fig. 9.8 and

$$s_{2_r} - s_{2^*} \equiv (\Delta s)_{i,P} = 0 \tag{9.92a}$$

when

$$\frac{T_w}{T_1} \approx 1 \tag{9.92b}$$

neglecting second order of magnitude terms. Consequently, the limiting case per Eq. (9.91b) is applicable and

$$(\Delta s)_{\text{irrev, tot}} \simeq \Delta s_f \tag{9.92c}$$

when Δs_f has a finite value $(\Delta s)_f > 0$. Note that this conclusion is fully consistent with the results of the thermodynamic analysis of the MIHES derived on the basis of the Gouy–Stodola theorem in the previous section.

9.3.4.3 Low Thermal Lift Simple MIHES. The simple MIHES defined in Section 9.1.5 can be interpreted in the light of the above linear laws as follows.

The end equilibrium temperature $T_{C,\max} = T_{H,\min} = T_\infty = T_w$, when $L_{\text{rel}} = \overline{\Delta\theta}_{\text{rel}} > 5$, coincides with the temperature obtained by mixing the two equal streams $\dot{m}_C = \dot{m}_H = \dot{m}$ of the same fluid, $(c_p)_C = (c_p)_H = c_p$, that is, $T_\infty = T_w = T_{1,C} + \frac{1}{2}(T_{1,H} - T_{1,C}) = T_{1,H} - \frac{1}{2}(T_{1,H} - T_{1,C})$ on the basis of the mass and energy conservation laws. It is important to note that this conclusion is valid regardless of the maximum total temperature difference $T_{1,H} - T_{1,C}$ between the hottest and coldest sides of the MIHES. This is shown graphically in Fig. 9.8 for $T_{1,H} = 305$ K and $T_{1,C} = 295$ K, or $T_{1,H} - T_{1,C} = 10$ K with $T_{C,\max} = T_{H,\min} = T_\infty = T_w = 300$ K and a corresponding maximum wall-to-fluid temperature difference on either side of the MIHES of $T_w - T_{1,C} = T_{1,H} - T_w = (\frac{1}{2})10 = 5$ K.

The origin of the entropy $s_o = 0$ in the temperature–entropy diagram, Fig. 9.8, has again been arbitrarily selected at $T_o = T_w = 300$ K and $P_o = P_{1,C} = P_{1,H}$ as in the case of the normalized $s_n - t_n$ graph shown in Fig. 9.5 for the limiting case $\Delta H_f = 0$ and therefore $\Delta s_f = 0$.

Because of the complete symmetry between the two sides of the simple (i.e., $R_M = 0$) MIHES, it is clear that T_w remains constant along the full length of the heat exchanger. Thus, the constant wall temperature at any axial position $0 \leq x \leq L$ is exactly equal to the mixing temperature resulting from the physical mixing of identical small "packets" of cold and hot fluid reaching the separating wall (i.e., as if the zero thermal resistance wall did not exist).

Focusing on the cold (fluid heating mode) side of the simple MIHES, we consider first the zero pressure drop limiting case when $\Delta P_f = 0$ or $\Delta s_f = 0$. It is evident that the locus of all conceivable end states 2_C^* is represented by the isobar P_1 when $\Delta P_f = 0$. In other words, the points 2_C^* must be located on the straight mixing line $\overline{1_C w}$ connecting the inlet state 1_C and the point w in Fig. 9.8. The actual end point 2_C^* on the cold side can thus be visualized as the resulting equilibrium state $(\text{ES})_{2_C^*}$ obtained by mixing at constant pressure a stream of fluid at T_w, $P_w = P_1$ flowing at a mass flow rate $\dot{m}_w = (\text{CF})\dot{m}$ with a second stream of the same fluid at $T_{1,C}$, P_1 flowing at a mass flow rate $(\text{BF})\dot{m}$. We thus imagine the total flowing mass $\dot{m}$ segregated in two streams $(\text{BF})\dot{m}$ and $(1 - \text{BF})\dot{m}$: the first fraction, the bypass fraction (BF), flowing through the channel without "reaching" the wall, as if "bypassed" around the heat exchanger, and the second fraction $\text{CF} = 1 - \text{BF}$ fully "contacting" the wall and acquiring the temperature of the wall T_w.[20] At a relative length of travel $L_{\text{rel}} = 1$ or relative residence time $\overline{\Delta\theta}_{\text{rel}} = 1$, the bypass fraction has a numerical value $\text{BF} = 1/e = 0.368$ with a corresponding mass flow fraction contacting the wall $\text{CF} = 1 - \text{BF} = 0.632$. At $L_{\text{rel}} = \overline{\Delta\theta}_{\text{rel}} > 5$ the bypassed fraction BF vanishes for all practical purposes and the corresponding fraction contacting the wall $\text{CF} = 1 - \text{BF} \rightarrow 1$.

Following the linear mixing rules outlined in the previous section, we see that the equilibrium state $(ES)_{2,C}$ for the real case $\Delta P_f > 0$ will be located on the linear isobar $P_2 = P_1 - \Delta P_f$ as illustrated in Fig. 9.8 for $T_w - T_{1,C} = \frac{1}{2}(T_{1,H} - T_{1,C}) = \frac{1}{2}(\Delta T)_{\max} = 5$ K. By reason of symmetry, the equilibrium state $(ES)_{2,H}$ on the hotter side of the simple MIHES will be located as shown in Fig. 9.8 for a maximum $(\Delta T)_{\max} = T_{1,H} - T_{1,C} = 10$ K and $T_{1,H} - T_w = 5$ K.

Within the linear domain it is evident that the mixing rules remain valid as stated regardless of the value of $(\Delta T)_{\max}$ and corresponding $T_w - T_{1,C} = T_{1,H} - T_w = \frac{1}{2}(\Delta T)_{\max}$. It is therefore permissible to normalize the analysis on the basis of $T_w - T_{1,C} = T_{1,H} - T_w = 1$ K and this automatically leads to the previous graphical solutions in normalized t_n versus s_n diagrams as shown in Fig. 9.5 for $\Delta P_f = 0$ and discussed more fully in the next numerical analysis for $\Delta P_f > 0$.

In conclusion, the proposed *global thermodynamic mixing model*[21] is valid since it *correctly duplicates the results* derived from conventional empirically validated heat transfer and friction correlations.

It should be noted that the application of the thermodynamic mixing model derived here for simple MIHES can be extended to treat the more general *nonmirror* heat exchanger cases in pure parallel flow [i.e., T_w = constant $\neq \frac{1}{2}(T_{1,H} + T_{1,C})$] as well as in pure counterflow ($T_w \neq$ constant) arrangements, with single- and two-phase fluids including finite wall thermal resistances [7.83, 9.4].

9.3.5 Graphical / Numerical Analysis of a Real MIHES

We now analyze a real MIHES under the following conditions: $T_w = 300$ K, normal atmospheric pressure, a velocity head $\frac{1}{2}\bar{V}^2 = 100$, that is, $\bar{V} = (200)^{1/2} = 14.14$ m/s, and otherwise identical conditions as in the previous conventional (first law) analysis highlighted in Table 9.1, lines (1)–(8).

Problem Statement

(a) Find ΔH_f and the corresponding normalized $(\Delta s_f)_n$ at $L_{\text{rel}} = \overline{\Delta\theta}_{\text{rel}} = 1$ for $T_w - T_{1,C} = T_{1,H} - T_w = 1$°C/K, noting that the maximum temperature differential on each side of the MIHES has been deliberately selected low so as to "magnify" the impact of ΔH_f in the normalized temperature–entropy diagram.

(b) Same as (a) for the values of L_{rel} shown in Table 9.1.

(c) Determine analytically the real exit state 2_C and 2_H at $T_w - T_{1,C} = 1$ K for $L_{\text{rel}} = 1$ and all L_{rel} in Table 9.1.

(d) Same as (c) graphically in a normalized s_n versus t_n chart for $L_{\text{rel}} = 1$ and $L_{\text{rel}} = 4.61$ at corresponding values of BF = 0.37 and 0.010, respectively.

(e) Reassess the above results assuming a fully rough wall channel (i.e., f_{fr} = constant independent of $\dot{m}$ at Re $\gg$ 10,000) under otherwise identical conditions but with hydrogen, instead of air on the basis of ideal gas laws. Note that, at 300 K, $(\mathrm{Pr})_{\mathrm{hydrogen}} = (\mathrm{Pr})_{\mathrm{air}}$, $(c_p)_{\mathrm{hydrogen}} = 14.27$ kJ/kg · K, and $(c_p)_{\mathrm{hydrogen}}/(c_p)_{\mathrm{air}} = 14.2$, whereas the molar specific heat ratio $(Mc_p)_{\mathrm{hydrogen}}/(Mc_p)_{\mathrm{air}} = 0.99 \simeq 1$.

Solution

(a) According to the definition of $(\mathrm{NVH})_f$, $\Delta H_f = \overline{V}^2/2(\mathrm{NVH})_f$; therefore, $\Delta H_f = 100(\mathrm{NVH})_f$ at $\overline{V}^2/2 = 100$. For $L_{\mathrm{rel}} = 1$, line (9), column (d), in Table 9.1, shows $(\mathrm{NVH})_f = 1.66$; therefore, $\Delta H_f = 100(1.66) = 166$ J/kg, the value shown on line (10). From Eq. (9.71) at $T_w - T_{1,C} = 1$ and $c_p = 1005$ J/kg · K, $(\Delta s_f)_n = \Delta H_f/c_p(1) = 166/1005(1) = 0.165$ as shown on line (11), column (d).

(b) Since $\Delta H_f \propto L \propto L_{\mathrm{rel}}$ at a fixed velocity $\overline{V}$, it is only necessary to multiply the above value $\Delta H_f = 166$ by L_{rel} to obtain the values of ΔH_f at any other L_{rel} as shown on line (10). Similarly, since Δs_f is proportional to ΔH_f, the values of $(\Delta s_f)_n$ shown on line (11) are obtained by multiplying the above value $(\Delta s_f)_n = 0.165$ at $L_{\mathrm{rel}} = 1$ by L_{rel}.

(c) We have already seen that the coordinates of the exit state 2* in Fig. 9.5 are $(-0.37, -0.37)$ for a BF = 0.37 in the idealized case $\Delta s_f = 0$. At the same BF = 0.37, corresponding to $L_{\mathrm{rel}} = 1$ and $(\Delta s_f)_n > 0$, the BF $= 1 - (\Delta t)_n =$ 0.37 remains unchanged; therefore, on the cold fluid side of the MIHES the ordinate $t_n = 0 - \mathrm{BF} = -0.37$ is the same as for the idealized frictionless case for perfect gases [in view of Eq. (9.22c)], whereas $(\Delta s)_n = (\Delta s_e)_n + (\Delta s_f)_n$ must increase by a nonnegative amount $(\Delta s_f)_n$ due to internal friction dissipation; that is,

$$(s_{2,C} - s_o)_n = -\mathrm{BF} + (\Delta s_f)_n \tag{9.93}$$

where s_o is the origin of the entropy at $T_o = T_w$ and $P_o = P_1$. Similarly, on the hot side of the MIHES, where $t_n = +\mathrm{BF}$ and since $(\Delta s_f)_n$ is always > 0, we obtain

$$(s_{2,H} - s_o)_n = +\mathrm{BF} + (\Delta s_f)_n \tag{9.94}$$

The results obtained from Eqs. (9.93) and (9.94) are shown on line (12) for the *cold* fluid and line (13) for the *hot* fluid.

Line (14) in Table 9.1 has been added to double check that all the entropy terms have been correctly accounted for, by verifying that the total entropy production $\dot{S}/\dot{m} = 2\,\Delta s_f = 2(\Delta s_f)_n$ for $T_w - T_{1,C} = T_{1,H} - T_w = 1$ per Eq. (9.74), that is, 2 × [value on line (11)] does agree (as it should) with the sum of lines (12) and (13).

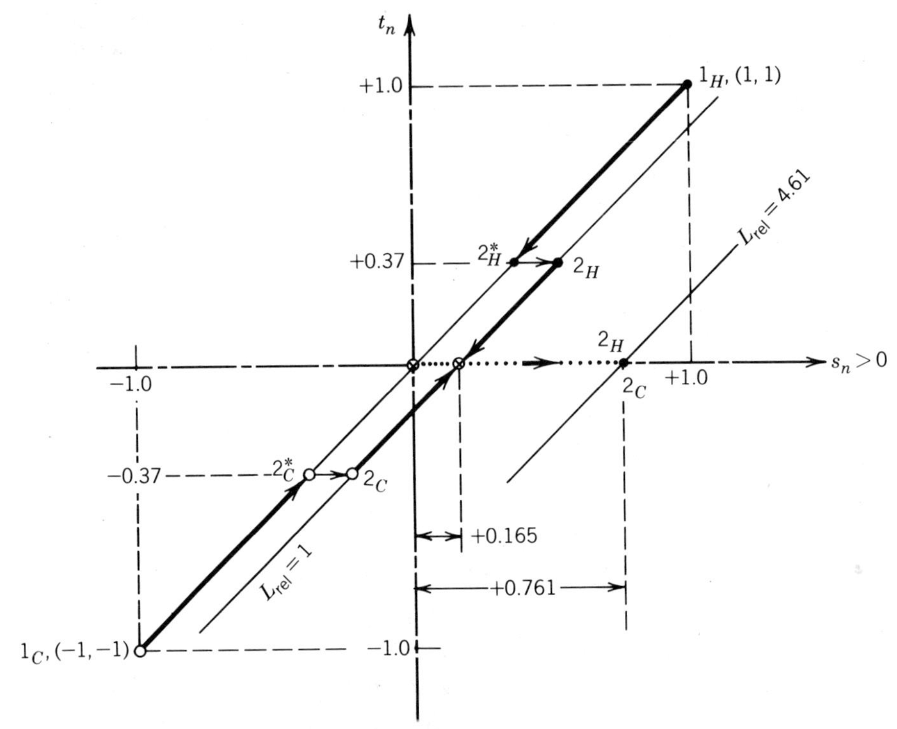

FIGURE 9.9. Real MIHES, $\Delta s_f > 0$, in normalized s_n versus t_n diagram. (2_C and 2_H for $L_{rel} = 1$ and 4.61 at BF = 0.37 and 0.01, respectively.) [From H. Soumerai, "Thermodynamic Considerations on a 'Mirror Image' Concept—Forced Convection Single-Phase Tube Flow Application," ASME Winter Annual Meeting, Miami, Florida, Nov. 1985.]

It is interesting to note the convergence toward $s_{2,C} - s_o = s_{2,H} - s_o = (\Delta s_f)_n$ indicated on lines (12) and (13) and Fig. 9.9 at $L_{rel} \gg 1$ when $T_w = T_{2,C} = T_{2,H}$. One can therefore visualize the full length MIHES as a succession of elementary MIHES connected in series with the *attractor state* initially at the origin of the coordinate system (0, 0) *continuously shifting in the direction of increasing (i.e., positive) entropy* as the normalized thermal driving force NTDF → 0 as L_{rel} is increased to $L_{rel} = \overline{\Delta\theta}_{rel} \geq 5$ in the case of perfect gases.

(d) Figure 9.9 illustrates the graphical representation of Eqs. (9.93) and (9.94) in the s_n versus t_n graph for $L_{rel} = 1$ at BF = 0.37 and $L_{rel} = 4.61$ at BF = 0.01.

(e) Since the Prandtl numbers for air and hydrogen are identical, lines (1)–(10) of Table 9.1 remain unchanged for hydrogen. On the other hand, since $(\Delta s_f)_n$ is proportional to $1/c_p$ according to Eq. (9.71), the ratio $(\Delta s_f)_{n,\text{hydrogen}}/(\Delta s_f)_{n,\text{air}} = (c_p)_{\text{air}}/(c_p)_{\text{hydrogen}} = 0.07$ and line (11) would be reduced by (1 − 0.07)100 = 93% in the case of hydrogen. Since the average velocity $\overline{V}$, BF, and CF = 1 − BF are identical at the specified conditions and since the Prandtl numbers as well as the fully rough friction factor f_{fr} are also identical for air and hydrogen, it follows from the conventional Eq. (9.21) that the same actual length of travel L will be required in both cases for the same bypass factor BF. We can therefore compute the total heat exchange capacity for the same channel geometry (D_h, L/D_h, $\bar{e}/D_h$ fully rough) for both gases with $\dot{m} = \rho \overline{V} \Omega_i$ according to

$$\dot{Q}_e = \overline{V}\Omega_i(1 - \mathrm{BF})\rho c_p(T_w - T_1) \tag{9.95}$$

or

$$\dot{Q}_e = \dot{\mathscr{V}}\rho c_p(1 - \mathrm{BF})(T_w - T_1) \tag{9.96}$$

where $\dot{\mathscr{V}} \equiv \Omega_i \overline{V}$ is the volumetric flow rate in $\mathrm{m^3/s}$ and $c_p\rho$ is the constant pressure specific heat per unit volume in $\mathrm{J/m^3 \cdot K}$. The above equations are valid for any single-phase constant specific heat incompressible fluid (ρ and c_p = constant) per conventional assumptions A.3 and A.5.

For real gases the specific mass can be expressed according to

$$\rho = \frac{1}{v} = \frac{M}{Z\mathscr{R}}\frac{P}{T} \tag{9.97}$$

where $\mathscr{R}$ is the universal gas constant, M is the molar mass, and Z is the compressibility factor, which can be expressed most generally in terms of reduced pressures and temperatures P^+ and T^+ for real gases (or highly superheated vapors) and in terms of P^+ only for vapors at saturation according to the corresponding states principle, as shown in Chapter 6.

Introducing Eq. (9.97) into Eq. (9.96) yields with Eq. (9.59a)

$$\dot{Q}_e = \dot{\mathscr{V}}\frac{(1 - \mathrm{BF})}{\mathscr{R}}\left(\frac{P}{Z}\right)Mc_p\frac{(T_w - T_1)}{T_w} \tag{9.98}$$

or

$$\dot{Q}_e = \dot{\mathscr{V}}\frac{(1 - \mathrm{BF})}{\mathscr{R}}\left(\frac{P}{Z}\right)M(s_w - s_1)_P \propto M(s_w - s_1)_P \tag{9.99}$$

where $M(s_w - s_1)_P$ is the entropy driving force per mole. Comparing the total capacity $\dot{Q}_e$ with hydrogen and air treated as quasi-ideal gases with $Z = 1$ at the same P, T_1, T_w, and BF, we obtain from Eq. (9.98) the ratio

$$\frac{\dot{Q}_{\text{hydrogen}}}{\dot{Q}_{\text{air}}} = \frac{(Mc_p)_{\text{hydrogen}}}{(Mc_p)_{\text{air}}} \tag{9.100}$$

or since $(Mc_p)_{\text{hydrogen}} = (Mc_p)_{\text{air}}$

$$\dot{Q}_{\text{hydrogen}} = \dot{Q}_{\text{air}} \tag{9.101}$$

Consequently, hydrogen is a much more efficient heat transfer medium, requiring roughly only 7% of the pumping power used with air, for the same cooling duty. Furthermore, it is easy to see by repeating the numerical analysis summarized in Table 9.1 that, for the same relative pumping power $(\dot{Q}_f/\dot{Q}_e)_{\text{hydrogen}} = (\dot{Q}_f/\dot{Q}_e)_{\text{air}}$, the ratios $(\Delta s_f/\Delta s_e)_{\text{hydrogen}}$ and $(\Delta s_f/\Delta s_e)_{\text{air}}$ must also be identical and as a corollary

$$\frac{(\dot{Q}_e)_{\text{hydrogen}}}{(\dot{Q}_e)_{\text{air}}} = \left[\frac{(c_p)_{\text{hydrogen}}}{(c_p)_{\text{air}}}\right]^{1/2} = (14.2)^{1/2} = 3.7.$$

Thus, for the same geometry, operating condition, and pumping power, the cooling capacity will be roughly 3.7 times greater with hydrogen. This explains why hydrogen is used in preference to air as a coolant in large electric generators, in spite of the added costs and complications of a complete hydrogen cooling system. The assumption of a fully rough wall is justified because the actual electrical generator cooling ducts are typically fully rough.

Comments

It is interesting to note that the expression for $\dot{Q}_e$ in terms of an entropy driving force per mole, according to Eq. (9.99), can be generalized to include saturated vapors and two-phase forced-convection heat exchangers (evaporators or condensers) on the basis of the ECSP in conjunction with the relationships proposed in Section 8.2.3.6 roughly as follows.

For *identical* P^+, channel geometry, volumetric flow rate $\dot{\mathscr{V}}$, and $(\text{BF})_{\text{tp}} \equiv (\text{AF})_{\text{tp}} \equiv y_2/y_1 = (1 - x_2)/(1 - x_1)$ per Eqs. (8.152) and (8.159), respectively, we obtain from Eq. (9.99) the ratio

$$\frac{\dot{Q}_e}{(\dot{Q}_e)_{\text{ref}}} = \frac{P/Z}{(P/Z)_{\text{ref}}} \frac{M(s_w - s_1)}{[M(s_w - s_1)]_{\text{ref}}} \tag{9.102}$$

Substituting $P_{\text{cr}}(P^+)$ and $Z_{\text{cr}}(Z^+)$ for P and Z in the above expression yields

$$\frac{\dot{Q}_e}{(\dot{Q}_e)_{\text{ref}}} = \frac{P_{\text{cr}}/Z_{\text{cr}}}{(P_{\text{cr}}/Z_{\text{cr}})_{\text{ref}}} \left(\frac{(P^+/Z^+)M(s_w - s_1)}{(P^+/Z^+)_{\text{ref}}[M(s_w - s_1)]_{\text{ref}}}\right) \tag{9.103}$$

TABLE 9.4. Two-Phase Heat Exchanger Capacity Ratio[a] $\dot{Q}_e/(\dot{Q}_e)_{R\text{-}22} = (P_{cr}/Z_{cr})/(P_{cr}/Z_{cr})_{R\text{-}22}$

Fluid	P_{cr}[b] (psia)	Z_{cr}[b]	(P_{cr}/Z_{cr}) (psia)	$\dot{Q}_e/(\dot{Q}_e)_{R\text{-}22}$
R-11	635	0.277	2292	0.85
R-12	596.2	0.278	2145	0.79
R-13	561	0.278	2018	0.75
R-113	495	0.274	1807	0.67
R-114	474	0.275	1724	0.64
R-115	453	0.274	1653	0.61
R-21	750	0.272	2757	1.02
R-22	721.9	0.267	2704	1.00
R-502	591	0.274	2157	0.80
R-717	1657	0.245	6763	2.50
R-718	3226	0.234	13786	5.10

[a] Valid for *identical* P^+, x_1, x_2, exit *volumetric* flow rate, and geometry under in / out SSSF conditions per Eq. (9.104b).

[b] From Soumerai [4.21, Table 1] except for refrigerant R-502 which is from Ref. 4.24.

which reduces to

$$\frac{\dot{Q}_e}{(\dot{Q}_e)_{\text{ref}}} = \frac{P_{\text{cr}}/Z_{\text{cr}}}{(P_{\text{cr}}/Z_{\text{cr}})_{\text{ref}}} \tag{9.104a}$$

since the terms in the numerator and denominator inside the large parentheses on the right-hand side of Eq. (9.103) are identical on the basis of the ECSP under the following design conditions.

$$\text{\textit{Identical} } P^+, x_1, x_2, \text{ geometry } (D_h, \Omega, L/D_h, \bar{e}/D_h), \text{ exit velocity } \overline{V}_2, \text{ and therefore identical volumetric flow rate } \dot{\mathcal{V}}_2 = \Omega \overline{V}_2 \text{ under in/out SSSF conditions} \tag{9.104b}$$

The heat exchanger capacity ratio $\dot{Q}_e/(\dot{Q}_e)_{\text{ref}}$ can thus be estimated according to Eq. (9.104a) for typical single-component fluids such as the similar halogenated refrigerants and even the dissimilar polar molecules, ammonia and water, listed in Soumerai [4.21, Table 1] using R-22 as the arbitrary reference fluid, as shown in Table 9.4.

The numerical values of $\dot{Q}_e/(\dot{Q}_e)_{\text{R-22}}$ summarized in Table 9.4 indicate that the two most dissimilar (polar) fluids, ammonia (R-717) and water (R-718), yield significantly higher capacity ratios of 2.50 and 5.10, respectively, whereas the thermodynamically similar halogenated refrigerants fall roughly within a rather narrow band of $\pm 20\%$ if we were to select R-12 as a reference instead of R-22. It should be kept in mind that the ratios $\dot{Q}_e/(\dot{Q}_e)_{ref}$ are based on the same reduced pressures P^+ and *identical volumetric* flow rates as stated in Eq. (9.104b).

9.4 ADVANTAGES OF SECOND LAW DRIVING FORCES

Some of the unique characteristics of entropy as the heat transfer driving force can now be stated more precisely.

9.4.1 First and Second Law Driving Forces

In the following assessment we differentiate the group of *first law* driving forces, or potentials, such as $(\Delta T)_w$, $(\Delta u)_w$, and $(\Delta h)_w$ per Eqs. (9.67)–(9.69) from the second law entropy (or more generally, entropy-based) driving forces. To identify the fundamental limitations and advantages of these alternative heat transfer driving forces, it is convenient to focus on a concrete case such as the simple MIHES in stable turbulent single-phase flow regimes, including applications beyond the validity limits of the linearized approximation for $|\Delta s_e|_{max}$ per Eq. (9.64). We recall that the linear approximation is valid within an accuracy of better than 5% (which is usually adequate in actual design practice) provided $0.90 < T_w/T_1 < 1.10$, that is, a maximum relative temperature variation from the end equilibrium temperature $T_w = T_\infty$ of 10% on either side of the MIHES. In other words, the linear approximation Eq. (9.64) is valid within a 5% accuracy when the degree of nonequilibrium of the incoming fluids in relation to $T_w = T_\infty$ does not exceed 10%. Under conditions *further* from equilibrium, the symmetry between the heating mode (cold fluid) and the cooling mode (hot fluid) is broken since $|\Delta s_e|_{max, C} \gg |\Delta s_e|_{max, H}$ at a prescribed large differential $T_{1, H} - T_{1, C}$ and it is necessary to consider alternate nonlinear driving forces (DF) for these more complex *nonlinear* applications, as highlighted in the concluding section of this chapter.

The major limitations of the first law potentials and the significant practical and theoretical advantages of the second law driving forces can be summarized as follows in the context of the previous analyses.

9.4.2 Limitations of First Law Driving Forces

All first law potentials are based on the axiomatic assumption (see conventional analytical methods assumption A.9) of *linearity in terms of* ΔT_w and do *not* recognize the significant impact of the *absolute temperature* level. These inherent limitations are compensated in practice by additional corrective empirical terms in the conventional correlations for the heat transfer coefficient α such as $(\mathrm{Pr})^\beta$ which is valid as shown in Eq. (9.21) only for moderate $T_w - T$ when the correction factor $M_{\Delta T}$ accounting for nonisothermal effects approaches unity. At less moderate $T_w - T$, the isothermal heat transfer coefficient is corrected by the empirical multiplier $M_{\Delta T}$, as per Eq. (8.3), for instance, in Chapter 8. The multiplier $M_{\Delta T}$ is so formulated that at the limit of a vanishingly small wall-to-fluid temperature differential $\Delta T_w \to 0$, the heat transfer coefficients in the heating and cooling modes are identical (i.e., the heat transfer coefficients are independent of the direction of heat flow).

No doubt the most flagrant limitation of the first law driving forces is the fact that the absolute temperature does *not* appear in ΔT_w and a very moderate ΔT_w of, say, 30 K at an absolute level $T_w = 300$ K is *perceived as equally moderate at a much lower absolute temperature* level $T_w = 30$ K. This does *not* reflect the physical reality at temperature levels significantly below or above normal ambient conditions, since it is clear that a ± 30 K variation at $T_w = 300$ K is not significant because it corresponds to an absolute temperature ratio range of only $0.9 < T_w/T_1 < 1.10$, whereas the same ± 30 K variation at $T_w = 30$ K corresponds to an absolute temperature ratio range of $\frac{1}{2} < T_w/T_1 < \infty$!

9.4.3 Advantages of Second Law Driving Forces

These can be summed up in one sentence: *The main liabilities of the conventional first law driving forces highlighted above define the basic advantages of the second law potentials.*

To be more specific, the second law potentials recognize the existence of nonlinear relationships between heat fluxes and driving forces, the significance of the absolute temperature levels, and have (on fundamental grounds) a much broader validity range than the conventional heat transfer driving forces.

9.4.3.1 Recognition of the Impact of Absolute Temperature. In contrast to the conventional ΔT_w and other first law potentials, the entropy $(\Delta s_e)_w$ and any second law potential involving the state property entropy recognize the central importance of the absolute temperature level and the fact that the *linearity* between heat fluxes and ΔT_w is *generally not* correct and becomes an acceptable approximation per Eq. (9.59) only *under conditions not too far from equilibrium* when the relative temperature differential $|\overline{T}_w - \overline{T}|/\overline{T} \simeq |\overline{T}_w - \overline{T}|/\overline{T}_w < 10\%$ Moreover, $(\Delta s_e)_{\max, C}$ being significantly larger in a MIHES than $|\Delta s_e|_{\max, H}$ beyond the linear domain, the driving force $|\Delta s_e|_w$ correctly diagnoses the effects of nonlinearity, namely, that the heat flux in the fluid heating mode should generally be larger than in the cooling mode for a *prescribed constant wall temperature* T_w when $T_w - T_{1,C} = T_{1,H} - T_w$ in the case of incompressible fluids.

Furthermore, the second law potential $|\Delta s_e|_w$ clearly recognizes that a fixed differential $|T_w - T_{1,C}| = +30$ K at an absolute level $T_w = 330$ K and $|T_w - T_{1,C}|/T_{1,C} = 30/300 = 0.1$ is *not* significant since the linear law Eq. (9.59) is applicable, but that the same temperature difference $|T_w - T_{1,C}| = +30$ K becomes immensely significant in the neighborhood of $T_w = +30$ K when the relative wall-to-fluid temperature difference $|T_w - T_{1,C}|/T_{1,C} =$

$30/0 \to \infty$ and the attractor state at $T_w = +30$ K becomes *immensely attractive to* to the fluid entering the MIHES at an absolute temperature $T_{1,C} \cong 0$ K, which is in line with some of the strange phenomena experienced with fluids at ultralow cryogenic temperatures (see also Table 9.2).

9.4.3.2 *Unified Thermodynamic Treatment: Single-Phase Fluids.* Another practical advantage illustrated in the previous graphical/numerical analysis is that the combined heat and fluid flow problems can be conveniently integrated in a *unified thermodynamic treatment* [6.5, 6.8] in a manner similar to the methods typically used in steam and gas turbine (or compressor) analyses where the irreversibility of the adiabatic expansion (or compression) process, $\Delta s_f > 0$, is fully recognized (in contrast to highly idealized models based on isentropic expansion or compression at $s =$ constant and $\Delta s = 0$) and the friction heat fully accounted for by a *reheat factor* in standard temperature–entropy or enthalpy–entropy diagrams. Similarly, Fig. 9.9 depicts the conditions encountered in real world heat exchangers when $\Delta s_f \gg 0$ and $\dot{Q}_e \gg 0$, in contrast to the idealized situation described in Fig. 9.5 at $\Delta s_f \to 0$ and $\dot{Q}_e \to 0$.

It is also clear that in forced-convection heat exchangers the entropy production due to the *friction at the wall*,[22] $(\Delta s_f)_{\text{wall}}$, *plays a constructive role* (in sharp contrast to thermal engines or compressors) in as much as a hypothetical *zero friction coating* at the wall (since under no-sticking conditions, $f \to 0$ and the coupled heat transfer coefficient $\alpha \to 0$) would be associated with a vanishingly small rate of heat exchange $\dot{Q}_e$ in typical stable turbulent flow regimes with the usual lengths of travel encountered in heat exchanger design practice.

9.4.3.3 *Unified Thermodynamic Treatment With Phase Changes.* The most important practical advantage of entropy, or entropy-based second law driving forces, is undoubtedly their successful application in complex two-phase flow situations as highlighted previously.[23] The phenomenological law, indicating that the molar entropy of common single-component fluids is a universal function of reduced pressure, provides a theoretical rationale for the empirically validated generalization methods based on the extended corresponding states principle. These thermodynamic generalization methods are (in principle) universally applicable regardless of the type of fluid under consideration, the types of flow regime, the flow arrangements (external or internal flows, natural and/or forced convection), heat exchanger geometry, or direction of heat flow (evaporation or condensation) under stable operating conditions near as well as far from equilibrium (i.e., linear and highly nonlinear situations such as critical heat flux generalizations in nuclear applications).

Conversely, the mere fact that complex coupled two-phase fluid and heat flow phenomena can be generalized successfully on the basis of reduced pressures (as outlined in Chapters 6 and 8) provides additional fundamental support for the validity of entropy or entropy-based second law driving forces.

In *low thermal lift* two-phase flow applications (i.e., in the absence of nucleate boiling regimes), the simplest second law driving force, namely, $(\Delta s_e)_w$, makes it possible to demonstrate the symmetry between the fluid evaporating and condensing modes as outlined previously and demonstrated somewhat more rigorously [7.83a, 7.83b] on the basis of a two-phase counter-flow mirror-image heat exchanger system" (TPC-MIHES). The abstract of Ref. 7.83a is reproduced here as Appendix D.

9.4.3.4 Universal Validity of Nonequilibrium Thermodynamic Laws. The second law thermodynamic treatment of the coupled fluid and heat flow phenomena described here (as well as in earlier publications cited) is fully consistent with the now generally accepted twentieth century theory of irreversible processes or nonequilibrium thermodynamics (NET). Indeed, the differentiation and designation of the two radically different contributions to the total entropy change Δs of a fluid, namely, Δs_e, due to *external* heat exchange, which can be positive or negative, and a nonnegative component $\Delta s_f \equiv \Delta s_i > 0$ due to *internal* (here friction) dissipation was inspired by publications from the Brussels School of Irreversible Thermodynamics [7.16–7.18, 7.63–7.66, 7.70].

Since NET laws are universally applicable (within stated validity limits), it is logical to explore complex situations beyond the current state of the art of conventional first law analytical methods on the basis of the twentieth century theory of NET. After a rather slow start, the practical application of NET tools in heat transfer engineering seems to be taking place at a faster pace,[24] judging by the increasing number of publications addressing complex engineering heat transfer problems on the basis of NET that have appeared in recent years [7.69, 7.71, 9.13, 9.14, and references cited therein].

9.5 BEYOND THE CURRENT STATE OF THE ART

The second law analytical methods discussed in this chapter, which focus especially on forced-convection turbulent flow heat exchanger applications, have in effect *extended* the practical validity of the *linear branch of NET theory to include nonlinear relationships between (stable turbulent) fluid flows and driving forces* as anticipated by the last paragraph of Guideline 20 in Chapter 7. It seems therefore appropriate to consider a number of potentially useful applications of this methodology beyond the "current state of the art."

9.5.1 Momentum, Heat, and Mass Transfer

In view of the analogy and interrelationship between momentum, heat, and mass transfer [9.15–9.19], the same basic methodology should also prove useful in the case of mass transfer and in applications involving simultaneous heat *and* mass transfer such as those typically encountered in air conditioning processes. In fact, the original analytical method proposed for forced-circulation air coolers by the U.S. pioneer in the science and art of air conditioning, W. H. Carrier [9.15, 9.16, 9.17], is based on an equivalent constant external (air side) coil surface temperature, called the *apparatus dew point* (ADP), in conjunction with a coil *bypass factor* (BF). The analogy with the methods used in the present text is quite evident in the context of the following excerpt [9.18]:

> **Coil "by-pass" factor (BF).** In the usual application of a cooling surface, where condensation of moisture from the air is taking place, the air mixture leaving the coil has a temperature and moisture content somewhat higher than that equivalent to the apparatus dewpoint. This is due to the effect of a certain amount of air by-passing the coil surface. By-passed air represents that portion of total air through the cooling coil which has not contacted the cooling surface and therefore has not been cooled or dehumidified to the apparatus dewpoint condition.
>
> $$\text{BF} = \frac{\text{By-passed air}}{\text{Total air through coil}}$$
>
> The by-pass factor is determined from test data and is dependent on many factors, such as depth of coil, type of surface, air velocity, etc. Values for high temperature conditioning coils, where 7 to 14 fins per inch are used, vary from 0.05 to 0.30 depending on fin spacing and number of rows deep. For product cooling and freezing applications, where 3 fin per inch tubing or prime surface pipe is used, by-pass factors range from 0.25 for 8 row finned coils to 0.59 for 10 row prime surface type coils. Stated in another way, the by-pass factor is a measure of the relation between the temperature of the air mixture leaving the coil and the "apparatus dewpoint."

9.5.2 Laminar and High-Velocity Single-Phase Flows

It is possible in the context of what was said previously in Chapter 7 (see, in particular, Variation Analyses, in Sections 7.9 and 7.10) to extend the unified thermodynamic treatment from stable turbulent tube flow regimes at $\text{Re} > 10{,}000$ down to Reynolds numbers below 2100 to include stable laminar flow regimes.

From the definition of the dimensionless total wall friction head loss,

$$(\text{NVH})_f \equiv \frac{\Delta H_f}{\overline{V}^2/2} \tag{9.105}$$

and the key *high-velocity parameter*

$$(\Delta T)_{\text{stagnation}} \equiv \frac{\overline{V}^2/2}{c_p} \tag{9.106}$$

defined by McAdams [7.87] to deal with high gas velocity (Mach number > 1) heat transfer applications, it follows after dividing the numerator and denominator on the right-hand side of Eq. (9.105) by c_p that

$$(\text{NVH})_f \equiv \frac{\Delta H_f / c_p}{(\bar{V}^2/2)/c_p} \equiv \frac{(\Delta T)_f}{(\Delta T)_{\text{stagnation}}} \tag{9.107}$$

where, $(\Delta T)_f$ is the apparent temperature rise due to wall friction dissipation per Eq. (7.98b).

Thus, the number of velocity heads $(\text{NVH})_f$ can also be visualized alternatively as a *number of stagnation temperature rise* due to friction dissipation at the wall, or $(\text{NSTR})_f$:

$$(\text{NVH})_f \equiv (\text{NSTR})_f \equiv \frac{\Delta T_f}{\Delta T_{\text{stagnation}}} \equiv \frac{\bar{T}(\Delta s)_f}{\bar{T}(\Delta s)_{\text{stagnation}}} = \frac{(\Delta s)_f}{(\Delta s)_{\text{stagnation}}} \tag{9.108}$$

It is clear that the proposed unified thermodynamic treatment of the coupled heat and fluid problem based on $(\text{NVH})_f \equiv (\text{NSTR})_f$ is fully compatible with the thermodynamic treatment recommended by McAdams for high gas velocities [7.87].

9.5.3 Regime Bifurcations: Single- and Two-Phase Flows

In some applications, such as direct electrical resistance heating or nuclear heating, the outlet equilibrium state $(\text{ES})_{2,C}$ of the coolant is known beforehand for a given entrance equilibrium state $(\text{ES})_{1,C}$ and mass flow rate $\dot{m}$ from an energy balance (regardless of channel geometry since the heat input *must* go through). The heat transfer problem consists in finding the unknown average (or local) wall temperature $\bar{T}_w$ (or T_w) for a given channel configuration (D_h, L/D_h, Ω, $\bar{e}/D_h$) and wall material or analyzing *instability phenomena* frequently experienced in single- and two-phase flow heat exchanger applications which result from sudden changes of flow regimes.

The oldest and simplest flow regime transition case, from laminar to turbulent flow regimes (and vice versa), can be analyzed on the basis of an entropy maximizing principle as highlighted in the abstracts of Refs. 9.19a and 9.19b reproduced here as Appendix E. The same basic second law methodology has been used in some complex two-phase flow situations [5.18, 9.5, 9.20, 9.21] to help shed some light on the possible choices of two-phase flow regimes and the hysteresis reported in the scientific literature for the inception of the transition from one regime to another under adiabatic conditions, that is, without external heat transfer. A similar thermodynamic approach yields useful insights into the far more complex two-phase flow regime transitions with an external heat input, such as the boiling crises associated with a sudden transition from wet to dry wall flow regimes experienced in direct-expansion refrigeration evaporators and at high heat fluxes in nuclear reactor channels using water as coolant.[25]

9.5.4 More Exotic Applications

Since the fundamental laws of thermodynamics have no known validity limits for systems satisfying the continuum hypothesis, it should be permissible to venture beyond the realm of normal engineering applications and consider some potentially fruitful avenues of investigation by raising such simple questions as: What are the fundamental implications of extreme extrapolations of key parameters like T_w, D_h, L/D_h, and $\overline{V}$ for the following situations, for example: (a) when $T_w \to 0$ and $(T_w - T_{1,C})/T_{1,C} \to \infty$ in ultralow cryogenic applications; (b) when the radiation component of $(\dot{Q}_e)_{tot}$ for certain types of real gases flowing inside fully enclosed long channels increases and becomes significant and ultimately controlling (recall that the basic laws of radiant heat transmission are also based on absolute temperatures and thermodynamic second law considerations); (c) when $\tilde{T}_w$ represents the time and surface integrated average of a *fluctuating wall temperature* field [4.4]; and (d) when D_h and Ω become extremely large while maintaining $L/D_h \gg 1$ in a zero gravity field?

Although some of these extreme extrapolations may rightfully seem rather wild to the cautious heat transfer engineer who has to "live" with earthbound designs, they are in fact benign compared to such exotic concepts as black hole entropy currently analyzed by astrophysicists to prove or disprove [9.22] the existence of black holes or, conversely, the universal validity of the second law of thermodynamics. Such exotic questions, however, are outside the scope of a text dedicated to normal engineering applications and we conclude this chapter with a more down to earth problem, namely, the formulation of nonlinear second law driving forces suitable for heat exchanger design applications in the nonlinear domain of nonequilibrium thermodynamics.

9.6 EXTENSION TO NONLINEAR DOMAIN

We first consider thermodynamically efficient heat exchangers designed to operate at small temperature differentials (to keep irreversible losses within acceptable limits), in the linear domain of NET, with special emphasis on forced-convection turbulent flow of gases in straight constant cross-section channels. The thermodynamic analysis is then extended to applications *far from equilibrium*, that is, to the *nonlinear* domain of twentieth century NET (see Chapter 7, Section 7.5.5, and particularly Guideline 20 and Fig. 7.2).

9.6.1 First Law Heat Transfer Driving Forces

We again focus on the archetype constant wall temperature boundary case of perfect gases flowing in turbulent regimes inside long ($L/D_H > 60$) straight constant cross-section channels, for the reasons stated in Section 9.1. Consequently, it is assumed that the perfect gas relationships per Eqs. (9.22a)–(9.22c) are applicable. It should be reemphasized, however, that the basic methods derived from these perfect gas analyses can be extended (in principle) to real gases, liquids, and two-phase fluids on the basis of general thermodynamic considerations such as the application of the extended corresponding states principle according to the methods outlined in Chapter 6.

The conventional temperature potential is based on Newton's (1701) cooling law,[26] which states that the heat transfer rate per unit area between a gas at bulk temperature T_b and a surrounding constant temperature channel wall T_w is proportional to the local driving force $\Delta T_w = T_w - T_b$. The proportionality constant is by definition the conventional heat transfer coefficient α_t in the well-known equation

$$q'' \equiv \alpha_t \, \Delta T_w \tag{9.109}$$

where the subscript t has been added to differentiate the conventional coefficient defined on the basis of a temperature driving force from those obtained on the basis of alternate driving forces.

Multiplying and dividing the right-hand side of Eq. (9.109) by the constant specific heat c_v yields, since $\Delta u = c_v \, \Delta T$, the following equivalent expression for q'':

$$q'' = \left(\frac{\alpha_t}{c_v} \right) \Delta u_w \equiv \alpha_u \, \Delta u_w \tag{9.110}$$

TABLE 9.5. Driving Forces (DF) and Corresponding Heat Transfer Coefficients α_{DF} for the Linear Domain When $q'' \propto |\Delta T_w|$ at $|y| = |\Delta T_w / T| \ll 1$

	DF	α_{DF}	α_{DF} / α_t	Equation
1.1	ΔT_w	α_t	1	(9.109)
1.2	Δu_w	α_u	$1 / c_v$	(9.110)
1.3	Δh_w	α_h	$1 / c_p$	(9.111)
2.1	$\Delta s_{w,p}$	$\alpha_{s,p}$	T / c_p	(9.114)
2.2	$\Delta s_{w,v}$	$\alpha_{s,v}$	T / c_v	(9.115)
2.3	$(\mathrm{SRRW})_p$	$\alpha_{\mathrm{SRRW},p}$	$(2T / c_p)^{1/2}$	(9.123)
2.4	$(\mathrm{SRRW})_v$	$\alpha_{\mathrm{SRRW},v}$	$(2T / c_v)^{1/2}$	(9.124)

where $\Delta u_w = c_v \Delta T_w$ is an alternative potential or driving force with an appropriate heat transfer coefficient $\alpha_u \equiv \alpha_t / c_v$. An enthalpy potential Δh_w can be obtained in a similar fashion with a corresponding α_h as shown on line (1.3) of Table 9.5. The three alternative driving forces, ΔT_w, Δu_w, and Δh_w, are grouped under lines 1.1–1.3 in Table 9.5 to distinguish these first law potentials from the second law driving forces addressed in the next section.

The essential common characteristic of Eqs. (9.109)–(9.111) is the linear relationship between the heat fluxes and driving forces:

$$q'' \propto \Delta T_w \propto \Delta u_w \propto \Delta h_w \tag{9.112}$$

9.6.2 Entropy-Based Driving Forces: Linear Domain

Within the linear domain $T_w / T_{1,b} \simeq 1 \pm 0.1$, the specific entropy change of a perfect gas from an inlet equilibrium state at $T_{1,b}$ to an exit equilibrium state at $T_w = T_{1,b} + (T_w - T_{1,b})$ at either constant specific volume, $(\Delta s_w)_v$, or constant pressure, $(\Delta s_w)_p$, can be expressed (as shown earlier in this chapter) according to

$$\frac{\Delta s_{w,v}}{c_v} = \frac{\Delta s_{w,p}}{c_p} = \frac{\Delta T_w}{T_{1,b}} \equiv y \tag{9.113}$$

Keep in mind the algebraic sign of the term $y \equiv \Delta T_w / T_{1,b}$. In the fluid heating mode, when $T_w / T_{1,b} > 1$, the term y is positive, whereas in the fluid cooling mode, when $T_w / T_{1,b} < 1$, the term y is negative.

Following the same procedure as indicated above for the first law driving forces Δu_w and Δh_w, we define two alternative entropy driving forces and corresponding heat transfer coefficients and express them in terms of the conventional α_t per Eqs. (9.114) and (9.115) shown on lines 2.1 and 2.2 of Table 9.5. Because the heat fluxes vary in direct proportion to the entropy driving forces, within the linear domain, Eq. (9.112) can be extended to include the entropy driving forces $(\Delta s_w)_v$ and $(\Delta s_w)_p$ as follows:

$$q'' \propto \Delta T_w \propto \Delta u_w \propto \Delta h_w \propto (\Delta s_w)_p \propto (\Delta s_w)_v \tag{9.116a}$$

Validity: Linear domain at $|y| = |\Delta T_w / T_{1,b}| \leq 0.1 \ll 1$ (9.116b)

9.6.3 Square Root Reversible Work Driving Forces

Our prime objective is to assess the merits of more widely applicable heat transfer driving forces combining the two independent thermodynamic state properties—T and s—introduced in [6.7] under the name maximum reversible

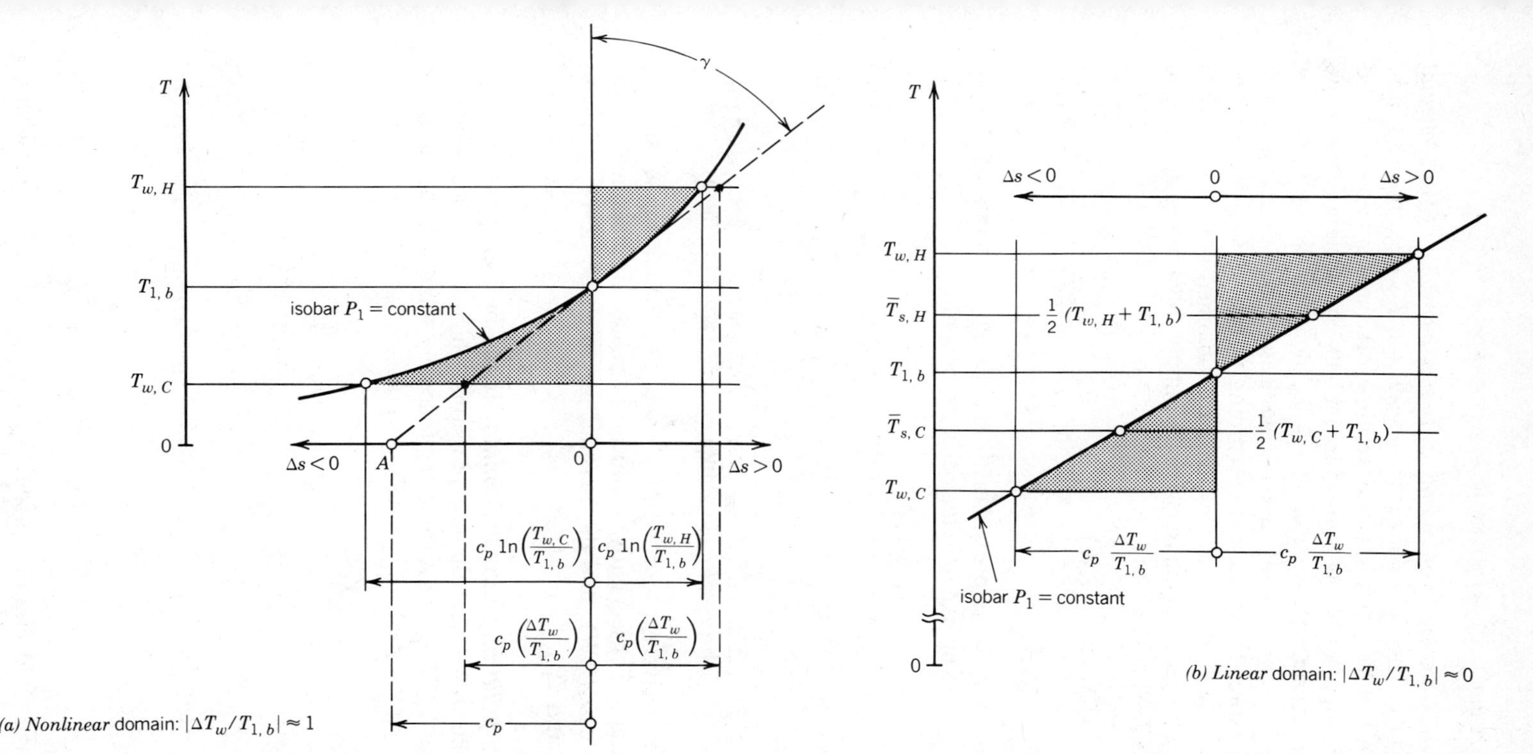

FIGURE 9.10. Reversible work potential $(RW)_p$ and $(RW)_v$ in *T-s* diagram. (a) *Nonlinear* domain: $T_w / T_{1,b}$ far from unity. (b) *Linear* domain: $T_w / T_{1,b} \simeq 1$. *Explanatory notes*: (1) Full logarithmic curve in part (*a*) is the $P_1 = P$ = constant isobar. (2) The origin of entropy $s_o = 0$ is selected arbitrarily at $T_{1,b}$ on the P_1 isobar. (3) The shaded areas represent $(RW)_p$ at prescribed $T_{1,b}$ for $T_{w,H} > T_{1,b}$ and $T_{w,C} < T_{1,b}$ in the fluid heating and cooling modes, respectively, shown here for $(T_{w,H} - T_{1,b})_{\text{heating}} = -(T_{w,C} - T_{1,b})_{\text{cooling}} \equiv |\Delta T_w|$. (4) The dashed straight line is tangent to the isobar at point $T_{1,b}$, P_1. The distance $\overline{AO} \equiv c_p(T_{1,b} / T_{1,b}) = c_p$ = constant, regardless of the absolute temperature level $T_{1,b}$. Consequently, angle γ varies from 90° to 0° as $T_{1,b}$ varies from 0 to ∞ K with a corresponding variation of $ds / dT \propto \tan\gamma$ from ∞ to 0. (5) In the linear domain, the isobar and tangent merge into a single straight line as shown in part (*b*). (6) Similar graphs are obtained for $(RW)_v$ when the logarithmic curve becomes the $v_1 = v$ = constant line, by substituting c_v for c_p. This is equivalent to multiplying the Δs scale by a constant factor c_v / c_p. [Adapted from H. Soumerai, "Entropy Based Heat Transfer Potentials—Forced Convection Single Phase," Fig. 1, 8th International Heat Transfer Conference, San Francisco, Aug. 1986.]

work potential, which we shall now simply call *reversible work* (RW) *potential* or driving force (DF). The shaded areas in the skeleton T-s diagrams shown in Fig. 9.10*a* quantify $(\mathrm{RW})_p$, that is, the maximum work that could be produced while raising one unit mass of gas from an initial bulk equilibrium temperature T_{1b} to the end equilibrium temperature $T_{2b} = T_w$ along the isobar $P_1 = P_2 =$ constant, in the absence of any irreversible losses. This reversible work potential can be expressed analytically according to Eqs. (9.49a) and (9.49b) for $P =$ constant, and similar expressions are obtained for $(\mathrm{RW})_v$ when $v = 1/\rho =$ constant (i.e., along an isochore) by simply substituting c_v for c_p in Eqs. (9.49a) and (9.49b). Therefore,

$$\frac{(\mathrm{RW})_p}{c_p} = \frac{(\mathrm{RW})_v}{c_v} = T_w \ln\left(\frac{T_w}{T_{1,b}}\right) - \Delta T_w \tag{9.117}$$

which can alternatively be expressed with Eq. (9.49b) in terms of $\overline{T}_s$, the equivalent constant thermodynamic temperature, according to

$$\frac{(\mathrm{RW})_p}{c_p} = \frac{(\mathrm{RW})_v}{c_v} = (T_w - \overline{T}_s)\ln\left(\frac{T_w}{T_{1,b}}\right) \tag{9.118a}$$

where

$$\overline{T}_s \equiv \frac{T_w - T_{1,b}}{\ln(T_w/T_{1,b})} = \frac{\Delta T_w}{\ln(1 + \Delta T_w/T_{1,b})} \tag{9.118b}$$

Note that $\overline{T}_s$ is exclusively a function of T_w and $T_{1,b}$ for ideal gases since c_v and c_p are constant. Consequently, identical $\overline{T}_s$ values are obtained for the constant volume case $(\mathrm{RW})_v$ and the constant pressure case $(\mathrm{RW})_p$ at identical values of T_w and $T_{1,b}$.

9.6.3.1 Near-Equilibrium Linear Domain. It can be seen by referring to Fig. 9.10*b* that the above logarithmic expressions, which are applicable at any temperature ratios $T_w/T_{1,b}$, converge to the following simpler equations in the linear domain when $0.9 \leq T_w/T_{1,b} \leq 1.10$ and $|\Delta T_w/T_{1,b}| = |y| \ll 1$:

$$\frac{(\mathrm{RW})_p}{c_p} = \frac{(\mathrm{RW})_v}{c_v} = (T_w - \overline{T}_{\mathrm{am}})\frac{\Delta T_w}{T_{1,b}} = \frac{(\Delta T_w)^2}{2T_{1,b}} \tag{9.119a}$$

where

$$\overline{T}_{\mathrm{am}} = \tfrac{1}{2}(T_{1,b} + T_w) = \overline{T}_s = \text{arithmetic mean temperature} \tag{9.119b}$$

or

$$T_w - \overline{T}_s = T_w - \overline{T}_{\mathrm{am}} = \tfrac{1}{2}(T_w - T_{1,b}) \tag{9.119c}$$

Note that, in the fluid heating mode, $T_w > T_{1,b}$ and $\Delta T_w > 0$, whereas

$T_w < T_{1,b}$ and $\Delta T_w < 0$ in the fluid cooling mode, while the $(\mathrm{RW})_p$ and $(\mathrm{RW})_v$ remain positive, as they should [6.7], since $(\Delta T_w)^2$ is always positive even in the cooling mode when ΔT_w is negative!

Since empirical evidence has amply confirmed that the heat fluxes q'' are proportional to the first power of the conventional driving force $|\Delta T_w|$ according to Eq. (9.116) in forced convection at $|y| = |\Delta T_w|/T_{1,b} \ll 1$, it is clear that the $(\mathrm{RW})_p$ and $(\mathrm{RW})_v$ driving forces per Eq. (9.119) are proportional to the second power of the heat flux q''; that is,

$$(q'')^2 \propto (\Delta T_w)^2 \propto (\mathrm{RW})_p \tag{9.120}$$

and

$$(q'')^2 \propto (\Delta T_w)^2 \propto (\mathrm{RW})_v \tag{9.121}$$

Taking the square root on both sides of Eqs. (9.120) and (9.121) yields

$$q'' \propto \Delta T_w \propto \left[(\mathrm{RW})_p\right]^{1/2} \equiv (\mathrm{SRRW})_{p,\,\mathrm{linear}} \tag{9.122a}$$

and

$$q'' \propto \Delta T_w \propto \left[(\mathrm{RW})_v\right]^{1/2} \equiv (\mathrm{SRRW})_{v,\,\mathrm{linear}} \tag{9.122b}$$

where $(\mathrm{SRRW})_{p,\,\mathrm{linear}}$ and $(\mathrm{SRRW})_{v,\,\mathrm{linear}}$ denote the *square root of the reversible work* at P = constant and v = constant, respectively. These relationships are valid *a priori* within the linear domain when $T_w/T_{1,b} \simeq 1$ or, with an adequate accuracy for heat transfer engineering applications, within the following practical limits:

$$\textit{Validity}:\quad 0.9 \le T_w/T_{1,b} \le 1.10 \tag{9.122c}$$

The corresponding expressions in terms α_t for the heat transfer coefficients $\alpha_{\mathrm{SRRW},\,p}$ and $\alpha_{\mathrm{SRRW},\,v}$, Eqs. (9.123) and (9.124), are shown on lines 2.3 and 2.4 of Table 9.5.

The following relationships are obtained from the square root of the ratio $(\mathrm{RW})_p/(\mathrm{RW})_v$:

$$\frac{(\mathrm{SRRW})_{p,\,\mathrm{linear}}}{(\mathrm{SRRW})_{v,\,\mathrm{linear}}} = \left\{\frac{(\mathrm{RW})_{p,\,\mathrm{exact}}}{(\mathrm{RW})_{v,\,\mathrm{exact}}}\right\}^{1/2} = \left(\frac{c_p}{c_v}\right)^{1/2} = \kappa^{1/2} \tag{9.125}$$

where κ is the isentropic exponent of the gas under consideration. Although outside the scope of this introductory text, it is interesting to note that $\kappa^{1/2}$ closely duplicates the empirical property term $1/(\mathrm{Pr})^{1/2} \equiv (\mathrm{Pr})^{-1/2}$ in the modified Chilton–Colburn heat transfer coefficient correlation for gases ($0.5 < \mathrm{Pr} < 1.0$) in turbulent channel flow regimes and that $(c_p/c_v)^{1/2} \simeq 1$ for most common gases [9.23], as can also be seen from the physical property data in Appendix A.

9.6.3.2 Far From Equilibrium Nonlinear Domain. The main purpose of the following analysis is to assess, qualitatively and quantitatively, the impact of large relative temperature differentials in both the fluid heating and cooling modes, on the basis of the SRRW driving forces, that is, exclusively from *theoretical* thermodynamic considerations, and to compare the results thus obtained with the correlations recommended by Petukhov [3.42].

Background

Conventional semiempirical correlations for fully developed turbulent tube flow regimes, valid for constant fluid physical properties, are generally expressed, as shown previously, in terms of Nu or $\mathrm{St} = \mathrm{Nu}/\mathrm{Re}\,\mathrm{Pr} = \Psi(\mathrm{Re}, \mathrm{Pr})$. To account for the effects of variable physical properties, a number of analytical and/or purely empirical correlations for gases have been proposed [3.42] which are based on a correction factor accounting for nonisothermal effects defined[27] as follows:

$$F \equiv \frac{(\mathrm{Nu})_b}{(\mathrm{Nu})_{ob}} = F(\theta) \tag{9.126}$$

where $\theta \equiv T_w/T_{1,b}$ is the absolute wall-to-gas bulk temperature ratio, $(\mathrm{Nu})_b$ is the Nusselt number based on bulk temperature properties at $\theta \neq 1$, and $(\mathrm{Nu})_{ob}$ is the reference value computed at bulk temperature when $\theta = 1$, that is, at vanishingly small temperature differentials and therefore constant physical properties.

Problem Statement

(a) Find on the basis of the SRRW driving force a theoretical expression for the correction factor F_{th} defined by Eq. (9.126) in terms of $y \equiv \Delta T_w/T_{1,b}$, for gases in the fluid heating mode and, by treating the relative temperature term y as an algebraic quantity < 0, in the fluid cooling mode.

(b) Compute F_{th} for the fluid heating mode in the range $0 < y < +1$ and for the fluid cooling mode in the range $-1 < y < 0$. Then compare these theoretical results with those reported for gases by Petukhov [3.42].

Solution

(a) Since the exact expression of the RW driving force, $(\mathrm{RW})_{\mathrm{exact}}$ per Eq. (9.117), converges toward the approximate linear expression $(\mathrm{RW})_{\mathrm{linear}}$ per Eq. (9.119) when $y \to 0$ (or $\theta \to 1$) in the $v = 1/\rho =$ constant as well as the

P = constant case, we simply set

$$F_{\text{th}} = \frac{(\text{Nu})_b}{(\text{Nu})_{ob}} = \frac{(\text{SRRW})_{\text{exact}}}{(\text{SRRW})_{\text{linear}}} = \left(\frac{(\text{RW})_{\text{exact}}}{(\text{RW})_{\text{linear}}} \right)^{1/2} \tag{9.127}$$

or, with Eqs. (9.117) and (9.119a),

$$F_{\text{th}} = \left(\frac{T_w \ln(T_w/T_{1,b}) - \Delta T_w}{(\Delta T_w)^2/2T_{1,b}} \right)^{1/2} \tag{9.128}$$

Dividing the numerator and denominator of the expression in the large parentheses on the right-hand side of Eq. (9.128) by $T_{1,b}$ yields, with

$$\theta \equiv \frac{T_w}{T_{1,b}} \equiv \frac{T_{1,b} + \Delta T_w}{T_{1,b}} = 1 + \frac{\Delta T_w}{T_{1,b}} = 1 + y \tag{9.129}$$

the following *fully dimensionless* equation *in terms of* y:

$$F_{\text{th}} = \left(\frac{2[(1+y)\ln(1+y) - y]}{y^2} \right)^{1/2} \tag{9.130}$$

Equation (9.130) is applicable in the fluid heating *and* cooling modes and also yields a positive *real* numerical value for the cooling mode when $-1 < y < 0$ with $\ln(1+y) < 0$ but $-y = +|y| > 0$ and $y^2 > 0$.

Equation (9.130) shows that F_{th} is defined uniquely in terms of the relative temperature differential y, as expected. Therefore, identical values of F_{th} are obtained at identical values of y regardless of the absolute temperature level. It is thus clear from this dimensionless expression that it is the *algebraic relative temperature differential* $y = \Delta T_w / T_{1,b}$ that is significant and *not* the thermodynamically meaningless *absolute temperature differential* $|\Delta T_w|$ as already emphasized [6.5].

(b) The theoretical values of F_{th} computed from Eq. (9.130) are shown on line (3) of Table 9.6 and are compared to the values of F reported by Petukhov [3.42] and shown on lines (4), (6), (8), and (9).

Qualitative Comparison: Gas Heating and Cooling Modes. The values of F_{th} shown in Table 9.6 correctly diagnose the empirically validated fact that $F > 1$ for the gas cooling mode, that is, the heat transfer is larger at $y < 0$ (or $\theta < 1$) than at $y \to 0$, and, conversely, $F < 1$ for the gas heating mode when $y > 0$ (or $\theta > 1$).

Quantitative Comparison: Gas Cooling Mode. Petukhov recommends[28] [3.42, Eq. (64)] the following correlation based on the most reliable published

TABLE 9.6. Theoretical Value of F_{th} Versus $F \equiv (Nu)_b / (Nu)_{ob}$ from Petukhov [3.42] for Gases

(1) $\theta \equiv T_w / T_{1,b}$	0	0.10	0.25	0.50	0.75	0.90	0.99	1.00	1.01	1.10	1.25	1.50	1.75	2.0
(2) $y \equiv \Delta T_w / T_{1,b}$	−1.00	−0.90	0.75	−0.50	−0.25	−0.10	−0.01	0	+0.01	+0.10	+0.25	+0.50	+0.75	+1.0
(3) F_{th} from Eq. (9.130)	$\sqrt{2} \simeq 1.41$	1.30	1.20	1.11	1.05	1.02	1.002	1	0.998	0.98	0.96	0.93	0.90	0.88
(4) F from Eq. (9.131)	—	$(1.39)^a$	$(1.24)^a$	1.135	1.068	1.027	1.0027	1						
(5) Deviation $(\%)^b$	—	$(-6.7)^a$	$(-4)^a$	−2.4	−2.0	−0.9	−0.1	0						
(6) $F_{analytical}$ = average values of H_2 and air from Deissler [3.42, Fig. 10]								1	∼1.00	0.98	0.97	0.93	0.89	0.87
(7) Deviation $(\%)^c$								0	←——	<	|1.2%|		——→	
(8) F_{max} = *maximum* values [3.42, Fig. 12]								1.06	1.04	1.05	1.00	0.91	0.80	0.77
(9) F_{min} = *minimum* values [3.42, Fig. 12]								0.96	0.96	0.91	0.84	0.77	0.68	0.66

[a] Approximate values extrapolated from line –··–··– [3.42, Fig. 12].
[b] % deviation ≡ 100[line (3) − line (4)] / line (4).
[c] % deviation ≡ 100[line (3) − line (6)] / line (6).

experimental data available at the time:

$$F = 1.27 - 0.27\theta \quad \text{valid for } 0.5 < \theta < 1 \tag{9.131}$$

The values of F computed according to Eq. (9.131), shown on line (4) in Table 9.6, are compared to F_{th} on line (3). The maximum deviation of less than 2.5% shown on line (5) demonstrates an exceptionally good agreement. It is also noteworthy that even the rather wild extrapolation to $(F_{\text{th}})_{\text{max}} = \sqrt{2} = 1.414$ at $T_w \to 0$ K, that is, absolute zero (conceivable for gases if $\rho = 1/v =$ constant when P decreases in direct proportion with T), suggesting a finite upper limit, seems to be well in line with the extrapolation to $\theta = 0.1 \simeq 0$ of the only set of empirical data shown [3.42, Fig. 12] that is sufficiently refined to indicate a definite variation of F with θ in the cooling mode [see values in parentheses on line (5) of Table 9.6].

Quantitative Comparison: Gas Heating Mode. Lines (6)–(9) of Table 9.6 show that the computed values of F_{th} fall within the total scatter/accuracy band of the data analyzed by Petukhov. A comparison on line (7) with the analytical results proposed by Deissler for hydrogen and air indicates a deviation of less than 1.2%. This is particularly remarkable for the heating mode when one recalls that Eq. (9.130) was derived from perfect gas laws, a rather extreme idealization for temperature levels as high as $T_w = 1000$–2000 K analyzed by Petukhov [3.42].

Comments

1. This analysis indicates that the theoretical expression of $F_{\text{th}} = (\text{Nu})_b/(\text{Nu})_{ob} = (\text{St})_b/(\text{St})_{ob}$ per Eq. (9.127), (9.128), or (9.130), derived from a square root reversible work driving force concept (and therefore exclusively from second law theoretical thermodynamic considerations), correctly diagnoses the direction of the effect of $y = \Delta T_w/T_{1,b} \neq 0$ and provides a fairly good (heating mode) to excellent (cooling mode) quantitative estimate of the effect of $y \neq 0$ within the limits investigated here.

2. It should be kept in mind that this analysis is based on perfect gases and the theoretical results compared to analytical/empirical correlations also obtained with quasi-ideal gases. These predictions and correlations are relatively simple because of the assumed constancy of the terms c_v, c_p, $c_p/c_v = \kappa$ for a given gas and the assumption that the compressibility factor $Z = 1$. By treating the specific heat c (either c_p or c_v) as a constant, regardless of the thermodynamic state, the first derivative $\delta s/\delta T = c/T \propto 1/T > 0$, and the second derivative becomes $\delta^2 s/\delta T^2 = -c/T^2 \propto -1/T^2 < 0$. This leads to the well-known logarithmic curves for $(s - s_o)_p$ and $(s - s_o)_v$ in a T-s diagram, characterized by a *continuously decreasing* slope $(\delta s/\delta T)$ as T increases since the second derivative remains negative at any (positive) ab-

solute temperature. The monotonous *decrease* of the first derivative $(\delta s/\delta T)$ from $+\infty$ to $+0$ as $T_{1,b}$ *increases* from 0 to ∞ K is highlighted in Fig. 9.10 by the variation of the angle γ from $\pi/4$ to 0 and a corresponding variation of the tangent of γ in the range $+\infty \leq \tan\gamma \leq 0$ (see explanatory note 4 in legend to Fig. 9.10). This perfect gas behavior, however, is *not* typical of real fluids, particularly in the neighborhood of the thermodynamic critical point (P_{cr}, T_{cr}) and/or near liquid–vapor phase transitions. Indeed, all fundamental studies of phase transitions focus on the *variations and discontinuities of the specific heat* [9.24]. It is clear, by referring to the graphical representation of RW in Fig. 9.10*a*, that the resulting changes in the shape of the $s = \varphi(T)$ curves can have a major impact on the variation of RW, SRRW, and F_{th} with ΔT_w and, consequently, on the response of the thermodynamic system to finite ΔT_w variations and/or fluctuations.

3. On the basis of the previous comment, major derivations from the results derived here for perfect gases can logically be expected in the case of real fluids such as saturated vapors and gases closer to the critical thermodynamic point, as well as liquids and single-component two-phase fluids. Fortunately, the problems associated with the absence of simple explicit analytical expressions for these more complex situations can be greatly simplified with the help of such classical engineering thermodynamic tools as the extended corresponding states principle (ECSP) and by fully capitalizing on the unique features of the temperature–entropy diagram to estimate the variation of SRRW graphically and/or numerically. The ECSP method of generalizations is particularly effective in highly complex evaporator and condenser design applications because the molar latent entropy change is for all practical purposes a universal function of reduced pressure for most common fluids [4.21, 4.61], as shown in Chapter 6.

4. It is also noteworthy that the graphical/analytical interpretations of the RW potential presented here and in Ref. 6.7 are fully consistent with the new concepts introduced in the 1970s by mathematical/experimental physicists [2.14] to deal with dissipative systems dynamics as well as the basic methods of twentieth century NET *near* and *far from equilibrium* [2.17, 7.18, 7.64]. More specifically, the *monotonous contraction (for ideal gases) of the RW area in a T-s diagram* from a maximum initial value $(\text{RW})_{\max}$ as the fluid enters the heat exchanger channel at $T_{1,b}$, at a relative length of travel or residence time $L_{rel} = \overline{\Delta\theta}_{rel} = 0$, to a vanishingly small value, RW $\to$ 0, at exit end equilibrium state $T_{2,b} \to T_w$ when $L_{rel} = \overline{\Delta\theta}_{rel} \geq 5$, *meets all the requirements of the stable attractors* defined in Chapter 6 of Bergé et al. [2.14].

9.6.4 Concluding Remarks

The NET concepts reviewed earlier for practical applications in simple linear heat flux heat exchangers operating not far from equilibrium have been extended here to deal with more complex situations farther from equilibrium in the nonlinear domain of twentieth century NET (see Fig. 7.2). The rather encouraging results obtained on the basis of reversible work driving forces, that is, purely theoretical second law considerations, further reinforce the most important general conclusions summarized in section 9.4, namely,

- The significant practical and theoretical advantages of second law driving forces.
- The relevancy and practical value of nonequilibrium thermodynamic considerations in many difficult heat transfer engineering applications.

In this context, it may be appropriate and reassuring to stress once more (see Section 7.2.3 and Guideline 11) the often forgotten fact that the Kirchhoff and the Stephan–Boltzmann laws of thermal radiation were derived in the nineteenth century from the second law of thermodynamics, and to recall the central role played by the concept of entropy in Planck's pioneering work on thermal radiation as clearly stated in Planck's scientific autobiography [7.30]. Second law thermodynamic considerations therefore have an excellent century-old track record in the science and, more importantly for the practicing heat transfer engineers, in the art of heat transmission.

It is hoped that this introductory text on a *unified* ***thermodynamic*** *treatment of momentum, heat, and mass*[29] *transfer* will encourage other engineers as well as applied researchers to consider the application of all equilibrium and nonequilibrium thermodynamic tools available today in the solution of complex heat transfer engineering problems.

NOTES

1. This chapter is essentially an expanded and updated consolidation of recent and forthcoming papers [4.4, 6.7, 6.8, 9.1–9.5]. To make this rather theoretical chapter more or less self-contained, a few key concepts discussed earlier have deliberately been highlighted here. These minor redundancies should prove helpful to the many engineers who prefer to read a technical book more or less "a la carte" without following the somewhat arbitrary chapter sequence selected by the author.
2. The reason for the use of the alternative term bypass factor (BF) in lieu of approach factor (AF) will become clearer later, when the thermodynamic mixing model is discussed.
3. See Chapter 7, Sections 7.5.5–7.5.7.
4. They do not satisfy by far Eq. (8.2b).
5. See also Guideline 9 in Chapter 4.
6. As discussed by Soumerai [4.4–4.6, and references cited therein].
7. Because of the *arbitrary* (though convenient) selection of a Reynolds number $(\mathrm{Re})_{max}$ based on $\overline{V}_{max}$ at Ω_{min}, the corresponding f obtained in typical ($\Omega \neq$ constant) compact heat exchangers cannot be compared with those obtained in long straight channels of constant cross section at the same numerical values of the Reynolds number. A more sensible comparison might be possible if the compact heat exchanger correlations were based on an "average" gas velocity $\overline{V}_{avg} < \overline{V}_{max}$ which would reduce the characteristic Reynolds number to lower values than those shown in the curves of Kays and London [9.7]. Clearly, a residence time based on $\overline{V}_{max}$ at Ω_{min} would be quite misleading, whereas an average residence time $(\overline{\Delta\theta})$ based on $\overline{V}_{avg}$ according to Eq. (7.71) would be physically more meaningful in compact heat exchangers with $\Omega \neq$ constant.
8. In contrast, the hydrodynamic flow regime is typically *not* fully developed in many compact heat exchanger design applications.
9. See Section 7.6.
10. See also Section 8.1.3.1.

11. Although a constant value $\beta < 0.5$ may be more accurate (see Section 8.1.4). However this would have no impact on the results shown in terms of L_{rel} in Fig. 9.1 and Table 9.1.
12. Although the temperature–entropy diagrams are not the most practical tools in actual thermal engineering design applications (the Mollier enthalpy–entropy and the log P–enthalpy charts are more popular in power generation and refrigeration applications, respectively), the T-s chart is a most effective heuristic and didactic tool, because it makes it possible to demonstrate graphically complex thermodynamic relationships quite simply (e.g., see the Clapeyron–Clausius equations in Chapter 4, Fig. 4.9, and later sections of this chapter). The T-s chart was also the traditional didactic tool in classical German language engineering thermodynamic textbooks [3.54, 9.11, 9.12]. It is thus no mere coincidence that the thermodynamic mixing model first proposed by Soumerai [6.8, 9.3] is based in large part on graphical methods described by Bosnjakovic [9.11] and Plank [9.12].
13. For an isothermal process at $T_1 = T_2 =$ constant, the positive work of expansion from P_1 to $P_2 < P_1$ is exactly equal to $T_2(\Delta s)_T$. This ideal (maximum) reversible work output is represented in the T-s diagram by the area of the rectangle in Fig. 9.3 confined between the horizontal line $T_2 =$ constant and the $T = 0$ K line, and the constant entropy vertical lines s_2^* and s_2.
14. For example, Plank [1.12, Table 1a] shows values of κ_o at 0°C ranging from 1.408 to 1.399 or $\kappa_o \simeq 1.4$ for H_2, N_2, O_2, and air, and a corresponding $1/\kappa_o \simeq 0.71$; therefore, $(1 - 1/\kappa_o) \equiv (\kappa_o - 1)/\kappa_o \simeq 0.29$. For the refrigerants R-11, R-12, and R-22, lower κ_o values are indicated—1.124, 1.143, and 1.19, respectively—or $\kappa_o \simeq 1.15$; therefore, $1/\kappa_o \simeq 0.87$ and $(\kappa_o - 1)/\kappa_o \simeq 0.13$. It is interesting to note that $1/\kappa_o \equiv c_v/c_p$ approximates rather closely the Prandtl numbers of these gases as shown in Appendix A and in Fig. 8.5 for the common halogenated refrigerant vapors over the normal range of reduced pressures encountered in heat pumping cycles.
15. This also complies with the terminology used by the Brussels School of Irreversible Thermodynamics [7.1, 7.17, 7.18, 7.63–7.66, 7.70].
16. The heat and mass transfer analogy is a standard and most useful tool in chemical engineering applications. For example, see a very recent application to automotive catalytic converters [9.25].
17. The sign of Δs being a matter of convention, Δs_f could just as well have been arbitrarily defined as negative rather than positive. The essential feature of Δs_f, in contrast to Δs_e, is that the sign of Δs_f can never change.
18. In his introduction Bosnjakovic [9.11] lists a few typical engineering applications involving thermodynamic mixing processes and makes the following very relevant observation: "Aber auch die wichtigen Erscheinungen der Turbulenz einer Stroemung sind eng mit Mischungsvorgaengen kleinster Teilchen verknuepft." This can be translated roughly as follows: "But the occurrence of the all important hydrodynamic turbulence is also closely associated with small particles mixing phenomena."
19. This is, of course, in line with the Clausius definition of a differential entropy change $(ds)_P = c_p(dT/T) = c_p d(\ln T)$ and simply means that for $\delta T/T \to 0$, the tangent to the logarithmic isobar line in a T-s diagram coincides with the curve itself.

20. This is in full agreement with conventional analytical methods which are also based on the fundamental assumption that at the ultimate wall–fluid interface the fluid has the same temperature as the wall. This basic assumption is always valid provided the continuum hypothesis of thermodynamics is satisfied. It is possible to provide a rigorous interpretation of the thermodynamic mixing model on the basis of statistical mechanics in terms of residence time (or $\overline{\Delta\theta}_{\text{rel}}$) in the light of the probabilistic interpretation of the entropy (see note 3, Chapter 3) at T_w, P_w, that is, at the ultimate end equilibrium or attractor state.
21. Recognizing the semantic hazards associated with all models of physical laws according to Di Francia [1.1, pp. 50–52] who comes to the following conclusion: "I believe that in this situation the term 'model,' which is often quite useless, should be used as little as possible." This is a far stronger recommendation than our Guideline 13 in Chapter 7.
22. The accent is *at the wall* since individual flow friction losses are fully counterproductive, as shown in Section 8.1.5.
23. See, for example, the thermodynamic interpretation of the Bo Pierre paradox in Section 8.2.3.
24. The increased interest among scientists in NET is evidenced by the new (1980s) Wiley series in *Nonequilibrium Problems in the Physical Sciences and Biology*.
25. However, in the latter case q'' and T have much higher values and T_w is closer to T_{cr}, so that such assumptions as c_p = constant may no longer be permissible (since as $T^+ = T/T_{cr} \to 1$, the liquid specific heat increases sharply and $ds/dT \to \infty$ before the liquid phase vanishes at $T^+ > 1$ as readily seen from standard H_2O temperature–entropy thermodynamic diagrams or tables).
26. See Section 7.2.1.
27. The definition of Petukhov's factor F is equivalent to the multiplier $M_{\Delta T}$ introduced in Section 8.1.1.
28. The following excerpt from Petukhov [3.42, p. 542] is of special interest in connection with the gas cooling mode (italics added to emphasize particularly relevant aspects):

    > In the case of *gas cooling*, ... the relation between $\text{Nu}_b/\text{Nu}_{ob}$ and θ, from experimental data, appears to be significantly weaker than that obtained analytically [based on Petukhov's conventional analytical methods]. *Further experiments and theoretical studies are necessary to explain these differences*. To obtain new data for practical calculations of heat transfer under the conditions of gas cooling, the simple equation suggested by Ivaschenko (63) from the treatment of Ilyin's experimental data (48) may be used: $\text{Nu}_b/\text{Nu}_{ob} = 1.27 - 0.27\theta$. This equation is valid when $0.5 < \theta < 1$.

    It is noteworthy that the above empirical correlation of experimental data agrees rather well with the *purely theoretical* F_{th} according to Eq. (9.130), derived exclusively from second law thermodynamic considerations as shown in our Table 9.6.
29. This is in view of the well-known analogy between heat and mass transfer. See also Section 9.5.1, footnote 16, and the related discussion following Eq. (9.69) in this chapter.

APPENDIX A

THERMODYNAMIC AND THERMOPHYSICAL PROPERTIES OF MATTER

TABLE A.1. Thermophysical Properties of Saturated Water

TEMPERATURE, T (K)	PRESSURE, P (bars)[b]	SPECIFIC VOLUME (m³/kg) $v_f \cdot 10^3$	v_g	HEAT OF VAPORIZATION, h_{fg} (kJ/kg)	SPECIFIC HEAT (kJ/kg·K) $c_{p,f}$	$c_{p,g}$	VISCOSITY (N·s/m²) $\mu_f \cdot 10^6$	$\mu_g \cdot 10^6$	THERMAL CONDUCTIVITY (W/m·K) $k_f \cdot 10^3$	$k_g \cdot 10^3$	PRANDTL NUMBER Pr_f	Pr_g	SURFACE TENSION, $\sigma_f \cdot 10^3$ (N/m)	EXPANSION COEFFICIENT, $\beta_f \cdot 10^6$ (K⁻¹)	TEMPERATURE, T (K)
273.15	0.00611	1.000	206.3	2502	4.217	1.854	1750	8.02	569	18.2	12.99	0.815	75.5	−68.05	273.15
275	0.00697	1.000	181.7	2497	4.211	1.855	1652	8.09	574	18.3	12.22	0.817	75.3	−32.74	275
280	0.00990	1.000	130.4	2485	4.198	1.858	1422	8.29	582	18.6	10.26	0.825	74.8	46.04	280
285	0.01387	1.000	99.4	2473	4.189	1.861	1225	8.49	590	18.9	8.81	0.833	74.3	114.1	285
290	0.01917	1.001	69.7	2461	4.184	1.864	1080	8.69	598	19.3	7.56	0.841	73.7	174.0	290
295	0.02617	1.002	51.94	2449	4.181	1.868	959	8.89	606	19.5	6.62	0.849	72.7	227.5	295
300	0.03531	1.003	39.13	2438	4.179	1.872	855	9.09	613	19.6	5.83	0.857	71.7	276.1	300
305	0.04712	1.005	27.90	2426	4.178	1.877	769	9.29	620	20.1	5.20	0.865	70.9	320.6	305
310	0.06221	1.007	22.93	2414	4.178	1.882	695	9.49	628	20.4	4.62	0.873	70.0	361.9	310
315	0.08132	1.009	17.82	2402	4.179	1.888	631	9.69	634	20.7	4.16	0.883	69.2	400.4	315
320	0.1053	1.011	13.98	2390	4.180	1.895	577	9.89	640	21.0	3.77	0.894	68.3	436.7	320
325	0.1351	1.013	11.06	2378	4.182	1.903	528	10.09	645	21.3	3.42	0.901	67.5	471.2	325
330	0.1719	1.016	8.82	2366	4.184	1.911	489	10.29	650	21.7	3.15	0.908	66.6	504.0	330
335	0.2167	1.018	7.09	2354	4.186	1.920	453	10.49	656	22.0	2.88	0.916	65.8	535.5	335
340	0.2713	1.021	5.74	2342	4.188	1.930	420	10.69	660	22.3	2.66	0.925	64.9	566.0	340
345	0.3372	1.024	4.683	2329	4.191	1.941	389	10.89	668	22.6	2.45	0.933	64.1	595.4	345
350	0.4163	1.027	3.846	2317	4.195	1.954	365	11.09	668	23.0	2.29	0.942	63.2	624.2	350
355	0.5100	1.030	3.180	2304	4.199	1.968	343	11.29	671	23.3	2.14	0.951	62.3	652.3	355
360	0.6209	1.034	2.645	2291	4.203	1.983	324	11.49	674	23.7	2.02	0.960	61.4	697.9	360
365	0.7514	1.038	2.212	2278	4.209	1.999	306	11.69	677	24.1	1.91	0.969	60.5	707.1	365
370	0.9040	1.041	1.861	2265	4.214	2.017	289	11.89	679	24.5	1.80	0.978	59.5	728.7	370
373.15	1.0133	1.044	1.679	2257	4.217	2.029	279	12.02	680	24.8	1.76	0.984	58.9	750.1	373.15
375	1.0815	1.045	1.574	2252	4.220	2.036	274	12.09	681	24.9	1.70	0.987	58.6	761	375
380	1.2869	1.049	1.337	2239	4.226	2.057	260	12.29	683	25.4	1.61	0.999	57.6	788	380
385	1.5233	1.053	1.142	2225	4.232	2.080	248	12.49	685	25.8	1.53	1.004	56.6	814	385
390	1.794	1.058	0.980	2212	4.239	2.104	237	12.69	686	26.3	1.47	1.013	55.6	841	390
400	2.455	1.067	0.731	2183	4.256	2.158	217	13.05	688	27.2	1.34	1.033	53.6	896	400
410	3.302	1.077	0.553	2153	4.278	2.221	200	13.42	688	28.2	1.24	1.054	51.5	952	410
420	4.370	1.088	0.425	2123	4.302	2.291	185	13.79	688	29.8	1.16	1.075	49.4	1010	420
430	5.699	1.099	0.331	2091	4.331	2.369	173	14.14	685	30.4	1.09	1.10	47.2		430
440	7.333	1.110	0.261	2059	4.36	2.46	162	14.50	682	31.7	1.04	1.12	45.1		440

450	9.319	1.123	0.208	2024	4.40	2.56	152	14.85	678	33.1	0.99	1.14	42.9		450
460	11.71	1.137	0.167	1989	4.44	2.68	143	15.19	673	34.6	0.95	1.17	40.7		460
470	14.55	1.152	0.136	1951	4.48	2.79	136	15.54	667	36.3	0.92	1.20	38.5		470
480	17.90	1.167	0.111	1912	4.53	2.94	129	15.88	660	38.1	0.89	1.23	36.2		480
490	21.83	1.184	0.0922	1870	4.59	3.10	124	16.23	651	40.1	0.87	1.25	33.9	—	490
500	26.40	1.203	0.0766	1825	4.66	3.27	118	16.59	642	42.3	0.86	1.28	31.6	—	500
510	31.66	1.222	0.0631	1779	4.74	3.47	113	16.95	631	44.7	0.85	1.31	29.3	—	510
520	37.70	1.244	0.0525	1730	4.84	3.70	108	17.33	621	47.5	0.84	1.35	26.9	—	520
530	44.58	1.268	0.0445	1679	4.95	3.96	104	17.72	608	50.6	0.85	1.39	24.5	—	530
540	52.38	1.294	0.0375	1622	5.08	4.27	101	18.1	594	54.0	0.86	1.43	22.1	—	540
550	61.19	1.323	0.0317	1564	5.24	4.64	97	18.6	580	58.3	0.87	1.47	19.7	—	550
560	71.08	1.355	0.0269	1499	5.43	5.09	94	19.1	563	63.7	0.90	1.52	17.3	—	560
570	82.16	1.392	0.0228	1429	5.68	5.67	91	19.7	548	76.7	0.94	1.59	15.0	—	570
580	94.51	1.433	0.0193	1353	6.00	6.40	88	20.4	528	76.7	0.99	1.68	12.8	—	580
590	108.3	1.482	0.0163	1274	6.41	7.35	84	21.5	513	84.1	1.05	1.84	10.5	—	590
600	123.5	1.541	0.0137	1176	7.00	8.75	81	22.7	497	92.9	1.14	2.15	8.4	—	600
610	137.3	1.612	0.0115	1068	7.85	11.1	77	24.1	467	103	1.30	2.60	6.3	—	610
620	159.1	1.705	0.0094	941	9.35	15.4	72	25.9	444	114	1.52	3.46	4.5	—	620
625	169.1	1.778	0.0085	858	10.6	18.3	70	27.0	430	121	1.65	4.20	3.5	—	625
630	179.7	1.856	0.0075	781	12.6	22.1	67	28.0	412	130	2.0	4.8	2.6	—	630
635	190.9	1.935	0.0066	683	16.4	27.6	64	30.0	392	141	2.7	6.0	1.5	—	635
640	202.7	2.075	0.0057	560	26	42	59	32.0	367	155	4.2	9.6	0.8	—	640
645	215.2	2.351	0.0045	361	90	—	54	37.0	331	178	12	26	0.1	—	645
647.3[c]	221.2	3.170	0.0032	0	∞	∞	45	45.0	238	238	∞	∞	0.0	—	647.3[c]

[b] 1 bar = 10^5 N/m^2.

[c] Critical temperature.

Source: F. P. Incropera and D. P. de Witt, *Fundamentals of Heat and Mass Transfer,* 2nd ed., Wiley, New York, 1985, pp. 774–775.

TABLE A.2. Thermophysical Properties of Saturated Fluids

Saturated Liquids

T (K)	ρ (kg/m^3)	c_p (kJ/kg · K)	$\mu \cdot 10^2$ (N · s/m^2)	$\nu \cdot 10^6$ (m^2/s)	$k \cdot 10^3$ (W/m · K)	$\alpha \cdot 10^7$ (m^2/s)	Pr	$\beta \cdot 10^3$ (K^{-1})
Engine Oil (unused)								
273	899.1	1.796	385	4,280	147	0.910	47,000	0.70
280	895.3	1.827	217	2,430	144	0.880	27,500	0.70
290	890.0	1.868	99.9	1,120	145	0.872	12,900	0.70
300	884.1	1.909	48.6	550	145	0.859	6,400	0.70
310	877.9	1.951	25.3	288	145	0.847	3,400	0.70
320	871.8	1.993	14.1	161	143	0.823	1,965	0.70
330	865.8	2.035	8.36	96.6	141	0.800	1,205	0.70
340	859.9	2.076	5.31	61.7	139	0.779	793	0.70
350	853.9	2.118	3.56	41.7	138	0.763	546	0.70
360	847.8	2.161	2.52	29.7	138	0.753	395	0.70
370	841.8	2.206	1.86	22.0	137	0.738	300	0.70
380	836.0	2.250	1.41	16.9	136	0.723	233	0.70
390	830.6	2.294	1.10	13.3	135	0.709	187	0.70
400	825.1	2.337	0.874	10.6	134	0.695	152	0.70
410	818.9	2.381	0.698	8.52	133	0.682	125	0.70
420	812.1	2.427	0.564	6.94	133	0.675	103	0.70
430	806.5	2.471	0.470	5.83	132	0.662	88	0.70
Ethylene Glycol [$C_2H_4(OH)_2$]								
273	1,130.8	2.294	6.51	57.6	242	0.933	617	0.65
280	1,125.8	2.323	4.20	37.3	244	0.933	400	0.65
290	1,118.8	2.368	2.47	22.1	248	0.936	236	0.65
300	1,111.4	2.415	1.57	14.1	252	0.939	151	0.65
310	1,103.7	2.460	1.07	9.65	255	0.939	103	0.65
320	1,096.2	2.505	0.757	6.91	258	0.940	73.5	0.65
330	1,089.5	2.549	0.561	5.15	260	0.936	55.0	0.65
340	1,083.8	2.592	0.431	3.98	261	0.929	42.8	0.65
350	1,079.0	2.637	0.342	3.17	261	0.917	34.6	0.65
360	1,074.0	2.682	0.278	2.59	261	0.906	28.6	0.65
370	1,066.7	2.728	0.228	2.14	262	0.900	23.7	0.65
373	1,058.5	2.742	0.215	2.03	263	0.906	22.4	0.65
Glycerin [$C_3H_5(OH)_3$]								
273	1,276.0	2.261	1,060	8,310	282	0.977	85,000	0.47
280	1,271.9	2.298	534	4,200	284	0.972	43,200	0.47
290	1,265.8	2.367	185	1,460	286	0.955	15,300	0.48
300	1,259.9	2.427	79.9	634	286	0.935	6,780	0.48
310	1,253.9	2.490	35.2	281	286	0.916	3,060	0.49
320	1,247.2	2.564	21.0	168	287	0.897	1,870	0.50
Freon (refrigerant-12) (CCl_2F_2)								
230	1,528.4	0.8816	0.0457	0.299	68	0.505	5.9	1.85
240	1,498.0	0.8923	0.0385	0.257	69	0.516	5.0	1.90
250	1,469.5	0.9037	0.0354	0.241	70	0.527	4.6	2.00
260	1,439.0	0.9163	0.0322	0.224	73	0.554	4.0	2.10

TABLE A.2. Continued.

Saturated Liquids

T (K)	ρ (kg/m^3)	c_p (kJ/kg·K)	$\mu \cdot 10^2$ (N·s/m^2)	$\nu \cdot 10^6$ (m^2/s)	$k \cdot 10^3$ (W/m·K)	$\alpha \cdot 10^7$ (m^2/s)	Pr	$\beta \cdot 10^3$ (K^{-1})
Freon (refrigerant-12) (CCl_2F_2) Continued								
270	1,407.2	0.9301	0.0304	0.216	73	0.558	3.9	2.25
280	1,374.4	0.9450	0.0283	0.206	73	0.562	3.7	2.35
290	1,340.5	0.9609	0.0265	0.198	73	0.567	3.5	2.55
300	1,305.8	0.9781	0.0254	0.195	72	0.564	3.5	2.75
310	1,268.9	0.9963	0.0244	0.192	69	0.546	3.4	3.05
320	1,228.6	1.0155	0.0233	0.190	68	0.545	3.5	3.5
Mercury (Hg)								
273	13,595	0.1404	0.1688	0.1240	8,180	42.85	0.0290	0.181
300	13,529	0.1393	0.1523	0.1125	8,540	45.30	0.0248	0.181
350	13,407	0.1377	0.1309	0.0976	9,180	49.75	0.0196	0.181
400	13,287	0.1365	0.1171	0.0882	9,800	54.05	0.0163	0.181
450	13,167	0.1357	0.1075	0.0816	10,400	58.10	0.0140	0.181
500	13,048	0.1353	0.1007	0.0771	10,950	61.90	0.0125	0.182
550	12,929	0.1352	0.0953	0.0737	11,450	65.55	0.0112	0.184
600	12,809	0.1355	0.0911	0.0711	11,950	68.80	0.0103	0.187

Saturated Liquid–Vapor, 1 atm[b]

FLUID	T_{sat} (K)	h_{fg} (kJ/kg)	ρ_f (kg/m^3)	ρ_g (kg/m^3)	$\sigma \cdot 10^3$ (N/m)
Ethanol	351	846	757	1.44	17.7
Ethylene Glycol	470	812	1,111[c]	—	32.7
Glycerin	563	974	1,260[c]	—	63.0[c]
Mercury	630	301	12,740	3.90	417
Refrigerant R-12	243	165	1,488	6.32	15.8
Refrigerant R-113	321	147	1,511	7.38	15.9

[a]Adapted from references 15 and 16.

[b]Adapted from references 8, 17, and 18.

[c]Property value corresponding to 300 K.

Source: F. P. Incropera and D. P. de Witt, *Fundamentals of Heat and Mass Transfer*, 2nd ed., Wiley, New York, 1985, pp. 772–773.

TABLE A.3. Thermophysical Properties of Gases at Atmospheric Pressure

T (K)	ρ (kg/m^3)	c_p (kJ/kg · K)	$\mu \cdot 10^7$ (N · s/m^2)	$\nu \cdot 10^6$ (m^2/s)	$k \cdot 10^3$ (W/m · K)	$\alpha \cdot 10^6$ (m^2/s)	Pr
Air							
100	3.5562	1.032	71.1	2.00	9.34	2.54	0.786
150	2.3364	1.012	103.4	4.426	13.8	5.84	0.758
200	1.7458	1.007	132.5	7.590	18.1	10.3	0.737
250	1.3947	1.006	159.6	11.44	22.3	15.9	0.720
300	1.1614	1.007	184.6	15.89	26.3	22.5	0.707
350	0.9950	1.009	208.2	20.92	30.0	29.9	0.700
400	0.8711	1.014	230.1	26.41	33.8	38.3	0.690
450	0.7740	1.021	250.7	32.39	37.3	47.2	0.686
500	0.6964	1.030	270.1	38.79	40.7	56.7	0.684
550	0.6329	1.040	288.4	45.57	43.9	66.7	0.683
600	0.5804	1.051	305.8	52.69	46.9	76.9	0.685
650	0.5356	1.063	322.5	60.21	49.7	87.3	0.690
700	0.4975	1.075	338.8	68.10	52.4	98.0	0.695
750	0.4643	1.087	354.6	76.37	54.9	109	0.702
800	0.4354	1.099	369.8	84.93	57.3	120	0.709
850	0.4097	1.110	384.3	93.80	59.6	131	0.716
900	0.3868	1.121	398.1	102.9	62.0	143	0.720
950	0.3666	1.131	411.3	112.2	64.3	155	0.723
1000	0.3482	1.141	424.4	121.9	66.7	168	0.726
1100	0.3166	1.159	449.0	141.8	71.5	195	0.728
1200	0.2902	1.175	473.0	162.9	76.3	224	0.728
1300	0.2679	1.189	496.0	185.1	82	238	0.719
1400	0.2488	1.207	530	213	91	303	0.703
1500	0.2322	1.230	557	240	100	350	0.685
1600	0.2177	1.248	584	268	106	390	0.688
1700	0.2049	1.267	611	298	113	435	0.685
1800	0.1935	1.286	637	329	120	482	0.683
1900	0.1833	1.307	663	362	128	534	0.677
2000	0.1741	1.337	689	396	137	589	0.672
2100	0.1658	1.372	715	431	147	646	0.667
2200	0.1582	1.417	740	468	160	714	0.655
2300	0.1513	1.478	766	506	175	783	0.647
2400	0.1448	1.558	792	547	196	869	0.630
2500	0.1389	1.665	818	589	222	960	0.613
3000	0.1135	2.726	955	841	486	1570	0.536
Ammonia (NH_3)							
300	0.6894	2.158	101.5	14.7	24.7	16.6	0.887
320	0.6448	2.170	109	16.9	27.2	19.4	0.870
340	0.6059	2.192	116.5	19.2	29.3	22.1	0.872
360	0.5716	2.221	124	21.7	31.6	24.9	0.872
380	0.5410	2.254	131	24.2	34.0	27.9	0.869
400	0.5136	2.287	138	26.9	37.0	31.5	0.853
420	0.4888	2.322	145	29.7	40.4	35.6	0.833
440	0.4664	2.357	152.5	32.7	43.5	39.6	0.826

TABLE A.3. Continued.

T (K)	ρ (kg/m³)	c_p (kJ/kg · K)	$\mu \cdot 10^7$ (N · s/m²)	$\nu \cdot 10^6$ (m²/s)	$k \cdot 10^3$ (W/m · K)	$\alpha \cdot 10^6$ (m²/s)	Pr
Ammonia (NH_3) Continued							
460	0.4460	2.393	159	35.7	46.3	43.4	0.822
480	0.4273	2.430	166.5	39.0	49.2	47.4	0.822
500	0.4101	2.467	173	42.2	52.5	51.9	0.813
520	0.3942	2.504	180	45.7	54.5	55.2	0.827
540	0.3795	2.540	186.5	49.1	57.5	59.7	0.824
560	0.3708	2.577	193	52.0	60.6	63.4	0.827
580	0.3533	2.613	199.5	56.5	63.8	69.1	0.817
Carbon Dioxide (CO_2)							
280	1.9022	0.830	140	7.36	15.20	9.63	0.765
300	1.7730	0.851	149	8.40	16.55	11.0	0.766
320	1.6609	0.872	156	9.39	18.05	12.5	0.754
340	1.5618	0.891	165	10.6	19.70	14.2	0.746
360	1.4743	0.908	173	11.7	21.2	15.8	0.741
380	1.3961	0.926	181	13.0	22.75	17.6	0.737
400	1.3257	0.942	190	14.3	24.3	19.5	0.737
450	1.1782	0.981	210	17.8	28.3	24.5	0.728
500	1.0594	1.02	231	21.8	32.5	30.1	0.725
550	0.9625	1.05	251	26.1	36.6	36.2	0.721
600	0.8826	1.08	270	30.6	40.7	42.7	0.717
650	0.8143	1.10	288	35.4	44.5	49.7	0.712
700	0.7564	1.13	305	40.3	48.1	56.3	0.717
750	0.7057	1.15	321	45.5	51.7	63.7	0.714
800	0.6614	1.17	337	51.0	55.1	71.2	0.716
Carbon Monoxide (CO)							
200	1.6888	1.045	127	7.52	17.0	9.63	0.781
220	1.5341	1.044	137	8.93	19.0	11.9	0.753
240	1.4055	1.043	147	10.5	20.6	14.1	0.744
260	1.2967	1.043	157	12.1	22.1	16.3	0.741
280	1.2038	1.042	166	13.8	23.6	18.8	0.733
300	1.1233	1.043	175	15.6	25.0	21.3	0.730
320	1.0529	1.043	184	17.5	26.3	23.9	0.730
340	0.9909	1.044	193	19.5	27.8	26.9	0.725
360	0.9357	1.045	202	21.6	29.1	29.8	0.725
380	0.8864	1.047	210	23.7	30.5	32.9	0.729
400	0.8421	1.049	218	25.9	31.8	36.0	0.719
450	0.7483	1.055	237	31.7	35.0	44.3	0.714
500	0.67352	1.065	254	37.7	38.1	53.1	0.710
550	0.61226	1.076	271	44.3	41.1	62.4	0.710
600	0.56126	1.088	286	51.0	44.0	72.1	0.707
650	0.51806	1.101	301	58.1	47.0	82.4	0.705
700	0.48102	1.114	315	65.5	50.0	93.3	0.702
750	0.44899	1.127	329	73.3	52.8	104	0.702
800	0.42095	1.140	343	81.5	55.5	116	0.705

TABLE A.3. Continued.

T (K)	ρ (kg/m^3)	c_p (kJ/kg·K)	$\mu \cdot 10^7$ (N·s/m^2)	$\nu \cdot 10^6$ (m^2/s)	$k \cdot 10^3$ (W/m·K)	$\alpha \cdot 10^6$ (m^2/s)	Pr
Helium (He)							
100	0.4871	5.193	96.3	19.8	73.0	28.9	0.686
120	0.4060	5.193	107	26.4	81.9	38.8	0.679
140	0.3481	5.193	118	33.9	90.7	50.2	0.676
160	—	5.193	129	—	99.2	—	—
180	0.2708	5.193	139	51.3	107.2	76.2	0.673
200	—	5.193	150	—	115.1	—	—
220	0.2216	5.193	160	72.2	123.1	107	0.675
240	—	5.193	170	—	130	—	—
260	0.1875	5.193	180	96.0	137	141	0.682
280	—	5.193	190	—	145	—	—
300	0.1625	5.193	199	122	152	180	0.680
350	—	5.193	221	—	170	—	—
400	0.1219	5.193	243	199	187	295	0.675
450		5.193	263	—	204	—	—
500	0.09754	5.193	283	290	220	434	0.668
550	—	5.193	—	—	—	—	—
600	—	5.193	320	—	252	—	—
650	—	5.193	332	—	264	—	—
700	0.06969	5.193	350	502	278	768	0.654
750	—	5.193	364	—	291	—	—
800	—	5.193	382	—	304	—	—
900	—	5.193	414	—	330	—	—
1000	0.04879	5.193	446	914	354	1400	0.654
Hydrogen (H_2)							
100	0.24255	11.23	42.1	17.4	67.0	24.6	0.707
150	0.16156	12.60	56.0	34.7	101	49.6	0.699
200	0.12115	13.54	68.1	56.2	131	79.9	0.704
250	0.09693	14.06	78.9	81.4	157	115	0.707
300	0.08078	14.31	89.6	111	183	158	0.701
350	0.06924	14.43	98.8	143	204	204	0.700
400	0.06059	14.48	108.2	179	226	258	0.695
450	0.05386	14.50	117.2	218	247	316	0.689
500	0.04848	14.52	126.4	261	266	378	0.691
550	0.04407	14.53	134.3	305	285	445	0.685
600	0.04040	14.55	142.4	352	305	519	0.678
700	0.03463	14.61	157.8	456	342	676	0.675
800	0.03030	14.70	172.4	569	378	849	0.670
900	0.02694	14.83	186.5	692	412	1030	0.671
1000	0.02424	14.99	201.3	830	448	1230	0.673
1100	0.02204	15.17	213.0	966	488	1460	0.662
1200	0.02020	15.37	226.2	1120	528	1700	0.659
1300	0.01865	15.59	238.5	1279	568	1955	0.655
1400	0.01732	15.81	250.7	1447	610	2230	0.650
1500	0.01616	16.02	262.7	1626	655	2530	0.643

TABLE A.3. Continued.

T (K)	ρ (kg/m^3)	c_p (kJ/kg · K)	$\mu \cdot 10^7$ (N · s/m^2)	$\nu \cdot 10^6$ (m^2/s)	$k \cdot 10^3$ (W/m · K)	$\alpha \cdot 10^6$ (m^2/s)	Pr
Hydrogen (H_2) Continued							
1600	0.0152	16.28	273.7	1801	697	2815	0.639
1700	0.0143	16.58	284.9	1992	742	3130	0.637
1800	0.0135	16.96	296.1	2193	786	3435	0.639
1900	0.0128	17.49	307.2	2400	835	3730	0.643
2000	0.0121	18.25	318.2	2630	878	3975	0.661
Nitrogen (N_2)							
100	3.4388	1.070	68.8	2.00	9.58	2.60	0.768
150	2.2594	1.050	100.6	4.45	13.9	5.86	0.759
200	1.6883	1.043	129.2	7.65	18.3	10.4	0.736
250	1.3488	1.042	154.9	11.48	22.2	15.8	0.727
300	1.1233	1.041	178.2	15.86	25.9	22.1	0.716
350	0.9625	1.042	200.0	20.78	29.3	29.2	0.711
400	0.8425	1.045	220.4	26.16	32.7	37.1	0.704
450	0.7485	1.050	239.6	32.01	35.8	45.6	0.703
500	0.6739	1.056	257.7	38.24	38.9	54.7	0.700
550	0.6124	1.065	274.7	44.86	41.7	63.9	0.702
600	0.5615	1.075	290.8	51.79	44.6	73.9	0.701
700	0.4812	1.098	321.0	66.71	49.9	94.4	0.706
800	0.4211	1.22	349.1	82.90	54.8	116	0.715
900	0.3743	1.146	375.3	100.3	59.7	139	0.721
1000	0.3368	1.167	399.9	118.7	64.7	165	0.721
1100	0.3062	1.187	423.2	138.2	70.0	193	0.718
1200	0.2807	1.204	445.3	158.6	75.8	224	0.707
1300	0.2591	1.219	466.2	179.9	81.0	256	0.701
Oxygen (O_2)							
100	3.945	0.962	76.4	1.94	9.25	2.44	0.796
150	2.585	0.921	114.8	4.44	13.8	5.80	0.766
200	1.930	0.915	147.5	7.64	18.3	10.4	0.737
250	1.542	0.915	178.6	11.58	22.6	16.0	0.723
300	1.284	0.920	207.2	16.14	26.8	22.7	0.711
350	1.100	0.929	233.5	21.23	29.6	29.0	0.733
400	0.9620	0.942	258.2	26.84	33.0	36.4	0.737
450	0.8554	0.956	281.4	32.90	36.3	44.4	0.741
500	0.7698	0.972	303.3	39.40	41.2	55.1	0.716
550	0.6998	0.988	324.0	46.30	44.1	63.8	0.726
600	0.6414	1.003	343.7	53.59	47.3	73.5	0.729
700	0.5498	1.031	380.8	69.26	52.8	93.1	0.744
800	0.4810	1.054	415.2	86.32	58.9	116	0.743
900	0.4275	1.074	447.2	104.6	64.9	141	0.740
1000	0.3848	1.090	477.0	124.0	71.0	169	0.733
1100	0.3498	1.103	505.5	144.5	75.8	196	0.736
1200	0.3206	1.115	532.5	166.1	81.9	229	0.725
1300	0.2960	1.125	588.4	188.6	87.1	262	0.721

TABLE A.3. Continued.

T (K)	ρ (kg/m^3)	c_p (kJ/kg · K)	$\mu \cdot 10^7$ (N · s/m^2)	$\nu \cdot 10^6$ (m^2/s)	$k \cdot 10^3$ (W/m · K)	$\alpha \cdot 10^6$ (m^2/s)	Pr
Water Vapor (steam)							
380	0.5863	2.060	127.1	21.68	24.6	20.4	1.06
400	0.5542	2.014	134.4	24.25	26.1	23.4	1.04
450	0.4902	1.980	152.5	31.11	29.9	30.8	1.01
500	0.4405	1.985	170.4	38.68	33.9	38.8	0.998
550	0.4005	1.997	188.4	47.04	37.9	47.4	0.993
600	0.3652	2.026	206.7	56.60	42.2	57.0	0.993
650	0.3380	2.056	224.7	66.48	46.4	66.8	0.996
700	0.3140	2.085	242.6	77.26	50.5	77.1	1.00
750	0.2931	2.119	260.4	88.84	54.9	88.4	1.00
800	0.2739	2.152	278.6	101.7	59.2	100	1.01
850	0.2579	2.186	296.9	115.1	63.7	113	1.02

Source: F. P. Incropera and D. P. de Witt, *Fundamentals of Heat and Mass Transfer*, 2nd ed., Wiley, New York, 1985, pp. 767–771.

TABLE A.4. Thermophysical Properties of Liquid Metals

COMPOSITION	MELTING POINT (K)	T (K)	ρ (kg/m^3)	c_p (kJ/kg · K)	$\nu \cdot 10^7$ (m^2/s)	k (W/m · K)	$\alpha \cdot 10^5$ (m^2/s)	Pr
Bismuth	544	589	10,011	0.1444	1.617	16.4	0.138	0.0142
		811	9,739	0.1545	1.133	15.6	1.035	0.0110
		1033	9,467	0.1645	0.8343	15.6	1.001	0.0083
Lead	600	644	10,540	0.159	2.276	16.1	1.084	0.024
		755	10,412	0.155	1.849	15.6	1.223	0.017
		977	10,140	—	1.347	14.9	—	—
Potassium	337	422	807.3	0.80	4.608	45.0	6.99	0.0066
		700	741.7	0.75	2.397	39.5	7.07	0.0034
		977	674.4	0.75	1.905	33.1	6.55	0.0029
Sodium	371	366	929.1	1.38	7.516	86.2	6.71	0.011
		644	860.2	1.30	3.270	72.3	6.48	0.0051
		977	778.5	1.26	2.285	59.7	6.12	0.0037
NaK, (45%/55%)	292	366	887.4	1.130	6.522	25.6	2.552	0.026
		644	821.7	1.055	2.871	27.5	3.17	0.0091
		977	740.1	1.043	2.174	28.9	3.74	0.0058
NaK, (22%/78%)	262	366	849.0	0.946	5.797	24.4	3.05	0.019
		672	775.3	0.879	2.666	26.7	3.92	0.0068
		1033	690.4	0.883	2.118	—	—	—
PbBi, (44.5%/55.5%)	398	422	10,524	0.147	—	9.05	0.586	—
		644	10,236	0.147	1.496	11.86	0.790	0.189
		922	9,835	—	1.171	—	—	—

Source: F. P. Incropera and D. P. de Witt, *Fundamentals of Heat and Mass Transfer*, 2nd ed., Wiley, New York, 1985, p. 776. Adapted (in part) from *ASHRAE Handbook* — 1981 Fundamentals.

TABLE A.5. ASHRAE Standard Designation of Refrigerants (ANSI / ASHRAE Standard 34-1978)

Refrigerant Number	Chemical Name	Chemical Formula
Halocarbon Compounds		
10	Carbontetrachloride	CCl_4
11	Trichlorofluoromethane	CCl_3F
12	Dichlorodifluoromethane	CCl_2F_2
13	Chlorotrifluoromethane	$CClF_3$
13B1	Bromotrifluoromethane	$CBrF_3$
14	Carbontetrafluoride	CF_4
20	Chloroform	$CHCl_3$
21	Dichlorofluoromethane	$CHCl_2F$
22	Chlorodifluoromethane	$CHClF_2$
23	Trifluoromethane	CHF_3
30	Methylene Chloride	CH_2Cl_2
31	Chlorofluoromethane	CH_2ClF
32	Methylene Fluoride	CH_2F_2
40	Methyl Chloride	CH_3Cl
41	Methyl Fluoride	CH_3F
50[a]	Methane	CH_4
110	Hexachloroethane	CCl_3CCl_3
111	Pentachlorofluoroethane	CCl_3CCl_2F
112	Tetrachlorodifluoroethane	CCl_2FCCl_2F
112a	Tetrachlorodifluoroethane	CCl_3CClF_2
113	Trichlorotrifluoroethane	CCl_2FCClF_2
113a	Trichlorotrifluoroethane	CCl_3CF_3
114	Dichlorotetrafluoroethane	$CClF_2CClF_2$
114a	Dichlorotetrafluoroethane	CCl_2FCF_3
114B2	Dibromotetrafluoroethane	$CBrF_2CBrF_2$
115	Chloropentafluoroethane	$CClF_2CF_3$
116	Hexafluoroethane	CF_3CF_3
120	Pentachloroethane	$CHCl_2CCl_3$
123	Dichlorotrifluoroethane	$CHCl_2CF_3$
124	Chlorotetrafluoroethane	$CHClFCF_3$
124a	Chlorotetrafluoroethane	CHF_2CClF_2
125	Pentafluoroethane	CHF_2CF_3
133a	Chlorotrifluoroethane	CH_2ClCF_3
140a	Trichloroethane	CH_3CCl_3
142b	Chlorodifluoroethane	CH_3CClF_2
143a	Trifluoroethane	CH_3CF_3
150a	Dichloroethane	CH_3CHCl_2
152a	Difluoroethane	CH_3CHF_2
160	Ethyl Chloride	CH_3CH_2Cl
170[a]	Ethane	CH_3CH_3
218	Octafluoropropane	$CF_3CF_2CF_3$
290[a]	Propane	$CH_3CH_2CH_3$
Cyclic Organic Compounds		
C316	Dichlorohexafluorocyclobutane	$C_4Cl_2F_6$
C317	Chloroheptafluorocyclobutane	C_4ClF_7
C318	Octafluorocyclobutane	C_4F_8

Refrigerant Number	Chemical Name	Chemical Formula
Azeotropes		
500	Refrigerants 12/152a (73.8/26.2)	CCl_2F_2/CH_3CHF_2
501	Refrigerants 22/12 (75/25)	$CHClF_2/CCl_2F_2$
502	Refrigerants 22/115 (48.8/51.2)	$CHClF_2/CClF_2CF_3$
503	Refrigerants 23/13 (40.1/59.9)	$CHF_3/CClF_3$
504	Refrigerants 32/115 (48.2/51.8)	$CH_2F_2/CClF_2CF_3$
505	Refrigerants 12/31 (78.0/22.0)	CCl_2F_2/CH_2ClF
506	Refrigerants 31/114 (55.1/44.9)	$CH_2ClF/CClF_2CClF_2$
Miscellaneous Organic Compounds		
Hydrocarbons		
50	Methane	CH_4
170	Ethane	CH_3CH_3
290	Propane	$CH_3CH_2CH_3$
600	Butane	$CH_3CH_2CH_2CH_3$
600a	Isobutane (2 methyl propane)	$CH(CH_3)_3$
1150[b]	Ethylene	$CH_2{=}CH_2$
1270[b]	Propylene	$CH_3CH{=}CH_2$
Oxygen Compounds		
610	Ethyl Ether	$C_2H_5OC_2H_5$
611	Methy l Formate	$HCOOCH_3$
Nitrogen Compounds		
630	Methy l Amine	CH_3NH_2
631	Ethyl Amine	$C_2H_5NH_2$
Inorganic Compounds		
702	Hydrogen (Normal and Para)	H_2
704	Helium	He
717	Ammonia	NH_3
718	Water	H_2O
720	Neon	Ne
728	Nitrogen	N_2
729	Air	$.21O_2, .78N_2, .01A$
732	Oxygen	O_2
740	Argon	A
744	Carbon Dioxide	CO_2
744A	Nitrous Oxide	N_2O
764	Sulfur Dioxide	SO_2
Unsaturated Organic Compounds		
1112a	Dichlorodifluoroethylene	$CCl_2{=}CF_2$
1113	Chlorotrifluoroethylene	$CClF{=}CF_2$
1114	Tetrafluoroethylene	$CF_2{=}CF_2$
1120	Trichloroethylene	$CHCl{=}CCl_2$
1130	Dichloroethylene	$CHCl{=}CHCl$
1132a	Vinylidene Fluoride	$CH_2{=}CF_2$
1140	Vinyl Chloride	$CH_2{=}CHCl$
1141	Vinyl Fluoride	$CH_2{=}CHF$
1150	Ethylene	$CH_2{=}CH_2$
1270	Propylene	$CH_3CH{=}CH_2$

[a] Methane, ethane, and propane appear in the Halocarbon section in their proper numerical order, but these compounds are not halocarbons.

[b] Ethylene and propylene appear in the Hydrocarbon section to indicate that these compounds are hydrocarbons, but are properly identified in the section Unsaturated Organic Compounds.

Source: Reproduced with permission from ASHRAE.

TABLE A.6. Physical Properties of Refrigerants

Refrigerant		Chemical Formula	Molecular Mass	Boiling Point (NBP),[k] °C	Freezing Point, °C	Critical Temperature, °C	Critical Pressure, kPa	Critical Volume, L/kg	Surface Tension, mN/m[a]	Refractive Index of Liquid[a,b]
No.	Name									
704	Helium	He	4.0026	−268.9	None	−267.9	228.8	14.43	0.12(−269)	1.021(NBP)5461Å
702n	Hydrogen (normal)	H_2	2.0159	−252.8	−259.2	−239.9	1315	33.21	2.31(−255)	1.097(NBP)5791 Å
702p	Hydrogen (para)	H_2	2.0159	−252.9	−259.3	−240.2	1292	31.82	2.172	1.09(NBP)[j]
720	Neon	Ne	20.183	−246.1	−248.6	−228.7	3397	2.070	5.50(−248)	...
728	Nitrogen	N_2	28.013	−198.8	−210	−146.9	3396	3.179	8.27(−193)	1.205(83K)5893 Å
729	Air	...	28.97	−194.3	...	−140.7	3772	3.048	Plait Point	
						−140.6	3764	3.126	Point of Contact	
740	Argon	A	39.948	−185.9	−189.3	−122.3	4895	1.867	13.2(−188)	1.233(TP)5893 Å
732	Oxygen	O_2	31.9988	−182.9	−218.8	−118.4	5077	2.341	13.2(−183)	1.221(92K)5893 Å
50	Methane	CH_4	16.04	−161.5	−182.2	− 82.5	4638	6.181		
14	Tetrafluoromethane	CF_4	88.01	−127.9	−184.9	− 45.7	3741	1.598		
1150	Ethylene	C_2H_4	28.05	−103.7	−169	9.3	5114	4.37	16.5(−104)[3]	1.363(−100)[3]
503	e	...	87.5	− 88.7	...	19.5	4182	2.035		
170	Ethane	C_2H_6	30.07	− 88.8	−183	32.2	4891	5.182		
744A[4]	Nitrous Oxide	N_2O	44.02	− 89.5	−102	36.5	7221	2.216	1.75(20)	
23	Trifluoromethane	CHF_3	70.02	− 82.1	−155	25.6	4833	1.942	9.5(−40)[5]	
13	Chlorotrifluoromethane	$CClF_3$	104.47	− 81.4	−181	28.8	3865	1.729	8.5(−40)[5]	1.146(25)[9]
744	Carbon Dioxide	CO_2	44.01	− 78.4	− 56.6[d]	31.1	7372	2.135	1.16(20)	1.195(15)
13B1	Bromotrifluoromethane	$CBrF_3$	148.93	− 57.75	−168	67.0	3962	1.342	3.8(26.7)[5]	1.239(25)[9]
504	f	...	79.2	− 57.2	...	66.4	4758	2.023		
1270	Propylene	C_3H_6	42.09	− 47.7	−185	91.8	4618	4.495	16.7(20)[4]	1.3640(−50)[3]
502[6]	g	...	111.64	− 45.4	...	82.2	4075	1.785		
290	Propane	C_3H_8	44.10	− 42.07	−187.7	96.8	4254	4.545		1.3397(−42)
22	Chlorodifluoromethane	$CHClF_2$	86.48	− 40.76	−160	96.0	4974	1.904	8.0(26.7)[5]	1.234(25)[9]
115	Chloropentafluoroethane	$CClF_2CF_3$	154.48	− 39.1	−106	79.9	3153	1.629	5.1(26.7)[5]	1.221(25)[9]
717	Ammonia	NH_3	17.03	− 33.3	− 77.7	133.0	11417	4.245	23.4(11.1)	1.325(16.5)
500	i	...	99.31	− 33.5	−159	105.5	4423	2.016		
12	Dichlorodifluoromethane	CCl_2F_2	120.93	− 29.79	−158	112.0	4113	1.792	8.9(26.7)[5]	1.288(25)[9]
152a	Difluoroethane	CH_3CHF_2	66.05	− 25.0	−117	113.5	4492	2.741		
40[4]	Methyl Chloride	CH_3Cl	50.49	− 12.4	− 97.8	143.1	6674	2.834	16.2(20)	
600a	Isobutane	C_4H_{10}	58.13	− 11.73	−160	135.0	3645	4.526		1.3514(−25)[3]
764[7]	Sulfur Dioxide	SO_2	64.07	− 10.0	− 75.5	157.5	7875	1.910		
142b	Chlorodifluoroethane	CH_3CClF_2	100.5	− 9.8	−131	137.1	4120	2.297		
630[7]	Methyl Amine	CH_3NH_2	31.06	− 6.7	− 92.5	156.9	7455		23(−20)	1.432(17.5)
C318	Octafluorocyclobutane	C_4F_8	200.04	− 5.8	− 41.4	115.3	2781	1.611	7.3(26.7)[5]	
600	Butane	C_4H_{10}	58.13	− 0.5	−138.5	152.0	3794	4.383		1.3562(−15)[3]
114	Dichlorotetrafluoroethane	$CClF_2CClF_2$	170.94	3.8	− 94	145.7	3259	1.717	12(26.7)[5]	1.294(25)[9]
21[8]	Dichlorofluoromethane	$CHCl_2F$	102.93	8.9	−135	178.5	5168	1.917	18(26.7)[5]	1.332(25)[9]
160[4]	Ethyl Chloride	C_2H_5Cl	64.52	12.4	−138.3	187.2	5267	3.028	21.2(5)[4]	
631[7]	Ethyl Amine	$C_2H_5NH_2$	45.08	16.6	− 80.6	183.0	5619		20.4(9.6)	
11	Trichlorofluoromethane	CCl_3F	137.38	23.82	−111	198.0	4406	1.804	18(26.7)[5]	1.362(25)[9]
611[7]	Methyl Formate	$C_2H_4O_2$	60.05	31.8	− 99	214.0	5994	2.866	25.08(20)	
610[7]	Ethyl Ether	$C_4H_{10}O$	74.12	34.6	−116.3	194.0	3603	3.790	17.01(20)	1.3526(20)
216	Dichlorohexafluoropropane	$C_3Cl_2F_6$	220.93	35.69	−125.4	180.0	2753	1.742		
30[7]	Methylene Chloride	CH_2Cl_2	84.93	40.2	− 97	237.0	6077		26.5(20)	1.4244(20)[h]
113	Trichlorotrifluoroethane	CCl_2FCClF_2	187.39	47.57	− 35	214.1	3437	1.736	19(26.7)[5]	1.357(25)[9]
1130[2]	Dichloroethylene	CHCl=CHCl	96.95	47.8	− 50	243.3	5478			
1120[7]	Trichloroethylene	$CHCl=CCl_2$	131.39	87.2	− 73	271.1	5016		29(30)[h]	1.4782(20)[h]
718[7]	Water	H_2O	18.02	100	0	374.2	22103	3.128	71.97(25)	

•The source of data is from Ref. 1 unless otherwise noted.
[a] The temperature of measurement (Celsius) is shown in parentheses. The data are from Ref. 2 unless otherwise noted.
[b] For the sodium D line.
[c] Sublimes.
[d] At 527 kPa.
[e] Refrigerants 23 and 13 (40.1/59.9% by mass).
[f] Refrigerants 32 and 115 (48.2/51.8% by mass).
[g] Refrigerants 22 and 115 (48.8/51.2% by mass).
[h] Data from Electrochemicals Department, E. I. duPont de Nemours & Co.
[i] Refrigerants 12 and 152a (73.8/26.2% by mass).
[j] Dielectric constant data.
[k] At standard atmospheric pressure (101.325 kPa).

Source: Reproduced with permission from ASHRAE.

TABLE A.7. Pressure–Enthalpy Diagram for Refrigerant 22 (Chlorodifluoromethane): Properties of Saturated Liquid and Saturated Vapor

Temp C	Pressure MPA	Volume Vapor m³/kg	Density Liquid kg/m³	Enthalpy Liquid kJ/kg	Enthalpy Vapor kJ/kg	Entropy Liquid kJ/kg K	Entropy Vapor kJ/kg K
-90	0.004748	3.6939	1542.8	98.575	364.20	0.55032	2.0006
-85	0.007084	2.5394	1529.9	104.54	366.63	0.58246	1.9754
-80	0.010308	1.7883	1516.8	110.42	369.06	0.61326	1.9524
-75	0.014662	1.2870	1503.6	116.21	371.49	0.64284	1.9312
-70	0.020424	0.94477	1490.3	121.92	373.91	0.67132	1.9117
-65	0.027914	0.70609	1476.7	127.58	376.32	0.69879	1.8938
-60	0.037491	0.53641	1463.1	133.18	378.72	0.72535	1.8773
-55	0.049556	0.41362	1449.2	138.74	381.10	0.75109	1.8621
-50	0.064549	0.32330	1435.2	144.27	383.45	0.77610	1.8479
-45	0.082947	0.25586	1421.0	149.77	385.77	0.80044	1.8348
-40.82	0.101325	0.21223	1408.9	154.37	387.69	0.82034	1.8246
-40	0.10527	0.20480	1406.5	155.26	388.06	0.82419	1.8227
-38	0.11542	0.18790	1400.7	157.46	388.96	0.83354	1.8180
-36	0.12632	0.17268	1394.8	159.66	389.86	0.84281	1.8135
-34	0.13801	0.15894	1388.9	161.86	390.75	0.85200	1.8091
-32	0.15053	0.14651	1382.9	164.06	391.64	0.86113	1.8049
-30	0.16391	0.13524	1376.9	166.26	392.52	0.87018	1.8007
-28	0.17821	0.12502	1370.9	168.46	393.39	0.87917	1.7967
-26	0.19346	0.11573	1364.8	170.67	394.25	0.88810	1.7927
-24	0.20969	0.10726	1358.7	172.89	395.10	0.89697	1.7889
-22	0.22696	0.09954	1352.6	175.10	395.95	0.90579	1.7851
-20	0.24531	0.09249	1346.4	177.33	396.79	0.91455	1.7815
-18	0.26477	0.08603	1340.1	179.56	397.62	0.92327	1.7779
-16	0.28540	0.08012	1333.8	181.79	398.43	0.93194	1.7744
-14	0.30724	0.07470	1327.5	184.04	399.24	0.94057	1.7710
-12	0.33034	0.06971	1321.1	186.29	400.04	0.94916	1.7677
-10	0.35474	0.06513	1314.6	188.55	400.83	0.95771	1.7644
-8	0.38049	0.06090	1308.1	190.82	401.61	0.96623	1.7612
-6	0.40763	0.05701	1301.5	193.10	402.37	0.97471	1.7581
-4	0.43622	0.05341	1294.9	195.39	403.12	0.98317	1.7550
-2	0.46630	0.05008	1288.2	197.69	403.87	0.99160	1.7520
0	0.49792	0.04700	1281.5	200.00	404.59	1.0000	1.7490
2	0.53113	0.04415	1274.7	202.32	405.31	1.0084	1.7461
4	0.56599	0.04150	1267.8	204.66	406.01	1.0167	1.7432
6	0.60254	0.03904	1260.8	207.01	406.70	1.0251	1.7404
8	0.64083	0.03675	1253.8	209.37	407.37	1.0334	1.7376
10	0.68091	0.03462	1246.7	211.74	408.03	1.0417	1.7349
12	0.72285	0.03263	1239.5	214.13	408.67	1.0500	1.7322
14	0.76668	0.03078	1232.3	216.54	409.29	1.0583	1.7295
16	0.81246	0.02905	1224.9	218.96	409.90	1.0665	1.7269
18	0.86025	0.02743	1217.5	221.40	410.49	1.0748	1.7243
20	0.91009	0.02592	1210.0	223.85	411.06	1.0831	1.7217
22	0.96205	0.02451	1202.4	226.32	411.61	1.0913	1.7191
24	1.0162	0.02318	1194.6	228.80	412.14	1.0996	1.7165
26	1.0725	0.02193	1186.8	231.31	412.65	1.1078	1.7140
28	1.1312	0.02076	1178.9	233.83	413.13	1.1160	1.7114
30	1.1921	0.01967	1170.8	236.38	413.60	1.1243	1.7089
32	1.2555	0.01863	1162.6	238.94	414.03	1.1325	1.7063
34	1.3213	0.01766	1154.3	241.52	414.45	1.1408	1.7038
36	1.3896	0.01674	1145.9	244.13	414.83	1.1490	1.7012
38	1.4605	0.01588	1137.3	246.75	415.19	1.1573	1.6987
40	1.5340	0.01506	1128.6	249.40	415.52	1.1656	1.6961
42	1.6102	0.01429	1119.7	252.07	415.82	1.1739	1.6934
44	1.6892	0.01356	1110.6	254.77	416.08	1.1822	1.6908
46	1.7710	0.01287	1101.4	257.49	416.31	1.1905	1.6881
48	1.8556	0.01221	1091.9	260.24	416.50	1.1989	1.6854
50	1.9432	0.01159	1082.3	263.02	416.65	1.2072	1.6826
52	2.0339	0.01101	1072.4	265.83	416.75	1.2156	1.6798
54	2.1276	0.01045	1062.3	268.67	416.81	1.2241	1.6769
56	2.2244	0.009915	1051.9	271.55	416.83	1.2326	1.6739
58	2.3245	0.009409	1041.3	274.46	416.79	1.2411	1.6709
60	2.4279	0.008927	1030.3	277.41	416.69	1.2497	1.6677
65	2.7015	0.007816	1001.3	284.98	416.16	1.2714	1.6594
70	2.9975	0.006819	969.68	292.88	415.14	1.2937	1.6500
75	3.3175	0.005917	934.38	301.22	413.46	1.3169	1.6393
80	3.6633	0.005086	893.89	310 18	410.88	1.3414	1.6265
85	4.0370	0.004301	845.17	320.13	406.90	1.3681	1.6104
90	4.4413	0.003517	780.60	331.96	400.28	1.3996	1.5877
95	4.8808	0.002547	660.94	350.67	384.95	1.4490	1.5421
*96.15	4.988	0.00195	513.	368.1	368.1	1.496	1.496

*Critical Point

Temp K	Viscosity, μPa · s			Thermal Conductivity, mW/m· K			Specific Heats, kJ/kg · K						Velocity of Sound, m/s		
	Sat. Liquid	Sat. Vapor	Gas (1 Atm.)	Sat. Liquid	Sat. Vapor	Gas (1 Atm.)	Sat. Liquid C_p	Sat. Liquid C_v	Sat. Vapor C_p	Sat. Vapor C_v	Gas (0 Atm.) C_p	Gas (0 Atm.) C_v	Sat. Liquid	Sat. Vapor	Gas (1 Atm.)
170	770		—	151.2		—	1.06		0.49		0.491				—
180	647		—	146.3		—	1.06		0.51		0.503				
190	554		—	141.4		—	1.07		0.52		0.515				
200	481	8.68	—	136.5	4.7	—	1.07		0.53		0.527				
210	424	9.12	—	131.6	5.4	—	1.08		0.55		0.539				—
220	378	9.56	—	126.6	6.0	—	1.08		0.58		0.551				
230	340	10.00	—	121.7	6.7	—	1.10		0.60		0.563				—
234.41[a]	325	10.19	10.19	119.5	7.0	7.0	1.10		0.61		0.568				—
240	309	10.43	10.43	116.7	7.4	7.39	1.11		0.63		0.575				
250	282.4	10.87	10.86	111.7	8.1	7.99	1.12		0.65		0.587				
260	260.2	11.32	11.29	106.8	8.7	8.59	1.14		0.67		0.599				
270	241.2	11.80	11.72	101.8	9.3	9.19	1.16		0.70		0.611				
280	224.8	12.33	12.14	96.8	9.9	9.79	1.19		0.74		0.623				
290	210.5	12.90	12.57	91.8	10.5	10.39	1.22		0.79		0.635				
300	198.0	13.50	12.99	86.8	11.1	10.98	1.26		0.85		0.647				
310	187.0	14.15	13.41	81.8	11.7	11.58	1.31		0.94		0.659				
320	177.2	14.85	13.82	76.9	12.3	12.18	1.37		1.04		0.671				
330	167.0	15.60	14.24	72.1	13.1	12.78	1.46		1.17		0.682				
340	150.0	16.40	14.65	66.8	14.1	13.38	1.57		1.32		0.693				
350	132.0	17.7	15.06	59.9	16.2	13.99	1.71		1.49		0.704				
360	105.0	19.9	15.47	50.9	20.1	14.59	1.9		1.75		0.715				
369.16[b]	30.5	30.5	15.87	∞	∞	15.14	∞		∞		0.724		0	0	
370	—	—	15.88	—	—	15.19	—		—		0.726		—	—	
380	—	—	16.28	—	—	15.79	—		—		0.737		—	—	
390	—	—	16.73	—	—	16.39	—		—		0.747		—	—	
400	—	—	17.08	—	—	16.98	—		—		0.757		—	—	
410	—	—	17.48	—	—	17.59	—		—		0.767		—	—	
420	—	—	17.87	—	—	18.20	—		—		0.777		—	—	
430	—	—	18.26	—	—	18.80	—		—		0.787		—	—	
440	—	—	18.65	—	—	19.40	—		—		0.796		—	—	
450	—	—	19.04	—	—	20.00	—		—		0.805		—	—	
460	—	—	19.42	—	—	20.60	—		—		0.814		—	—	
470	—	—	19.80	—	—	20.20	—		—		0.823		—	—	
480	—	—	20.18	—	—	21.80	—		—		0.832		—	—	
490	—	—	20.56	—	—	22.40	—		—		0.840		—	—	
500	—	—	20.93	—	—	23.00	—		—		0.848		—	—	

[a]Normal boiling point. [b]Critical point. [c]Very large. [d]Large. [e]Small.

Source: *ASHRAE Handbook, Fundamentals*, 1985. Reproduced with permission of the American Society of Heating, Refrigerating, and Air-Conditioning Engineers.

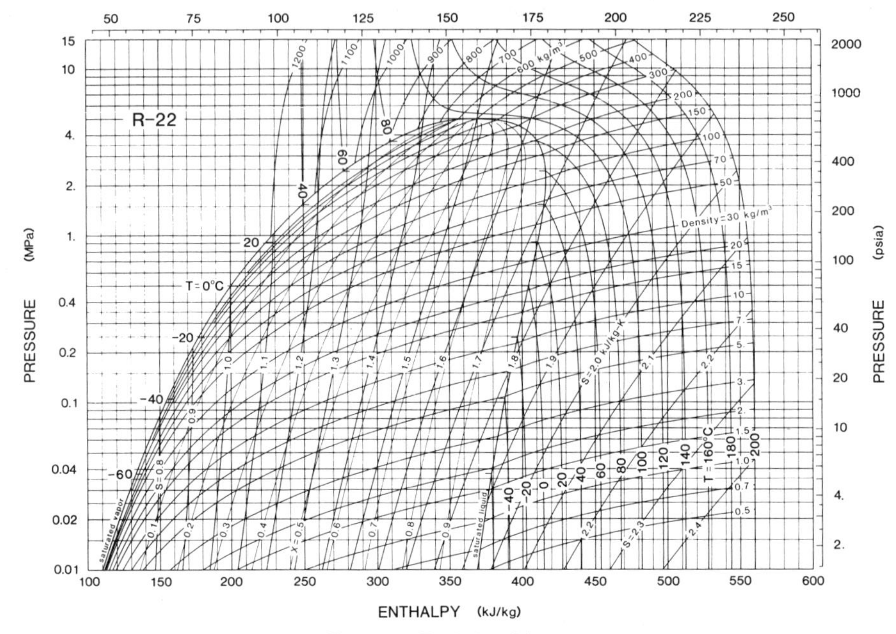

Pressure-Enthalpy Diagram

TABLE A.8. Thermophysical Properties of Selected Metallic Solids

		Properties at 300 K				Properties at various temperatures, K							
						k, W/m·K				c_p, J/kg·K			
Composition	Melting point, K	ρ, kg/m³	c_p, J/kg·K	k, W/m·K	$\alpha \times 10^6$, m²/s	100	200	400	600	100	200	400	600
Aluminum													
Pure	933	2,702	903	237	97.1	302	237	240	231	482	796	949	1,033
Alloy 2024–T6 (4.5% Cu, 1.5% Mg, 0.6% Mn)	775	2,770	875	177	73.0	65	163	186	186	473	787	925	1,042
Alloy 195, cast (4.5% Cu)	—	2,790	883	168	68.2	—	—	174	185	—	—	—	—
Chromium	2,118	7,160	449	93.7	29.1	159	111	90.9	80.7	192	384	484	542
Copper													
Pure	1,358	8,933	385	401	117	482	413	393	379	252	356	397	417
Commercial bronze (90% Cu, 10% Al)	1,293	8,800	420	52	14	—	42	52	59	—	785	460	545
Phosphor gear bronze (89% Cu, 11% Sn)	1,104	8,780	355	54	17	—	41	65	74	—	—	—	—
Cartridge brass (70% Cu, 30% Zn)	1,188	8,530	380	110	33.9	75	95	137	149	—	360	395	425
Constantan (55% Cu, 45% Ni)	1,493	8,920	384	23	6.71	17	19	—	—	237	362	—	—
Iron													
Pure	1,810	7,870	447	80.2	23.1	134	94.0	69.5	54.7	216	384	490	574
Armco (99.75% pure)	—	7,870	447	72.7	20.7	95.6	80.6	65.7	53.1	215	384	490	574

TABLE A.8. Continued.

Composition	Melting point, K	Properties at 300 K				Properties at various temperatures, K							
		ρ, kg/m^3	c_p, J/kg·K	k, W/m·K	$\alpha \times 10^6$, m^2/s	k, W/m·K				c_p, J/kg·K			
						100	200	400	600	100	200	400	600
Carbon steels													
Plain carbon (Mn ≤ 1%, Si ≤ 0.1%)	—	7,854	434	60.5	17.7	—	—	56.7	48.0	—	—	487	559
A1S1 1010	—	7,832	434	63.9	18.8	—	—	58.7	48.8	—	—	487	559
Carbon–silicon (Mn ≤ 1%, 0.1% < Si ≤ 0.6%)	—	7,817	446	51.9	14.9	—	—	49.8	44.0	—	—	501	582
Carbon—manganese—silicon (1% < Mn ≤ 1.65%, 0.1% < Si ≤ 0.6%)	—	8,131	434	41.0	11.6	—	—	42.2	39.7	—	—	487	559
Chromium (low) steels													
½ Cr–¼ Mo–Si (0.18% C, 0.65% Cr), 0.23% Mo, 0.6% Si)	—	7,822	444	37.7	10.9	—	—	38.2	36.7	—	—	492	575

1 Cr–½ Mo (0.16% C, 1% Cr, 0.54% Mo,0.39% Si)	—	7,858	442	42.3	12.2	—	—	42.0	39.1	—	—	492	575
1 Cr–V (0.2% C, 1.02% Cr, 0.15% V)	—	7,836	443	48.9	14.1	—	—	46.8	42.1	—	—	492	575
Stainless steels													
A1S1 302	—	8,055	480	15.1	3.91	—	—	17.3	20.0	—	—	512	559
A1S1 304	1,670	7,900	477	14.9	3.95	9.2	12.6	16.6	19.8	272	402	515	557
A1S1 316	—	8,238	468	13.4	3.48	—	—	15.2	18.3	—	—	504	550
A1S1 347	—	7,978	480	14.2	3.71	—	—	15.8	18.9	—	—	513	559
Lead	601	11,340	129	35.3	24.1	39.7	36.7	34.0	31.4	118	125	132	142
Magnesium	923	1,740	1,024	156	87.6	169	159	153	149	649	934	1,074	1,170
Molybedenum	2,894	10,240	251	138	53.7	179	143	134	126	141	224	261	275
Nickel													
Pure	1,728	8,900	444	90.7	23.0	164	107	80.2	65.6	232	383	485	592
Nichrome (80% Ni, 20% Cr)	1,672	8,400	420	12	3.4	—	—	14	16	—	—	480	525
Inconel X–750 (73% Ni, 15% Cr, 6.7% Fe)	1,665	8,510	439	11.7	3.1	8.7	10.3	13.5	17.0	—	372	473	510

Source: Reproduced from Frank W. Schmidt et al., *Introduction to Thermal Sciences*, Wiley, New York, 1984, pp. 427–429.

TABLE A.9. Normal Total Emissivity

Substance	Metals Surface temperature, K	ε_η
Cast iron, freshly turned	310	0.44
Iron plate, rusted	293	0.61
Cast, iron, rough and strongly oxidized	310–530	0.95
Platinum, polished	500–900	0.054–0.104
Silver, polished	310–810	0.01–0.03
Stainless steel		
Type 310, smooth	1090	0.39
Type 316, polished	480–1310	0.24–0.31
Tins, polished	310	0.05
Tungsten, filament	3590	0.39
	Nonmetals	
Asbestos		
Paper	310	0.93
Board	310	0.96
Brick		
White refractory	1370	0.29
Red, rough	310	0.93
Carbon, lampsoot	310	0.95
Concrete, rough	310	0.94
Ice, smooth	273	0.966
Marble, white	310	0.95
Paint		
Oil, all colors	373	0.92–0.96
Lead, red	370	0.93
Plaster	310	0.91
Rubber, hard	293	0.92
Snow	270	0.82
Water, deep	273–373	0.96
Wood		
Oak	295	0.90
Beech	340	0.94

Source: Frank W. Schmidt et al., *Introduction to Thermal Sciences*, Wiley, New York, 1984, pp. 372–373.

APPENDIX B

HEAT EXCHANGERS[1]

11.1 HEAT EXCHANGER TYPES

Heat exchangers are typically classified according to *flow arrangement* and *type of construction*. The simplest heat exchanger is one for which the hot and cold fluids move in the same or opposite directions in a *concentric tube* (or *double-pipe*) construction. In the *parallel-flow* arrangement of Figure 11.1*a*, the hot and cold fluids enter at the same end, flow in the same direction, and leave at the same end. In the *counterflow* arrangement of Figure 11.1*b*, the fluids enter at opposite ends, flow in opposite directions, and leave at opposite ends.

Alternatively, the fluids may move in *cross flow* (perpendicular to each other), as shown by the *finned* and *unfinned* tubular heat exchangers of Figure 11.2. The two configurations differ according to whether the fluid moving over the tubes is *unmixed* or *mixed*. In Figure 11.2*a*, the fluid is said to be unmixed because the fins prevent motion in a direction (y) that is transverse to the mainflow direction (x). In this case the fluid temperature varies with x and y. In contrast, for the unfinned tube bundle of Figure 11.2*b*, fluid motion, hence mixing, in the transverse direction is possible, and temperature variations are primarily in the main-flow direction. Since the tube flow is unmixed, both fluids are unmixed in the finned exchanger, while one fluid is mixed and the other unmixed in the unfinned exchanger. The nature of the mixing condition can significantly influence heat exchanger performance.

Another common configuration is the *shell-and-tube* heat exchanger. Specific forms differ according to the number of shell and tube passes, and the simplest

[1] Excerpts presented in this appendix are from F. P. Incropera and D. P. de Witt, *Fundamentals of Heat and Mass Transfer*, Chap. 11, Wiley, New York, 1985. Section and figure numbers of the original article have been retained.

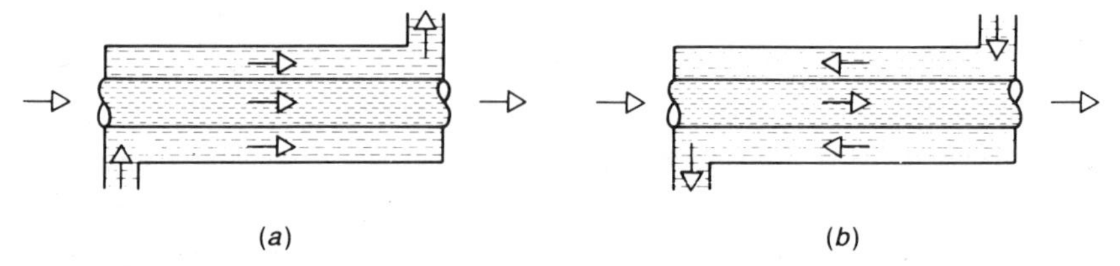

FIGURE 11.1. Concentric tube heat exchangers. (*a*) Parallel flow. (*b*) Counterflow.

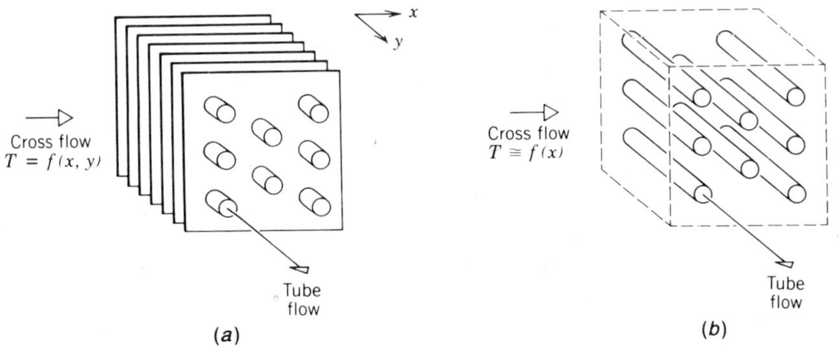

FIGURE 11.2. Cross-flow heat exchangers. (*a*) Finned with both fluids unmixed. (*b*) Unfinned with one fluid mixed and the other unmixed.

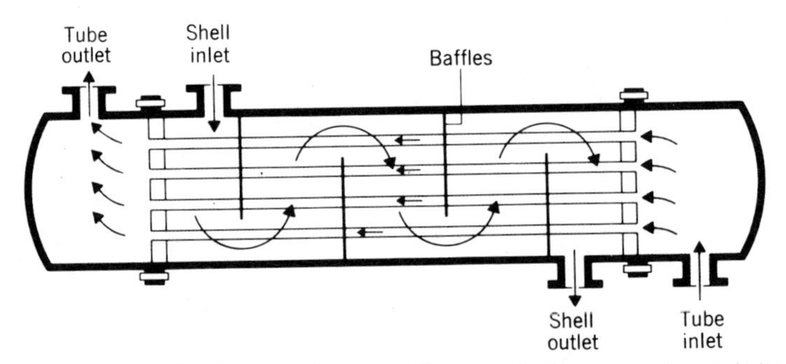

FIGURE 11.3. Shell-and-tube heat exchanger with one shell pass and one tube pass (cross-counterflow mode of operation).

form, which involves single tube and shell *passes*, is shown in Figure 11.3. Baffles are usually installed to increase the convection coefficient of the shell-side fluid by inducing turbulence and a cross-flow velocity component. Baffled heat exchangers with one shell pass and two tube passes and with two shell passes and four tube passes are shown in Figures 11.4*a* and 11.4*b*, respectively.

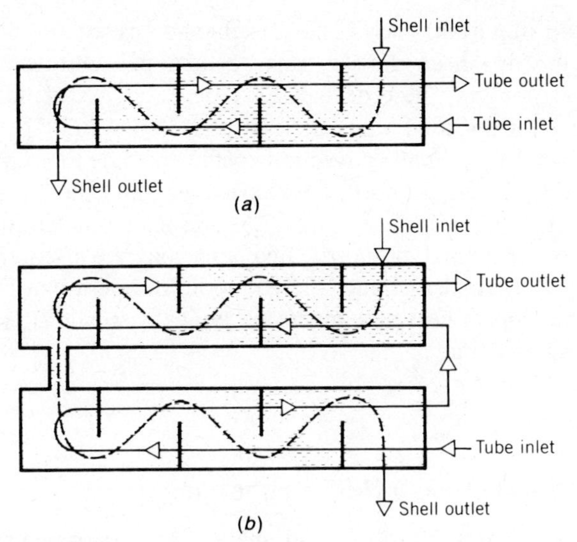

FIGURE 11.4. Shell-and-tube heat exchangers. (*a*) One shell pass and two tube passes. (*b*) Two shell passes and four tube passes.

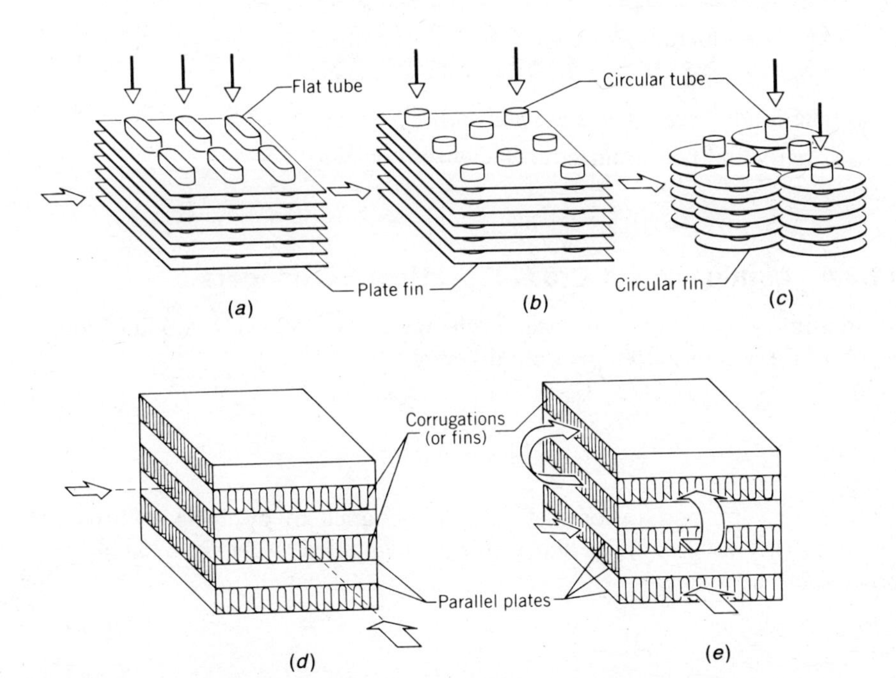

FIGURE 11.5. Compact heat exchanger cores. (*a*) Fin–tube (flat tubes, continuous plate fins). (*b*) Fin–tube (circular tubes, continuous plate fins). (*c*) Fin–tube (circular tubes, circular fins). (*d*) Plate–fin (single pass). (*e*) Plate–fin (multipass).

A special and important class of heat exchangers is used to achieve a very large ($\gtrsim 700\ \mathrm{m^2/m^3}$) heat transfer surface area per unit volume. Termed *compact heat exchangers*, these devices have dense arrays of finned tubes or plates and are typically used when at least one of the fluids is a gas, and is hence characterized by a small convection coefficient. The tubes may be *flat* or *circular*, as in Figures 11.5a and 11.5b, respectively, and the fins may be *plate* or *circular*, as in Figures 11.5a, b and 11.5c, respectively. Parallel plate heat exchangers may be finned or corrugated and may be used in single-pass (Figure 11.5d) or multipass (Figure 11.5e) modes of operation. Flow passages associated with compact heat exchangers are typically small ($D_h \lesssim 5$ mm), and the flow is usually laminar.

11.3.1 The Parallel-Flow Heat Exchanger

The energy balances and the subsequent analysis are performed subject to the following assumptions.

1. The heat exchanger is insulated from its surroundings, in which case the only heat exchange is between the hot and cold fluids.
2. Axial conduction along the tubes is negligible.
3. Potential and kinetic energy changes are negligible.
4. The fluid specific heats are constant.
5. The overall heat transfer coefficient is constant.

11.3.4 Multipass and Cross-Flow Heat Exchangers

In multipass and cross-flow heat exchangers, the following modification is made to the log mean temperature difference.

$$\Delta T_{\mathrm{lm}} = F \Delta T_{\mathrm{lm,\,CF}} \tag{11.18}$$

That is, the appropriate form of ΔT_{lm} is obtained by applying a correction factor to the value of ΔT_{lm} that would be computed *under the assumption of counterflow conditions*.

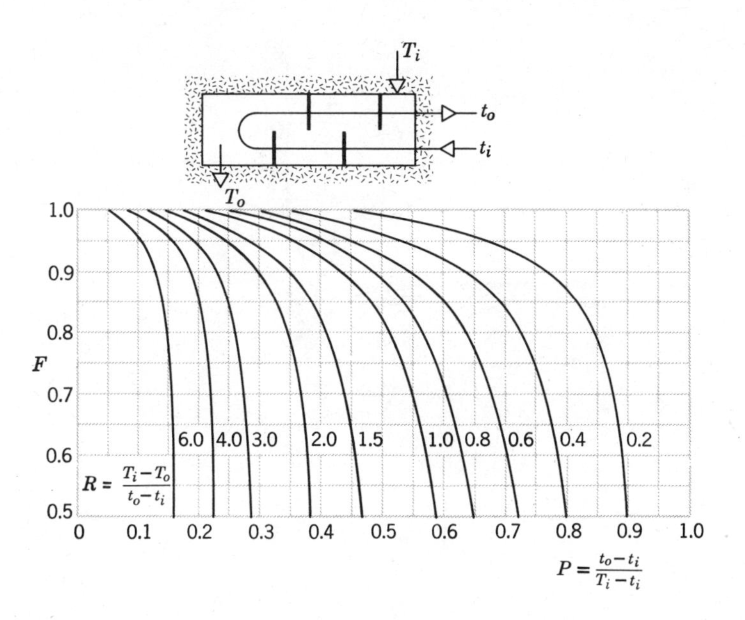

FIGURE 11.10. Correction factor for a shell-and-tube heat exchanger with one shell and any multiple of two tube passes (two, four, etc. tube passes).

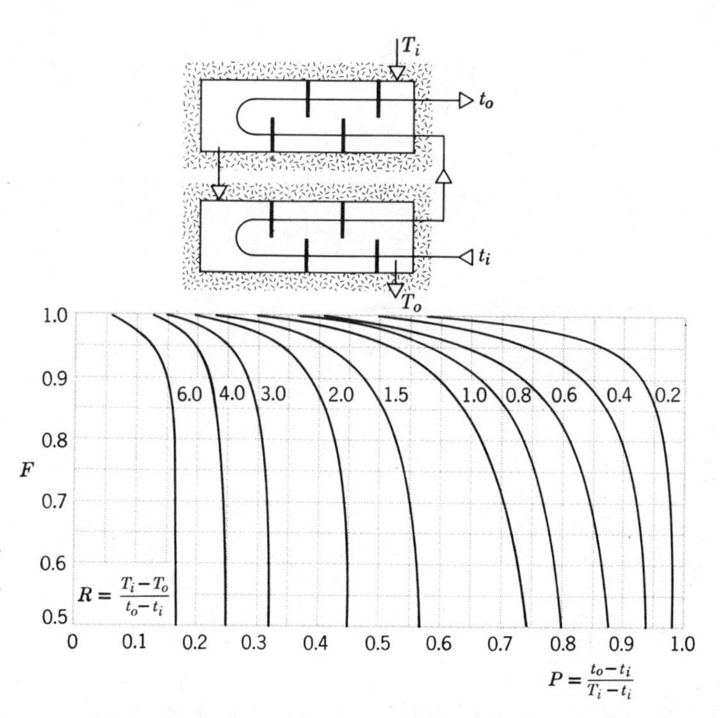

FIGURE 11.11. Correction factor for a shell-and-tube heat exchanger with two shell passes and any multiple of four tube passes (four, eight, etc. tube passes).

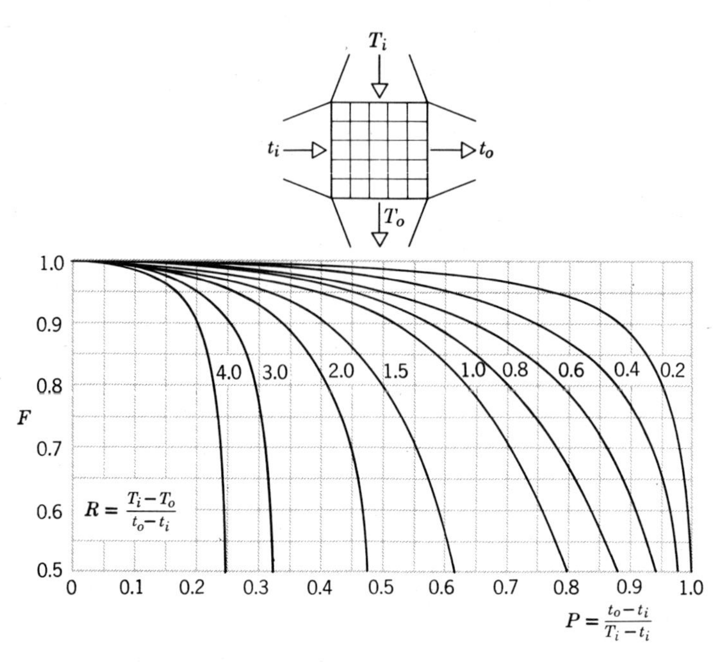

FIGURE 11.12. Correction factor for a single-pass, cross-flow heat exchanger with both fluids unmixed.

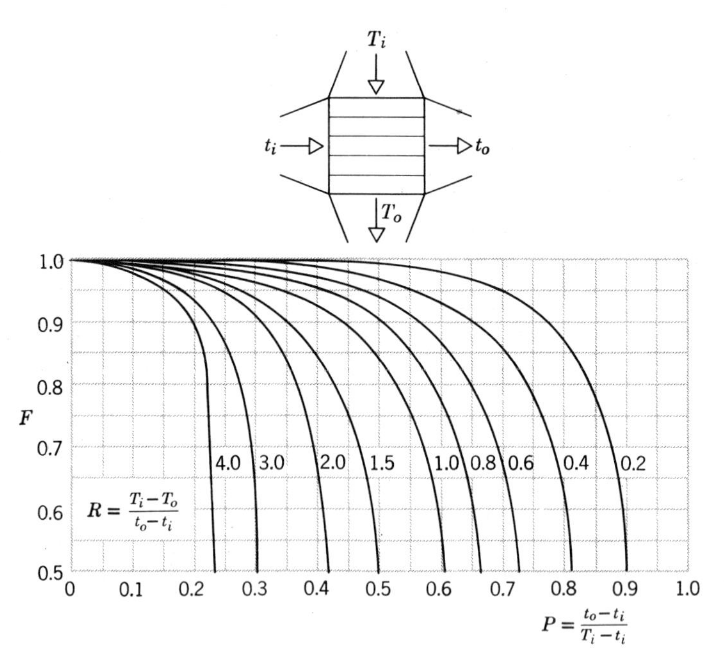

FIGURE 11.13. Correction factor for a single-pass, cross-flow heat exchanger with one fluid mixed and the other unmixed.

11.4 HEAT EXCHANGER ANALYSIS: THE EFFECTIVENESS-NTU METHOD

Representative results are summarized in Table 11.3, where C_r, is the *heat capacity ratio* $C_r \equiv C_{min}/C_{max}$. In deriving Equation 11.32a, it is assumed that the total NTU is equally distributed between shell passes of the same arrangement, $\text{NTU} = n(\text{NTU})_1$. Hence, when using ε_1 with this expression, NTU is replaced by $(\text{NTU})/n$ in Equation 11.31a. Note that for $C_r = 0$, as in a boiler or condenser ε is given by Equation 11.36a for *all flow arrangements*. Hence, for that special case, it again follows that heat exchanger behavior is independent of flow arrangement. For the cross-flow heat exchanger with both fluids unmixed, Equation 11.33 is exact only for $C_r = 1$. However, it may be used to an excellent approximation for all $0 < C_r \leq 1$. For $C_r = 0$, Equation 11.36a must be used.

In heat exchanger design calculations, it is more convenient to work with ε–NTU relations of the form

$$\text{NTU} = f\left(\varepsilon, \frac{C_{min}}{C_{max}}\right)$$

Explicit relations for NTU as a function of ε and C_r are provided in Table 11.4. Note that Equation 11.33 may not be manipulated to yield a direct relationship for NTU as a function of ε and C_r. Note also, that in using

TABLE 11.3. Heat Exchanger Effectiveness Relations

Flow Arrangement	Relation	
Concentric tube		
Parallel flow	$\varepsilon = \dfrac{1 - \exp[-\text{NTU}(1 + C_r)]}{1 + C_r}$	(11.29a)
Counterflow	$\varepsilon = \dfrac{1 - \exp[-\text{NTU}(1 - C_r)]}{1 - C_r\exp[-\text{NTU}(1 - C_r)]}$	(11.30a)
Shell and tube		
One shell pass (2, 4, ... tube passes)	$\varepsilon_1 = 2\left\{1 + C_r + (1 + C_r^2)^{1/2}\dfrac{1 + \exp[-\text{NTU}(1 + C_r^2)^{1/2}]}{1 - \exp[-\text{NTU}(1 + C_r^2)^{1/2}]}\right\}^{-1}$	(11.31a)
n Shell passes ($2n, 4n, \ldots$ tube passes)	$\varepsilon = \left[\left(\dfrac{1 - \varepsilon_1 C_r}{1 - \varepsilon_1}\right)^n - 1\right]\left[\left(\dfrac{1 - \varepsilon_1 C_r}{1 - \varepsilon_1}\right)^n - C_r\right]^{-1}$	(11.32a)
Cross flow (single pass)		
Both fluids unmixed	$\varepsilon = 1 - \exp[(1/C_r)(\text{NTU})^{0.22}\{\exp[-C_r(\text{NTU})^{0.78}] - 1\}]$	(11.33)
C_{max} (mixed), C_{min} (unmixed)	$\varepsilon = (1/C_r)(1 - \exp\{-C_r[1 - \exp(-\text{NTU})]\})$	(11.34a)
C_{min} (mixed), C_{max} (unmixed)	$\varepsilon = 1 - \exp(-C_r^{-1}\{1 - \exp[-C_r(\text{NTU})]\})$	(11.35a)
All exchangers ($C_r = 0$)	$\varepsilon = 1 - \exp(-\text{NTU})$	(11.36a)

TABLE 11.4. Heat Exchanger NTU Relations

Flow Arrangement	Relation	
Concentric tube		
Parallel flow	$\mathrm{NTU} = -\dfrac{\ln[1-\varepsilon(1+C_r)]}{1+C_r}$	(11.29b)
Counterflow	$\mathrm{NTU} = -\dfrac{1}{C_r-1}\ln\left(\dfrac{\varepsilon-1}{\varepsilon C_r-1}\right)$	(11.30b)
Shell and tube		
One shell pass	$\mathrm{NTU} = -(1+C_r^2)^{-1/2}\ln\left(\dfrac{E-1}{E+1}\right)$	(11.31b)
(2, 4, ... tube passes)	$E = \dfrac{2/\varepsilon_1-(1+C_r)}{(1+C_r^2)^{1/2}}$	(11.31c)
n Shell passes	Use Equations 11.31b and 11.31c with	
($2n, 4n, \ldots$ tube passes)	$\varepsilon_1 = \dfrac{F-1}{F-C_r}, \quad F = \left(\dfrac{\varepsilon C_r-1}{\varepsilon-1}\right)^{1/n}$	(11.32b, c)
Cross flow (single pass)		
C_{max} (mixed), C_{min} (unmixed)	$\mathrm{NTU} = -\ln[1+(1/C_r)\ln(1-\varepsilon C_r)]$	(11.34b)
C_{min} (mixed), C_{max} (unmixed)	$\mathrm{NTU} = -(1/C_r)\ln[C_r\ln(1-\varepsilon)+1]$	(11.35b)
All exchangers ($C_r = 0$)	$\mathrm{NTU} = -\ln(1-\varepsilon)$	(11.36b)

Equations 11.32b, c with 11.31b, c, it is the NTU *per shell pass* that is computed from Equation 11.31b. This result is multiplied by n to obtain the NTU for the entire exchanger.

The foregoing expressions are represented graphically in Figures 11.14 to 11.19. For Figure 11.19 the solid curves correspond to C_{min} mixed and C_{max} unmixed, while the dashed curves correspond to C_{min} unmixed and C_{max} mixed. Note that for $C_r = 0$, all heat exchangers have the same effectiveness, which may be computed from Equation 11.36a. Moreover, if NTU $\lesssim 0.25$, all heat exchangers have the same effectiveness regardless of the value of C_r, and ε may again be computed from Equation 11.36a. More generally, for $C_r > 0$ and NTU $\gtrsim 0.25$, the counterflow exchanger is the most effective. For any exchanger, maximum and minimum values of the effectiveness are associated with $C_r = 0$ and $C_r = 1$, respectively.

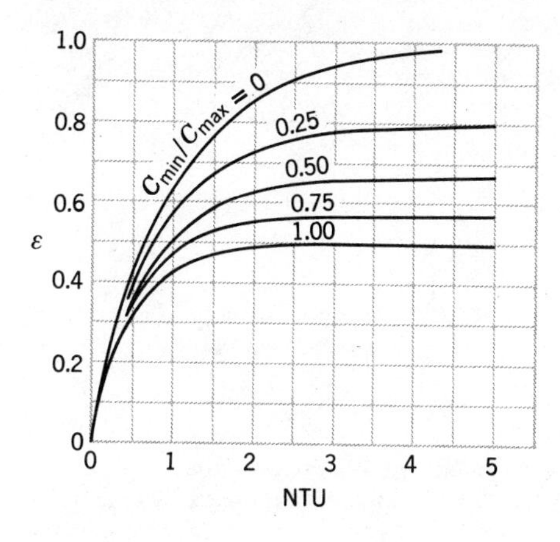

FIGURE 11.14. Effectiveness of a parallel-flow heat exchanger (Equation 11.29).

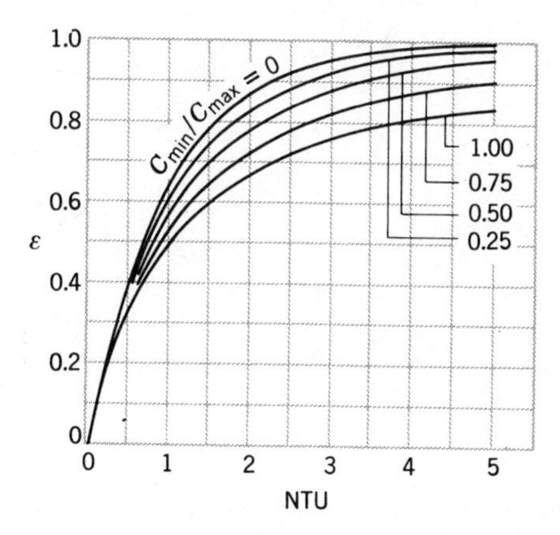

FIGURE 11.15. Effectiveness of a counter-flow heat exchanger (Equation 11.30).

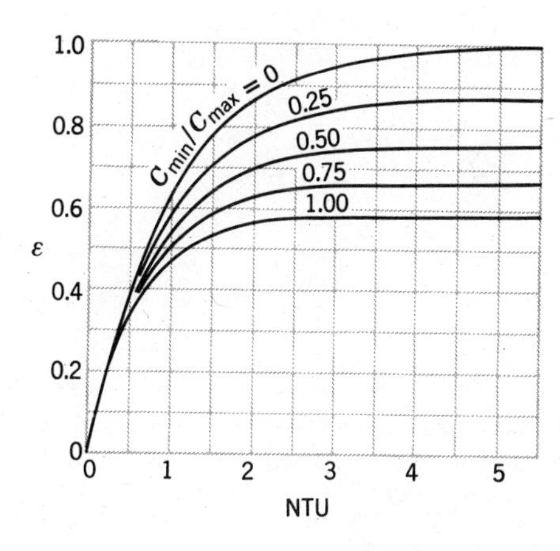

FIGURE 11.16. Effectiveness of a shell-and-tube heat exchanger with one shell and any multiple of two tube passes (two, four, etc. tube passes) (Equation 11.31).

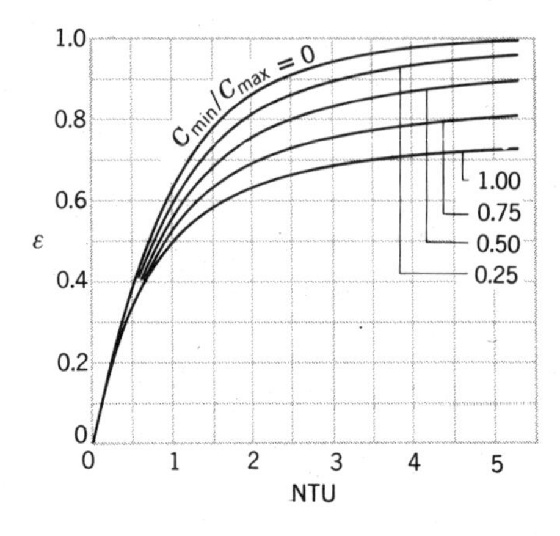

FIGURE 11.17. Effectiveness of a shell-and-tube heat exchanger with two shell passes and any multiple of four tube passes (four, eight, etc. tube passes) (Equation 11.32 with $n = 2$).

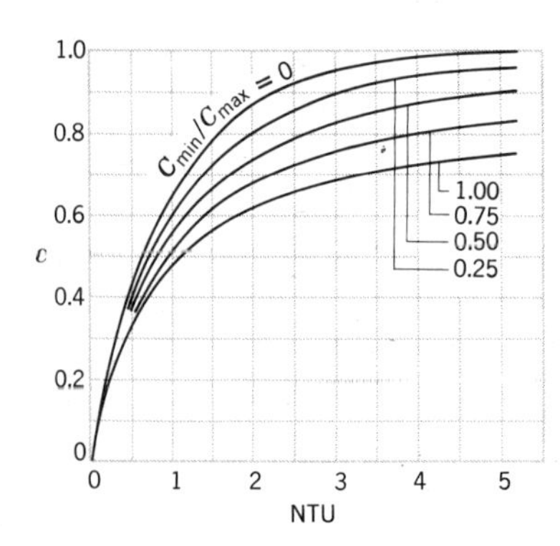

FIGURE 11.18. Effectiveness of a single-pass, cross-flow heat exchanger with both fluids unmixed (Equation 11.33).

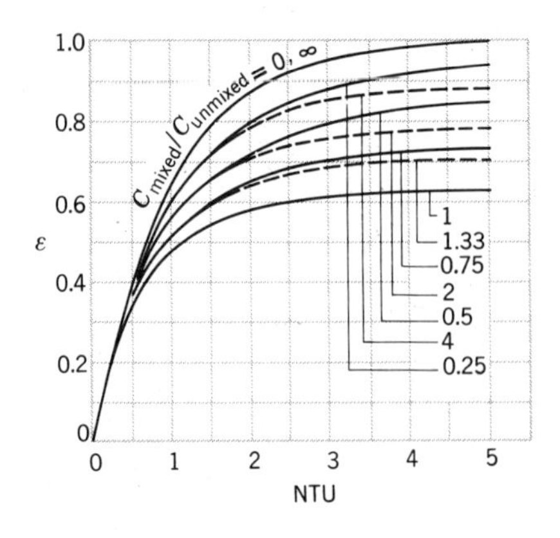

FIGURE 11.19. Effectiveness of a single-pass, cross-flow heat exchanger with one fluid mixed and the other unmixed (Equations 11.34, 11.35).

APPENDIX C

A SIMPLE METHOD OF RATING WATER-COOLED CONDENSERS[1]

ABSTRACT

This is a method of presenting the performance of a series of condensers on a single graph. This shows at a glance the effect of the different variables for a given type of tube, refrigerant and design fouling factor. The same performance curves can be used at different scale factors simply by multiplying the actual length of travel by a correction factor. This length correction factor can be plotted in terms of scale deposit factor at different water flow rates.

The condenser is the "heat sink" of the refrigerating system. Head pressure, motor loading as well as cooling capacity of a mechanical refrigeration system depend on the ability of the condenser to dissipate the proper amount of heat at selected operating conditions. In designing a system it is important to determine beforehand the performance of the condenser, that is, the heat rejection capacity, over the complete range of conditions that can be expected in the field.

Performance Effected. Several factors besides the physical proportions of the condenser will effect its performance. In the case of water cooled condensers the heat rejection capacity will vary with water flow, water temperature, condensing temperature and with the degree of scale deposit on the water side of the tubes. The effect of fouling is especially noticeable in Freon* condensers using finned tubes with large external to internal surface ratios A_o/A_i.

[1]By Henri Soumerai, from *Refrigerating Engineering*, ASRE, Oct. 1955. Reproduced with permission of ASHRAE.

*Freon is the registered trademark of the du Pont Co. for dichlorofluoromethane.

Although it is difficult to foresee exactly the degree of fouling for a particular type of material and water, the application engineer should nevertheless check the performance of the condenser at the maximum anticipated scale deposit factor. Failure to do so can result in too frequent nuisance trip outs on high pressure cutout, motor overloading or abnormally frequent tube cleaning and general customer dissatisfaction.

BASIC EQUATIONS

The basic equations will be briefly reviewed for the case of a single circuit shell type condenser. These equations can readily be applied to multicircuit condensers of the shell and tube type (see Nomenclature).

Neglecting heat losses, the heat rejected by the condensing refrigerant is entirely absorbed by the condenser water:

$$q = 500\,(\text{gpm})(T_L - T_E) \tag{1}$$

Referring to Fig. 1, we can define a contact factor, CF, indicating how closely the water temperature rise $(T_L - T_E)$ approaches the potential temperature rise, $(T_C - T_E)$,

$$\frac{T_L - T_E}{T_C - T_E} = \text{CF} \tag{2}$$

In line with present practice it is assumed that the average overall heat transfer coefficient based on the log mean temperature difference between refrigerant at saturation temperature and water inlet and outlet temperatures can be correlated in terms of water velocity; then

$$q = U(L)\frac{(T_L - T_E)}{\ln\left(\dfrac{T_C - T_E}{T_C - T_L}\right)} \tag{3}$$

From (1) and (3) follows:

$$\frac{T_C - T_L}{T_C - T_E} = e^{-x} = \text{AF} \tag{4}$$

$$\frac{T_L - T_E}{T_C - T_E} = 1 - e^{-x} = \text{CF} \tag{5}$$

with

$$x = \frac{L_T}{500(\text{gpm}/U)} \tag{6}$$

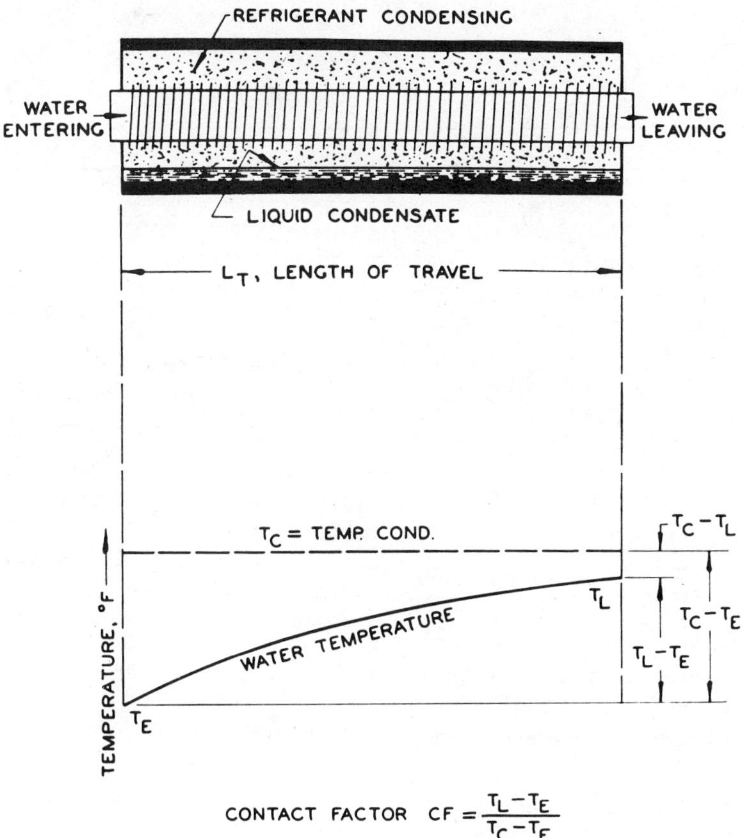

FIGURE 1. Schematic single-circuit water-cooled condenser.

The water temperature rise can be expressed in terms of entering temperature difference:

$$T_L - T_E = \text{CF}(T_C - T_E) \tag{7}$$

Introducing (7) in (1) we obtain the following equation for the heat rejection:

$$q = 500(\text{gpm})(\text{CF})(T_C - T_E) \tag{8}$$

Since the heat rejection is proportional to the entering temperature difference it is possible to express the heat rejection in Btu/h · °F entering temperature difference:

$$\text{Specific heat rejection} = \frac{q}{T_C - T_E} = 500(\text{CF})(\text{gpm}) \tag{9}$$

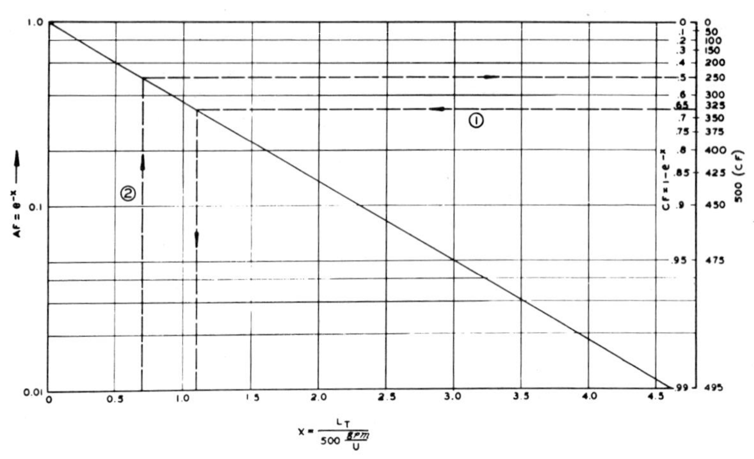

FIGURE 2. Contact factor versus exponent x.

For practical use of equation (9) the terms CF and 500(CF) have been plotted on exponential paper in Fig. 2.

Condenser calculations can be reduced to the following typical problems:

1. *Determine the Required Length of Travel at Given Design Conditions.* In this case the contact factor, CF, is known. From Fig. 2, we read x then

$$L_T = x\left(500\frac{\text{gpm}}{U}\right) \tag{10}$$

Allowable water pressure drop and maximum water velocities (approximately 8 ft/s) impose certain limitations on the condenser design.

2. *Performance of Given Condenser.* In this case L_T, gpm, and U are known. The term x can be computed and the contact factor CF read from Fig. 2.

It is then possible to determine either:

(a) The heat rejection for a given initial temperature difference $(T_C - T_E)$ from equation (8).

(b) The condensing temperature for a given heat rejection

$$T_C - T_E = \frac{q}{500(\text{CF})(\text{gpm})} \tag{11}$$

then

$$T_C = T_E + (T_C - T_E) \tag{12}$$

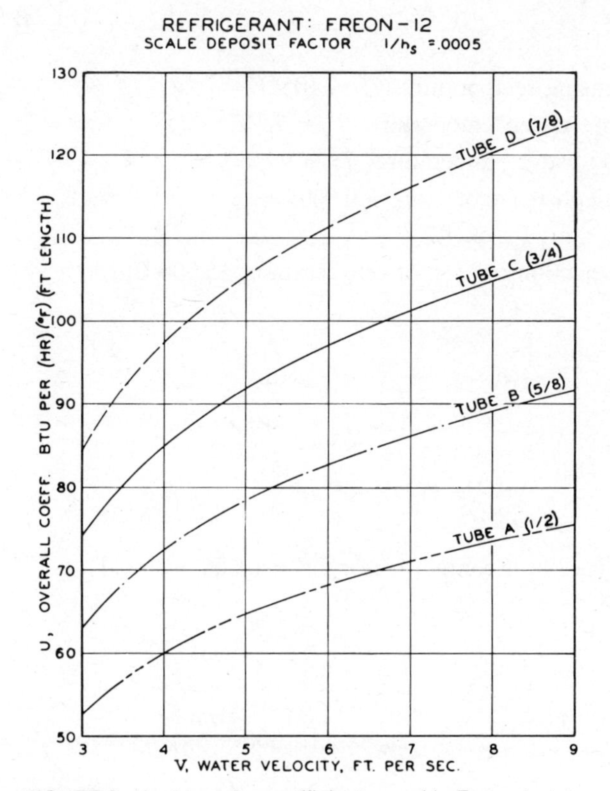

FIGURE 3. Heat transfer coefficients used in Example 1.

Example 1: Selection of Tube Size

Given the average heat transfer coefficient (Fig. 3) and physical dimensions (Table 1) of four different sizes of finned tubes, determine the application range of each tube size, that is, the practical range of capacity in tons of refrigeration, length of travel, and water pressure drop at the following design

TABLE 1. Physical Dimensions of Finned Tubes Having 12 fins / in., $\frac{1}{8}$ in. High

Tube	Tube OD (in.)	Tube ID (in.)	Ai (ft^2/ft)	Ao (ft^2/ft)	Diameter Over Fins (in.)
A	$\frac{1}{2}$	0.430	0.112	0.720	$\frac{3}{4}$
B	$\frac{5}{8}$	0.555	0.145	0.849	$\frac{7}{8}$
C	$\frac{3}{4}$	0.680	0.178	0.981	1
D	$\frac{7}{8}$	0.805	0.210	1.111	$1\frac{1}{8}$

conditions:

- Condensing temperature: $T_C = 105°F$
- Entering water temperature: $T_E = 75°F$
- Leaving water temperature: $T_L = 95°F$
- Scale deposit factor: $1/h_s = 0.0005$
- Refrigerant: Freon 12
- Heat rejection per ton of refrigeration: 15,000 Btu/h

Solution

$$T_L - T_E = 20$$

$$T_C - T_E = 30$$

$$\text{CF} = \frac{20}{30} = 0.666$$

Entering Fig. 2 at a contact factor CF = 0.666, we read

$$x = 1.10$$

From (10)

$$L_T = 1.10\left(500\frac{\text{gpm}}{U}\right) \tag{13}$$

From (1)

$$q = 500(20)(\text{gpm})$$

Dividing by 15,000 Btu/h per ton of refrigeration, we obtain

$$\text{TON} = \left(\frac{2}{3}\right)\text{gpm} \tag{14}$$

For repetitive calculations such as these it is advantageous to compute the terms 500(gpm$/U$) and plot the results in term of water velocity and water flow per circuit (Fig. 4). Note that the term 500(gpm$/U$) has the dimension of a length. Equation (10) shows that 500(gpm$/U$) represents the length of travel for $x = 1$, which corresponds to a contact factor of 63%.

With the help of Fig. 4 the capacity and corresponding length of travel can readily be computed at different water velocities from (13) and (14). The water pressure drop in a length of tube L_T is calculated at the corresponding velocity in the usual manner [1, 2].

Capacity Range Shown. The results are plotted on Fig. 5, which shows the refrigeration capacity range as well as the water pressure drop in terms of

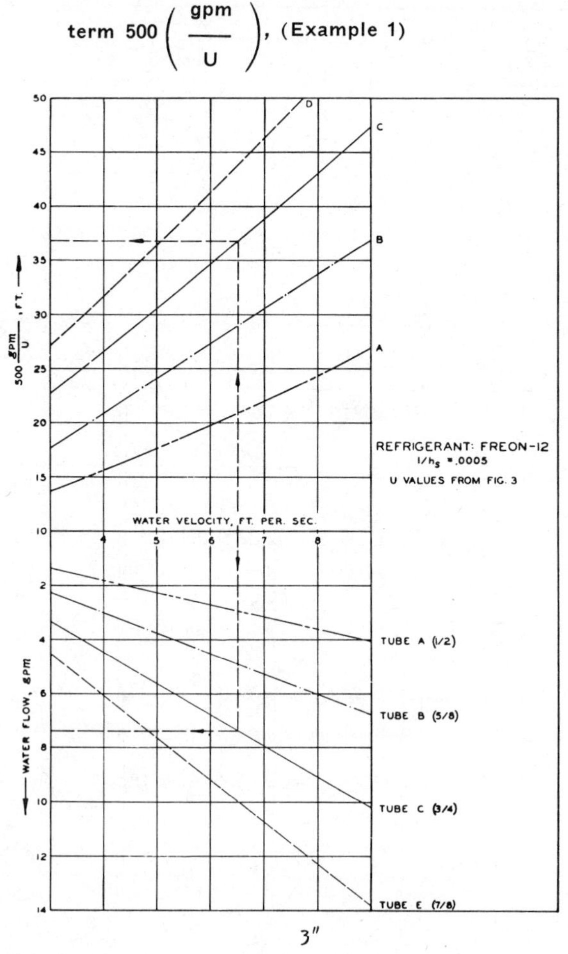

FIGURE 4. Effect of water velocity or gpm on the term 500 (gpm/U) (Example 1).

length of travel for the different tube sizes. Curves showing the length of tube per ton of refrigeration have been added for cost comparison purposes.

The maximum allowable pressure drop per circuit will set a capacity limit for each tube size. Let us assume that the pressure drop, excluding additional losses in water boxes, bends, and fittings is to be limited to 8 psi. Entering Fig. 5 at a pressure drop, $\Delta p = 8$ psi, we read the maximum length of travel and loadings per circuit listed on Table 2.

Condensers can be designed for larger capacities with any of the tubes shown on Table 2 simply by connecting several circuits in parallel. This leads automatically to the shell and tube condenser design.

Most Economical Tube. Curves such as those on Fig. 5 are valuable in choosing the most economical tube for a given application. It should be noted

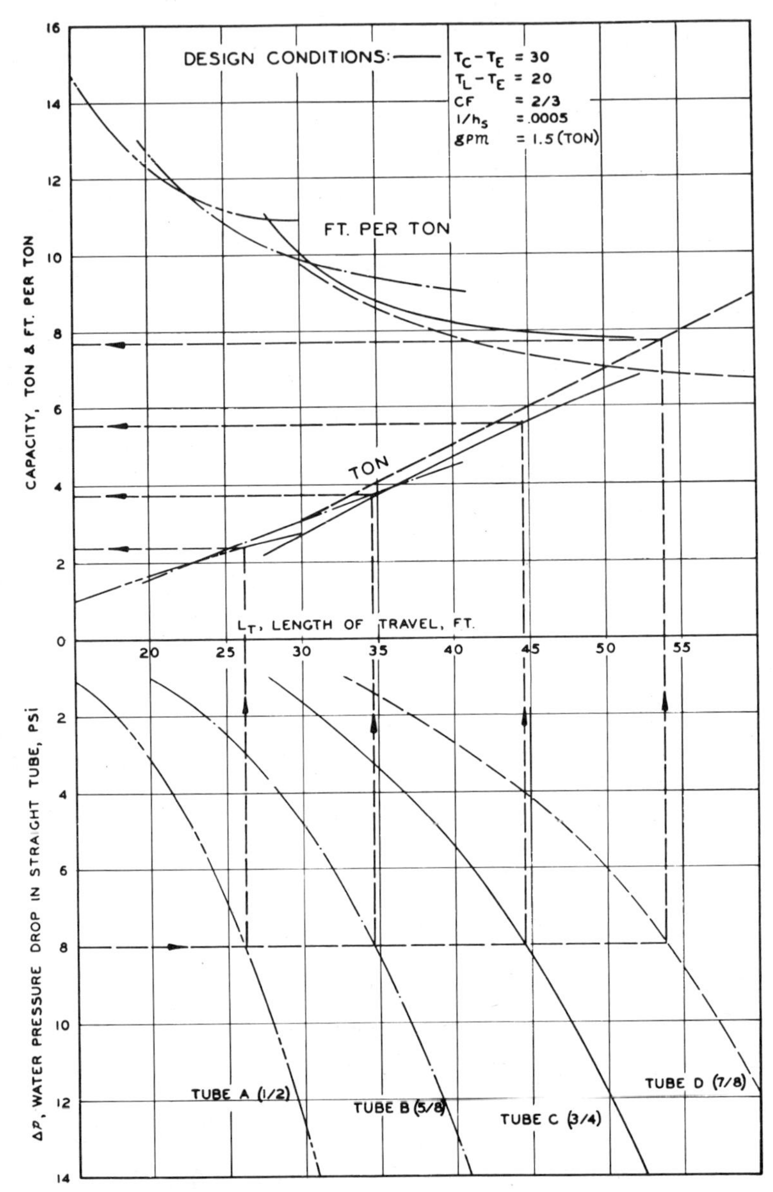

FIGURE 5. Application range of four different tubes (Example 1).

however that the cost of the tubes is just one factor to consider. Smaller, low fin tubes usually result in more compact shell dimensions, that is, lower shell cost, but more tubes are required, that is, the labor cost is higher than is the case with the larger diameter tubes. Complete cost studies of alternate designs will determine the most economical tube size. The optimum tube type depends to a large extent on manufacturing methods, as well as local labor and material costs.

TABLE 2. Maximum Tube Loading (from Example 1)

Tube	Tube OD (in.)	Maximum Length of Travel (ft)	Maximum Capacity per Circuit (ton)
A	$\frac{1}{2}$	26.0	2.35
B	$\frac{5}{8}$	34.5	3.60
C	$\frac{3}{4}$	44.75	5.50
D	$\frac{7}{8}$	54.00	7.75

HEAT REJECTION CURVES

Assume that the $\frac{3}{4}$ in. OD finned tube has been found to be the most economical. The next step is to determine the performance of this tube over the normal application range of water flow at different lengths of travel.

The same procedure as in the preceding example can be applied to determine at a given water flow rate, the length of travel required for different contact factors and the corresponding specific heat rejection. The results are plotted on Fig. 6 which shows the effect of length of travel on heat rejection at different water flow rates. These curves were computed with the scale deposit factor most widely used for copper tubes, that is, $1/h_s = 0.0005$. The constant contact factor lines were obtained by connecting points on the heat rejection curves having equal contact factor.

Application of heat rejection curves.

1. Condenser selection

(a) For a given contact factor, read the lengths of travel at different water flow rates and corresponding pressure drops along a constant CF line.
(b) For a given specific heat rejection, enter ordinate at given specific heat rejection, read along horizontal line the required length of travel at different water gpm and corresponding pressure drop.

2. Performance of given condenser.

Enter abscissa at given length of travel, read on vertical line the specific heat rejection and pressure drop at different flow rates.

Example 2

The performance curves of a Freon-12 compressor show a heat rejection of 600,000 Btu/h at design conditions. Select a standard condenser from the tabulation on Fig. 6 to balance the system at 105°F condensing temperature.

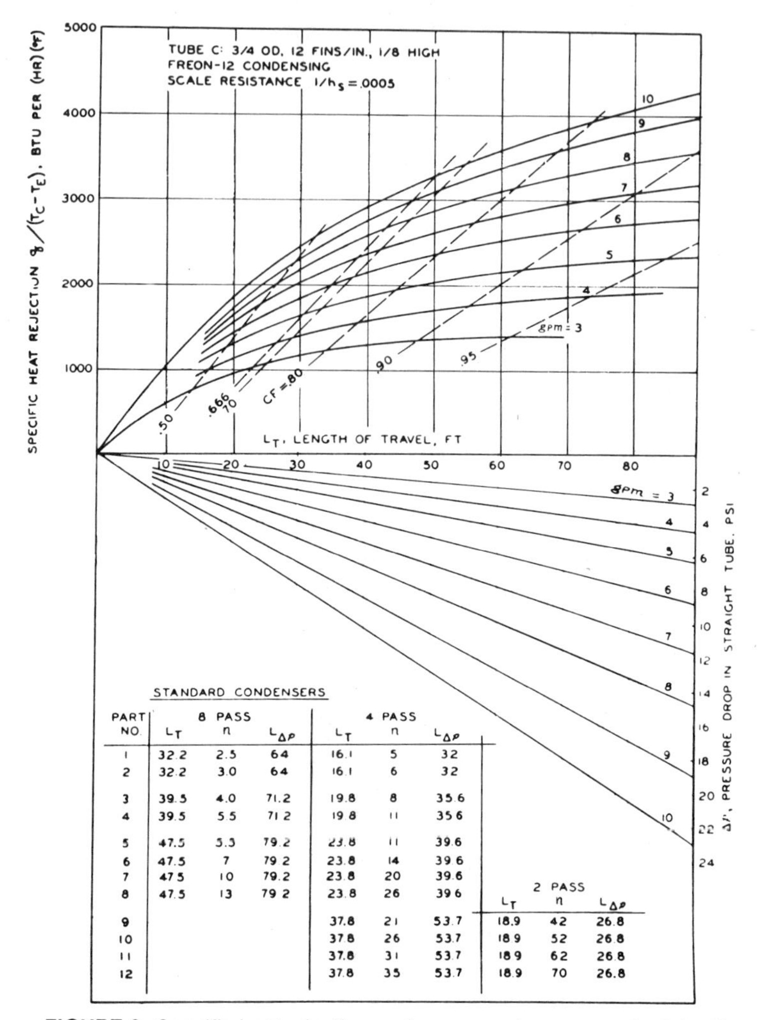

PART NO.	8 PASS L_T	8 PASS n	8 PASS $L_{\Delta P}$	4 PASS L_T	4 PASS n	4 PASS $L_{\Delta P}$	2 PASS L_T	2 PASS n	2 PASS $L_{\Delta P}$
1	32.2	2.5	64	16.1	5	32			
2	32.2	3.0	64	16.1	6	32			
3	39.5	4.0	71.2	19.8	8	35.6			
4	39.5	5.5	71.2	19.8	11	35.6			
5	47.5	5.3	79.2	23.8	11	39.6			
6	47.5	7	79.2	23.8	14	39.6			
7	47.5	10	79.2	23.8	20	39.6			
8	47.5	13	79.2	23.8	26	39.6			
9				37.8	21	53.7	18.9	42	26.8
10				37.8	26	53.7	18.9	52	26.8
11				37.8	31	53.7	18.9	62	26.8
12				37.8	35	53.7	18.9	70	26.8

FIGURE 6. Specific heat rejection and pressure drop curves for tube C.

Fifty (50) gpm of water are available at an entering temperature of 75°F. The pressure drop must not exceed 8 psi.

Solution

$$T_C - T_E = 30$$

From (1)

$$T_L - T_E = \frac{600{,}000}{500(50)} = 24$$

$$\text{CF} = \frac{24}{30} = 0.80$$

Note that the standard condensers tabulation on Fig. 6 lists the different condenser part numbers in order of increasing cost with the following data for different pass arrangements: length of travel L_T, number of circuits, and the effective length of straight tube used to determine the total pressure drop through the condenser. This effective length, $L_{\Delta p}$, is obtained by adding the water-box pressure losses expressed in equivalent feet of tubes to the actual length of travel.

Choice Limited. At a CF of 0.80, Fig. 6 shows that the length of travel should be equal or greater than 33.5 ft. This immediately limits the choice to one of the 8-pass condensers #3 to 8 or the 4-pass condensers #9 to 12. With a length of travel of 39.5 the flow rate per circuit at a CF of 0.80 is 4 gpm. Obviously condensers #3 and 4 have too few circuits for a total of 50 gpm. With a length of travel of 47.5, the flow rate per circuit required to maintain a CF of 0.80 is 5.5 gpm; therefore

$$n = \frac{50}{5.5} = 9.09$$

Select the next larger size, that is, condenser #7 with $n = 10$ resulting in a flow of 5 gpm per circuit.

To determine the water pressure drop enter abscissa at

$$L_{\Delta p} = 79.2$$

Read at 5 gpm

$$\Delta p = 5.6 < 8 \text{ psi}$$

EFFECT OF SCALE DEPOSIT

Scale deposits have a greater effect on finned tubes than on bare tubes. This is shown on Fig. 7 where the overall heat transfer coefficients per unit length of a $\frac{3}{4}$ in. diameter finned tube and of two bare tubes are plotted as a function of the scale deposit factor $1/h_s$.

The heat transfer coefficients per unit length of the finned tube C in Fig. 7 as well as Fig. 3 were computed from "clean" heat transfer coefficient values and the following equation:

$$\frac{1}{U_{1/h_s}} = \frac{1}{U_{1/h_s=0}} + \frac{1}{h_s A_i} \tag{15}$$

For the bare $\frac{3}{4}$ and $\frac{7}{8}$ in. diameter copper tubes the following formula was

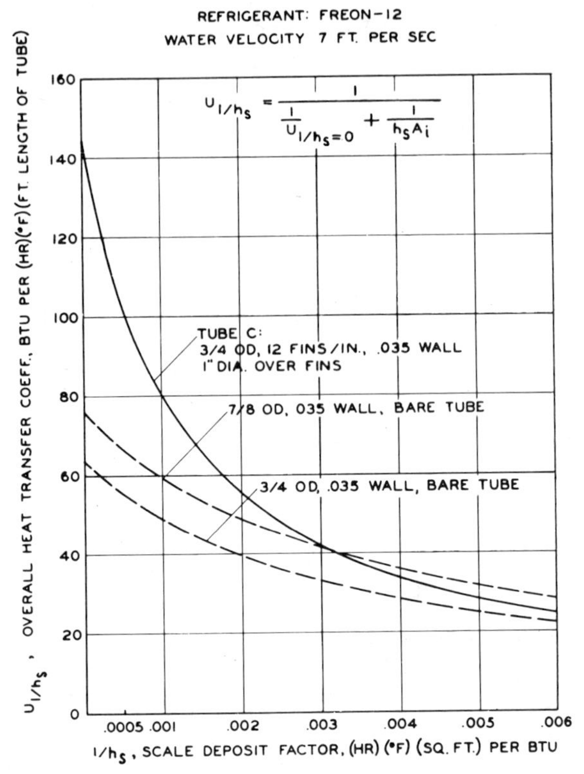

FIGURE 7. Effect of scale deposit on overall heat transfer coefficient per unit length.

used to determine the "clean" heat transfer coefficients:

$$\frac{1}{U_{1/h_s=0}} = \frac{1}{h_i A_i} + \frac{1}{h_o A_o}$$

with h_i calculated at 85°F average water temperature and a velocity of 7 ft/s from Ref. 3:

$$h_i = 150(1 + 0.011T)\left(\frac{V^{0.8}}{D_i^{0.2}}\right)$$

and an average heat transfer coefficient for Freon-12 condensing

$$h_o \simeq 350$$

Figure 7 shows that the use of external fins with abnormally hard waters becomes questionable. At any rate the effect of fouling cannot be neglected in the case of finned tube Freon condensers. Rather than prepare a series of heat rejection curves at different scale deposit factors it is possible to determine the performance of condensers at any fouling factor with the heat rejection curves computed at $1/h_s = 0.0005$ by entering Fig. 6 at an effective length of travel.

$$L_{Te} = M(L_T)$$

where

$$M = \frac{U_{1/h_s}}{U_{1/h_s=0.0005}} \tag{16}$$

To prove this it is only necessary to regroup the terms used to determine the exponent x. At any scale deposit factor $1/h_s$,

$$x = \frac{L_T(U_{1/h_s})}{500 \text{ gpm}}$$

Multiplying and dividing by $U_{1/h_s=0.0005}$ yields

$$x = \frac{L_T M}{500 \text{ gpm}/U_{1/h_s=0.0005}}$$

This means that the exponent x and contact factor, CF, at a given gpm, scale deposit factor $1/h_s$ and an actual length of travel L_T have the same value as at a scale factor $1/h_s = 0.0005$ and an effective length of travel $L_{Te} = ML_T$. In other words the heat rejection curves can be used with this equivalent length of travel to determine the performance of a condenser at different fouling factors.

Results Are Plotted. The length correction factor M for the finned tube C was computed from (15) and (16). The results are plotted on Fig. 8 in terms of $1/h_s$ at different water gpm. Taking a flow of 6 gpm as an average design value we read at $1/h_s = 0$ a length correction factor $M = 1.36$. This means that the "clean" tube will have an effective length 36% greater than at a fouling factor $1/h_s = 0.0005$. The effective length is already reduced to half the actual length of travel at a scale factor of about 0.0024 and the length correction factor reaches a value of 0.25 at $1/h_s = 0.006$.

The actual change in heat rejection will depend on the original condenser selection. From Fig. 6 it is apparent that the effect of water side fouling is less pronounced if the condenser has been selected on the flat portion of the heat rejection curves, that is, at high contact factors, than if the condenser is selected at high water velocities and short length of travel, that is, at low contact factors.

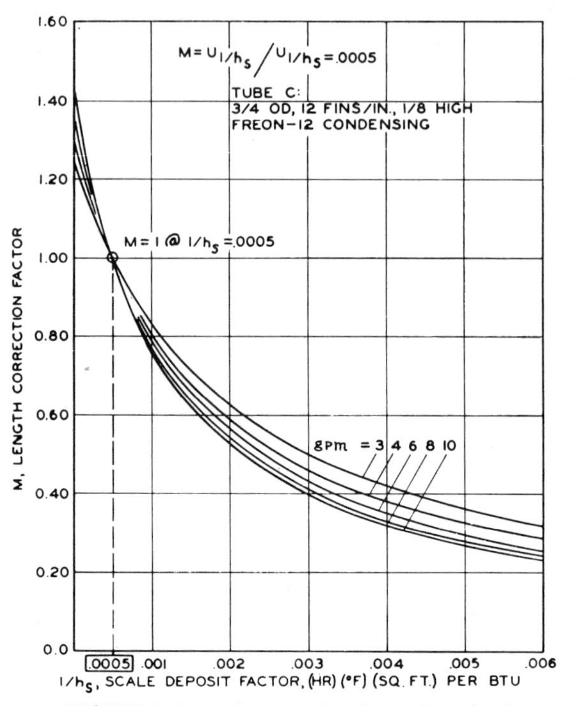

FIGURE 8. Length correction factor for tube C.

Example 3

Assuming the same condenser, heat rejection, water flow, and inlet temperature as in Example 2, determine the condensing temperature at a scale deposit factor $1/h_s = 0.004$.

Solution. At a water flow of 5 gpm per circuit and $1/h_s = 0.004$, we read from Fig. 8 that

$$M = 0.37$$

Then, with the length of travel of 47.5 ft selected in Example 2,

$$L_{Te} = 0.37(47.5) = 17.6$$

From Fig. 6 at 17.6 ft and 5 gpm

$$\frac{q}{T_C - T_E} = 1200$$

with a total heat rejection of 600,000 Btu/h, that is, 60,000 per circuit:

$$T_C - T_E = \frac{60,000}{1200} = 50°\text{F}$$

then

$$T_C = 75 + 50 = 125°\text{F}.$$

Example 4

The condenser selected in Example 2 has been in operation in the field for some time. At design water flow, inlet temperature, and total heat rejection, the system balances at 115°F condensing temperature. Determine approximately the degree of fouling assuming that the system is properly charged and free of non-condensibles.

Solution

$$T_C - T_E = 115 - 75 = 40$$

$$\frac{q}{T_C - T_E} = \frac{60{,}000}{40} = 1500$$

From Fig. 6 we read at a specific heat rejection of 1500 Btu/h · °F and 5 gpm that

$$L_{Te} = 25$$

The length correction factor is

$$M = \frac{25}{47.5} = 0.576$$

At a correction factor $M = 0.576$ and 5 gpm, Fig. 8 shows a scale deposit factor of

$$\frac{1}{h_s} \cong 0.002$$

BIBLIOGRAPHY

1. *Air Conditioning-Refrigerating Data Book*, Design volume, 8th ed. The American Society of Refrigerating Engineers, Chap. 3, 10, and 21.
2. Crane Company, Flow of Fluids through valves, fittings and pipe, Tech. Paper No. 409, May 1942.
3. Report No. 4—Wolverine Tube division. Condensation of Vapors on Finned Tubes Having 19 fins per inch.

NOMENCLATURE

T_C	= condensing temperature	°F
T_E	= entering water temperature	°F
T_L	= leaving water temperature	°F
V	= water velocity	ft/s
Δp	= water pressure drop	psi
A_o	= total external surface per foot length	ft^2/ft
A_i	= internal surface per foot length	ft^2/ft
h_o	= heat transfer coefficient for refrigerant vapor condensing outside tube	$Btu/h \cdot °F \cdot ft^2$
h_i	= heat transfer coefficient for water inside tube	$Btu/h \cdot °F \cdot ft^2$
$1/h_s$	= scale deposit factor	$h \cdot °F \cdot ft^2/Btu$
U	= overall heat transfer coefficient per foot of tube	$Btu/h \cdot °F \cdot ft$

U_{1/h_s}	= overall heat transfer coefficient per foot of tube including scale factor $1/h_s$	Btu/h · °F · ft

For Single Circuit Condensers

gpm	= water flow	gpm
q	= heat rejection	Btu/h
$q/(T_C - T_E)$	= specific heat rejection	Btu/h · °F
L_T	= actual length of travel	ft
L_{Te}	= effective length of travel	ft
e	= base of natural logarithm	
x	= exponent = $L_T/500(\text{gpm})/U$	
CF	= contact factor = $(T_L - T_E)/(T_C - T_E) = 1 - e^{-x}$	
AF	= approach factor = $(T_C - T_L)/(T_C - T_E) = e^{-x}$	
M	= $U_{1/h_s}/U_{1/h_s=0.0005}$ = Length of travel correction factor	

For Multicircuit Condensers

n	= number of circuits, in parallel	
P	= number of passes	
$N = nP$	= total number of tubes in condenser	
L	= length of tube	ft
$L_T = LP$	= length of travel	ft
$L_{\Delta P}$	= equivalent length of straight tube used in computing pressure drop through condenser	ft
GPM	= n (gpm) = total water flow	gpm
Q	= $n(q)$ = total heat rejection	Btu/h

APPENDIX D

THERMODYNAMIC CONSIDERATIONS ON A TWO-PHASE COUNTERFLOW MIRROR-IMAGE HEAT EXCHANGER SYSTEM[1]

ABSTRACT

The "mirror-image" concept for single-phase tube flow applications is extended in the present paper to two-phase forced-convection applications with heat addition (evaporation) and rejection (condensation). The two-phase counterflow mirror-image heat exchanger system (TPC-MIHES) is defined and its key characteristics are identified with special emphasis on typical conditions encountered in thermodynamically efficient refrigeration cycle applications:

(a) Stable, wet wall, annular-spray flow regimes in long channels (aspect ratio $L/D \gg 1$) of constant cross section Ω.

(b) Low average relative wall-to-fluid temperature differentials,

$$\frac{\bar{T}_w - \bar{T}_{sat}}{\bar{T}_{sat}} \ll 1$$

(c) Low relative total friction dissipation,

$$\frac{\Delta P_{tp}}{\bar{P}} \ll 1$$

[1] This appendix is the abstract of Henri Soumerai, "Thermodynamic Considerations on a Two-Phase Counterflow Mirror Image Heat Exchanger System—Forced Convection Channel Flow Applications with Bulk Evaporation and Condensation" approved for presentation at the XVIII International Symposium on Heat and Mass Transfer in Cryoengineering and Refrigeration in Dubrovnik, Yugoslavia, September 1986. As the author could not attend this meeting, this paper was not presented, and will therefore not appear in the symposium proceedings. However, a shorter version of this paper entitled: "Forced convection Channel Flows With Bulk Evaporation or Condensation: The Two-Phase Mirror Image Concept" will appear in the proceedings of the XVIIth Congress of the International Institute of Refrigeration, Vienna, 1987.

The class of TPC-MIHES thus defined is analyzed in the light of conventional methods as well as classical thermodynamic considerations with the help of a temperature–entropy diagram. The main conclusions derived from this analysis can be summarized as follows:

1. Within the typical operating conditions defined and investigated here, the local (or length integrated average) two-phase heat transfer coefficients α_{tp} (or $\bar{\alpha}_{tp}$) and wall-to-fluid temperature differentials $|T_w - T_{sat}|$ (or $|\bar{T}_w - \bar{T}_{sat}|$) are identical in the fluid heating (evaporating) and fluid cooling (condensing) modes.
2. The results *corroborate* and extend the analytical and empirical conclusions reported previously for single- and two-phase heat exchanger applications, including the global *thermodynamic mixing model.*
3. These technical thermodynamic considerations provide a simple bridge with the new (twentieth century) theory of *linear* irreversible thermodynamics and demonstrate the relevancy of nonequilibrium thermodynamics in heat, mass, and fluid flow engineering applications.

A number of useful practical applications of the mirror-image concept beyond the scope of this presentation are highlighted and the paper concludes with a brief assessment of the value of modern thermodynamic tools in the solution of heat, mass, and fluid flow engineering problems.

APPENDIX E

THERMODYNAMIC ASPECTS OF ADIABATIC AND DIABATIC TUBE FLOW REGIME TRANSITIONS: SINGLE-PHASE FLUIDS[1]

ABSTRACT

The transition between stable steady-state laminar and turbulent flow regimes in long channels ($L/D_H \gg 1$) *of constant cross section is first analyzed according to conventional methods under both adiabatic and diabatic conditions, and then in the light of classical second law considerations. It is shown that the ultimate choice of stable steady-state flow regime is always associated with* either *a minimum* or *a maximum irreversible entropy production per unit time, depending on external constraints. Thus, the choice of flow regime is* ***most generally*** *associated with an* ***extremum*** *of entropy production per unit time rather than a necessary minimum entropy production according to a fundamental theorem of the "linear branch" of irreversible thermodynamics.*

[1]Abstract reproduced from Henri Soumerai, "Thermodynamic Aspects of Adiabatic and Diabatic Tube Flow Regime Transitions: Single Phase Fluids." Presented at the AIAA/ASME 4th Joint Thermophysics and Heat Transfer Conference in Boston, June 2nd, 1986 at Convective Heat Transfer session.

ON THE APPLICATION OF AN "ENTROPY MAXIMIZING PRINCIPLE" IN FLOW REGIME PREDICTIONS[2]

ABSTRACT

The validity of the "theorem" of minimum entropy production (per unit time) is assessed for the well documented case of single-phase fluid tube flow bifurcations from laminar to turbulent regimes, with and without external heat exchange. The analysis leads to the following main conclusions. (1): *This "theorem" is not generally valid.* (2): *The stable type of flow regime is associated with a maximum specific (per unit mass) entropy change. The second conclusion provides additional empirical support for the analytical methods based on an "entropy maximizing principle" proposed by several researchers to predict stable flow regimes and/or void fractions in highly complex two-phase flow situations.*

[2]Abstract reproduced from Henri Soumerai, "On the Application of an 'Entropy Maximizing Principle' in Flow Regime Predictions," International Communications in Heat and Mass Transfer **14** (3), 303–312 (1987). The nonadiabatic aspects of bifurcations which were highlighted rather superficially in the earlier AIAA paper (see previous abstract) are treated more rigorously and in detail in this more recent communication.

APPENDIX F

CONVERSION FACTORS AND PHYSICAL CONSTANTS

CONVERSION FACTORS

Acceleration	1 m/s^2	= 4.2520 × 10^7 ft/h^2
Area	1 m^2	= 1550.0 $in.^2$
		= 10.764 ft^2
Energy	1 J ≡ 1 N · m	= 9.4787 × 10^{-4} Btu
Force	1 N	= 0.22481 lb_f
Heat transfer rate	1 W	= 3.4123 Btu/h
Heat flux	1 W/m^2	= 0.3171 Btu/h · ft^2
Heat generation rate	1 W/m^3	= 0.09665 Btu/h · ft^3
Heat Transfer coefficient	1 $W/m^2 \cdot K$	= 0.17612 Btu/h · ft^2 · °F
Kinematic viscosity and diffusivities	1 m^2/s	= 3.875 × 10^4 ft^2/h
Latent heat	1 J/kg	= 4.2995 × 10^{-4} Btu/lb_m
Length	1 m	= 39.370 in.
		= 3.2808 ft
	1km	= 0.62137 mile
Mass	1 kg	= 2.2046 lb_m
Mass density	1 kg/m^3	= 0.062428 lb_m/ft^3
Mass flow rate	1 kg/s	= 7936.6 lb_m/h
Mass transfer coefficient	1 m/s	= 1.1811 × 10^4 ft/h
Pressure and stress[1]	1 N/m^2 ≡ J/m^3	= 0.020886 lb_f/ft^2
		= 1.4504 × 10^{-4} $lb_f/in.^2$
		= 4.015 × 10^{-3} in. water
		= 2.953 × 10^{-4} in. Hg
	1.0133 × 10^5 N/m^2	= 1 standard atmosphere
	1 × 10^5 N/m^2	= 1 bar

[1] The SI name for the quantity pressure is pascal (Pa) having units N/m^2 or $kg/m \cdot s^2$.

Specific heat	1 J/kg · K	$= 2.3886 \times 10^{-4}$ Btu/lb_m · °F
Temperature	K	= (5/9)°R
		= (5/9)(°F + 459.67)
		= °C + 273.15
Temperature difference	1 K	= 1°C
		= (9/5)°R = (9/5)°F
Thermal conductivity	1 W/m · K	= 0.57782 Btu/h · ft · °F
Thermal resistance	1 K/W	= 0.52750 °F · h/Btu
Viscosity (dynamic)[2]	1 N · s/m^2	= 2419.1 lb_m/ft · h
		$= 5.8016 \times 10^{-6}$ lb_f · h/ft^2
Volume	1 m^3	$= 6.1023 \times 10^4$ $in.^3$
		= 35.314 ft^3
		= 264.17 gal
Volume flow rate	1 m^3/s	$= 1.2713 \times 10^5$ ft^3/h
		$= 2.1189 \times 10^3$ ft^3/min
		$= 1.5850 \times 10^4$ gal/min

PHYSICAL CONSTANTS

Universal Gas Constant:

$\mathcal{R} = 8.205 \times 10^{-2}$ m^3 · atm/kmol · K
$= 8.314 \times 10^{-2}$ m^3 · bar/kmol · K
= 8.315 kJ/kmol · K or J/mol · K
= 1545 ft · lb_f/lbmole · °R
= 1.986 Btu/lbmole · °R

Avogadro's Number:

$\mathcal{N} = 6.024 \times 10^{23}$ molecules/mol

Planck's Constant:

$h = 6.625 \times 10^{-34}$ J · s/molecule

Boltzmann's Constant:

$k = \mathcal{R}/\mathcal{N} = 1.380 \times 10^{-23}$ J/K · molecule

Speed of Light in Vacuum:

$c_o = 2.998 \times 10^8$ m/s

Stefan-Boltzmann Constant:

$\sigma = 5.670 \times 10^{-8}$ W/m^2 · K^4
$= 0.1714 \times 10^{-8}$ Btu/h · ft^2 · $°R^4$

Gravitational Acceleration (Sea Level):

$g = 9.807$ m/s^2

Normal Atmospheric Pressure:

$p = 101{,}325$ N/m^2 or Pa ≅ 101.3 kPa

[2]Also expressed in equivalent units of kg/s · m.

Adapted with permission from Frank P. Incropera and D. P. de Witt, Fundamentals of Heat and Mass Transfer, 2nd ed., Wiley, New York, 1985.

REFERENCES

CHAPTER 1

1.1 G. Toraldo Di Francia, *The Investigation of the Physical World*, English translation, Cambridge University Press, Cambridge, 1981.

1.2 W. H. McAdams, *Heat Transmission*, 3rd ed., McGraw-Hill, New York, 1954.

1.3 D. Q. Kern, *Process Heat Transfer*, McGraw-Hill, New York, 1950.

1.4 P. Grassmann, *Physikalische Grundlagen der Chemie-Ingenieur-Technik*, Verlag H. R. Sauerlaender & Co., Aarau, Switzerland, 1961.

1.5 F. W. Schmidt et al., *Introduction to Thermal Sciences*, Sec. 1.2, p. 2, Wiley, New York, 1984.

1.6 R. E. Sonntag and Gordon J. Van Wylen, *Introduction to Thermodynamics, Classical and Statistical*, 2nd ed., Chap. 2, p. 16, Wiley, New York, 1982.

1.7 Sadi Carnot, *Reflexions sur la Puissance Motrice du Feu et sur les Machines Propres a developper cette puissance*, Chez Bachelier, Libraire, Paris, 1824.

1.8 Actes de la Table Ronde, *Sadi Carnot et l'Essor de la Thermodynamique*, sous le patronage du Centre National de La Recherche Scientifique et de l'Ecole Polytechnique, Editions du Centre National de La Recherche Scientifique, Paris, 1976.

1.9 W. Diamant, Redacteur en Chef, Revue Generale de Thermique, Private communication, Paris, Aug. 1985.

1.10 R. Clausius, Ueber verschiedene fuer die Anwendung bequeme Formen der Hauptgleichungen der mechanischen Waermetheorie, *Pogg. Ann.*, **125**, 353–400 (1865).

1.11 T. E. Stanton, On the Passage of Heat between Metal Surfaces and Liquids in Contact with Them, *Philos. Trans. A*, **190**, 67–88 (1897).

1.12 Rudolf Plank, *Handbuch der Kaeltetechnik*, Zweiter Band, *Thermodynamische Grundlagen*, Springer-Verlag, Berlin, 1953.

1.13 *Dictionary of National Bibliography*, Sir Sidney Lee (Ed.), 2nd Suppl. (1901–1911), pp. 508–510, Smith, Elder, London, 1912.

CHAPTER 2

2.1 E. T. Bell, *Men of Mathematics*, Vols. 1 and 2. See Vol. 1, Chap 6, On the Seashore, Newton (1642–1727), pp. 97–126; Chap. 8, Nature or Nurture? The Bernoullis (17th and 18th centuries), pp. 143–150; Chap. 9, Analysis Incarnate, Euler (1707–1783), pp. 151–166; Chap. 11, From Peasant to Snob, Laplace (1749–1827), pp. 188–199; Chap. 12, Friends of an Emperor, ... Fourier (1768–1830), pp. 209–225; Penguin Books Ltd., Middlesex, England, 1965.

2.2 D. Vischer, Schweizer Pioniere der Hydraulik—Zum 200. Todestag von Leonhard Euler, *Schweiz. Ing. Architekt*, **48**, 1129–1134 (1983).

2.3 D. Vischer, Daniel Bernoulli zum 200 Todestag, *Zeitschrift Wasser, Energie, Luft*, **5/6**, (1982).

2.4 H. Rouse and S. Ince, *History of Hydraulics*, Dover, New York, 1963.

2.5 D. Speiser, Daniel Bernoulli (1700–1783), *Helv. Phys. Acta*, **55**, (1982).

2.6 Naturforschende Gesellschaft in Basel, verschiedene Autoren: *Die Werke von Daniel Bernoulli*, Birkhaeuser-Verlag, Basel, 1982.

2.7 I. Szabo, Ueber die sog. Bernoullische Gleichung der Hydromechanik; die Stromfadentheorie Daniel und Johann Bernoulli. *Technikgeschichte*, Vol. 37 (1), VDI-Verlag, Duesseldorf, 1970.

2.8 Euler-Kommission der Schweizerischen Naturforschenden Gesellschaft; verschiedene Autoren: *Leonhard Euler—Opera Omnia, Birkhaeuser-Verlag, Basel*, 1911.

2.9 F. W. Schmidt et al. *Introduction to Thermal Sciences* Chap. 6.3.2, pp. 174–176, Wiley, New York, 1984.

2.10 *Kent's Mechanical Engineers' Handbook*, Power Volume, 12th ed., pp. 1-13 to 1-21; 5-21 to 5-22, Wiley, New York, 1950.

2.11 *Mark's Mechanical Engineers' Handbook*. 6th ed., pp. 3-60 to 3-67; 16-18 to 16-20, McGraw-Hill, New York, 1958.

2.12 G. Toraldo Di Francia, *The Investigation of the Physical World*, Chap. 3.1, p. 152, English Translation, Cambridge University Press, Cambridge, 1981.

2.13 I. Prigogine and I. Stengers, *Order Out of Chaos*, Chapter II, The Identification of the Real: The Language of Dynamics, p. 68; Chapter IV, Energy and the Industrial Age: Heat Engines and the Arrow of Time, p. 111; Bantam Books New York, 1984.

2.14 P. Bergé et al., *l'Ordre dans le Chaos—Vers une approche deterministe de la turbulence*," Hermann, Paris, 1984.

2.15 R. W. Haywood, *Equilibrium Thermodynamics*, Wiley, Chichester, 1980.

2.16 J. Ackeret, The Role of Entropy in the Aerospace Sciences, Daniel & Florence Guggenheim Memorial Lecture, *J. Aerospace Sci.*, **28**(2), 81 (1961).

2.17 J. Ackeret, Die Entwicklung des Entropiebegriffes, *Schweiz. Bauzeitung—77*, **20**, 316 (May 14, 1959).

2.18 H. J. Schroeder, *Die Entschleierte Entropie*, Technischer Verlag Resch KG, Muenchen, 1982.

2.19 F. Schultz-Grunow, J. Ackeret: Persoenliche Erinnerungen, *Schweiz. Ing. Architekt*, **21**, 587 (1983); N. Rott, Jakob Ackeret und die Geschichte der Machschen Zahl, *Schweiz. Ing. Architekt*, **21**, 591 (1983); 50 Jahre Institut fuer Aerodynamik an der ETH Zurich, Gedenkfeier fuer Jakob Ackeret, *Schweiz. Ing. Architekt*, **21**, 587 (1983).

2.20 James Watt and the Steam Engine, *Heat Transfer Engineering*, **6**(1) 14 (1985).

2.21 T. D. Spearman, Mathematical and Theoretical Physics, in *The Royal Irish Academy—A Bicentennial History 1785–1985*, Royal Irish Academy, Dublin, 1985.

2.22 *Modern Power Station Practice*, 2nd ed., Vol. 3, p. 57, Pergamon Press, Oxford, 1971.

2.23 E. R. G. Eckert, Experiments on Energy Separation in Fluid Streams, *Mech. Eng.*, **106**(10), 58–65 (1984).

CHAPTER 3

3.1 E. T. Bell, *Men of Mathematics*, Vol. 1, Chap. 6, pp. 97–126, Penguin Books Ltd., Middlesex, England, 1965.

3.2 W. M. Rohsenow and H. Choi, *Heat, Mass, and Momentum Transfer*, Prentice-Hall, Englewood Cliffs, NJ, 1961.

3.3 A. Bejan, *Convection Heat Transfer*, Chap. 1, Wiley, New York, 1984.

3.4 H. Soumerai, Pressure Drop of Refrigerant Boiling in Horizontal Internally Finned Annulus, *ASHRAE Trans.*, **72**, Pt. I, 43 (1966).

3.5 O. Reynolds, On the Experimental Investigation of the Circumstances which Determine Whether the Motion of Water Shall Be Direct or Sinuous and the Law of Resistance in Parallel Channels, *Philos. Trans. R. Soc.*, **174**, 935–982 (1883).

3.6 P. Bergé et al., *l'Ordre dans le Chaos—Vers une approche deterministe de la turbulence*, Hermann, Paris, 1984.

3.7 O. Reynolds, On the Two Manners of Motion of Water, *Proc. R. Inst. Great Britain*, 1884.

3.8 *Initiation aux Transfer Thermiques*, J. F. Sacadura (Coordinateur de l'ouvrage), Chap. IV, Sec. 1.3, p. 204, Technique et Documentation, Paris, 1980.

3.9 W. M. Rohsenow and H. Choi, *Heat, Mass, and Momentum Transfer*, Chap. 3.7, pp. 47–53, Prentice-Hall, Englewood Cliffs, NJ, 1961.

3.10 S. S. Kutateladze, *Fundamentals of Heat Transfer*, Chap. IV, Academic Press, New York, 1963.

3.11 P. Bergé et al., *l'Ordre dans le Chaos—Vers une approche deterministe de la turbulence*, Introduction, Hermann, Paris, 1984.

3.12 G. Hagen, Ueber die Bewegung des Wassers in engen zylindrischen Roehren, *Pogg Ann.*, **46**, 423 (1839).

3.13 J. Poiseuille, Recherches expérimentales sur le mouvement des liquides dans les tubes de très petits diametres, *Comptes Rendus*, **11**, 961, 1041 (1840).

3.14 W. H. McAdams, *Heat Transmission*, 3rd ed., Chap. 6, McGraw-Hill, New York, 1954.

3.15 *ASHRAE Handbook*, *Fundamentals*, Chap. 2, ASHRAE, Atlanta, Georgia, 1985.

3.16 L. Z. Graetz, *Math. Phys.*, **25**, 316, 375 (1880).

3.17 A. G. Greenhill, *Proc. London Math. Soc.*, **13**, 43 (1881).

3.18 J. Boussinesq, *Théorie analytique de la chaleur*, Vol. III, Gauthier-Villars, Paris, 1903; *J. Math. Pures Appl.*, **1**, 285 (1905).

3.19 H. Lamb, *Hydrodynamics*, 5th ed., pp. 555, 556, Cambridge University Press, London, 1924.

3.20 A. Bejan, *Convection Heat Transfer*, Chap. 3, Wiley, New York, 1984.

3.21 W. M. Rohsenow and H. Choi, *Heat, Mass, and Momentum Transfer*, Chap. 17, pp. 427–443, Prentice-Hall, Englewood Cliffs, NJ, 1961.

3.22 W. H. McAdams, *Heat Transmission*, 3rd ed., Chap. 5, pp. 126–139, McGraw-Hill, New York, 1954.

3.23 S. S. Kutateladze, *Fundamentals of Heat Transfer*, Chap. VI, pp. 41–56, Academic Press, New York, 1963.

3.24 J. H. Lienhard, *A Heat Transfer Textbook*, Chap. 4.3, p. 113, Prentice-Hall, Englewood Cliffs, NJ, 1981.

3.25 H. L. Langhaar, *Dimensional Analysis and Theory of Models*, Wiley, New York, 1951.

3.26 A. A. Guckhman, *Introduction to the Theory of Similarity*, Academic Press, New York, 1965.

3.27 E. Buckingham, *Phys. Rev.*, **4**, 345 (1914); *Trans. ASME*, **35**, 262 (1915).

3.28 G. E. Barnes, Dimensional Analysis, Sec. 5, Art. 2, pp. 5-04 to 5-08, in *Kent's Mechanical Engineers' Handbook*, Power Volume, 12th ed., Wiley, New York, 1950.

3.29 F. P. Incropera and D. P. de Witt, *Fundamentals of Heat and Mass Transfer*, 2nd ed., p. 368–369, Wiley, New York, 1985.

3.30 H. L., Langhaar, *J. Appl. Mech.*, **64**, A-55 (1942).

3.31 W. M. Kays and M. E. Crawford, *Convective Heat and Mass Transfer*, McGraw-Hill, New York, 1980.

3.32 H. Soumerai, "Single Component Two-Phase Annular-Dispersed Flow in Horizontal Tubes without/with Heat Addition from a Constant or Variable Temperature Source," Dissertation No. 4322, ETH, Zurich, 1969.

3.33 D. E. Andrews et al., "The Prediction of Pressure Loss During Two-Phase Horizontal Flow in Two Inch Pipes," ASME Paper No. 66-PET-13, 1966.

3.34 W. H. McAdams, *Heat Transmission*, 3rd ed., pp. 155–158, McGraw-Hill, New York, 1954.

3.35 (a) E. C. Koo, thesis in Chemical Engineering, Massachusetts Institute of Technology, 1932. (b) T. B. Drew et al., *Trans. Am. Inst. Chem. Eng.*, **28**, 56–72 (1932); *Trans. Am. Inst. Chem. Eng.*, **32**, 17–19 (1936).

3.36 W. Froessel, *Forsch. Gebiete Ing.* **7**, 75–84 (1936).

3.37 J. H. Keenan and E. P. Neumann, *Natl. Advisory Comm. Aeronaut., Tech. Note*, 963, 1945.

3.38 L. F. Moody, *Trans. ASME*, **66**, 671–684 (1944); *Mech. Eng.*, **69**, 1005–1006 (1947).

3.39 S. S. Kutateladze, *Fundamentals of Heat Transfer*, Chap. X, Fig. 47, p. 174, Academic Press, New York, 1963.

3.40 J. Nikuradze, *Forschungsheft*, **361**, 1–22 (1933); *Pet. Eng.*, **11**(6), 164–166 (1940); **11**(8), 75–82 (1940); **11**(9), 124–130 (1940); **11**(11), 38–42 (1940); **11**(12), 83 (1940).

3.41 T. J. von Karman, *Aeronaut. Sci.*, **1**, 1–20 (1934); *Engineering*, **148**, 210–213 (1939); The analogy between fluid friction and heat transfer, *Trans. ASME*, **61**, 705–710 (1939).

3.42 B. S. Petukhov, Heat Transfer and Friction in Turbulent Pipe Flow with Variable Physical Properties, in *Advances in Heat Transfer*, J. P. Hartnett and T. F. Irvine, Jr. (Eds.), Vol. 6, pp. 503–564, Academic Press, New York, 1970.

3.43 J. H. Lienhard, *A Heat Transfer Textbook*, p. 323, Prentice-Hall, Englewood Cliffs, NJ, 1981.

3.44 R. K. Pefley and R. I. Murray, *Thermofluid Mechanics*, Chap. 6-2, pp. 103–111; Chap. 14-6, pp. 324–326, McGraw-Hill, New York, 1966.

3.45 *ASHRAE Handbook*, *Fundamentals*, Chap. 2, p. 2.9, ASHRAE, Atlanta, Georgia, 1985.

3.46 E. Kemler, *Trans. Hyd.* **55**, 7–32 (1933).

3.47 R. J. S. Pigott, *Mech. Eng.*, **55**, 497–501 (1933).

3.48 W. M. Rohsenow and H. Choi, *Heat, Mass, and Momentum Transfer*, Chap. 4, Prentice-Hall, Englewood Cliffs, NJ, 1961.

3.49 H. Soumerai, Discussion of "On the Presentation of Performance Data for Enhanced Tubes Used in Shell-and-Tube Heat Exchangers, by W. J. Marner et al.," *ASME J. Heat Transfer*, **106**, 908–909 (Nov. 1984).

3.50 W. Bober and R. A. Kenyon, *Fluid Mechanics*, Wiley, New York, 1980.

3.51 W. M. Rohsenow and H. Choi, *Heat, Mass, and Momentum Transfer*, Chap. 20, Prentice-Hall, Englewood Cliffs, NJ, 1961.

3.52 W. H. McAdams, *Heat Transmission*, 3rd ed., pp. 216–217, McGraw-Hill, New York, 1954.

3.53 W. H. McAdams, *Heat Transmission*, 3rd ed., Chap. 9, p. 129, McGraw-Hill, New York, 1954.

3.54 G. Eichelberg, *Thermodynamik I*, ETH, Zurich, 1943. *Thermodynamik II*, Course in German language, Chap. 11.5, p. 7; Chap. 13, p. 17; Chap. 13.3, p. 20, ETH, Zurich, 1944.

3.55 M. M. Bolstad and R. C. Jordan, Theory and Use of the Capillary Tube Expansion Device, *J. ASRE Refrig. Eng.*, **56**(6), 519 (Dec. 1948).

3.56 J. H. Keenan, Friction Coefficients for the Compressible Flow of Steam, *J. Appl. Mech.*, (Mar. 1939).

3.57 A. Stodola, *Steam and Gas Turbines*, McGraw-Hill, New York, 1927.

3.58 R. Plank, *Handbuch der Kaeltetechnik*, 2 Band, *Thermodynamische Grundlagen*, Chap. 1, Sec. 4, pp. 56–59, Springer-Verlag, Berlin, 1953.

3.59 R. Plank, "The Concept of Entropy," The Deans' Lecture Series in honor of Dean Emeritus Andrey A. Potter, Purdue University, 1962.

3.60 J. S. Doolittle and F. J. Hale, *Thermodynamics for Engineers*, Chap. 7, pp. 151–152, Wiley, New York, 1984.

3.61 R. E. Sonntag and G. J. Van Wylen, *Introduction to Thermodynamics, Classical and Statistical*, 2nd ed., Chap. 17, Sec. 17.3, pp. 602–607, Wiley, New York, 1982.

3.62 F. Bosnjakovic, *Waermelehre und Waermewirtschaft in Einzeldarstellungen*, Vol. 11, *Technische Thermodynamik*, Chap. IX, pp. 114–122, Verlag Theodor Steinkopff, Dresden, 1965.

CHAPTER 4

4.1 *ASHRAE Handbook*, *Fundamentals*, Chap. 4, p. 4.16; Chap. 11, Fig. 2, p. 11.2, ASHRAE, Atlanta, Georgia, 1981.

4.2 *ASHRAE Handbook*, *Equipment*, Chaps. 10 and 11, ASHRAE, Atlanta, Georgia, 1983.

4.3 S. Nukiyama, *J. Soc. Mech. Eng.* (*Japan*), **37**, 367–374, S53–54 (1934).

4.4 H. Soumerai, "Method of Computing Total Heat Transfer Between Boiling Refrigerant Inside Horizontal Tubes and an External Variable Temperature Source," IIR, Commission 3, Prague, 1965.

4.5 H. Soumerai, Pressure Drop of Refrigerant Boiling in Horizontal Internally Finned Annulus," *ASHRAE Trans.*, **72**, Pt. I, 43 (1966).

4.6 H. Soumerai, "Single Component Two-phase Annular-Dispersed Flow in Horizontal Tubes without/with Heat Addition from a Constant or Variable Temperature Source," Dissertation No. 4322, ETH, Zurich, 1969.

4.7 H. Soumerai, Compression Refrigeration, in *Air Conditioning Refrigerating Data Book*, Design Volume, 10th ed., pp. 4-01 to 4-25. The American Society of Refrigerating Engineers (ASRE) (now ASHRAE), New York, 1957.

4.8 *ASHRAE Handbook*, *Fundamentals*, Chap. 1, Part II, pp. 1.8–1.21; Chaps. 16 and 17, ASHRAE, Atlanta, Georgia, 1981.

4.9 *ASHRAE Handbook*, *Equipment*, Chap. 12, Part I, pp. 12-1 to 12-20; Chaps. 16 and 17; Chap. 18, Part II, pp. 18-5 to 18-8; Chap. 18, Part IV, pp. 18-13 to 18-16, ASHRAE, Atlanta, Georgia, 1983.

4.10 K. J. Bell, Heat Exchangers with Phase Change, in *Heat Exchangers: Theory and Practice*, J. Taborek, G. F. Hewitt, and N. Afgan (Eds.), pp. 3–17, Hemisphere Publishing, New York, 1983.

4.11 J. W. Palen et al., Comments to the Application of Enhanced Boiling Surfaces in Tube Bundles, in *Heat Exchangers: Theory and Practice*, J. Taborek, G. F. Hewitt, and N. Afgan (Eds.), pp. 193–203, Hemisphere Publishing, New York, 1983.

4.12 *Second Law Aspects of Thermal Design*, HTD-Vol. 33, ASME, 1984. Presented at the 22nd National Heat Transfer Conference and Exhibition, Niagara Falls, NY, A. Bejan and R. L. Reid (Eds.), August 1984.

4.13 A. Bejan, *Entropy Generation Through Heat and Fluid Flow*, Wiley, New York, 1982.

4.14 R. Thevenot, History, in *Refrigeration in France*, pp. 6–12, XVIth International Congress of Refrigeration, Paris, 1983.

4.15 R. Thevenot, *A History of Refrigeration Throughout the World*, J. C. Fidler (Trans.), International Institute of Refrigeration, Paris, 1979.

4.16 E. Clapeyron, Memoire sur la Puissance Motrice de la Chaleur, *J. l'Ecole R. Polytech.* **XIV**, 153–190 (1833).

4.17 G. Lorentzen, Energy and Refrigeration, *Int. J. Refrig.*, **6**(5/6), 262–273 (1983).

4.18 H. Soumerai, "Design and Operation of Modern Two-Pole Hermetic Screw Package Chillers," XIIth IIR Congress, Madrid, Spain, 1967.

4.19 H. Soumerai, Large Screw Compressors for Refrigeration, *ASHRAE J.*, **9** (Mar. 1967).

4.20 *ASHRAE Handbook*, *Equipment*, Chap. 12, Part IV, section on Features of Helical Screw Compressors, p. 12.16, ASHRAE, Atlanta, Georgia, 1983.

4.21 H. Soumerai, Application of Thermodynamic Similitude, Part I: Pressure Drop, *ASHRAE J.*, **8**, 78 (June 1966); Part II; Heat and Mass Transfer, *ASHRAE J.*, **8**, 38 (July 1966); Part III: Comments on Applications of Thermodynamic Similitude, *ASHRAE J.*, **8** (Sept. 1966).

4.22 H. Soumerai, "Generalized Design Criteria for Single-Component, Two-Phase Flow inside Tubes—Evaporation or Condensation, Part II, Diabatic Flow," p. 389, Condensation in Annular-Dispersed Regimes, 12th IIR Congress, Madrid, Spain, 1967.

4.23 *ASHRAE Handbook*, *Fundamentals*, Chaps. 2 and 4, ASHRAE, Atlanta, Georgia, 1981.

4.24 *ASHRAE Handbook*, *Fundamentals*, Chap. 4, p. 4.7, ASHRAE, Atlanta, Georgia, 1985.

4.25 G. H. Green and F. G. Furse, Effect of Oil on Heat Transfer from a Horizontal Tube to Boiling Refrigerant 12–Oil Mixtures, *ASHRAE J.*, 63 (Oct. 1963).

4.26 P. Worsoe-Schmidt, *ASME Trans.*, 197 (Aug. 1960).

4.27 Bo Pierre, Flow Resistance with Boiling Refrigerants, *ASHRAE J.*, **6**, 58, 73. Translated from *Kylteknisk Tidskrift*, No. 6 (Dec. 1957).

4.28 W. M. Rohsenow, Heat Transfer with Boiling, *ASHRAE Trans.*, **72**, Pt. I, 7 (1966).

4.29 S. W. Anderson et al., Evaporation of Refrigerant 22 in a Horizontal 3/4-in. OD Tube, *ASHRAE Trans.*, **72**, Pt. I, 28 (1966).

4.30 W. R. Zahn, Flow Conditions when Evaporating Refrigerant 22 in Air-Conditioning Coils, *ASHRAE Trans.*, **72**, Pt. I, 82 (1966).

4.31 J. B. Chaddock and J. A. Noerager, Evaporation of Refrigerant 12 in a Horizontal Tube with Constant Wall Heat Flux, *ASHRAE Trans.*, **72**, Pt. I, 90 (1966).

4.32 S. W. Gouse, Jr. and A. J. Dickson, Heat Transfer and Fluid Flow Inside a Horizontal Tube Evaporator: Phase II, *ASHRAE Trans.*, **72**, Pt. I, 104 (1966).

4.33 J. J. Kowalczewski, Vapor Slip in Two-Phase Fluid Flow, *ASHRAE Trans.*, **72**, Pt. I, 115 (1966).

4.34 F. W. Staub and N. Zuber, Void Fraction Profiles, Flow Mechanisms and Heat Transfer Coefficients for Refrigerant 22 Evaporating in a Vertical Tube, *ASHRAE Trans.*, **72**, Pt. I, 130 (1966).

4.35 *ASHRAE Handbook*, *Equipment*, ASHRAE, Atlanta, Georgia, 1983.

4.36 P. Bancel and W. E. Schaffnit, Jr., Steam Condensation, in *Mark's Mechanical Engineers' Handbook*, Design Volume, 6th ed., pp. 9-97, 9-98, McGraw-Hill, New York, 1958.

4.37 K. S. Brundige, Condensers, Sec. 9, Figs. 9 and 10, and section on Condenser Design, p. 9-11, in *Kent's Mechanical Engineers' Handbook*, Power Volume, 12th ed., Wiley, New York, 1950.

4.38 G. Oplatka and H. Lang, Theory and Design of "Church-Window" Condensers for Large Steam Turbines, pp. 326–336, in *Brown Boveri Review*, Vol. 60, 7/8, Brown, Boveri & Co., Baden, Switzerland, 1973.

4.39 H. Lang, Investigations and Measurements on "Church-Window" Condensers, pp. 337–344, in *Brown Boveri Review*, Vol. 60, 7/8, Brown, Boveri & Co., Baden, Switzerland, 1973.

4.40 G. Baumann and G. Oplatka, A New Concept on Condenser Design, in *Brown Boveri Review*, Vol. 54, 10/11, pp. 675–681, Brown, Boveri & Co., Baden, Switzerland, 1967.

4.41 V. L. Eriksen et al., Design of Gas Turbine Exhaust Heat Recovery Boiler Systems, in *Boiler Circulation*, p. 7, 84-GT-126, ASME, New York, 1984.

4.42 W. L. Badger, Evaporators and Evaporation, in *Kent's Mechanical Engineers' Handbook*, Power Volume, 12th ed., pp. 3-71 to 3-82, Wiley, New York, 1950.

4.43 *Chemical Engineers' Handbook*, McGraw-Hill, New York, 1950.

4.44 J. Taborek, Evolution of Heat Exchanger Design Techniques, *Heat Transfer Eng.*, **1**(1), 15 (July/Sept. 1979); see Fig. 3, p. 23; Fig. 5, p. 27.

4.45 W. M. Rohsenow, Why Laminar Flow Heat Exchangers Can Perform Poorly, pp. 1057–1071, in *Heat Exchangers*, Kakac, Bergles, and Mayinger (Eds.), Hemisphere Publishing, New York, 1981.

4.46 G. R. Putnam and W. M. Rohsenow, Viscosity Induced Non-Uniform Flow in Laminar Flow Heat Exchangers, *Int. J. Heat and Mass Transfer*, in press.

4.47 G. Lorentzen, How to Design Piping for Refrigerant Recirculation, *Heating, Piping and Air Conditioning*, (June 1965).

4.48 W. H. McAdams, *Heat Transmission*, 3rd ed., Chap. 10, Sec. IV, see Fig. 10.22, pp. 276–280, McGraw-Hill, New York, 1954.

4.49 D. Q. Kern, *Process Heat Transfer*, Chap. 7, pp. 127–174, McGraw-Hill, New York, 1950.

4.50 H. Soumerai, "Thermodynamic Considerations on a 'Mirror Image' Concept—Forced Convection Two-Phase Tube Flow Applications," unpublished.

4.51 F. Tippets, Analysis of the Critical Heat-Flux Condition in High-Pressure Boiling Water Flows, *ASME J. Heat Transfer*, (Feb. 1964).

4.52 *ASHRAE Handbook*, *Fundamentals*, Chap. 4, Table 2, pp. 4.6-4.7, ASHRAE, Atlanta, Georgia, 1985.

4.53 *ASHRAE Handbook*, *Fundamentals*, Chap. 4, pp. 4.1–4.14, ASHRAE, Atlanta, Georgia, 1985.

4.54 K. Kawaji et al., Reflooding With Steady and Oscillatory Injection: Part 1—Flow Regimes, Void Fraction, and Heat Transfer, *ASME J. Heat Transfer*, **107**, 670–678 (Aug. 1985).

4.55 A. Sharan et al., Corresponding States Correlations for Pool and Flow Boiling Burnout, *ASME J. Heat Transfer*, **107**, 392–397 (May 1985).

4.56 M. G. Cooper, Correlations for Nucleate Boiling-Formulation Using Reduced Properties, *PhysicoChemical Hydrodynamics*, **3**(2), 89–111 (1982).

4.57 M. G. Cooper, "Saturation Nucleate Pool Boiling—A Simple Correlation," First U.K. National Heat Transfer Conference, July 3–5, Leeds, 1984.

4.58 M. G. Cooper, Heat Flow Rates in Saturated Nucleate Pool Boiling—A Wide-Ranging Examination Using Reduced Properties, in *Advances in Heat Transfer*, J. P. Hartnett and T. F. Irvine, Jr. (Eds.), Vol. 16, Academic Press, New York, 1984.

4.59 H. Soumerai, Generalizations of Forced Convection Using Reduced Properties —Single- and Two-Phase Flow, *PhysicoChemical Hydrodynamics*, **6**(3), 339–342, (1985).

4.60 H. Soumerai, Discussion of "The Latent Heat of Vaporization of a Widely Diverse Class of Fluids, by S. Torquato and P. Smith," *ASME J. Heat Transfer*, **107**, 499–500 (May 1985).

4.61 H. Soumerai, Thermodynamic Generalizations of Heat and Fluid Flow Data—With and Without Phase Changes, *Int. Commun. Heat and Mass Transfer*, **12**(1), 101 (1985).

4.62 I. I. Novikov, The Generalized Relation for the Critical Thermal Load, *Zh. Atomnaya Energiya*, **7**(3), (Sept. 1959).

4.63 W. Borishansky et al., Generalization of Experimental Data for the Heat Transfer Coefficient in Nucleate Boiling, *Leningrad Phys. J.*, (Dec. 1962).

4.64 V. M. Borishansky et al., "Use of Thermodynamic Similarity in Generalizing Experimental Data of Heat Transfer," Paper No. 56, International Heat Transfer Conference, University of Colorado, Boulder, Aug. 1961; *Int. Dev. Heat Transfer*, *ASME*, 475–482 (1963).

4.65 J. H. Lienhard and V. E. Schrock, The Effect of Pressure, Geometry, and the Equation of State upon the Peak and Minimum Boiling Heat Fluxes, *ASME J. Heat Transfer*, **85**(3), 261–272 (1963).

4.66 G. Danilova, Influence of Pressure and Temperature on Heat Exchange in the Boiling of Halogenated Hydrocarbons, *Kholodilnaya Teknika*, No. 2, (1965). English abstract in *Modern Refrig.*, (Dec. 1965).

4.67 H. Soumerai, "Generalized Design Criteria for Single-Component, Two-Phase Flow inside Tubes—Evaporation or Condensation, 12th IIR Congress, Madrid, Spain, 1967.

4.68 H. Soumerai, "Generalization of Heat Pump Evaporator Design Data on the Basis of Reduced Pressures," Commission E2, Heat Pumps and Energy Recovery, International Institute of Refrigeration, Trondheim, Norway, June 1985.

4.69 H. Soumerai, Predicting Heat Pump Performance by Thermodynamic Generalizations of Fluid Flow and Heat Transfer Data, *Int. J. Refrig.*, **9**(2), 113 (Mar. 1986).

4.70 H. Soumerai, "Thermodynamic Generalization of Heat Transfer and Fluid Flow Data on the Basis of Reduced Pressures," ASHRAE Semi-Annual Meeting, San Francisco, Jan. 1986.

4.71 ASHRAE Research 1985-86, *ASHRAE J.*, **27**, 4–8 (Oct. 1985).

4.72 W. H. McAdams, *Heat Transmission*, 3rd ed., Chap. 13, p. 325, McGraw-Hill, New York, 1954.

4.73 D. Q. Kern, *Process Heat Transfer*, Chap. 12, p. 252, McGraw-Hill, New York, 1950.

4.74 Modern Power Station Practice, Vol. 8, *Nuclear Power Generation*, Pergamon Press, Oxford, 1971.

4.75 H.-J. Thomas, *Thermische Kraftanlagen*, Springer-Verlag, Berlin, 1974.

4.76 Vereinigung der Grosskesselbetreiber E. V., Essen, VGB-Kernkraftwerks-Seminar, 1970.

4.77 *Steam, Its Generation and Use*, The Babcock & Wilcox Company, 1978.

4.78 Cours de Post-Formation en Genie Nucleaire, École Polytechnique Federale de Lausanne, Eidg. Technische Hochschule, Zurich, Eidg. Institut Fuer Reaktorforschung, Wuerenlingen, Suisse, Autumn 1977.

CHAPTER 5

5.1 H. H. Brehm, *Kaeltetechnik*, Sec. 23, Item 4, p. 65, Schweizer Druck- und Verlagshaus, Zurich, 1947.

5.2 H. Soumerai, "Single Component Two-Phase Annular-Dispersed Flow in Horizontal Tubes without/with Heat Addition from a Constant or Variable Temperature Source," Dissertation No. 4322, ETH, Zurich, 1969, Chap. 5, pp. 89–119; see also item 4, pp. 110–111.

5.3 "Discussions" on pp. 41–42, 60–61, 89, 101–103, 113–114, 129, 146, Proceedings of the ASHRAE Semiannual Meeting, *ASHRAE Trans.*, **72**, Pt. I (Jan. 1966).

5.4 *ASHRAE Handbook*, *Fundamentals*, pp. 91–98, ASHRAE, New York, 1967.

5.5 C. Boling et al., Heat Transfer of Evaporating Freon with Inner-Fin Tubing, *Refrig. Eng.*, (Feb. 1954).

5.6 C. Ashley, The Heat Transfer of Evaporating Freon, *Refrig. Eng.*, (Feb. 1942).

5.7 H. Soumerai, "Single Component Two-Phase Annular-Dispersed Flow in Horizontal Tubes without/with Heat Addition from a Constant or Variable Temperature Source," Dissertation No. 4322, ETH, Zurich, 1969, Chap. 5.4, and more specifically Sec. 5.4.1, pp. 111–114.

5.8 L. S. Tong, *Boiling Heat Transfer and Two-Phase Flow*, Wiley, New York, 1965.

5.9 Bo Pierre, The Coefficient of Heat Transfer for Boiling Freon-12 in Horizontal Tubes, *Heating and Air Treatment Eng.*, (1956).

5.10 Bo Pierre, *S. F. Review*, **2**(1), 55 (1955).

5.11 *ASHRAE Handbook*, *Fundamentals*, Chap. 4, p. 4.16, ASHRAE, Atlanta, Georgia, 1981.

5.12 R. C. Martinelli et al., Isothermal Pressure Drop for Two-Phase Two-Component Flow in a Horizontal Pipe, *ASME Trans.*, **66**, 139 (1944).

5.13 R. C. Martinelli and D. B. Nelson, Prediction of Pressure Drop During Forced-Circulation Boiling of Water, *ASME Trans.*, **70**, 695 (1945).

5.14 R. W. Lockhart and R. C. Martinelli, Proposed Correlation of Data for Isothermal Two-Phase Two-Component Flow in Pipes, *Chem. Eng. Prog.*, **45**, 39 (1949).

5.15 J. Starczewski, Generalized Design of Evaporation Heat Transfer to Nucleate Boiling Liquids, *Bri. Chem. Eng.*, (Aug. 1965).

5.16 P. Worsoe-Schmidt, Some Characteristics of Flow-Pattern and Heat Transfer of Freon-12 Evaporating in Horizontal Tubes, *Ingenieren*, *Int. Ed.*, **3**(3), (Sept. 1959).

5.17 G. H. Green, Influence of Oil on Boiling Heat Transfer and Pressure Drop in Refrigerants 12 and 22, *ASHRAE J.*, **7**, 57 (Dec. 1965).

5.18 S. M. Zivi, "Estimate of the Steady State Steam Void Fraction by Means of the Principle of Minimum Entropy Production," Paper 63-HT-16, April 23, 1963, ASME, JHT, May 1964.

5.19 J. Wicks and A. Dukler, *AIChE J.*, **6**, 463 (1960).

5.20 P. Magiros and A. Dukler, Entrainment and Pressure Drop in Concurrent Gas–Liquid Flow, in *Developments in Mechanics*, Lay and Malvern (Eds.), Plenum Press, New York, 1961.

5.21 G. Wallis, *One-Dimensional Two-Phase Flow*, McGraw-Hill, New York, 1969.

5.22 G. Wallis, Annular Two-phase Flow, Part I: A Simple Theory, *J. Basic Eng.*, *ASME Trans.*, **92D**, 59 (1970).

5.23 G. Wallis, Annular Two-Phase Flow, Part II, Additional Effect, *J. Basic Eng.*, *ASME Trans.*, **92D**, 73 (1970).

5.24 S. W. Gouse, Jr., "An Index to the Two-Phase Gas–Liquid Flow Literature, Part 1," MIT Engineering Projects Laboratory Report DSR-8734-1, Mechanical Engineering Department, Massachusetts Institute of Technology, Cambridge, MA, 1963.

5.25 A. J. Stepanoff, *Pumps and Blowers*, *Two-Phase Flow*, Chap. 12, pp. 246–262; see also Chap. 9, Sec. 9.3, p. 164, Wiley, New York, 1965.

5.26 A. Bergelin, Flow of Gas–Liquid Mixtures, *Chem. Eng.*, 104–106 (May 1949).

5.27 R. Jenkins, Ch.E. thesis, University of Delaware, 1947.

5.28 *ASHRAE Handbook*, *Fundamentals*, pp. 93–94, ASHRAE, New York, 1967.

5.29 R. M. Hatch and R. B. Jackobs, Prediction of Pressure Drop in Two-Phase Single Component Fluid Flow, *Am. Inst. Chem. Eng. J.*, **8**, 18 (1962).

5.30 M. Altmann et al., Local and Average Heat Transfer and Pressure Drop for Refrigerants Evaporating in Horizontal Tubes, *ASME Trans. J. Heat Transfer*, **82**, 189 (Aug. 1960).

5.31 H. Soumerai, Application of Thermodynamic Similitude, Part I: Pressure Drop, *ASHRAE J.*, **8**, 87 (June 1966).

5.32 *ASHRAE Handbook*, *Fundamentals*, Chap. 5, Example 3, pp. 94–96, ASHRAE, New York, 1967.

5.33 A. J. Stepanoff, *Pumps and Blowers*, *Two-Phase Flow*, Chap. 12, Sec. 12.7, p. 252, Wiley, New York, 1965.

5.34 Durand, "Basic Relationships of the Transportation of Solids in Pipes," Proceedings of the Minnesota International Hydraulic Convention, Sept. 1953.

5.35 Condolios and Chapus, Transporting Solid Materials in Pipelines, *Chem. Eng.*, (June 24, July 8, July 22, 1963).

CHAPTER 6

6.1 *Directory of Certified Unitary Heat Pumps*, Air Conditioning and Refrigeration Institute.

6.2 *ASHRAE Handbook*, *Equipment*, Chap. 44, ASHRAE, Atlanta, Georgia, 1983.

6.3 Was leisten Waermepumpen wirklich? (What is the Real Performance of Heat pumps?), *Schweiz. Ing. Architekt*, **15**, 309 (1985).

6.4 H. Soumerai, Compression Refrigeration, in *Air Conditioning and Refrigeration Data Book*, Chap. 4, ASRE (now ASHRAE), New York, 1957.

6.5 H. Soumerai, "A Unified Thermodynamic Theory of Heat, Mass, and Momentum Exchange, Part B: Extensions Beyond the Current State of the Art," 16th International Congress on Refrigeration, Paris, Comm. B1, 1983.

6.6 H. Auracher, Advances and Future Aspects of Heat Transfer Relating to Refrigeration and Heat Pumps, *Int. J. Refrig.*, **7**(5), 333 (1984).

6.7 H. Soumerai, A Unified Thermodynamic Theory of Heat, Mass, and Momentum Exchange, *ASHRAE Trans.*, **90**, Pt. 2 (1984).

6.8 H. Soumerai, Second Law Thermodynamic Treatment of Heat Exchangers, in *Second Law Aspects of Thermal Design*, ASME, HTD-Vol. 33, Bk. No. G00249, pp. 11–18, ASME, New York, 1984.

6.9 H. Soumerai, Evaluation of Refrigerant 502 in Integral Horsepower Commercial Refrigeration Compressors, *ASHRAE J.*, **6**, 31 (Jan. 1964).

6.10 I. P. Basarow, *Thermodynamik*, Sec. 9.3, p. 182, Veb Deutscher Verlag der Wissenschaften, Berlin, 1964. (Translation of the updated 1961 textbook in Russian.)

6.11 T. W. Leland, Jr. and P. S. Chappelear, The Corresponding States Principle, A Review of Current Theory and Practice, *Ind. Eng. Chem.*, **60**(7), 15–42 (July 1968).

6.12 E. A. Guggenheim, *J. Chem. Phys.*, **13**, 253 (1945).

6.13 H. Soumerai, Applications of Thermodynamic Similitude, Part III: Comments on Applications of Thermodynamic Similitude, *ASHRAE J.*, **8** (Sept. 1966).

6.14 E. Holfond and S. A. Rice, Principle of Corresponding States for Transport Properties, *J. Chem. Phys.*, **32**(G), (June 1960).

6.15 P. E. Liley, Survey of Recent Work on the Viscosity, Thermal Conductivity and Diffusion of Gases and Liquefied Gases Below 500 K, in *Progress in Interna-*

tional Research on Thermodynamic and Transport Properties, ASME publication, New York, January 1962.

6.16 H. Eyring and D. Henderson, Thermodynamic and Transport Properties of Liquids, in *Progress in International Research on Thermodynamic and Transport Properties*, ASME publication, New York, January 1962.

6.17 L. I. Stiel and G. Thodos, The Prediction of the Transport Properties of Pure Gaseous and Liquid Substances, in *Progress in International Research on Thermodynamic and Transport Properties*, ASME publication, New York, January 1962.

CHAPTER 7

7.1 I. Prigogine and I. Stengers, *Order Out of Chaos*, Chap. 4, p. 109, Bantam Books, New York, 1984.

7.2 R. Plank, *Handbuch der Kaeltetechnik, 2 Band, Die Temperatur*, p. 3, Springer-Verlag, Berlin, 1953.

7.3 F. Burckhardt, *Die Erfindung des Thermometers*, Basel, 1867—Die wichtigsten Thermometer des 18 Jahrhunderts, Basel, 1871.

7.4 G. Amontons, *Hist. et Mémoires de l'Académie des Sciences*, Paris, 1699, 1702, and 1703.

7.5 J. H. Lambert, "Pyrometrie oder vom Masse des Feuers und der Waerme," erschienen in Berlin nach dem Tode Lamberts, 1799.

7.6 R. De Reaumur, *Hist. et Mémoires de l'Acádemie de Paris*, 1730, 1731.

7.7 D. G. Fahrenheit, *Philos. Trans. R. Soc. London*, **30**, 1 (1724); **33**, 78 (1724).

7.8 A. Celsius, *Abh.d.Schwedischen Akad.*, **4**, 197 (1742).

7.9 I. Newton, Tabula Quantitatum et gradium caloris, *Philos. Trans.* (1701), first published anonymously.

7.10 L. C. Burmeister, *Convective Heat Transfer*, Wiley, New York, 1983.

7.11 J. P. Holman, *Heat Transfer*, 4th ed. McGraw-Hill, New York, 1976.

7.12 F. P. Incropera and D. P. de Witt, *Fundamentals of Heat and Mass Transfer*, 2nd ed., Wiley, New York, 1985.

7.13 J. S. Doolittle and F. J. Hale, *Thermodynamics for Engineers*, Wiley, New York, 1984.

7.14 F. W. Schmidt et al., *Introduction to Thermal Sciences*, Wiley, New York, 1984.

7.15 J. B. J. Fourier, *Théorie analytique de la chaleur*, Gauthier-Villars, Paris, 1822; English translation by Freeman, Cambridge, 1878.

7.16 I. Prigogine and I. Stengers, *Order Out of Chaos*, Chap. II, p. 68; Chap. IV, p. 111, Bantam Books, New York, 1984.

7.17 I. Prigogine, *From Being to Becoming*, Freeman, San Francisco, 1980.

7.18 P. Glansdorff and I. Prigogine, *Thermodynamic Theory of Structure, Stability, and Fluctuations*, Wiley, New York, 1971.

7.19 M. Planck, *Ann. Phys.*, **4**, 533 (1901).

7.20 J. Stefan, *Sitzungsber. Kais. Akad. Wiss. Wien Kl. Math. Naturwiss.*, **79**, 391 (1879).

7.21 L. Boltzmann, *Wied. Ann.*, **22**, 291 (1884).

7.22 W. M. Rohsenow and H. Choi, *Heat, Mass, and Momentum Transfer*, Chap. 13, p. 332, Prentice-Hall, Englewood Cliffs, NJ, 1961.

7.23 F. P. Incropera and D. P. de Witt, *Fundamentals of Heat and Mass Transfer*, 2nd ed., Chap. 12, p. 545; Chap. 13, p. 623, Wiley, New York, 1985.

7.24 W. McAdams, *Heat Transmission*, 3rd ed., Chap. 4, p. 55, McGraw-Hill, New York, 1954.

7.25 A. P. Fraas and M. N. Ozisik, *Heat Exchanger Design*, Chap. 3, p. 31, Wiley, New York, 165.

7.26 D. Q. Kern, *Process Heat Transfer*, Chap. 4, p. 62, McGraw-Hill, New York, 1950.

7.27 E. U. Schlünder (Ed.), *Heat Exchanger Design Handbook*, Vol. 2, *Fluid Mechanics and Heat Transfer*, Hemisphere Publishing, New York, 1983.

7.28 *Kent's Mechanical Engineers' Handbook*, Power Volume, 12th ed., Radiation, Sec. 3, No. 7, pp. 3-21 to 3-26, Wiley, New York, 1950.

7.29 F. Hell, *Grundlagen der Wärmeübertragung*, p. 165, VDI-Verlag, GmbH, Dusseldorf, 1982.

7.30 Max Planck, *Wissenschaftliche Selbstbiographie*, Zweite Auflage, Johann Ambrosius Barth-Verlag, Leipzig, 1948.

7.31 A. Kastler, Radiation and Entropy, Coherence and Negentropy, in *Synergetics*, pp. 18–25, H. Haken (Ed.), Springer-Verlag, Berlin, 1977.

7.32 A. Kastler, Cohérence des Vibrations Lumineuses et Négentropie, in *Sadi Carnot et l'Essor de la Thermodynamique*, p. 281–285, Éditions du Centre National de la Recherche Scientifique, Paris, 1976.

7.33 J. Bruhat, *Thermodynamique* (Edition entièrement remaniée par A. Kastler), 6th ed., p. 672, Masson, Paris, 1968.

7.34 W. McAdams, *Heat Transmission*, 3rd ed., pp. 43–51, Chap. 2, McGraw-Hill, New York, 1954.

7.35 W. McAdams, *Heat Transmission*, 3rd ed., Chap. I, pp. 1–6 and Table 1.1, McGraw-Hill, New York, 1954.

7.36 W. L. Badger and W. L. McCabe, *Elements of Chemical Engineering*, McGraw-Hill, New York, 1936.

7.37 O. Reynolds, "On the Extent and Action of the Heating Surface of Steam Generators," in *Proceedings of Literary and Philosophical Society*, Manchester, Vol. 14, pp. 81–85, Session 1874–1875.

7.38 A. P. Colburn, A Method of Correlating Forced Convection Heat-Transfer Data and a Comparison with Fluid Friction, *Int. J. Heat Mass Transfer*, **7**, 1359–1384 (1964). (Reprinted from *Trans. Am. Inst. Chem. Eng.*, **29**, (1933).

7.39 L. Prandtl, *Z. Phys.*, **11**, 1072 (1910); **29**, 487–489 (1928).

7.40 G. I. Taylor, *Br. Advisory Comm. Aeronaut., Rept. Mem. 272*, **31**, 423–429 (1916).

7.41 W. H. McAdams, *Heat Transmission*, 3rd ed., Part B, p. 208–218, McGraw-Hill, New York, 1954.

7.42 W. M. Rohsenow and H. Choi, *Heat, Mass, and Momentum Transfer*, Preface, pp. ix–xi, Prentice-Hall, Englewood Cliffs, NJ, 1961.

7.43 R. C. L. Bosworth, *Transport Processes in Applied Chemistry*, Wiley, New York, 1956.

7.44 Report on Engineering Sciences, *ASEE J. Eng. Educ.*, **49**, 1 (Oct. 1958).

7.45 R. B. Bird et al., *Transport Phenomena*, Wiley, New York, 1961.

7.46 A. S. Foust et al., *Principles of Unit Operations*, Part II, Wiley, New York, 1960.

7.47 A. V. Lykov and Y. A. Mikhaylov, *Theory of Energy and Mass Transfer*, Prentice-Hall, Englewood Cliffs, NJ, 1961.

7.48 M. Gaston Darboux, *Les Oeuvres de Joseph Fourier*, Gauthier-Villars et Fils, Paris, 1888–1890.

7.49 I. Grattan-Guinness (in collaboration with J. R. Ravetz), *Joseph Fourier 1768–1820*, MIT Press, Cambridge, MA, 1972.

7.50 J. Herivel, *Joseph Fourier*, *Face Aux Objections Contre sa Théorie de la Chaleur*, Lettres inédites 1808–1816, Bibliothèque Nationale, Paris, 1980.

7.51 A.-T. Dulong and P.-L. Petit, Sur la Mesure des Température sur les Lois de la Communication de la Chaleur, *Anal. Chim. Phys.*, **V11**, 113, 225 (1817).

7.52 *ASHRAE Handbook*, *Fundamentals*, Chap. 3, Table 5, p. 3.13; Table 6, pp. 3.15–3.16, ASHRAE, Atlanta, Georgia, 1985.

7.53 *ASHRAE Handbook*, *Fundamentals*, Chap. 4, Table 1, pp. 4.3–4.4, ASHRAE, Atlanta, Georgia, 1985.

7.54 E. R. G. Eckert, Pioneering Contributions to Our Knowledge of Convective Heat Transfer, *J. Heat Transfer*, **103**, 409–414 (1981).

7.55 A. Schack, *Der Industrielle Wärmeubergang*, 3 Auflage, Verlag Stahleisen, Dusseldorf, 1948.

7.56 E. Adiutori, *The New Heat Transfer*, Ventuno Press, Cincinnati, 1974.

7.57 E. Clapeyron, Mémoire sur la puissance motrice de la chaleur, *J. l'École Polytech.*, **14**, 153–190 (1834). English translation, pp. 73–105, see especially p. 101, E. Mendoza (Ed.), New York, 1960.

7.58 M. J. Klein, Closing the Carnot Cycle, in *Sadi Carnot et l'Essor de la Thermodynamique*, p. 214, Éditions du Centre National de la Recherche Scientifique, Paris, 1976.

7.59 W. Thomson, An Account of Carnot's Theory of the Motive Power of Heat; with Numerical Results Deduced from Regnault's Experiments on Steam, *Trans. Edinburgh R. Soc.*, 16 (1849). (Reprinted in W. Thomson, *Mathematical and Physical Papers*, Vol. 1, pp. 113–115, Cambridge, University Press, Cambridge 1882.)

7.60 S. Carnot, *Reflexions sur la Puissance Motrice de Feu et sur les Machines Propres a Developper cette puissance*. (English translation by R. H. Thurston, reprinted in S. Carnot, *Reflections on the Motive Power of Fire and Other Papers on the Second Law of Thermodynamics*, E. Mendoza (Ed.), New York, 1960.)

7.61 R. K. Pefley and R. I. Murray, *Thermofluid Mechanics*, Chap. 2, Sec. 2-1, p. 12, McGraw-Hill, New York, 1966.

7.62 E. Fermi, *Thermodynamics*, Dover, New York, 1956. (Originally published by Prentice-Hall, Englewood Cliffs, NJ, 1937.)

7.63 P. Glansdorff, private communication, "Remarques sur un exposé de H. Soumerai," Forced Convection Thermodynamics," May 1985.

7.64 P. Glansdorff, *Thermodynamics in Contemporary Dynamics*, Springer-Verlag, New York, 1972.

7.65 I. Prigogine and P. Glansdorff, *Acad. R. Belg., Bul. Cl. des Sci. Ser.*, 5, **LIX**, 672 (1973-8).

7.66 P. Glansdorff and I. Prigogine, Non-Equilibrium Stability Theory, *Physica*, **46**(3), 334 (1970).

7.67 (a) R. Haase, *Grundzüge der Physikalischen Chemie, Band III, Transportvorgange*, Dr. Dietrich Steinkopff Verlag, Darmstadt, 1973. (b) *Thermodynamik der irreversiblen Prozesse*, Dr. Dietrich Steinkopff Verlag, Darmstadt, 1963. (Translation: *Thermodynamics of Irreversible Processes*, Addison-Wesley, Reading, MA, 1969.)

7.68 S. R. de Groot and P. Mazur, *Non-Equilibrium Thermodynamics*, North Holland, Amsterdam, 1962.

7.69 G. N. Bochkov et al., Stochastic Models in the Theory of Combustion and Explosion, and Fluctuation–Dissipation Models of Mass Transfer in Systems with Chemical Reactions, in *International Communications in Heat and Mass Transfer*, Vol. 12, No. 1, Pergamon Press, New York, 1985.

7.70 I. Prigogine, *Acad. R. Belg. Bull. Cl. Sci.* **31**, 600 (1945).

7.71 A. V. Luikov and Yu. A. Mikhailov, *Theory of Energy and Mass Transfer*, Pergamon Press, New York, 1965.

7.72 P. Glansdorff and I. Prigogine, Non-Equilibrium Stability Theory, *Physica*, **46**(3), 345/346 (1970).

7.73 R. W. Haywood, *Equilibrium Thermodynamics*, Wiley, New York, 1980.

7.74 R. W. Haywood, A Critical Review of the Theorems of Thermodynamic Availability, With Concise Formulations, Part 1. Availability, *J. Mech. Eng. Sci.*, **16**(3), 160–173 (1974).

7.75 R. W. Haywood, A Critical Review of the Theorems of Thermodynamic Availability, With Concise Formulations, Part 2. Irreversibility, *J. Mech. Eng. Sci.*, **16**(4), 258–267 (1974).

7.76 J. E. Ahern, *The Exergy Method of Energy Systems Analysis*, Wiley, New York, 1980.

7.77 A. L. London, Economics and the Second Law: An Engineering View and Methodology, *Int. J. Heat Mass Transfer*, **14**(6), 743–751 (1982).

7.78 A. L. London and R. K. Shah, Costs of Irreversibilities in Heat Exchanger Design, *Heat Transfer Eng.*, **4**(2), 59–73 (Apr.–June, 1983).

7.79 T. H. Fehring and R. A. Gaggioli, Economics of Feedwater Heater Replacement, *J. Eng. Power, Trans. ASME*, 482–488 (July 1977).

7.80 T. A. Brzustowski and P. J. Golem, Second-Law Analysis of Energy Processes, Part I: Exergy—An Introduction, *Trans. CSME*, **4**(4), (1976–77). Part II: The Performance of Simple Heat Exchangers, *Trans. CSME*, **4**(4), 209–226 (1976–77).

7.81 *Second Law Aspects of Thermal Design*, A. Bejan and R. L. Reid (Eds.), presented at the 22nd National Heat Transfer Conference and Exhibition, Niagara Falls, NY, Aug. 5–8, ASME, HTD-Vol. 33, 1984.

7.82 "Symposium on Energy Systems Methodologies," AES-1 to 5, presented at the ASME Technical Conference, Winter Annual Meeting, Florida, Nov. 1985.

7.83 (a) H. Soumerai, "Thermodynamic Considerations on a Counterflow Mirror Image Heat Exchanger System (CMIHES)—Two-Phase Flow Applications," XVII International Symposium on Heat and Mass Transfer in Cryoengineering and Refrigeration, Dubrovnik, Yugoslavia, Sept. 1986 (see note in Appendix D); (b) H. Soumerai, "Forced Convection Channel Flows With Bulk Evaporation or Condensation: The Two-Phase Mirror Image Concept," XVII International Congress of Refrigeration, Vienna, Aug. 1987.

7.84 W. H. McAdams, Heat Transmission by Conduction and Convection (see section entitled "Graphical Method of Interpreting Over-all Coefficients of Heat Transfer"), in *Chemical Engineers' Handbook*, 3rd ed., p. 470, McGraw-Hill, New York, 1950.

7.85 P. Grassmann, *Physikalische Grundlagen der Chemie-Ingenieur-Technik* pp. 492–493, Verlag H. R. Sauerlaender & Co., Aarau, Switzerland, 1961.

7.86 P. Le Goff, *Energetique Industrielle*, Vol. 1, Chap. 7, p. 223, Technique et Documentation, Paris, 1979.

7.87 W. H. McAdams, *Heat Transmission*, 3rd ed., Chap. 12, p. 309, McGraw-Hill, New York, 1954.

7.88 L. C. Burmeister, *Convective Heat Transfer*, p. 607, Wiley, New York, 1983.

7.89 W. H. McAdams, *Heat Transmission*, 3rd ed., Chap. 9, Sec. IV, pp. 241–250, McGraw-Hill, New York, 1954.

7.90 *ASHRAE Handbook*, *Fundamentals*, Chap. 3, pp. 3.14–3.15, ASHRAE, Atlanta, Georgia, 1985.

7.91 A. Bejan, *Convection Heat Transfer*, Chap. 3, p. 78, Wiley, New York, 1984.

7.92 H. Soumerai, A Simple Method of Rating Water Cooled Condensers, *Refrig. Eng.*, 52–59 (Oct. 1955).

7.93 S. J. Kline and F. A. McClintock, Describing Uncertainties in Single-Sample Experiments, *Mech. Eng.*, **75**, 3–8 (Jan. 1953).

7.94 A. Bejan, *Convection Heat Transfer*, Fig. A.1, p. 461, Wiley, New York, 1984.

7.95 P. Grassmann, *Physikalische Grundlagen der Verfahrenstechnik*, 3rd ed., Verlag H. R. Sauerlaender & Co., Aarau, Switzerland, 1982.

7.96 J. Alfred Kastler, L'Oeuvre Posthume de Sadi Carnot, pp. 195–198 in Actes de la Table Ronde, *Sadi Carnot et l'Essor de la Thermodynamique*, Editions du Centre National de la Recherche Scientifique, Paris, 1976.

7.97 I. Prigogine, *From Being to Becoming*, pp. 88–89, Freeman, San Francisco, 1980.

7.98 I. Prigogine and I. Stengers, *Order Out of Chaos*, p. 142, Bantam Books, New York, 1984.

7.99 A. Bejan, *Convection Heat Transfer*, Chap. 5, section entitled "Fluid Layers Heated From Below," pp. 185–188, Wiley, New York, 1984.

7.100 "Navier, C. L. M. H.," *Dictionary of Scientific Biography*, Vol. X, pp. 2–5, Scribner's, New York, 1974.

7.101 "G. G. Stokes," *Dictionary of National Bibliography*, S. Lee (Ed.), 2nd Suppl., pp. 421–424, 1901–1911, Smith, Elder, London, 1912.

7.102 H. Kangro, pp. 229–245, Actes de la Table Ronde, *Sadi Carnot et l'Essor de la Thermodynamique*, Editions du Centre National de la Recherche Scientifique, Paris, 1976.

7.103 G. Toraldo Di Francia, *The Investigation of the Physical World*, English translation, p. 71, Cambridge University Press, Cambridge, 1981.

7.104 M. R. Cohen, *Reason and Nature*, Dover, New York, 1978. Republication of 2nd ed. by Free Press, Glencoe, IL 1953 (originally published by Harcourt, Brace and Co., New York, 1931).

7.105 P. Cotti, Naturwissenschaft und Philosophie—gestern und heute, *Schweiz. Ing. Architekt*, **43**, 831–836 (1984).

Classical Group

7.106 A. Bejan, *Entropy Generation Through Heat and Fluid Flow*, Wiley, New York, 1982.

7.107 J. E. Ahern, *The Exergy Method of Energy Systems Analysis*, Wiley, New York, 1980.

7.108 R. W. Haywood, *Equilibrium Thermodynamics*, Wiley, New York, 1980.

7.109 C. Caratheodory, Untersuchungen uber die Grundlagen der Thermodynamik, *Mathematische Annalen*, Vol. 67, pp. 355–386, A. Clebsch and C. Neumann, Eds., Druck und Verlag von B. G. Teubner, Leipzig, 1909.

7.110 Max Planck, *Vorlesungen ueber Thermodynamik*, Achte Auflage, de Gruyter, Berlin, 1927.

7.111 *Thermodynamik I* nach der Vorlesung von G. Eichelberg, (bearbeitet von stud. ing. E. Jaeckel), 1943; *Thermodynamik II* (bearbeitet von stud. ing. B. Friedrich und stud. ing. Th. Erismann), Verlag des Akademischen Machinen-Ingenieur-Vereins der ETH, Zurich, 1944.

7.112 *Chemical Engineers' Handbook*, 3rd ed., Sec. 4, pp. 287–358, McGraw-Hill, New York, 1950 .

7.113 *Kent's Mechanical Engineers' Handbook*, Power Volume, 12th ed., "Thermodynamics of Gases at High Velocity," p. 3-63, Wiley, New York, 1950.

7.114 R. Plank, *Handbuch Der Kaeltetechnik*, 2 Band, *Thermodynamische Grundlagen*, Springer-Verlag, Berlin, 1953.

7.115 E. Fermi, *Thermodynamics*, Dover, New York, 1956 (originally published by Prentice-Hall, Englewood Cliffs, NJ, 1937).

7.116 I. P. Basarow, *Thermodynamik*, Veb Deutscher Verlag der Wissenschaften, Berlin, 1964.

7.117 R. K. Pefley and R. I. Murray, *Thermofluid Mechanics*, McGraw-Hill, New York, 1966.

7.118 J. Bruhat, *Thermodynamique* (Edition entierement remaniée par A. Kastler), 6th ed., Masson, Paris, 1968.

7.119 Fr. Bosnjakovic, *Technische Thermodynamik*, Vol. I. Pt. 4, Auflage, Verlag Theodor Steinkopff, Dresden, 1965; Vol. II, Pt. 5, durchgesehene Auflage, 1971.

7.120 *Modern Power Station Practice*, Vol. 3, Published for and on behalf of the Central Electricity Generating Board by Pergamon Press, Oxford, 1971.

7.121 H. J. Thomas, *Thermische Kraftanlagen*, Springer-Verlag, Berlin, 1974.

7.122 C. Bory, *La Thermodynamique*, 4th ed., Presse Universitaires de France, Paris, 1981.

7.123 J. Bougard, *Thermodynamique Technique*, 4th ed., Presses Universitaires de Bruxelles, 1982–83.

7.124 L. Borel, *Thermodynamique et Energetique*, Presses Polytechniques Romandes, Lausanne, Switzerland, 1984.

7.125 F. W. Schmidt et al., *Introduction to Thermal Sciences*, Wiley, New York, 1984.

7.126 *ASHRAE Handbook*, *Fundamentals*, Sec. 1, Chap. 1, ASHRAE, Atlanta, Georgia, 1985.

Statistical Group

7.127 *Handbook of Physics*, Part 5, Heat and Thermodynamics, McGraw-Hill, New York, 1958.

7.128 M. Tribus, *Thermostatics and Thermodynamics*, Van Nostrand, Princeton, NJ, 1961.

7.129 F. Reif, *Fundamentals of Statistical and Thermal Physics*, Int. Student Ed., McGraw-Hill, New York, 1965.

7.130 W. G. V. Rosser, *An Introduction to Statistical Physics*, Ellis Horwood Ltd., Chichester, U.K., 1982.

Classical and Statistical

7.131 R. E. Sonntag and G. J. Van Wylen, *Introduction to Thermodynamics*, *Classical and Statistical*, 2nd ed., Wiley, New York, 1982.

7.132 J. S. Doolittle and F. J. Hale, *Thermodynamics for Engineers*, SI Version, Wiley, New York, 1984.

CHAPTER 8

8.1 D. Q. Kern, *Process Heat Transfer*, section entitled "The Reynolds Analogy," p. 54, McGraw-Hill, New York, 1950.

8.2 P. L. Geiringer, *Handbook of Heat Transfer Media*, Reinhold, New York, 1962.

8.3 *ASHRAE Handbook*, *Fundamentals*, ASHRAE, Atlanta, Georgia, 1985.

8.4 C. Boling et al., Heat Transfer of Evaporating Freon with Inner-Fin Tubing, *Refrig. Eng.*, (Feb. 1954).

8.5 H. Soumerai, Discussion of paper by W. J. Marner et al., On the Presentation of Performance Data for Enhanced Tubes Used in Shell-and-Tube Heat Exchangers, *ASME J. Heat Transfer*, **106**, 908–909 (1984).

8.6 H. Soumerai, "Thermodynamic Considerations on a Mirror Image concept—Forced Convection Single-Phase Tube Flow Applications," ASME Winter Annual Meeting, Florida, 1985.

8.7 S. Inada et al., Liquid–Solid Contact State in Subcooled Pool Transition Boiling System, *Trans. ASME J. Heat Transfer*, **108**, 219–221 (Feb. 1986).

8.8 *ASHRAE Handbook*, *Fundamentals*, Chap. 4, pp. 4.1–4.7, ASHRAE, Atlanta, Georgia, 1985.

8.9 E. Hahne and J. Muller, "Boiling From Finned Tube Bundles—The Effect of Enhanced Convection," presented at the 16th International Congress of Refrigeration, Paris, 1983.

8.10 E. U. Schluender et al., Mittlere Wärmeübergangszahl auf der Kältemittelseite eines Verdampferblockes, *Kältetechnik Klimatisierung*, **4** (1968).

8.11 D. Steiner, "Pressure Drop in Horizontal Flows," in *Proceedings of the International Workshop on Two-Phase Flow Fundamentals*, NBS-Gaithersburg, U.S.A., Sept. 1985.

8.12 J. C. Chen, "A Correlation for Boiling Heat Transfer to Saturated Fluids in Convective Flow," ASME Paper 63-HT-34, 1963.

8.13 D. W. Schmitt, "Waermeuebergang beim Blasensieden von Mehrkomponentengemischen," GVC, VDI April 24/25, 1986, Bad Saeckingen, W. Germany.

8.14 *Kent's Mechanical Engineers' Handbook*, Power Volume, 12th ed., pp. 4-06, 4-07, and 8-95 to 8-99, Wiley, New York, 1950.

8.15 R. Pictet, *Ann. Chim. Phys.*, **9**(5), 180 (1876).

8.16 Rudolf Plank, *Handbuch der Kaeltetechnik*, Sec. B, p. 131, Springer-Verlag, Berlin, 1953.

CHAPTER 9

9.1 H. Soumerai, "Thermodynamic Considerations on a 'Mirror Image' Concept—Forced Convection Single-Phase Tube Flow Applications," ASME Winter Annual Meeting, Miami, Florida, Nov. 1985.

9.2 H. Soumerai, "Entropy Based Heat Transfer Potentials—Forced Convection Single-Phase," 8th International Heat Transfer Conference, San Francisco, Aug. 1986.

9.3 H. Soumerai, Forced Convection Thermodynamics, *Int. Commun. in HMT*, submitted for publication.

9.4 H. Soumerai, "Thermodynamic Considerations on the Symmetry Between Forced Convection Cooling and Heating at Small Relative Temperature Differentials—Single-Phase/Two-Phase Tube Flow Applications," unpublished.

9.5 H. Soumerai, "Thermodynamic Aspects of Adiabatic and Diabatic Tube Flow Transition Regimes—Single-Phase/Two-Phase," unpublished.

9.6 P. Grassmann, *Physikalische Grundlagen der Chemie-Ingenieur-Technik*, Verlag H. R. Sauerlaender, Aarau, Switzerland, 1961.

9.7 W. M. Kays and A. L. London, *Compact Heat Exchangers*, McGraw-Hill, New York, 1958.

9.8 W. McAdams, *Heat Transmission*, 3rd ed., Chap. 13, pp. 356–365, McGraw-Hill, New York, 1954.

9.9 *ASHRAE Handbook*, *Fundamentals*, Sec. 1, Chap. 5, pp. 5.1–5.16, ASHRAE, Atlanta, Georgia, 1985.

9.10 A. Bejan, *Entropy Generation Through Heat and Fluid Flow*, section entitled "The Gouy–Stodola Theorem," pp. 21–24, Wiley, New York, 1982.

9.11 F. Bosnjakovic, *Technische Thermodynamik*, Pt. I, 4 ed., Sec. XIII.D, p. 276, Verlag Theodor Steinkopff, Dresden, 1965. (English translation: *Technical Thermodynamics*, Holt, Rinehart, and Winston, New York, 1965.)

9.12 R. Plank, *Handbuch der Kaeltetechnik*, 2. Die Vorgaenge im Mischraum, p. 365; 3. Der isotherme Drosseleffekt, p. 201; XVII. Die Drosselung, p. 88, Springer-Verlag, Berlin, 1953.

9.13 K. Kawashimo and K. Morikawa, Irreversible Thermodynamical Structure of Heat Conduction [English], *Bull. Jap. Soc. Mech. Eng.*, **27**(231), 1932–1937, (1984/09), *Bull. Int. Inst. Refrig.*, **LXV**(4), 410 (1985).

9.14 H. Lang, Entropieerzeugung, Wärme- und Stofftransport in Zweiphasenströmung (Entropy Production, Heat and Mass Transfer in Two-Phase Flows), presented at *Thermodynamik-Kolloquium 1986 der VDI-Gesellschaft Energietechnik*, Hinterzarten, 6/7 October 1986.

9.15 W. H. Carrier, The Contact-Mixture Analogy Applied to Heat Transfer With Mixtures of Air and Water Vapor, *Trans. ASME*, (Jan. 1937).

9.16 W. H. Carrier et al., *Modern Air Conditioning, Heating and Ventilating*, Pitman Publishing, Marshfield, MA, 1950.

9.17 L. N. Brown, Part II. Forced Circulation Air Coolers, in *Air Conditioning Refrigerating Data Book*, Design Volume, 10th ed., Chap. 22, pp. 22-13 to 22-27, The American Society of Refrigerating Engineers, New York, 1957.

9.18 L. N. Brown, Part II. Forced Circulation Air Coolers, in *Air Conditioning Refrigerating Data Book*, Design Volume, 10th ed., Chap. 22, p. 22-19, The American Society of Refrigerating Engineers, New York, 1957.

9.19 (a) H. Soumerai, "Thermodynamic Aspects of Adiabatic and Diabatic Tube Flow Regime Transitions—Single Phase," presented at AIAA/ASME 4th Joint Thermophysics and Heat Transfer Conference, Boston, June 2–4, 1986, American Institute of Aeronautics and Astronautics, New York. (b) H. Soumerai, "On the Application of an 'Entropy Maximizing Principle,' in Flow Regime Predictions", International Communications in Heat and Mass Transfer **14**(3), 303–312 (1987).

9.20 W. Yuen, Analysis of One-Component Two-Phase Flow using an Entropy Maximizing Principle, *Heat Transfer*, *AICHE Symp. Ser.*, **79**, 200–208 (1983).

9.21 F. J. Moody, Second Law Thinking-Example Applications in Reactor and Containment Technology, in section entitled "System Tendency Toward Stable States," *Second Law Aspects of Thermal Design*, HTD-Vol. 33, p. 7, ASME, New York, 1984.

9.22 D. Wilkins, On the Equilibrium of a Black Hole in a Radiation-Filled Cavity, *General Relativity and Gravitation*, **11**(1), 59–70 (1979).

9.23 H. Soumerai, Application of Thermodynamic Similitude, Part II: Heat and Mass Transfer, *ASHRAE J.*, **8**, 38 (July 1966).

9.24 N. Boccara, *Symetries Brisées—Théorie des transitions avec parametre d'ordre* Publié avec le concours du Centre National de la Recherche Scientifique, Hermann, Paris, 1976.

9.25 J. R. Mondt, "Adapting the Heat and Mass Transfer Analogy to Model Performance of Automotive Catalytic Converters," ASME Paper No. 85-WA/HT-33.

INDEX

Note: The following abbreviations are used: t = table, f = figure, n = note, Gl = Guideline.